Respiratory Medicine

More information about this series at http://www.springer.com/series/7665

David E. Griffith
Editor

Nontuberculous Mycobacterial Disease

A Comprehensive Approach to Diagnosis and Management

Editor
David E. Griffith
Professor of Medicine
University of Texas Health Science Center,
Tyler, TX
USA

ISSN 2197-7372 ISSN 2197-7380 (electronic)
Respiratory Medicine
ISBN 978-3-319-93472-3 ISBN 978-3-319-93473-0 (eBook)
https://doi.org/10.1007/978-3-319-93473-0

Library of Congress Control Number: 2018955583

Printed on acid-free paper

This Humana Press imprint is published by the registered company Springer Nature Switzerland AG
The registered company address is: Gewerbestrasse 11, 6330 Cham, Switzerland

The volume is dedicated to my colleague, mentor, and friend, Richard J. Wallace, Jr., whose contributions to the study of NTM diseases are unsurpassed. His profound and enduring influence on everyone currently working in the field is also unsurpassed. It is my incalculable good fortune to have worked with him over the last 40 years.

Fig. 1 Richard J. Wallace, Jr., Emanuel Wolinsky, Barbara Brown-Elliott circa 1990

Contents

Contributors

Jennifer Adjemian, PhD Epidemiology Unit, Laboratory of Clinical Immunology and Microbiology, Division of Intramural Research, National Institute of Allergy and Infectious Diseases, National Institutes of Health, Bethesda, MD, USA

United States Public Health Service, Commissioned Corps, Rockville, MD, USA

Timothy R. Aksamit, MD Mayo Clinic, Pulmonary Disease and Critical Care Medicine, Rochester, MN, USA

Sarah K. Brode, MD, MPH Division of Respirology, Department of Medicine, Toronto Western Hospital, University Health Network and West Park Healthcare Centre, University of Toronto, Toronto, ON, Canada

Barbara A. Brown-Elliott, MS, MT(ASCP)SM Department of Microbiology, Mycobacteria/Nocardia Research Laboratory, The University of Texas Health Science Center, Tyler, TX, USA

Robert Burkes, MD The Division of Pulmonary Medicine, University of North Carolina at Chapel Hill, Chapel Hill, NC, USA

James A. Caccitolo, MD Cardiothoracic Surgery, CHRISTUS Trinity Clinic, University of Texas Health Science Center Tyler, Tyler, TX, USA

Edward D. Chan, MD Pulmonary Section, Denver Veterans Affairs Medical Center, Denver, CO, USA

Program in Cell Biology and Department of Academic Affairs, National Jewish Health, Denver, CO, USA

Division of Pulmonary Sciences and Critical Care Medicine, University of Colorado Anschutz Medical Campus, Aurora, CO, USA

Jeremy M. Clain, MD Mayo Clinic, Pulmonary Diseases and Critical Care Medicine, Rochester, MN, USA

Samantha Cooray, MBiochem, MBBS, PhD Department of Respiratory Medicine, St Thomas' Hospital, Guys and St Thomas' NHS Foundation Trust, London, UK

Andrea T. Cruz, MD, MPH Department of Pediatrics, Baylor College of Medicine, Houston, TX, USA

Charles L. Daley, MD Division of Mycobacterial and Respiratory Infections, National Jewish Health, Denver, CO, USA

Shelby Daniel-Wayman, BA Epidemiology Unit, Laboratory of Clinical Immunology and Microbiology, Division of Intramural Research, National Institute of Allergy and Infectious Diseases, National Institutes of Health, Bethesda, MD, USA

David E. Griffith, MD University of Texas Health Science Center, Tyler, TX, USA

Emily Henkle, PhD, MPH OHSU-PSU School of Public Health, Oregon Health and Science University, Portland, OR, USA

Michael R. Holt, MD Division of Mycobacterial and Respiratory Infections, National Jewish Health, Denver, CO, USA

Won-Jung Koh, MD Division of Pulmonary and Critical Care Medicine, Department of Medicine, Samsung Medical Center, Sungkyunkwan University School of Medicine, Seoul, South Korea

Leah Lande, MD Division of Pulmonary and Critical Care Medicine, Lankenau Medical Center, Wynnewood, PA, USA

Lankenau Institute for Medical Research, Wynnewood, PA, USA

Marc Lipman, MD UCL Respiratory, University College London & Royal Free London NHS Foundation Trust, London, UK

Theodore K. Marras, MD, MSc Division of Respirology, Department of Medicine, Toronto Western Hospital, University Health Network, University of Toronto, Toronto, Canada

Kozo Morimoto, MD Division of Clinical Research, Fukujuji Hospital, Japan Anti-Tuberculosis Association, Tokyo, Japan

Peadar G. Noone, MD, FCCP, FRCPI The Division of Pulmonary Medicine, University of North Carolina at Chapel Hill, Chapel Hill, NC, USA

Anne E. O'Donnell, MD Division of Pulmonary, Critical Care and Sleep Medicine, Georgetown University Medical Center, Washington, DC, USA

Julie V. Philley, MD University of Texas Health Science Center, Tyler, TX, USA

D. Rebecca Prevots, PhD, MPH Epidemiology Unit, Laboratory of Clinical Immunology and Microbiology, Division of Intramural Research, National Institute of Allergy and Infectious Diseases, National Institutes of Health, Bethesda, MD, USA

Felix C. Ringshausen, MD Department of Respiratory Medicine, Hannover Medical School and German Center for Lung Research, Hannover, Germany

Jeffrey R. Starke, MD Department of Pediatrics, Baylor College of Medicine, Houston, TX, USA

Rachel Thomson, MBBS Grad Dip PhD FRACP Gallipoli Medical Research Institute, University of Queensland, Brisbane, Australia

Jakko van Ingen, MD, PhD Department of Medical Microbiology, Radboud University Medical Center, Nijmegen, The Netherlands

Dirk Wagner, MD Division of Infectious Diseases, Department of Internal Medicine II, Medical Center – University of Freiburg, Faculty of Medicine, Freiburg, Germany

Richard J. Wallace Jr., MD Department of Microbiology, Mycobacteria/Nocardia Laboratory, The University of Texas Health Science Center, Tyler, TX, USA

Kevin L. Winthrop, MD, MPH OHSU-PSU School of Public Health, Oregon Health and Science University, Portland, OR, USA

Nontuberculous Mycobacterial Disease: An Introduction and Historical Perspective

David E. Griffith

I want to begin this volume on nontuberculous mycobacterial (NTM) disease with a plea and an admonition. First, many aspects of NTM disease are difficult to understand, counterintuitive, and even paradoxical. I am repeatedly reminded about how poorly the nuances and idiosyncrasies of NTM disease are generally understood through years of interactions with clinicians who seek my advice about NTM disease management. Many aspects of NTM disease defy easy explanations and require sometimes detailed background information to build an adequate context for interpretation and comprehension. The reader is strongly encouraged to use this volume as more than a quick reference or handbook on NTM disease management. Rather, each chapter should be read in its entirety to promote an in-depth understanding of NTM disease with all of the attendant complexities, contradictions, and knowledge gaps. There are no shortcuts.

Second, many aspects of NTM disease defy the kind of evidence-based analysis and conclusions that would support rigorous or robust evidence-based recommendations. The necessary accumulation of information to achieve that goal is simply not yet available. In the absence of a better evidence base, many recommendations for NTM management have their origin in "expert opinion," and many recommendations in this volume reflect that reality. Clinicians faced with difficult NTM management decisions still require guidance, even the imperfect guidance of expert opinion. Controversial areas where strong opinions are offered will be evident to the reader who will be savvy enough to judge the merits of those opinions and to seek alternative opinions.

Interest in the "nontuberculous mycobacteria" or NTM is a relatively recent phenomenon that has now reached unprecedented levels. Although NTM were identified more than a century ago, their role as human pathogens was generally perceived

D. E. Griffith
University of Texas Health Science Center, Tyler, TX, USA
e-mail: david.griffith@uthct.edu

D. E. Griffith (ed.), *Nontuberculous Mycobacterial Disease*, Respiratory Medicine,
https://doi.org/10.1007/978-3-319-93473-0_1

as minor, even inconsequential during most of that time. For the purposes of this volume, the NTM are comprised of species in the genus *Mycobacterium* excluding species in the *M. tuberculosis* complex and *M. leprae.*

The term "nontuberculous mycobacteria" (NTM) is now in common use but is not universally endorsed as the collective term for these organisms. Alternative names such as "atypical mycobacteria," "mycobacteria other than tuberculosis" (MOTT), or "environmental mycobacteria" have been championed with variable penetrance into the NTM vernacular. "Atypical mycobacteria" is probably the most commonly used alternative label and presumably referred to isolation of a mycobacterial species other than "typical" *M. tuberculosis.* It seems inappropriate now because strictly from the perspective of isolation frequency, the "atypical" mycobacteria far outnumber "typical" *M. tuberculosis* isolates in major mycobacteriology laboratories in the United States. While "environmental mycobacteria" is appealing from taxonomic and pathophysiologic standpoints, the label NTM is now so firmly entrenched it cannot be easily displaced and is our preferred, if imperfect, term for this group of organisms.

For most of its history in the United States, NTM disease knowledge and understanding was impeded by the difficulty separating NTM pathogens and associated clinical disease syndromes from disease caused by *M. tuberculosis.* Clinical NTM isolates, especially those from respiratory specimens, were often regarded as contaminants and dismissed as clinically insignificant. It was also generally assumed that NTM pathogens and disease would respond favorably to antituberculosis antimicrobials leading to inevitable and understandable frustration when they did not. The lack of therapeutic response was probably also an unintentional disincentive to aggressively recognize and diagnose NTM disease, especially NTM lung disease. Clearly, the identification of NTM pathogens and recognition of their clinical significance have markedly improved. However, an easy separation between NTM disease and tuberculosis continues to be an ongoing and evolving process especially in the developing world, where due to a lack of available resources to isolate, identify, or treat NTM pathogens, mycobacterial disease is often initially assumed to be caused by *M. tuberculosis.*

The emergence of NTM pathogens and disease as subjects of serious interest in the United States can be dated roughly to the publication in 1980 of a state-of-the-art review in the American Review of Respiratory Disease by Dr. Emanuel Wolinsky titled, *Nontuberculous Mycobacteria and Associated Diseases* [1]. This highly influential manuscript, published almost 40 years ago, was the first comprehensive and more importantly widely read NTM disease review and represents a clear watershed moment in the recognition and appreciation of NTM disease. Progress in the NTM disease realm has been nothing short of remarkable since then. This brief introduction highlights some important milestones in that progress with chapter references to guide the reader to more detailed information and discussion about specific NTM disease aspects.

At the time of the Wolinsky manuscript, there were approximately 40 recognized NTM species that were identified utilizing insensitive phenotypic and biochemical characteristics including colony morphology and patterns of nutrient metabolism

[1, 2]. A widely adopted early NTM classification system based on this approach was eponymously labeled the Runyon classification system after Dr. Ernest H. Runyon [3]. The speed and accuracy of mycobacterial species identification dramatically improved first with high-performance liquid chromatography (HPLC), closely followed by the introduction of molecular laboratory methods, including DNA probes and gene sequencing techniques [4–7]. High-performance liquid chromatography and DNA probes are rapid and widely available but are restricted to identification of some commonly isolated NTM species including *Mycobacterium avium* complex (MAC), *M. kansasii*, and *M. gordonae*.

Nontuberculous mycobacterial species identification expanded in an almost explosive manner with the widespread application of 16S rRNA gene sequencing, a gene thought to be highly preserved within NTM species [7]. Utilizing this and other molecular-based techniques, the number of recognized NTM species continues to expand and has grown to approximately 200 [8]. It is now apparent that the 16S rRNA gene analysis, by itself, does not always satisfactorily discriminate between all NTM species and/or subspecies [7, 9–11]. The process of NTM organism identification has become sufficiently complex that discriminating between some NTM species and subspecies requires either multigene sequencing or whole-genome sequencing [7, 9–11]. Even then, controversy persists about the degree of difference between NTM isolates that is necessary for species versus subspecies determination and differentiation [10, 11]. Overall, however, molecular methods have revolutionized the microbiologic evaluation of NTM including rapid and accurate methods for clinical NTM isolate identification, molecular epidemiology investigations and discovery of innate NTM resistance mechanisms [12–14]. This important and rapidly changing field is discussed in detail in chapters "The Modern Mycobacteriology Laboratory and Its Role in NTM Disease Diagnosis and Management" and "In Vitro Drug Susceptibility Testing for NTM and Mechanisms of NTM Drug Resistance" with focused discussion in several other chapters.

Genotyping environmental and clinical NTM isolates has provided invaluable insights into the identification of NTM environmental niches and possible routes of NTM pathogen acquisition [12, 15–17]. This approach is a necessary element for developing disease prevention strategies which are discussed in chapters "Environmental Niches for NTM and Their Impact on NTM Disease" and "Healthcare Associated NTM Outbreaks and Pseudo-outbreaks". Genotyping clinical *Mycobacterium avium* complex (MAC) isolates from MAC lung disease patients also allows discrimination between true disease relapse isolates and (presumed) reinfection isolates which is discussed in chapter "*Mycobacterium avium* Complex Disease" [18, 19].

Ironically, the new molecular laboratory methods have so radically changed our view and understanding of NTM pathogens and disease that the advances have outpaced the capability of most mycobacterial laboratories to adopt and perform these invaluable services. Currently, most clinicians do not have access to laboratories utilizing these invaluable methods which have proven indispensable for optimal NTM patient management.

At the time of Dr. Wolinsky's manuscript, there was little data and limited understanding about the epidemiology of NTM disease. Initial efforts to estimate NTM disease prevalence in the United States suggested that it was 1–2 cases/100,000 population based on NTM isolation prevalence calculated at the Centers for Disease Control and Prevention (CDC) [20, 21]. Because NTM disease was not reportable, the NTM isolates received by the CDC were not part of a comprehensive national survey of NTM isolates or disease. Currently a minority of states within the United States have mandatory NTM reporting with variable requirements for the information that is collected.

Aside from the absence of a national or universal reporting requirement, the second major impediment to accurate determination of NTM disease prevalence is that in contrast to tuberculosis, a single NTM isolate is not necessarily an indication of active NTM disease, especially lung disease [22–26]. Unlike tuberculosis, NTM isolation prevalence from respiratory specimens does not equate to actual NTM lung disease prevalence. This frustrating observation is primarily due to the possibility that clinical specimens can be contaminated by NTM from environmental sources. Patients with suspected NTM lung disease, therefore, must meet a set of diagnostic criteria that are difficult to apply retrospectively in epidemiologic investigations without obtaining detailed information from the patient's medical record. The challenges of NTM disease diagnosis are discussed in detail in chapters "Epidemiology of NTM Disease: United States" and "Epidemiology of NTM Disease: Global".

Some investigators have undertaken the tedious analysis that is necessary for accurate NTM case definition [27–30]. Other investigators have utilized alternative epidemiologic tools such as querying extensive insurance-based patient databases utilizing diagnostic codes [31, 32]. While estimates of NTM disease prevalence vary, the available data consistently show that NTM prevalence in the United States is increasing. Mandatory NTM disease reporting would provide more accurate estimates of prevalence but would also facilitate new insights into incidence, which has been an elusive goal so far. Chapter "Epidemiology of NTM Disease: United States" discusses in detail the current understanding of NTM disease epidemiology in the United States.

Investigators outside the United States are providing a clearer picture of global NTM disease epidemiology [33–35]. Nontuberculous mycobacterial infections in the developed world have broadly comparable epidemiology to that in the United States although some important differences, particularly in Western Europe exist. Why those differences exist is unclear, but their investigation offers opportunities for better understanding of multiple NTM disease aspects beyond epidemiology.

Unfortunately, NTM disease epidemiology remains poorly described in large areas of the developing world. Even in these areas, however, NTM epidemiologic information is becoming more accessible in part because of the expanding availability of rapid and accurate tuberculosis diagnostic tools such as the Cepheid GeneXpert TB/RIF technology [36, 37]. This platform gives a first approximation of NTM disease prevalence by identifying patients whose respiratory specimens are acid-fast bacilli (AFB) smear positive but nucleic acid amplification negative for tuberculosis. As with NTM disease prevalence in developed countries, it is highly likely that the extent of NTM disease in the developing world will be much higher than is currently appreciated with an inevitable attendant demand on limited resources for treating the expanding number of NTM disease patients. Global NTM disease epidemiology is discussed in detail in chapter "Epidemiology of NTM Disease: Global".

When the Wolinsky manuscript was published, NTM lung disease pathophysiology was assumed to be analogous to tuberculosis with the notable exception that NTM lung disease pathogens had not been demonstrated to be transmissible between humans. It was also known that NTM were environmental organisms that inhabited specific niches including natural water sources inviting speculation that NTM lung disease might be the consequence of organism inhalation after naturally occurring aerosolization of the organism [15, 38–41].

Recently there have been multiple reports describing isolation of NTM respiratory pathogens from environmental sources including household or municipal water, and with the aid of organism genotyping, it has also been shown that some clinical NTM respiratory isolates are genotypically identical to household water NTM isolates [15, 42, 43]. These observations are strong evidence that municipal water is the source of NTM respiratory pathogens, especially *Mycobacterium avium* complex (MAC), for some patients with NTM lung disease. Municipal water is also a known environmental niche for NTM respiratory pathogens such as *M. kansasii* and *M. xenopi* as well as nosocomially acquired pathogens such as *M. abscessus* and *M. chimaera* [12, 15, 43]. The demonstration of NTM acquisition from specific environmental sources is a necessary prerequisite for developing NTM disease prevention strategies. The environmental acquisition of NTM is discussed in chapters "Environmental Niches for NTM and Their Impact on NTM Disease" and "Healthcare Associated NTM Outbreaks and Pseudo-outbreaks" including recommendations for the investigation of nosocomial NTM outbreaks and pseudo-outbreaks.

When the Wolinsky manuscript was published, NTM lung disease was regarded as clinically and radiographically similar to tuberculosis, and clearly NTM lung disease does sometimes present radiographically with upper lobe fibrocavitary changes similar to reactivation tuberculosis [1, 23, 44]. Currently, however, in the United States NTM lung disease, especially MAC lung disease, is now more commonly associated with nodular and bronchiectatic radiographic changes [23–25, 45]. Recognition of this shift has influenced the way that many NTM experts view NTM lung disease pathophysiology. Specifically, there is growing consensus that many (perhaps most) NTM lung disease patients not only require exposure to NTM but also must have a vulnerability or susceptibility to NTM infection such as structural lung abnormalities associated with bronchiectasis or obstructive lung disease [46, 47]. For many NTM lung disease patients, the infection is the consequence of the underlying anatomic lung abnormality or predisposition rather than a primary event. Recent work suggests that some patients with "idiopathic" bronchiectasis have polygenic mutations, the sum of which predispose to bronchiectasis and NTM infection [46, 47]. The role of NTM in cystic fibrosis, a disease associated with severe and progressive bronchiectasis, is discussed in chapter "NTM Disease Associated with Cystic Fibrosis". The management of bronchiectasis, which is an essential element in the comprehensive treatment of the NTM lung disease patient, is discussed in chapter "Management of Lung Diseases Associated with NTM Infection".

The identification of both genetic and acquired factors predisposing to NTM infection is rapidly expanding and is discussed in chapters "Vulnerability to NTM Lung Disease or Systemic Infection Due to Genetic /Heritable Disorders" and "Acquired immune Dysfunction and NTM Disease". Nontuberculous mycobacterial lung infection has, in general, not been found to be associated with systemic immune deficiency, although extrapulmonary and disseminated NTM disease is usually a consequence of systemic immune dysfunction or suppression [25, 48]. The role of NTM infection in children who represent another special and vulnerable host is discussed in chapter "NTM Disease in Pediatric Populations".

When the Wolinsky manuscript was published, NTM treatment was based on the principles of tuberculosis therapy. There was a limited armamentarium of antituberculosis drugs whose use was guided by in vitro susceptibility test breakpoints established for *M. tuberculosis* but not validated for NTM [1, 23]. One study suggested that MAC lung disease treatment success depended on the number of antituberculosis drugs used (up to five or six), including second-line TB drugs such as ethionamide and cycloserine [49]. The limitations of this approach were recognized at the time although few studies were done that critically evaluated the use of traditional antituberculosis medications in NTM disease [50].

In the mid-1980s, a seismic shift occurred in NTM disease with the advent of the acquired immunodeficiency syndrome (AIDS) epidemic and the emergence of MAC as a lethal pathogen [51–53]. These catastrophic events created a sense of urgency in the effort to find effective MAC therapy. Multiple antibiotics and combinations of antibiotics were tried with the new macrolide drugs, clarithromycin and azithromycin, emerging as the foundation of effective disseminated MAC therapy and prophylaxis [54–56]. It is noteworthy that this once feared AIDS-related infection is now infrequently encountered due to the success of antiretroviral therapy for AIDS.

Over the subsequent three decades, multiple studies demonstrated the utility of macrolide-based regimens for treating MAC lung infections [19, 57–62]. Regrettably, MAC lung disease therapy has stagnated with almost no significant innovations since the introduction of macrolide-based regimens. The recent introduction of an inhaled liposomal amikacin suspension (ALIS) for treatment of pulmonary MAC disease may prove to be an important exception to this generally bleak picture [63, 64]. While treatment outcomes have been generally favorable, MAC treatment success is still not comparable to the almost universally favorable TB treatment outcomes. Additionally, many other NTM respiratory pathogens such as *M. xenopi*, *M. malmoense*, *M. abscessus*, and *M. simiae* remain even more difficult to treat than MAC [25, 65]. The many challenges for treating MAC and other NTM pathogens as well as suggested treatment strategies are discussed in detail in chapters "In Vitro Drug Susceptibility Testing for NTM and Mechanisms of NTM Drug Resistance", "General Management Principles for NTM Lung Disease",

"*Mycobacterium avium* complex Disease", "NTM disease caused by *M. kansasii, M. xenopi, M. malmoense* and Other Slowly Growing NTM", and "*Mycobacterium abscessus* Disease and Disease Caused by Other Rapidly Growing NTM".

Since the publication of the Wolinsky manuscript, the presence of a particularly troublesome and frustrating aspect of NTM therapy has been repeatedly confirmed. For reasons that are not yet well understood, in vitro antibiotic susceptibility results for multiple NTM pathogens may not be predictive of treatment success or failure with a specific antibiotic [66, 67]. For MAC, for instance, the only antibiotic agents where in vitro susceptibility predicts in vivo response are macrolides and amikacin [25, 66]. Understanding the nuances and limitations of in vitro susceptibility testing for NTM is of such importance that the topic is covered in two chapters in this volume (chapters "The Modern Mycobacteriology Laboratory and Its Role in NTM Disease Diagnosis and Management" and "In Vitro Drug Susceptibility Testing for NTM and Mechanisms of NTM Drug Resistance"). The reader will note that the two chapters approach NTM in vitro susceptibility testing from different perspectives and with different areas of emphasis, but practical management considerations largely coincide between the two chapters. Both perspectives are valuable and instructive, and the reader is strongly encouraged to read both chapters in detail.

Molecular laboratory techniques have provided tools for investigating paradoxical NTM antibiotic resistance and have made us aware of multiple factors possessed by NTM that are associated with innate or natural drug resistance [66, 67]. These innate resistance factors may not be reflected in the MIC of the organism for specific drugs. This is the most vexing and counterintuitive characteristic of NTM lung disease for clinicians and the area where experience with tuberculosis is least helpful. Probably the best known example of this phenomenon is the inducible macrolide resistance gene, or *erm* gene, present in *M. abscessus* subsp. *abscessus* and subsp. *bolletii* as well as other mycobacterial species, such as *M. fortuitum* and even *M. tuberculosis* [13]. The activity of this gene can only be detected in vitro by preincubation of the organism in the presence of macrolide. While *erm* gene activity is only one mechanism of innate NTM drug resistance, its recognition has been transformative for how we approach patients with *M. abscessus* respiratory disease (chapters "In Vitro Drug Susceptibility Testing for NTM and Mechanisms of NTM Drug Resistance" and "*Mycobacterium abscessus* Disease and Disease Caused by Other Rapidly Growing NTM").

Ultimately, the future of NTM lung disease therapy will be guided by recognition of innate antibiotic resistance mechanisms and the discovery of ways to overcome them. The complexities of in vitro susceptibility testing for treating NTM disease are discussed in chapters "Laboratory Diagnosis and Antimicrobial Susceptibility Testing of Nontuberculous Mycobacteria" and "In Vitro Drug Susceptibility Testing for NTM and Mechanisms of NTM Drug Resistance" as well as multiple other chapters. For successful management of NTM infections, clinicians must become familiar with the idiosyncratic behavior of NTM pathogens. There is no substitute for having this knowledge.

Unfortunately, the discussion of NTM antibiotic drug resistance does not end with innate drug resistance. Many NTM pathogens including MAC and *M. abscessus* subsp. *abscessus* are also vulnerable to acquired mutational drug resistance, a mechanism for acquired drug resistance well known to clinicians who treat tuberculosis. For instance, macrolides must be protected by effective companion drugs in MAC treatment regimens to avoid the emergence of macrolide resistance through selection of organisms with a 23S rRNA mutation. This occurrence is associated with poor treatment response and poor overall outcome [68]. Acquired mutational drug resistance occurs with other NTM pathogens, notably the *rpoβ* gene and acquired *M. kansasii* rifamycin resistance. This type of antibiotic resistance is both predictable and avoidable if the clinician is aware of the risk for specific NTM pathogens and the necessary steps to avoid it. Again, there are no shortcuts and no substitutes for this knowledge. The management of NTM pathogens in the context of both innate and acquired drug resistance mechanisms is discussed in chapters ""In Vitro Drug Susceptibility Testing for NTM and Mechanisms of NTM Drug Resistance", "General Management Principles for NTM Lung Disease", "*Mycobacterium avium* Complex Disease", "NTM disease caused by *M. kansasii, M. xenopi, M. malmoense* and Other Slowly Growing NTM", and "*Mycobacterium abscessus* Disease and Disease Caused by Other Rapidly Growing NTM".

In large part because of antimicrobial resistance, surgical intervention is important for management of both pulmonary and extrapulmonary-pulmonary NTM disease and is discussed in chapter "Surgical Management of NTM Diseases" as well as chapters discussing treatment of specific NTM pathogens. Surgical resection of diseased lung has consistently been shown to be effective for selected NTM lung disease patients [13]. Surgery is a sufficiently important potential adjunct to medical therapy for NTM lung disease that it should be considered whenever possible. Surgical debridement of diseased tissue is absolutely essential for successful therapy of NTM skin, soft tissue, and bone infections.

Since 1990, there have been three NTM statements sponsored or co-sponsored by the American Thoracic Society [23–25]. These documents summarized contemporary knowledge about NTM with recommendations for treating specific NTM pathogens. As much as anything else, they focused attention on the numerous and persistent NTM disease knowledge gaps and the sparse evidence base for making NTM disease management recommendations. The NTM statements did, however, provide treatment recommendations based on the limited evidence base and expert opinion. The MAC lung disease recommendations proved to be effective if imperfect and less reliably effective than TB therapy. In that context, it is instructive that two studies have shown that there is poor adherence to the published treatment guidelines worldwide which may account for some of the frustration experienced by clinicians related to ineffective therapy [69, 70].

Unquestionably, many weaknesses and gaps in our knowledge of NTM disease remain. We need better understanding of environmental niches and mechanisms of organism acquisition. For NTM lung disease especially, we need markers of disease activity so that we can predict which patients will have progressive disease and require therapy. That type of marker would allow eliminating the confusing and the sometimes insensitive and nonspecific NTM disease diagnostic criteria. Equally important we need the ability to identify those patients with NTM lung disease who are likely to relapse after successful therapy. Overall, we need more efficient ways to define and predict the course of NTM lung disease. We need better ways to determine NTM disease prevalence and ultimately incidence. Making NTM disease uniformly reportable in the United States and globally would go a long way toward accomplishing those goals, although, without tools that improve diagnostic accuracy, even universal case reporting would probably still entail considerable inaccuracies. The most pressing need is for new and more effective antimicrobial agents, a process that will be driven by improved understanding of NTM drug resistance mechanisms. We will need new approaches to NTM disease prevention, a process only possible with early identification of patients at risk for developing NTM lung disease and better understanding of NTM environmental niches and mechanisms of organism acquisition from these niches.

The reader is once again strongly encouraged to read each chapter for a comprehensive overview of the complexities, subtleties, and paradoxes of NTM disease and its treatment. The understanding of NTM disease is clearly nascent, but we are experiencing an exciting acceleration in the pace of discovery and knowledge. It is also remarkable that progress so far has been accomplished largely without extramural funding from national (the United States) and international funding agencies, although that bleak scenario may be gradually improving. Convincing extramural funding sources that NTM disease, especially lung disease, is a growing international health burden and that committing research dollars to this field will yield important and widely applicable results are major priorities and challenges. A vital element in this task is procuring funding for prospective treatment trials which are necessary not only for critical evaluation of current treatment strategies but also to establish optimal study designs for testing new drugs as they become available [63, 64].

Since the publication of the Wolinsky manuscript, the study of NTM disease has been completely transformed. A fledgling field in 1980 has acquired legitimacy and momentum with a sound footing in clinical and basic science. There are many reasons to be optimistic about continued and accelerating progress with NTM disease. First among them is the proliferation of investigators around the world including the very talented investigators who contributed to this volume. I am immensely grateful to them for their excellent contributions. I am also impressed, humbled, and inspired by the superb quality of their innovative work. It is clear to me that over the next 40 years, there will be further exponential expansion of NTM disease knowledge and understanding. The inevitable result will be achievement of the ultimate goal, improved outcomes for our patients.

Bibliography

1. Wolinsky E. Nontuberculous mycobacteria and associated diseases. Am Rev Respir Dis. 1979;119(1):107–59.
2. Runyon EH. Identification of mycobacterial pathogens utilizing colony characteristics. Am J Clin Pathol. 1970;54(4):578–86.
3. Runyon EH. Atypical mycobacteria: their classification. Am Rev Respir Dis. 1965;91:288–9.
4. Jost KC Jr, Dunbar DF, Barth SS, Headley VL, Elliott LB. Identification of Mycobacterium tuberculosis and M. avium complex directly from smear-positive sputum specimens and

BACTEC 12B cultures by high-performance liquid chromatography with fluorescence detection and computer-driven pattern recognition models. J Clin Microbiol. 1995;33(5):1270–7.
5. Louro AP, Waites KB, Georgescu E, Benjamin WH Jr. Direct identification of Mycobacterium avium complex and Mycobacterium gordonae from MB/BacT bottles using AccuProbe. J Clin Microbiol. 2001;39(2):570–3. 8: Pauls RJ, Turenne CY, Wolfe JN, Kabani A. A high proportion of novel mycobacteria species identified by 16S rDNA analysis among slowly growing AccuProbe-negative strains in a clinical setting. Am J Clin Pathol. 2003;120(4):560–6
6. Cloud JL, Carroll KC, Cohen S, Anderson CM, Woods GL. Interpretive criteria for use of AccuProbe for identification of Mycobacterium avium complex directly from 7H9 broth cultures. J Clin Microbiol. 2005;43(7):3474–8.
7. Griffith DE, Brown-Elliott BA, Benwill JL, Wallace RJ Jr. Mycobacterium abscessus. "Pleased to meet you, hope you guess my name...". Ann Am Thorac Soc. 2015;12(3):436–9.
8. LPSN. bacterio.net web. List of prokaryotic names with standing in nomenclature. http://www.bacterio.net/-classification.html
9. Tettelin H, Davidson RM, Agrawal S, Aitken ML, Shallom S, Hasan NA, Strong M, de Moura VC, De Groote MA, Duarte RS, Hine E, Parankush S, Su Q, Daugherty SC, Fraser CM, Brown-Elliott BA, Wallace RJ Jr, Holland SM, Sampaio EP, Olivier KN, Jackson M, Zelazny AM. High-level relatedness among Mycobacterium abscessus subsp. massiliense strains from widely separated outbreaks. Emerg Infect Dis. 2014;20(3):364–71.
10. Tortoli E, Kohl TA, Brown-Elliott BA, Trovato A, Leão SC, Garcia MJ, Vasireddy S, Turenne CY, Griffith DE, Philley JV, Baldan R, Campana S, Cariani L, Colombo C, Taccetti G, Teri A, Niemann S, Wallace RJ Jr, Cirillo DM. Emended description of Mycobacterium abscessus, Mycobacterium abscessus subsp. abscessus and Mycobacteriumabscessus subsp. bolletii and designation of Mycobacterium abscessus subsp. massiliense comb. nov. Int J Syst Evol Microbiol. 2016;66(11):4471–9.
11. Adekambi T, Sassi M, van Ingen J, Drancourt M. Reinstating Mycobacterium massiliense and Mycobacterium bolletii as species of the Mycobacterium abscessus complex. Int J Syst Evol Microbiol. 2017;67(8):2726–30.
12. van Ingen J, Kohl TA, Kranzer K, Hasse B, Keller PM, Katarzyna Szafrańska A, Hillemann D, Chand M, Schreiber PW, Sommerstein R, Berger C, Genoni M, Rüegg C, Troillet N, Widmer AF, Becker SL, Herrmann M, Eckmanns T, Haller S, Höller C, Debast SB, Wolfhagen MJ, Hopman J, Kluytmans J, Langelaar M, Notermans DW, Ten Oever J, van den Barselaar P, Vonk ABA, Vos MC, Ahmed N, Brown T, Crook D, Lamagni T, Phin N, Smith EG, Zambon M, Serr A, Götting T, Ebner W, Thürmer A, Utpatel C, Spröer C, Bunk B, Nübel U, Bloemberg GV, Böttger EC, Niemann S, Wagner D, Sax H. Global outbreak of severe Mycobacterium chimaera disease after cardiac surgery: a molecular epidemiological study. Lancet Infect Dis. 2017;17(10):1033–41.
13. Nash KA, Brown-Elliott BA, Wallace RJ Jr. A novel gene, erm(41), confers inducible macrolide resistance to clinical isolates of Mycobacterium abscessus but is absent from Mycobacterium chelonae. Antimicrob Agents Chemother. 2009;53(4):1367–76.
14. van Ingen J, Boeree MJ, van Soolingen D, Mouton JW. Resistance mechanisms and drug susceptibility testing of nontuberculous mycobacteria. Drug Resist Updat. 2012;15(3):149–61.
15. Wallace RJ Jr, Iakhiaeva E, Williams MD, Brown-Elliott BA, Vasireddy S, Vasireddy R, Lande L, Peterson DD, Sawicki J, Kwait R, Tichenor WS, Turenne C, Falkinham JO 3rd. Absence of Mycobacterium intracellulare and presence of Mycobacterium chimaera in household water and biofilm samples of patients in the United States with Mycobacterium avium complex respiratory disease. J Clin Microbiol. 2013;51(6):1747–52.
16. Bryant JM, Grogono DM, Rodriguez-Rincon D, Everall I, Brown KP, Moreno P, Verma D, Hill E, Drijkoningen J, Gilligan P, Esther CR, Noone PG, Giddings O, Bell SC, Thomson R, Wainwright CE, Coulter C, Pandey S, Wood ME, Stockwell RE, Ramsay KA, Sherrard LJ, Kidd TJ, Jabbour N, Johnson GR, Knibbs LD, Morawska L, Sly PD, Jones A, Bilton D, Laurenson I, Ruddy M, Bourke S, Bowler IC, Chapman SJ, Clayton A, Cullen M, Daniels T, Dempsey O, Denton M, Desai M, Drew RJ, Edenborough F, Evans J, Folb J, Humphrey H,

Isalska B, Jensen-Fangel S, Jönsson B, Jones AM, Katzenstein TL, Lillebaek T, MacGregor G, Mayell S, Millar M, Modha D, Nash EF, O'Brien C, O'Brien D, Ohri C, Pao CS, Peckham D, Perrin F, Perry A, Pressler T, Prtak L, Qvist T, Robb A, Rodgers H, Schaffer K, Shafi N, van Ingen J, Walshaw M, Watson D, West N, Whitehouse J, Haworth CS, Harris SR, Ordway D, Parkhill J, Floto RA. Emergence and spread of a human-transmissible multidrug-resistant nontuberculous mycobacterium. Science. 2016;354(6313):751–7.
17. van Ingen J, Boeree MJ, Dekhuijzen PN, van Soolingen D. Environmental sources of rapid growing nontuberculous mycobacteria causing disease in humans. Clin Microbiol Infect. 2009;15(10):888–93.
18. Wallace RJ Jr, Zhang Y, Brown BA, Dawson D, Murphy DT, Wilson R, Griffith DE. Polyclonal Mycobacterium avium complex infections in patients with nodular bronchiectasis. Am J Respir Crit Care Med. 1998;158(4):1235–44.
19. Wallace RJ Jr, Brown-Elliott BA, McNulty S, Philley JV, Killingley J, Wilson RW, York DS, Shepherd S, Griffith DE. Macrolide/Azalide therapy for nodular/bronchiectatic mycobacterium avium complex lung disease. Chest. 2014;146(2):276–82.
20. O'Brien RJ, Geiter LJ, Snider DE Jr. The epidemiology of nontuberculous mycobacterial diseases in the United States. Results from a national survey. Am Rev Respir Dis. 1987;135(5):1007–14.
21. O'Brien RJ. The epidemiology of nontuberculous mycobacterial disease. Clin Chest Med. 1989;10(3):407–18.
22. Griffith DE, Brown-Elliott BA, Wallace RJ Jr. Diagnosing nontuberculous mycobacterial lung disease. A process in evolution. Infect Dis Clin North Am. 2002;16(1):235–49.
23. Diagnosis and treatment of disease caused by nontuberculous mycobacteria. Am Rev Respir Dis. 1990;142(4):940–53. Erratum in: Am Rev Respir Dis 1991 Jan;143(1):204.
24. Diagnosis and treatment of disease caused by nontuberculous mycobacteria. This official statement of the American Thoracic Society was approved by the Board of Directors, March 1997. Medical section of the American Lung Association. Am J Respir Crit Care Med. 1997;156(2 Pt 2):S1–25.
25. Griffith DE, Aksamit T, Brown-Elliott BA, Catanzaro A, Daley C, Gordin F, Holland SM, Horsburgh R, Huitt G, Iademarco MF, Iseman M, Olivier K, Ruoss S, von Reyn CF, Wallace RJ Jr, Winthrop K, ATS Mycobacterial Diseases Subcommittee, American Thoracic Society, Infectious Disease Society of America. An official ATS/IDSA statement: diagnosis, treatment, and prevention of nontuberculous mycobacterial diseases. Am J Respir Crit Care Med. 2007;175(4):367–416. Review. Erratum in: Am J Respir Crit Care Med 2007 Apr 1;175(7):744–5. Dosage error in article text.
26. Jankovic M, Sabol I, Zmak L, Jankovic VK, Jakopovic M, Obrovac M, Ticac B, Bulat LK, Grle SP, Marekovic I, Samarzija M, van Ingen J. Microbiological criteria in non-tuberculous mycobacteria pulmonary disease: a tool for diagnosis and epidemiology. Int J Tuberc Lung Dis. 2016;20(7):934–40.
27. Cassidy PM, Hedberg K, Saulson A, McNelly E, Winthrop KL. Nontuberculous mycobacterial disease prevalence and risk factors: a changing epidemiology. Clin Infect Dis. 2009;49(12):e124–9.
28. Winthrop KL, McNelley E, Kendall B, Marshall-Olson A, Morris C, Cassidy M, Saulson A, Hedberg K. Pulmonary nontuberculous mycobacterial disease prevalence and clinical features: an emerging public health disease. Am J Respir Crit Care Med. 2010;182(7):977–82.
29. Winthrop KL, Varley CD, Ory J, Cassidy PM, Hedberg K. Pulmonary disease associated with nontuberculous mycobacteria, Oregon, USA. Emerg Infect Dis. 2011;17(9):1760–1.
30. Henkle E, Hedberg K, Schafer S, Novosad S, Winthrop KL. Population-based incidence of pulmonary nontuberculous mycobacterial disease in Oregon 2007 to 2012. Ann Am Thorac Soc. 2015;12(5):642–7.
31. Adjemian J, Olivier KN, Seitz AE, Holland SM, Prevots DR. Prevalence of nontuberculous mycobacterial lung disease in U.S. Medicare beneficiaries. Am J Respir Crit Care Med. 2012;185(8):881–6.

32. Prevots DR, Shaw PA, Strickland D, Jackson LA, Raebel MA, Blosky MA, Montes de Oca R, Shea YR, Seitz AE, Holland SM, Olivier KN. Nontuberculous mycobacterial lung disease prevalence at four integrated health care delivery systems. Am J Respir Crit Care Med. 2010;182(7):970–6.
33. Hoefsloot W, van Ingen J, Andrejak C, Angeby K, Bauriaud R, Bemer P, Beylis N, Boeree MJ, Cacho J, Chihota V, Chimara E, Churchyard G, Cias R, Daza R, Daley CL, Dekhuijzen PN, Domingo D, Drobniewski F, Esteban J, Fauville-Dufaux M, Folkvardsen DB, Gibbons N, Gómez-Mampaso E, Gonzalez R, Hoffmann H, Hsueh PR, Indra A, Jagielski T, Jamieson F, Jankovic M, Jong E, Keane J, Koh WJ, Lange B, Leao S, Macedo R, Mannsåker T, Marras TK, Maugein J, Milburn HJ, Mlinkó T, Morcillo N, Morimoto K, Papaventsis D, Palenque E, Paez-Peña M, Piersimoni C, Polanová M, Rastogi N, Richter E, Ruiz-Serrano MJ, Silva A, da Silva MP, Simsek H, van Soolingen D, Szabó N, Thomson R, Tórtola Fernandez T, Tortoli E, Totten SE, Tyrrell G, Vasankari T, Villar M, Walkiewicz R, Winthrop KL, Wagner D, Nontuberculous Mycobacteria Network European Trials Group. The geographic diversity of nontuberculous mycobacteria isolated from pulmonary samples: an NTM-NET collaborative study. Eur Respir J. 2013;42(6):1604–13.
34. Ito Y, Hirai T, Fujita K, Maekawa K, Niimi A, Ichiyama S, Mishima M. Increasing patients with pulmonary Mycobacterium avium complex disease and associated underlying diseases in Japan. J Infect Chemother. 2015;21(5):352–6.
35. Hu P, Bai L, Liu F, Ou X, Zhang Z, Yi S, Chen Z, Gong D, Liu B, Guo J, Tan Y. Evaluation of the Xpert MTB/RIF assay for diagnosis of tuberculosis and rifampin resistance in county-level laboratories in Hunan province, China. Chin Med J (Engl). 2014;127(21):3744–5025.
36. Wu J, Zhang Y, Li J, Lin S, Wang L, Jiang Y, Pan Q, Shen X. Increase in nontuberculous mycobacteria isolated in Shanghai, China: results from a population-based study. PLoS One. 2014;9(10):e109736.
37. Shao Y, Chen C, Song H, Li G, Liu Q, Li Y, Zhu L, Martinez L, Lu W. The epidemiology and geographic distribution of nontuberculous mycobacteria clinical isolates from sputum samples in the eastern region of China. PLoS Negl Trop Dis. 2015;9(3):e0003623.
38. Wendt SL, George KL, Parker BC, Gruft H, Falkinham JO 3rd. Epidemiology of infection by nontuberculous mycobacteria. III. Isolation of potentially pathogenic mycobacteria from aerosols. Am Rev Respir Dis. 1980;122(2):259–63.
39. Collins CH, Grange JM, Yates MD. Mycobacteria in water. J Appl Bacteriol. 1984;57(2):193–211.
40. Parker BC, Ford MA, Gruft H, Falkinham JO 3rd. Epidemiology of infection by nontuberculous mycobacteria. IV. Preferential aerosolization of Mycobacterium intracellulare from natural waters. Am Rev Respir Dis. 1983;128(4):652–6.
41. Meissner PS, Falkinham JO 3rd. Plasmid DNA profiles as epidemiological markers for clinical and environmental isolates of Mycobacterium avium, Mycobacterium intracellulare, and Mycobacterium scrofulaceum. J Infect Dis. 1986;153(2):325–31.
42. Fujita K, Ito Y, Hirai T, Maekawa K, Imai S, Tatsumi S, Niimi A, Iinuma Y, Ichiyama S, Mishima M. Genetic relatedness of Mycobacterium avium-intracellulare complex isolates from patients with pulmonary MAC disease and their residential soils. Clin Microbiol Infect. 2013;19(6):537–41.
43. Falkinham JO 3rd. Environmental sources of nontuberculous mycobacteria. Clin Chest Med. 2015;36(1):35–41.
44. Iseman MD, Corpe RF, O'Brien RJ, Rosenzwieg DY, Wolinsky E. Disease due to Mycobacterium avium-intracellulare. Chest. 1985;87(2 Suppl):139S–49S.
45. Prince DS, Peterson DD, Steiner RM, Gottlieb JE, Scott R, Israel HL, Figueroa WG, Fish JE. Infection with Mycobacterium avium complex in patients without predisposing conditions. N Engl J Med. 1989;321(13):863–8.
46. Szymanski EP, Leung JM, Fowler CJ, Haney C, Hsu AP, Chen F, Duggal P, Oler AJ, McCormack R, Podack E, Drummond RA, Lionakis MS, Browne SK, Prevots DR, Knowles M, Cutting G, Liu X, Devine SE, Fraser CM, Tettelin H, Olivier KN, Holland SM. Pulmonary

nontuberculous mycobacterial infection. A multisystem, multigenic disease. Am J Respir Crit Care Med. 2015;192(5):618–28.
47. Chen F, Szymanski EP, Olivier KN, Liu X, Tettelin H, Holland SM, Duggal P. Whole exome sequencing identify the 6q12-q16 linkage region and a candidate gene TTK for pulmonary nontuberculous mycobacterial disease. Am J Respir Crit Care Med. 2017;196:1599.
48. Holland SM, Pierce VM, Shailam R, Glomski K, Farmer JR. Case 28-2017. A 13-month-old girl with pneumonia and a 33-year-old woman with hip pain. N Engl J Med. 2017;377(11):1077–91.
49. Horsburgh CR Jr, Mason UG 3rd, Heifets LB, Southwick K, Labrecque J, Iseman MD. Response to therapy of pulmonary Mycobacterium avium-intracellulare infection correlates with results of in vitro susceptibility testing. Am Rev Respir Dis. 1987;135(2):418–21.
50. Research Committee of the British Thoracic Society. First randomised trial of treatments for pulmonary disease caused by M avium intracellulare, M malmoense, and M xenopi in HIV negative patients: rifampicin, ethambutol and isoniazid versus rifampicin and ethambutol. Thorax. 2001;56(3):167–72.
51. Hawkins CC, Gold JW, Whimbey E, Kiehn TE, Brannon P, Cammarata R, Brown AE, Armstrong D. Mycobacterium avium complex infections in patients with the acquired immunodeficiency syndrome. Ann Intern Med. 1986;105(2):184–8.
52. Mycobacterioses and the acquired immunodeficiency syndrome. Joint position paper of the American Thoracic Society and the Centers for Disease Control. Am Rev Respir Dis. 1987;136(2):492–6.
53. Horsburgh CR Jr, Selik RM. The epidemiology of disseminated nontuberculous mycobacterial infection in the acquired immunodeficiency syndrome (AIDS). Am Rev Respir Dis. 1989;139(1):4–7.
54. Chin DP, Reingold AL, Stone EN, Vittinghoff E, Horsburgh CR Jr, Simon EM, Yajko DM, Hadley WK, Ostroff SM, Hopewell PC. The impact of Mycobacterium avium complex bacteremia and its treatment on survival of AIDS patients--a prospective study. J Infect Dis. 1994;170(3):578–84.
55. Shafran SD, Singer J, Zarowny DP, Phillips P, Salit I, Walmsley SL, Fong IW, Gill MJ, Rachlis AR, Lalonde RG, Fanning MM, Tsoukas CM. A comparison of two regimens for the treatment of Mycobacterium avium complex bacteremia in AIDS: rifabutin, ethambutol, and clarithromycin versus rifampin, ethambutol, clofazimine, and ciprofloxacin. Canadian HIV Trials Network Protocol 010 Study Group. N Engl J Med. 1996;335(6):377–83.
56. Chaisson RE, Keiser P, Pierce M, Fessel WJ, Ruskin J, Lahart C, Benson CA, Meek K, Siepman N, Craft JC. Clarithromycin and ethambutol with or without clofazimine for the treatment of bacteremic Mycobacterium avium complex disease in patients with HIV infection. AIDS. 1997;11(3):311–7. PubMed PMID: 9147422.
57. Wallace RJ Jr, Brown BA, Griffith DE, Girard WM, Murphy DT. Clarithromycin regimens for pulmonary Mycobacterium avium complex. The first 50 patients. Am J Respir Crit Care Med. 1996;153(6 Pt 1):1766–72.
58. Wallace RJ Jr, Brown BA, Griffith DE, Girard WM, Murphy DT, Onyi GO, Steingrube VA, Mazurek GH. Initial clarithromycin monotherapy for Mycobacterium avium-intracellulare complex lung disease. Am J Respir Crit Care Med. 1994;149(5):1335–41.
59. Griffith DE, Brown BA, Girard WM, Murphy DT, Wallace RJ Jr. Azithromycin activity against Mycobacterium avium complex lung disease in patients who were not infected with human immunodeficiency virus. Clin Infect Dis. 1996;23(5):983–9.
60. Jeong BH, Jeon K, Park HY, Kim SY, Lee KS, Huh HJ, Ki CS, Lee NY, Shin SJ, Daley CL, Koh WJ. Intermittent antibiotic therapy for nodular bronchiectatic Mycobacterium avium complex lung disease. Am J Respir Crit Care Med. 2015;191(1):96–103.
61. Kobashi Y, Abe M, Mouri K, Obase Y, Miyashita N, Oka M. Clinical usefulness of combination chemotherapy for pulmonary Mycobacterium avium complex disease. J Infect. 2010;
62. Kobashi Y, Matsushima T. The microbiological and clinical effects of combined therapy according to guidelines on the treatment of pulmonary Mycobacterium avium complex disease in Japan – including a follow-up study. Respiration. 2007;74(4):394–400.

63. Olivier KN, Griffith DE, Eagle G, JP MG 2nd, Micioni L, Liu K, Daley CL, Winthrop KL, Ruoss S, Addrizzo-Harris DJ, Flume PA, Dorgan D, Salathe M, Brown-Elliott BA, Gupta R, Wallace RJ Jr. Randomized trial of liposomal amikacin for inhalation in nontuberculous mycobacterial lung disease. Am J Respir Crit Care Med. 2017;195(6):814–23.
64. Griffith DE, Eagle G, Thomson RM, Aksamit TA, Hasegawa N, Morimoto K et al. Amikacin Liposome Inhalation Suspension for Treatment-Refractory Lung Disease Caused by *Mycobacterium avium* Complex. Am J Respir and Crit Care Med, In Press, 2018.
65. van Ingen J, Ferro BE, Hoefsloot W, Boeree MJ, van Soolingen D. Drug treatment of pulmonary nontuberculous mycobacterial disease in HIV-negative patients: the evidence. Expert Rev Anti Infect Ther. 2013;11(10):1065–77.
66. Brown-Elliott BA, Nash KA, Wallace RJ Jr. Antimicrobial susceptibility testing, drug resistance mechanisms, and therapy of infections with nontuberculous mycobacteria. Clin Microbiol Rev. 2012;25(3):545–82.
67. Brown-Elliott BA, Iakhiaeva E, Griffith DE, Woods GL, Stout JE, Wolfe CR, Turenne CY, Wallace RJ Jr. In vitro activity of amikacin against isolates of Mycobacterium avium complex with proposed MIC breakpoints and finding of a 16S rRNA gene mutation in treated isolates. J Clin Microbiol. 2013;51(10):3389–94.
68. Griffith DE, Brown-Elliott BA, Langsjoen B, Zhang Y, Pan X, Girard W, Nelson K, Caccitolo J, Alvarez J, Shepherd S, Wilson R, Graviss EA, Wallace RJ Jr. Clinical and molecular analysis of macrolide resistance in Mycobacterium avium complex lung disease. Am J Respir Crit Care Med. 2006;174(8):928–34.
69. Adjemian J, Prevots DR, Gallagher J, Heap K, Gupta R, Griffith D. Lack of adherence to evidence-based treatment guidelines for nontuberculous mycobacterial lung disease. Ann Am Thorac Soc. 2014;11(1):9–16.
70. van Ingen J, Wagner D, Gallagher J, Morimoto K, Lange C, Haworth CS, Floto RA, Adjemian J, Prevots DR, Griffith DE, NTM-NET. Poor adherence to management guidelines in nontuberculous mycobacterial pulmonary diseases. Eur Respir J. 2017;49(2). pii: 1601855.

Laboratory Diagnosis and Antimicrobial Susceptibility Testing of Nontuberculous Mycobacteria

Barbara A. Brown-Elliott

Introduction

Prior to the 1990s, clinical mycobacteriology laboratories used phenotypic cultural characteristics and conventional biochemical testing for a large portion of nontuberculous mycobacteria (NTM) species identification [1].

Following the "biochemical era" (and even today), some larger reference laboratories, especially public health laboratories, and some research laboratories relied upon species identification based on chromatographic/chemotaxonomic methods including high-performance liquid chromatography (HPLC) of cell wall mycolic acids, thin-layer chromatography (TLC), and gas-liquid chromatography (GLC).

Beginning in the 1990s, the advent of molecular testing by PCR and gene sequencing for species and subspecies identification of NTM marked a new era for NTM with the subsequent explosion of more than 100 species being described compared to approximately 55 species identified from 1880 to 1990! The definition of a "species" is somewhat of a "moving target" requiring constant modification as newer diagnostic methodologies are introduced. Gene sequencing has now become the accepted reference method for the identification of NTM [2]. As with the previous methodologies, the accuracy and quality

B. A. Brown-Elliott
Department of Microbiology, Mycobacteria/Nocardia Research Laboratory, The University of Texas Health Science Center, Tyler, TX, USA
e-mail: Barbara.Elliott@uthct.edu

D. E. Griffith (ed.), *Nontuberculous Mycobacterial Disease*, Respiratory Medicine,
https://doi.org/10.1007/978-3-319-93473-0_2

of available databases are fundamental to the success of this technology [3]. If the database is inadequate or poorly curated, the results, which rely upon it, will also be poor. In other words, the results are only as good as the database from which they are derived.

More recently, matrix-assisted laser desorption-ionization time-of-flight mass spectrometry (MALDI-TOF-MS) has been introduced as a novel method for rapid and less expensive per test (after the initial purchase of the mass spectrometry instrument), but the full story of its efficacy and discriminatory power for mycobacterial species and subspecies remains to be seen, and as is important with the previous identification methodologies, the completeness of the organism database is paramount to the utility of the system. These databases are just now being developed, and there is no public database to use as an additional reference to the commercial companies that market MALDI instrumentation. It should also be noted that unlike MALDI of bacteria, mycobacterial isolates require specialized extraction methods which increase the hands-on time for preparation prior to loading into the instrument.

In the last few years, whole genome sequencing (WGS) has emerged as the newest approach, not only to determine the overall genomic relatedness of species and subspecies of NTM but also to help elucidate antimicrobial susceptibility and resistance patterns for specific antimicrobials as related to species or subspecies. Indeed, this method may well represent the basis of future identification algorithms, although currently only a few (mostly research or reference) laboratories are able to implement this methodology due to cost and the specialized education (i.e., bioinformatics) required to interpret and analyze sequences that, so far, is beyond the current capability of most clinical laboratories. The population databases for most NTM currently do not exist.

Antimicrobial susceptibility testing (AST) of NTM began with agar dilution methods utilized for the *Mycobacterium tuberculosis* complex (MTBC). Due to its laborious nature and inability to test large numbers of isolates, the agar dilution method for NTM has been abandoned except in a few research laboratories. The agar disk diffusion method for rapidly growing mycobacteria was introduced but was never validated or recommended by the Clinical and Laboratory Standards Institute (CLSI) and has suffered from lack of reproducibility and intra- and interlaboratory subjectivity of the reads. In contrast, the agar disk elution method utilizing commercial susceptibility disks has been successfully used and recommended by the CLSI but only for the AST of fastidious species such as *Mycobacterium haemophilum*. Despite the popularity of the gradient agar MIC method (i.e., E-test), a commercial system that utilizes a combination of the MIC and agar methodology, it has proven to be more difficult to get reproducible endpoints with mycobacterial species. The introduction of broth macrodilution and microdilution proved to be the most reliable AST for NTM species, and, to date, only the broth microdilution method has been recommended by the CLSI for testing of both slowly and rapidly growing species of NTM.

Laboratory Identification of Nontuberculous Mycobacteria (NTM)

Conventional Biochemical Testing

Prior to the advent and increased usage of molecular technologies, NTM were routinely identified using large batteries of biochemical tests. This conventional approach for RGM included arylsulfatase, nitrate reduction, sodium chloride tolerance, iron uptake, catalase production, and growth on single carbon sources such as citrate, mannitol, and inositol, along with growth rate, pigmentation, and colonial morphology. Currently only the latter three properties, morphologic, and phenotypic features, continue to be important or essential even in the current molecular era [4].

These batteries of key conventional biochemicals were developed and standardized over several decades and were used in most mycobacteriology laboratories for identification of NTM. The major disadvantages were that they required actual growth of the organisms and thus were slow in providing clinically relevant information in a timely manner, thus preventing rapid diagnosis and treatment of patients. Not only was this methodology slow, since it involved the actual growth of the organism in various substrates, but it also proved to lack discriminatory power to separate many newer species and subspecies and was generally poorly reproducible.

Carbohydrate utilization testing was also previously used to aid in identification of RGM, but the method was cumbersome, difficult to quality control, and no commercially standardized substrates were available making the method tedious and requiring in-house preparation of the media with time-consuming specific test validation. Moreover, as genetic techniques have shown, conventional testing results alone, even with the addition of carbohydrate utilization, are inconsistent and unable to differentiate many species of NTM accurately. Newer species were/are difficult to identify with this method, and fewer and fewer biochemical test results are provided with newer species. Thus, ultimately conventional biochemical methods have been replaced by more definitive molecular methodologies, recognizing that not all countries are yet able to make this transition.

High-Performance Liquid Chromatography (HPLC) Identification

Mycobacterial mycolic acid patterns of the cell wall vary within the species of NTM. Three methods, including thin-layer chromatography (TLC), gas-liquid chromatography (GLC), and high-performance liquid chromatography (HPLC), have been used in the identification of NTM species. The former two methods were mainly performed in research laboratories outside the United States of America

(USA), while HPLC gained popularity in the USA. Concurrent with conventional biochemical testing and also following afterward, HPLC analysis of mycobacterial cell wall mycolic acid content became popular, especially in larger reference laboratories including public health department laboratories.

Although HPLC is still being used, mostly in public health laboratories, this technology based on separation of mycolic acids by carbon length and charge has been shown to be limited for the identification of most species of NTM. Species are characterized by arrangement of major peaks from the eluting of the mycolics, the height of the peaks, and the retention time [2]. Most mycobacterial experts agree that this method can help to categorize the more common NTM species such as *M. avium* complex and *M. kansasii*, but it is not specific enough to identify most species of RGM accurately, and it suffers from a low discriminatory power among closely related species of slowly and rapidly growing mycobacteria. Limited data is also available for many of the newer species that are identified by molecular technology.

AST as a Taxonomic Tool

Prior to the molecular era antimicrobial susceptibility patterns, especially those of the RGM were used to provide a preliminary "screening test" for the identification of the most commonly encountered species. Although genetic sequencing is necessary for definitive species identification, AST provides useful taxonomic help especially for the *M. fortuitum* group and the *M. chelonae-M. abscessus* complex. Even the non-validated (for diagnostic AST) agar disk diffusion method can be used taxonomically to easily differentiate the *M. fortuitum* group from the *M. chelonae-M. abscessus* complex by susceptibility (defined as any size zone of inhibition) to polymyxin B [5]. Isolates of the latter complex show no zone of inhibition to polymyxin B in contrast to isolates of the former group which exhibit partial to full zones of inhibition with polymyxin B [5, 6].

The sulfonamides [typically trimethoprim-sulfamethoxazole (TMP-SMX)] also offer another taxonomic clue in that almost all isolates of the *M. fortuitum* group (except rare isolates that have been treated with sulfonamides in long-term regimens) show zones of inhibition or low MICs (≤2/38 μg/mL) as compared to isolates of the *M. chelonae-M. abscessus* complex which rarely exhibit zones of inhibition and have TMP-SMX MICs >2/38 μg/mL. Less than 10% of the *M. chelonae-M. abscessus* complex exhibit susceptibility either by agar disk diffusion (i.e., zones) or by low MICs (broth microdilution) [7].

The *M. chelonae-M. abscessus* complex can also be differentiated by susceptibility to cefoxitin and tobramycin [5]. Isolates of *M. chelonae* exhibit cefoxitin MICs >128 μg/mL and tobramycin MICs ≤4 μg/mL in contrast to isolates of the *M. abscessus* complex which have modal cefoxitin MICs 32–64 μg/mL and tobramycin MICs ≥8 μg/mL [8, 9].

Caution must be exercised with isolates showing cefoxitin MICs >128 μg/mL and tobramycin MICs ≥8 μg/mL as these characteristics are also shared by isolates of *M. immunogenum*. By agar disk diffusion, isolates of *M. immunogenum* have small to no zones of inhibition with tobramycin and equivalent amikacin and kanamycin zones in contrast to the isolates of *M. chelonae-M. abscessus* complex which almost always have larger zones of inhibition with kanamycin than amikacin [5, 10].

In general, isolates of the *M. fortuitum* group show more overall susceptibility to antimicrobials than isolates of the *M. chelonae-M. abscessus* complex [5, 11]. Additionally, untreated isolates of *M. fortuitum* group are almost always susceptible to fluoroquinolones. Rarely patient isolates of this group after long-term quinolone treatment may show fluoroquinolone resistance [12]. However, most often, quinolone susceptibility provides another helpful marker for the *M. fortuitum* group, although some isolates of *M. chelonae* (less than 40%) may also show susceptibility to this class of antibiotics.

Isolates of the *M. mucogenicum* group (*M. mucogenicum*, *M. phocaicum*, *M. aubagnense*) are the only nonpigmented species of RGM which show susceptibility to cephalothin [5, 13]. Additionally, like isolates of *M. immunogenum*, these isolates exhibit equivalent kanamycin and amikacin zones and are polymyxin resistant.

Molecular Identification

Commercial Nucleic Acid Probes for NTM

For more than 25 years, clinical mycobacteriology laboratories worldwide, but especially in the USA, have relied upon commercial single-stranded DNA nucleic acid probes (Accuprobe, Hologic Gen-Probe, San Diego, CA) to rapidly detect and identify MTBC and NTM including *M. avium* complex (MAC as a combined probe), *M. intracellulare* and *M. avium* (separate species probes), *M. kansasii*, and *M. gordonae* [14, 15].

The Accuprobe employs acridinium ester-labeled oligonucleotide probes complementary to 16S rRNA in the target organisms. Colonies on solid media or in broth cultures provide the target nucleic acid for these assays. Extraction of nucleic acids following lysis with appropriate buffers and heat inactivation ensures safety (i.e., non-viability) during the procedure. The hybridization results are measured with a luminometer [16, 17]. Generally, species accuracy and sensitivity have been good, although some cross-reactivity has been reported, and it should be noted that most laboratories do not sequence isolates to confirm their probe identification once a probe is positive for one of the aforementioned species so the accuracy may be overestimated [18]. Although the probes can be costly, the decreased labor and shortened turnaround time can offset the expense. Moreover, more than half of the species commonly encountered in the clinical laboratory fall into one of these groups, and liquid cultures, which are typically becoming positive before solid media cultures, can be probed early without waiting for growth on solid media [19–22].

Despite the utility of the commercial probes, they are not able to identify all pathogenic and nonpathogenic mycobacteria, and thus other methods of identification must be used.

Line Probe Assays

INNO-LiPA (Innogenetics, Ghent, Belgium)

To identify a wider variety of species of NTM, INNO-LiPA genetic probe strip techniques based on the application of PCR plus reverse-hybridization DNA have also been developed. In general, the target sequences from culture growth on solid or in liquid media are amplified using PCR and biotinylated primers. Subsequently, the amplified PCR products are hybridized to nitrocellulose membrane-bound species-specific fragments on a strip followed by an enzyme-mediated color reaction. The species-specific banding patterns are visually analyzed following a colorimetric conjugation step by comparison to a commercially available chart which is coded to the NTM species identification.

Using this technology, there are currently three commercial testing systems available. Until approximately 10–15 years ago, the INNO-LiPA multiplex probe reverse hybridization based on nucleotide differences in the 16S-23S rRNA gene spacer region was used only outside the USA. Although, like the other two systems, it is still not US Food and Drug Administration (FDA) cleared, it is available in the USA and offers a method of identifying more species than the Accuprobe system [15]. In fact, this system has the capability of identifying both RGM and slowly growing NTM species by a single probe, and unlike Accuprobe, the technologist does not need to select the appropriate probe.

Limitations of the INNO-LiPA include the cross-reactivity with the species within the *M. fortuitum* group and several less commonly isolated species such as *M. thermoresistibile*, *M. agri, and M. alvei* [23] Furthermore, discrimination of some closely related species such as *M. chelonae and M. abscessus* can be problematic. The INNO-LiPA separates the *M. chelonae-M. abscessus* complex into three different groups, and in one study 20/21 isolates of *M. chelonae* were confirmed by *rpo*B sequence. In the same study, 24/38 isolates of *M. abscessus* were by INNO-LiPA and 14/38 were *M. abscessus* subsp. *bolletii* by *rpo*B sequence [24]. To address these issues, the manufacturer has included an additional probe specific for the *M. fortuitum-M. peregrinum* complex and also *M. smegmatis* [23, 25, 26]. The system also is capable of identifying several commonly isolated slowly growing NTM including *M. kansasii*, *M. gordonae*, *M. simiae*, *M. marinum*, and *M. avium* complex. It can also identify several fastidious species including *M. xenopi*, *M. genavense*, *M. malmoense*, and *M. haemophilum* [27]. In a 2014 study by de Zwaan et al., in the Netherlands, 417/455 isolates of NTM could be identified beyond the genus level, and 348/417 showed similarity to *rpoB* sequence results [24].

The INNO-LiPA MAIS probe may also cross-react with *M. arosiense*, *M. mantenii*, *M. heidelbergense*, *M. nebraskense*, *M. parascrofulaceum*, and *M. paraffinicum*, but importantly so far the system has not misidentified MTBC as NTM [18].

GenoType Mycobacterium CM/AS (Hain Lifescience, Nehren, Germany)

The second-line probe assay method is based on the detection of species-specific sequence in the 23S rRNA gene. This system utilizes two strips (CM, for "common mycobacteria," and AS, for "additional species") [15, 28] and provides probes for simultaneous identification of *M. chelonae* with specific probes for *M. peregrinum*, *M. fortuitum*, and *M. phlei* [28, 29]. The GenoType assay also identifies slowly growing NTM including the common species such as *M. avium*, *M. gordonae*, *M. interjectum*, and *M. kansasii* but is unable to discriminate between the following NTM groups: *M. intracellulare/M. chimaera*, *M. scrofulaceum/M. paraffinicum/M. parascrofulaceum*, *M. malmoense/M. haemophilum/M. palustre/M. nebraskense*, MTBC/*M. xenopi*, and *M. marinum/M. ulcerans* [27]. Using the AS strip, NTM species including *M. simiae*, *M. celatum* (types 1 and 3 results should only be reported if obtained from solid cultures with typical colony morphology and growth rate), *M. lentiflavum*, *M. heckeshornense*, *M. kansasii*, *M. ulcerans*, *M. gastri*, *M. asiaticum*, and *M. shimoidei* can be reported but with *M. genavense/M. triplex*, *M. szulgai/M. intermedium*, and *M. haemophilum/M. nebraskense* sharing the same patterns [27].

Recent studies also show that the GenoType *M. intracellulare* probe cross-hybridizes with several other NTM including *M. arosiense*, *M. chimaera*, *M. colombiense, and M. mantenii* in the "MAC X" complex and *M. saskatchewanense* [18]. Two rare NTM species, *M. riyadhense* and the non-validated species *M. simulans*, have been incorrectly identified as MTBC by the GenoType method [18]. However, similar to the INNO-LiPA, no MTBC isolates have been misidentified as NTM using the GenoType [18].

Speed-Oligo Mycobacteria (Vincel, Granada, Spain)

Recently a third-line probe assay has been developed based on targeting both 16S rRNA and 16S–23S rRNA regions [30]. This system has a limited number of species which it is capable of identifying compared to the other two-line probe assays. It also is only able to identify some species to only the group or complex level. So far, only limited reports of its use have been published. A major advantage of the system is its speed and ease to differentiate between NTM species and MTBC even in direct testing of respiratory samples [31, 32].

The Speed-oligo mycobacteria system can identify *M. avium*, *M. gordonae*, *M. interjectum*, *M. intracellulare*, *M. kansasii*, *M. malmoense*, *M. xenopi, and M. scrofulaceum* in addition to *M. marinum/M. ulcerans* which share the same pattern [27]. A major misidentification has been reported with *M. marinum* identified as *M. kansasii* [27].

Polymerase Chain Reaction (PCR) Restriction Fragment Length Analysis (PRA)

The initial shift toward molecular identification of NTM involved the use of PCR by amplification of a 441-bp segment of the 65 kDa heat shock protein (*hsp*) and a subsequent restriction fragment analysis method as described by Telenti et al. [33] and later modified by Steingrube et al. [34]. Moreover, unlike some gene sequence methodology, PRA will differentiate *M. chelonae* from *M. abscessus*, and additional enzymes even allow for the discrimination of *M. abscessus* subsp. *abscessus*, subsp. *massiliense*, and subsp. *bolletii*. Using this method, the amplified products of the PCR were digested using specific restriction enzymes (BstEII and HaeIII) and subsequently analyzed by agarose gel electrophoresis [27]. The resulting gel patterns are compared to an algorithm as prepared with the use of an in-house database as there is no commercial database available. Many of the patterns were specific for clinically significant species [35]. An algorithm describing 34 species, including new subgroups of *M. kansasii and also several RGM species* (*M. porcinum*, *M. senegalense*, *and M. peregrinum*), was proposed by Devallois et al. [35]. This simple and rapid method system was widely used for several years because a viable culture of the NTM was not necessary and the method was more rapid than conventional biochemical testing and less expensive than gene sequencing. However, technical difficulties such as small size differences between the fragments and the finding of some closely related NTM species with non-distinguishable patterns along with the need to establish large in-house databases [3] became the major disadvantages of this method which has subsequently been replaced by more specific gene sequencing procedures.

Algorithms for identification of NTM using PRA along with a web accessible database have been published [36].

Currently the 441-bp of the *hsp* 65 gene (Telenti fragment) remains the most useful sequence used for PRA. However, many newly described species of NTM have not been extensively studied using PRA.

Amplification Followed by DNA Sequencing

With the increasing numbers of NTM being described and the need for accurate identification to species level, analysis of gene sequences is currently considered the gold standard for taxonomic identification of NTM species. The nucleic acid source for sequencing can either be a colony or cell pellet from broth cultures. Sequencing identification is accomplished by amplification of a selected segment of the mycobacterial gene and determination of the nucleotide sequence traditionally by Sanger sequencing.

Once sequence data is obtained, comparison to a public, commercial, or in-house-validated database is an important step. Not to be overemphasized, integrity of the database is dependent upon the quality of sequences added to the library. Each database should be constantly monitored and should include new data as it becomes available. Rigorous quality control and curating of each entry is critical. As with any other identification method, sequencing results should correlate with key phenotypic characteristics. Mixed cultures can produce uninterpretable sequences. Development of an in-house algorithm to identify NTM requires expertise and a quality controlled database which should include supplementation with reference (especially type) strains.

Consistent with implementation of other laboratory methodology, the procedures should undergo careful evaluation, rigorous validation, verification, and quality control including appropriate sequencing controls with each run.

A major caveat for single gene target sequencing is that because many NTM are closely related and may be indistinguishable, additional molecular and/or phenotypic properties may need to be assessed to discriminate species [37].

Partial 16S rRNA Gene Sequence Analysis

For NTM, the identification focuses on two important hypervariable regions, known as region A and region B, which are located at the 5′ end of the 16S rRNA gene. These regions correspond to *Escherichia coli* positions 129–267 and 430–500, respectively. Region A contains most of the species-specific or "signature sequences" in mycobacteria [2]. This information stretch of approximately 500-bp usually allows taxonomic identification of many RGM and slowly growing NTM species and is the basis for the only commercially available sequencing method known as MicroSEQ (Life Technologies, Foster City, CA) [38–40]. The premade amplification and sequencing master mixes significantly decrease technologist time and make quality control easier [40]. This system uses universal primers for the amplification and sequencing of the first 500-bp sequence and subsequently compares to a commercially prepared database composed primarily of type strains (one per species). The major limitation of the partial (500-bp) sequencing method is the inability to distinguish between species that have identical hypervariable regions A and B or an identical complete 16S rRNA gene sequence such as the slowly growing NTM species: *M. marinum/M. ulcerans*, *M. kansasii/M. gastri*, and *M. genavense/M. simiae*. Interestingly, the complete, approximately 1500-bp sequence, among the RGM, is identical for the three subspecies of *M. abscessus*, *M. chelonae*, and *M. franklinii* and among the slow growers that are identical for both *M. genavense* and *M. simiae* and also *M. kansasii* and *M. gastri* and as stated previously *M. marinum* and *M. ulcerans*. Thus, most clinical laboratories have supplemented the commercial database with additional sequences from their own in-house databases or other sequence libraries such as RIDOM (Wurzburg, Germany) or GenBank. Caution must be used, however, with public databases as often they may contain errors [41]. It is critical to use a quality-controlled database or at a minimum to evaluate and confirm the sequences in a non-quality-controlled database such as GenBank [3].

Table 1 Algorithm for identification of 16S rRNA gene sequences[a]

% Sequence identity	Report	Example
100	"*Mycobacterium* + species"	*Mycobacterium fortuitum*
99–99.9	"Most closely related to *Mycobacterium* + species"	Most closely related to *Mycobacterium fortuitum*
95–98.9	"Unable to identify by 16S rRNA gene sequence; most closely related to *Mycobacterium* sp."	Unable to identify; most closely related to *Mycobacterium* sp.

[a]Modified from CLSI MM18-A, 2008 [41] using a minimum sequencing target of 300 bases

For accurate identification of species with the same partial 16S rRNA gene sequence, sequencing outside the first 500-bp region such as for *M. chelonae-M. abscessus* which differs in the 3′ region and can thus be exploited for differentiation by sequencing that region or targeting another gene for sequencing is necessary. To ensure accuracy, at least 300-bp of a quality sequence should be compared between the reference and the query segments and should cover at least one domain of the gene where variations should be expected to occur. For this reason, most clinical laboratories sequence between 450- and 480-bp [41].

The 2007 CLSI interpretation criteria recommend guidelines to ensure more consistency in reporting among clinical laboratories. Currently the cutoff recommendations for 16S rRNA gene sequences do not reflect strict taxonomical classification (see Table 1) [41].

Complete 16S rRNA Gene Sequence

Sequencing of the complete 16S rRNA gene (approximately 1500-bp) is not a practical technique for routine identification of NTM, but in some cases as previously mentioned, it is necessary for differentiation of closely related species that differ by only a few base pairs. Such is the case in the discrimination of two pigmented RGM, i.e., *M. goodii* and *M. smegmatis*, which differ in the complete 16S rRNA gene by only 4-bp [42]. Among slowly growing NTM, as previously mentioned, *M. marinum* and *M. ulcerans* and sequence variants I and IV of *M. kansasii* and *M. gastri* have identical complete 16S rRNA gene sequences.

rpoB Gene Sequencing

As previously discussed, the 16S rRNA gene sequence lacks discriminatory power to identify some closely related groups of NTM, especially the RGM in the *M. fortuitum* and *M. chelonae-M. abscessus* complex [43]. Thus, sequencing of the *rpoB* gene an almost 3600-bp gene which encodes the *B*-subunit of RNA polymerase has been suggested as a secondary target for the identification of NTM including non-pigmented and late-pigmented RGM [44] and species within the MAC [45]. A major problem is that there are no CLSI-established cutoff values based on species population studies for genes other than the 16S rRNA gene. Several different regions to the *rpoB* have been proposed for amplification and sequence determination among species of NTM. Kim et al. [46] used a 306-bp segment (Region III), while Lee et al. [47] amplified a 360-bp fragment for NTM including slowly growing species. In contrast, Adékambi targeted an approximately 760-bp segment (Region V) mostly for pre-identification of RGM including two newer species *M. phocaicum* and *M. aubagnense.* Adékambi et al. noted a 98.2% similarity between *M. fortuitum* and *M. houstonense.* Likewise, Ben Salah et al. found a 99.2% similarity between *M. intracellulare* and *M. chimaera* which suggests that *M. chimaera* could be a subspecies of *M. intracellulare.* It has also been suggested that if *rpoB* sequencing is used alone, horizontal gene transfer may cause misidentification of the subspecies *M. abscessus* [37]. However, it has also been noted that other gene targets should also be affected [48].

A second major limitation of *rpoB* sequencing is the lack of sequence databases for reference strains, especially of the less commonly recovered NTM species, and as for all molecular-based techniques, the absence of high-quality public sequence databases result in incomplete or incorrect sequences which can in turn result in misidentification or at best difficulty in comparing sequences. Multiple problems including ambiguous base designations and submissions labeled only as *Mycobacterium* sp. emphasize the need to evaluate and update databases and to use public databases with caution.

A recent study by de Zwaan et al., in the Netherlands, revealed a higher discriminatory power of the *rpoB* sequencing as compared to a reverse line probe method (INNO-LiPA) even with the addition of 16S rRNA gene sequence. (See discussion in Line Probe Assay (INNO-LiPA) section.) The investigators found that the greatest difference between *rpoB* gene sequencing as compared to the INNO-LiPA/16S rRNA gene sequence was the grouping of species by the latter method, whereas the *rpoB* sequence was able to provide species level identification of the NTM. The *rpoB* gene sequencing was able to identify more known species (among the MAC, *M. vulneris*, *M. colombiense*, *M. timonense*, and *M. yongonense* and *M. porcinum*, *M. houstonense*, *M. septicum*, *M. peregrinum*, and *M. setense* within the *M. fortuitum* group). It also identified less common species including *M. mantenii*, *M. heidelbergense*, *M. novocastrense*, *M. alvei*, and the non-validated species *M. tilburgii* [24]. Interestingly, four groups of species showed debatable mutual similarities including *M. mantenii*, *M. intracellulare* and *M. chimaera*, *M. fortuitum* and *M. houstonense*, *M. peregrinum* and *M. alvei*, and the *M. abscessus* subsp. *massiliense*, subsp. *bolletii*, and subsp. *abscessus* using strictly applied cutoff values [24]. This study definitively illustrated the utility of the *rpoB* gene sequencing for the differentiation of both RGM and slowly growing NTM.

Heat Shock Protein (hsp65) Gene Sequencing

The 65 kDA *hsp* gene is less conserved among mycobacterial species and thus shows more interspecies and intraspecies polymorphisms compared to those in the 16S rRNA gene sequence [49]. This variability has been exploited in the identification of closely related species of NTM using a 441-bp sequence known as the Telenti fragment [33]. The *hsp*65 gene allows for differentiation of several RGM species closely related by the 16S rRNA gene including *M. peregrinum*, *M. septicum*, *M. houstonense*, *M. fortuitum*, *M. porcinum*, and *M. senegalense.* Also in contrast to partial 16S rRNA gene sequencing, *M. chelonae* and *M. abscessus* can be easily discriminated based on almost 30-bp differences in the 441-bp segment, as can the three subspecies of *M. abscessus*. Additionally, among the slowly growing NTM, *M. marinum* is differentiated from *M. ulcerans, M. kansasii* from *M. gastri* and using the more variable 3′ region, subsets including *M. avium subsp. avium* (bird strains), *M. avium* subsp. *paratuberculosis* (bovine), *M. avium* subsp. *paratuberculosis* (ovine) and *M. avium* subsp. *"hominissuis"*(porcine and human) can be discerned [50].

A web-based database for NTM sequenced by *hsp*65 gene is also available [36]. As for other sequencing targets, development of an in-house-validated database is critical for clinical use of the sequencing of the *hsp*65 gene.

Erythromycin Ribosomal Resistance Methylase (erm) Gene Sequence

The presence of a functional rRNA methylase (erythromycin resistance) *erm* gene by methylation of the 23S rRNA binding of macrolides to the ribosomes induces macrolide resistance in most common species of RGM and two subspecies of *M. abscessus* (most *M. abscessus* subsp. *abscessus and M. abscessus* subsp. *bolletii*). This gene is harbored in several pathogenic RGM as shown in Table 2.

Treatment of *M. abscessus* subsp. *abscessus* has been hampered by its resistance to standard antituberculous agents and other antimicrobials including the

Table 2 Presence of functional *erm* gene types harbored in rapidly growing mycobacterial species

Erm type	Species
erm(38):	*M. goodii*, *M. smegmatis*
erm(39):	*M. fortuitum*, *M. houstonense*, *M. porcinum*, *M. neworleansense*
erm(40):	*M. mageritense*, *M. wolinskyi*
erm(41):	*M. abscessus* subsp. *bolletii* and *M. abscessus* subsp. *abscessus* but not *M. abscessus* subsp. *massiliense* which has a large deletion and inactive (non-functional) *erm* (gene)

References: [51–57]

recent discovery that approximately 80% of the isolates of *M. abscessus* subsp. *abscessus* recovered in the USA harbor a functional inducible methylase gene, *erm*(41), which confers macrolide resistance [57]. There are at least seven functional sequevars associated with macrolide resistance. There are also at least three known nonfunctional sequevars that contain a T28G substitution causing a change in protein structure, resulting in a nonfunctional *erm* gene; the latter isolates should be assumed to be macrolide susceptible unless a mutation in the 23S rRNA gene is confirmed [51, 57]. Studies by Kim et al. used the *erm*(41) gene sequences to differentiate *M. abscessus* subsp. *abscessus* from *M. abscessus* subsp. *massiliense* (truncated nonfunctional *erm*) and *M. abscessus* subsp. *bolletii* [52]. This finding has been illustrated in a study by Koh et al., in which patients with *M. abscessus* isolates with a nonfunctional *erm* gene (i.e., those with *M. abscessus* subsp. *massiliense* (a subspecies that harbors a truncated nonfunctional *erm* gene)) had an approximately 90% therapeutic response rate to macrolide-containing regimen compared to those patients with *M. abscessus* subsp. *abscessus* who only had an approximate 25% response rate [53].

Two other RGM groups consist of species with functional *erm* genes and thus are considered macrolide resistant. This includes the *M. fortuitum* group *erm*(39) with related species *M. fortuitum*, *M. porcinum*, *M. houstonense*, and *M. neworleansense*, the *M. smegmatis* group (*M. smegmatis* and *M. goodii*) *erm*(38), and *M. wolinskyi* and *M. mageritense erm*(40). In contrast to the macrolide-susceptible species, *M. peregrinum*, *M. senegalense*, *M. chelonae*, *M. immunogenum*, and the *M. mucogenicum* complex (*M. mucogenicum*, *M. phocaicum*, *M. aubagnense*) do not harbor functional *erm* genes and to date have no sequencing evidence for the presence of any *erm* genes [51, 54, 58].

Recent in vitro studies have shown that sequencing of the *erm* gene or its absence among isolates of RGM can predict macrolide (clarithromycin) resistance. These studies have also demonstrated the absence of a functional *erm* gene in *M. chelonae*, *M. immunogenum*, and the *M. mucogenicum* group [58]. Thus, these RGM species should be macrolide susceptible unless a mutation is present as indicated by an initial (3–5 days) clarithromycin MIC >8 μg/mL. Moreover, *M. abscessus* subsp. *massiliense* as previously mentioned has a truncated *erm* gene that renders the subspecies macrolide susceptible. A proposal to the CLSI has been made (but not yet approved) that this sequencing information combined with the 3–5-day clarithromycin MIC to exclude mutational resistance could be substituted for extended (up to 14 days) incubation to detect a functional *erm* gene, allowing for a more rapid turnaround time and more expeditious selection of antimicrobial therapy [58].

Further studies are necessary to evaluate the utility of sequencing of the *erm* gene in other RGM species including other species within the *M. fortuitum* and the *M. smegmatis* and the *M. wolinskyi/mageritense* groups. Therefore, until such evaluations are completed, extended incubation is required to assess macrolide resistance unless the isolate has a 23S rRNA gene mutation evident by a 3–5-day Clarithromycin MIC >8 μg/mL.

Sequencing of Other Gene Targets

Other gene targets including the 16S-23S rRNA internal transcribed spacer (ITS) region, *dnaJ*, *sec* A1, and *rec* A, have been proposed for taxonomic identification of NTM [2, 44, 59, 60]. However most of the research to date suggests that these targets are more variable than *hsp*65, and thus they have not been widely utilized. As always, the updated and accurate sequence databases for each of the targets is lacking, especially for newer species. Currently, most investigators have recommended a multigene target approach for taxonomic evaluation of the NTM not identified by partial 16S rRNA gene sequencing [61].

A 2011 report by Macheras et al. recommended a multigenic target approach using eight housekeeping gene sequences including *arg*H, *cya*, *glp*K, *gnd*, *mur*C, *pgm*, *pta*, and *pur*H [37]. However, for the routine clinical laboratory, this algorithm is too laborious. This work also questions the previously accepted >3% *rpoB* (Region V) sequence divergence cutoff between RGM species [44, 62].

The sequence of the ITS 1 region is a 200–330-bp target region that separates the 16S and 23S rRNA genes. Several primer sites for amplification and sequencing of this segment have been proposed [63]. The ITS 1 sequence has a high variability that can be used for species discrimination especially with the RGM; some slowly growing species such as *M. simiae* and *M. xenopi* have two or more sequence variants and thus cannot be differentiated using this technique [64]; however, ITS can be useful for other slowly growing species including the MAC [65, 66].

As previously mentioned, the *erm* sequence along with *rpoB* sequencing has recently been proposed for identification of many RGM, especially the *M. abscessus* complex [52, 67]. These results have been in agreement with multilocus sequence analysis including *rpoB*, *hsp65*, *sod A*, and the 16S-23S ITS although some investigators report that *sod A* has little utility in species identification [62].

Pyrosequencing

Pyrosequencing (Biotage, Uppsala, Sweden) differs from Sanger sequencing in that it employs short-segment (20–30-bp) nucleic acid sequencing of hypervariable region A of the 16S rRNA gene, and instead of chain termination with dideoxynucleotides (i.e., Sanger), it relies on detection of pyrophosphate during DNA synthesis [68–70]. During sequencing, visible light that is proportional to the number of nucleotides is produced [27].

Although pyrosequencing is not as discriminatory as Sanger sequencing and may require additional sequencing due to the short sequence length generated by this method, it has advantages in being rapid, relatively simple, and a less expensive molecular method than traditional sequencing approaches [23, 27, 70]. A 2010 study by Bao et al. showed that pyrosequencing could identify 114/117 (97.4%)

isolates of NTM to species level including slowly growing NTM, other than *M. scrofulaceum* and *M. simiae* [68]. In an earlier 2005 study by Tuohy and colleagues, 40/50 RGM isolates had agreement by pyrosequencing to traditional sequencing results. Closely related species within the *M. fortuitum* group were problematic [70].

A recent proposal to use a 60-bp segment of the *rpoB* gene along with pyrosequencing successfully identified 99/100 isolates of *M. abscessus and M. chelonae* [71]. Currently, pyrosequencing is primarily used only in research and large reference laboratories until more extensive studies are completed.

Guidelines for Comparison and Interpretation of Gene Sequences

In 2010, following the publication of CLSI guidelines for interpretation of 16S rRNA gene sequencing in 2008, Tortoli proposed a standard operating procedure and recommendation for optimal identification of NTM using the 16S rRNA gene [15, 41, 48].

1. Annotations accompanying the sequence in the database provide useful information for evaluation and comparison of sequences.
   - Source of strain (clinical or environment strains are often less reliable than established reference strains, especially type strains)
   - Reputation of submitter (a more established submitter suggests a higher reliability for the sequence)
   - Noting the year of the submission to determine if the sequences were derived using outdated techniques which would suggest less reliability than the use of updated techniques
2. Comparison to a reference strain (especially a type strain) is optimal even if the similarity percentage is lower than for other entries. Careful evaluation of the presence of ambiguous nucleotides in the database sequence or incorrect interpretation of the electropherogram should select the best choice for identity.
3. The presence of multiple entries assigned to the same species increases the confidence level of the identification.
4. Note bibliographic references if present.
5. Cautiously compare old sequence data. The lack of close similarity between a reference strain and the confirmed type strain should be viewed with suspicion.
6. If the sequence is derived from a non-validated species, the test isolate cannot be assigned to a species and remains unidentified.
7. Organism sequences should be ordered according to the highest similarity score using the "maximum identity" column.
8. Recovery of a single isolate is inadequate to establish a new species identity. Comment should "suggest" the "possibility" of undescribed species.

As noted previously, misidentified sequences and submission errors are not uncommon. Although a test sequence may be 100% identical to the sequence in the database, it is important to confirm the identification using the measures cited above. BLAST (Basic Local Alignment Search Tool) is a routinely used program that can be used to locate sequences similar to the test reference [15]. Although a quality electropherogram is important, the ultimate accuracy of the database and the ability of the user to carefully interpret and compare sequences are also critical.

Matrix-Assisted Laser Desorption-Ionization Time-Of-Flight (MALDI-TOF) Mass Spectrometry (MS)

MALDI-TOF-MS is one of the newest techniques introduced for the identification of NTM [1]. This method bases identification of organisms on the creation of a unique species-specific spectral profile (fingerprint) produced by extracted ribosomal proteins with specific mass to charge ratios [72, 73]. Inactivated mycobacterial cells are extracted, applied to a target mass spectrometer plate, and overlaid with a matrix solution. Additionally, safety (to rule out MTBC) must be considered so that the specimens should be rendered nonviable before analysis, and the extraction should be performed within a biosafety cabinet. The accuracy of the MALDI-TOF-MS is dependent upon obtaining quality spectra which can be challenging due to the complex cell walls of mycobacteria. One of the greatest advantages of this method is its speed with traditional bacterial identification. However, for mycobacteria a pre-extraction procedure usually involving bead beating or vortexing in formic acid, ethanol, and acetonitrile is necessary which adds to the hands-on time [74]. The application of a colony directly onto the target plate (whole cell preparation) is not feasible for mycobacteria. After extraction, the proteins are then spotted onto the target plate. The mass spectrometer is operated under a high vacuum, and the laser fires randomly within each well to produce spectra of sufficient intensity. The spectra are recorded, and the instrument software processes and compares the collected spectra and determines a standard duration for peak intensity and placement. The patterns (mass spectra) generated are compared to a consensus profile using all spectra within statistically significant threshold values [73, 75–77].

There are currently two commercially available MALDI-TOF-MS systems. A major concern to laboratories implementing the MALDI-TOF-MS is the initial expense (almost twice the price of a genetic analyzer), although some cost may be offset by less expensive consumables and more rapid turnaround time [15, 73]. Since 2011, several studies have been published comparing the MALDI-TOF-MS method to gene sequencing-based identification. As is important for all identification systems, validation of the method and creation of a database usually augmented with in-house data to supplement the commercial libraries are critical and are major challenges especially for the identification of NTM [72]. Additionally, it has been noted that the robustness of the spectral database with MALDI-TOF is improved if the same extraction method is used for the development of the database and the analysis of unknown test isolates.

Previous studies have shown MALDI-TOF to be useful for the identification of both RGM and slowly growing NTM. A major caveat is the use of pure cultures for accurate identification. For organisms grown on solid media, a single well-isolated colony should be used when possible. After selection is made, the isolation plate should be reincubated. Upon extended incubation, if multiple colony types are noted, each colony morphotype should be isolated and analyzed by MALDI-TOF-MS.

Limitations of the MALDI-TOF-MS with NTM include lack of discrimination with some clinically important closely related species and subspecies including *M. abscessus* subsp. *abscessus*, *M. abscessus* subsp. *massiliense*, and *M. abscessus* subsp. *bolletii*; *M. avium*, *M. avium* subsp. *avium*, *M. avium* subsp. *paratuberculosis*, and *M. avium* subsp. *silvaticum*; *M. chelonae* and *M. abscessus*; *M. intracellulare* and *M. chimaera*; and *M. mucogenicum* and *M. phocaicum* and the *M. fortuitum* group [73, 75–78].

As with all identification methods, the final instrument result should also correlate with known phenotypic characteristics including growth rate, pigmentation, and colony morphology. General consensus among experts is that organism identifications unfamiliar to the testing laboratory should be confirmed using these key phenotypic properties. If no identification is achieved with a good quality spectrum, this may mean that the organism is not represented in the database being applied. In contrast, if the spectral pattern is poor, there may be a mixed culture present, and colony purity should be investigated. In almost all failures, the first action step is to repeat the analysis. Most experts recommend supplemental molecular testing methods, or referral to a qualified reference laboratory may be required to discriminate between closely related species if the isolate is considered clinically significant [i.e., important to public health implications (usually not relevant with NTM but rather with MTBC)], linked to serious co-morbidities, carries specific prognostic implications, and/or resistant to commonly used antimicrobials. If the isolate is not considered clinically relevant, a report to genus, group, or complex level may be a reasonable alternative.

Table 3 provides an overview of the most commonly used methods of laboratory identification for the nontuberculous mycobacteria.

Molecular Strain Typing Systems

Pulsed-Field Gel Electrophoresis (PFGE)

The most widely used method for molecular strain typing is PFGE. The principle of PFGE is based on the premise that a periodic alternating electrical field causes DNA fragments to change directions and allows large molecules to be separated. By using restriction enzymes, larger DNA fragments >50 Kb can be separated.

Table 3 Laboratory methods for identification of nontuberculous mycobacteria

Method	Comment
Conventional biochemicals	Not useful for definitive species identification
	Should be replaced by molecular or proteomic method
High-performance liquid chromatography (HPLC)	Not suitable for definitive species identification
Commercial nucleic acid probes	Only useful for a few species:
	Mycobacterium avium, *M. intracellulare*, *M. gordonae*, *M. kansasii*
	May be cross-reactivity between other *Mycobacterium avium* complex and other mycobacteria
Polymerase chain reaction (PCR) Amplification followed by restriction enzyme analysis (PRA or REA)	Replaced by sequencing. Not useful for definitive species identification especially with newer species/subspecies
Line probe assay	Useful for species identification but there can be cross-reactivity with similar species
Gene sequencing	Useful for definitive species identification for most clinically relevant species. Specificity depends upon selection of gene target (16S rRNA, *rpoB*, *hsp*65, etc.). Sequence dependent upon updated and accurate database
Matrix-assisted laser desorption Time-of-flight mass spectrometry (MALDI-TOF MS)	Useful for species identification of many clinically relevant species. Cannot differentiate between subspecies of *M. abscessus*. Also cannot differentiate among species within the MAC, *M. fortuitum* and *M. mucogenicum* groups.
	Quality of results dependent upon updated and accurate database
	No public database

Generally the method requires an actively growing liquid culture and an average of 3–4 weeks for processing to prepare genomic DNA. A suspension of NTM is first embedded in agarose plugs, cells are lysed, and chromosomal DNA is digested with restriction enzymes. The plugs are then loaded into the wells of the gel and placed on the instrument. Problems with the technique include cell clumping or irregular amounts of DNA causing uneven loading.

Although the method has never been standardized for RGM or slowly growing NTM, most investigators use the criteria set forth by Tenover et al., in 1995, for interpretation of chromosomal DNA restriction patterns by PFGE with bacterial isolates with some modifications for mycobacterial species [79, 80].

2–3 band differences = "closely related"
4–6 bands = "possibly related"
>7 bands = "genetically different"

Zhang et al. modified the PFGE procedure with the addition of thiourea to enable more reliable results with all species of RGM, including those species (especially isolates of *M. abscessus*) that were originally affected by DNA degradation [81]. A 2013 study by Howard and colleagues correlated the previously described DNA smear pattern with the presence of *dnd*, a DNA degradation gene conferring DNA phosphorothioation [82].

PFGE is expensive and requires skill and experience. The method requires large amounts of high-quality DNA which may not be easy to obtain with some NTM species due to their lack of growth in liquid media. Another disadvantage is that only a limited number of isolates can be tested at one time. Adequate growth of some NTM species in liquid culture may require extended incubation on a shaker incubator, adding to the turnaround time for the test. However, one major advantage of PFGE is that it is generic and can be used with any NTM species or subspecies.

The PFGE method has been applied with both slowly growing mycobacteria and RGM in the investigations of mycobacterial outbreaks, pseudo-outbreaks, and epidemics [13, 83–85]. Moreover, the strategy has been used to analyze patients with chronic MAC lung disease in order to ascertain new infection versus relapse infections [80, 86, 87] and has also been used in population studies [81] and as confirmation for the previous typing of isolates of *M. abscessus* complex already typed using other methods such as rep-PCR [88].

Repetitive-Unit Sequence-Based PCR (rep-PCR)

The rep-PCR technology is the first commercially available method for DNA strain typing of mycobacteria. The method is a high-throughput automated system; the method, although proprietary, is available through the DiversiLab System (bioMérieux, Durham, NC) with a web-based database of sequences also available. Rep-PCR uses repetitive elements interspersed throughout the genome. The system electrophoretically separates repetitive-sequence-based PCR amplicons on microfluidic chips to generate a computerized "graphic" representation of bands [17, 89]. As the fragments migrate over the chip, their size and fluorescence intensity are measured by a laser computer. The discriminatory power has been reported to equal or exceed that of standard restriction fragment length polymorphism (RFLP) analysis for some NTM with a more rapid turnaround time and a smaller sample size than for standard PFGE [88, 89]. The system has been used in previous outbreak investigations of NTM but has not been validated against more standard methods such as PFGE [88].

Random Amplified Polymorphic DNA PCR (RAPD-PCR)

The RAPD-PCR method utilizes one to three arbitrary primers and low stringency conditions so that the primer anneals to both strands of template DNA, resulting in a match or partial match of strain-specific multiband heterogeneous DNA profiles [90]. Guidelines have been published for interpretation of RAPD compared to PFGE [90]. The RAPD-PCR, also known as arbitrarily primed PCR (AP-PCR), is considered to have less discriminatory power than PFGE. Nevertheless the method has been used to confirm several observations from prior RGM outbreaks including a nosocomial outbreak of *M. abscessus* in children with otitis media as well as a post-cardiac surgery outbreak and most recently *M. chimaera* (a species within the *M. avium* complex), in air cultures and the heater-cooler units following cardiac surgeries [90–92]. The most important disadvantage of RAPD-PCR is the lack of reproducibility. Some investigators believe that pattern variation is dependent upon priming efficiency during amplifications and these variations are in turn dependent on template concentration and purity, primer/template ratio, or thermal ramping rates used [17].

Variable Number Tandem Repeat (VNTR) and Mycobacterial Interspersed Repetitive Units (MIRU)

VNTRs are scattered throughout the genome of *Mycobacterium* and are known to be widely distributed among MTBC and NTM including MAC, *M. gordonae*, *M. kansasii*, *M. marinum*, *M. szulgai*, and *M. ulcerans* [17]. DNA strain typing using VNTR/MIRU is accomplished by assessing the number and length of tandem repeats at each locus of each isolate. The variability of the particular loci often is dependent on the sample collection, the geographical source, and the natural strain variability. The protocol for typing uses PCR amplification of each locus using specific primers complementary to the flanking regions and analysis of the resulting amplicons which are separated by gel electrophoresis. The size of the amplicons corresponds to the number of tandem repeat units which is determined in reference to the known size of the repeat unit within each targeted locus [17].

Recently VNTR/MIRU has been applied to strain typing of members of the MAC. Although VNTR loci are usually species-specific, the reproducibility, the smaller volume requirement, and the speed at which it is performed make it an attractive possible alternative method for PFGE comparison especially with isolates of MAC [93–95]. A recent report by Wallace and colleagues applying VNTR methodology revealed the presence of *M. chimaera* and the absence of *M. intracellulare* in household water and biofilm samples from patients with MAC lung disease throughout the USA [93, 94, 96]. The application of VNTR typing to other species including *M. abscessus* has recently been described by Wong and colleagues using 18 tandem repeats in the *M. abscessus* genomic sequence, with 100% typeability, 100% locus stability, and 100% reproducibility [97].

Multi-locus Sequence Typing (MLST)

MLST schemes analyze intergenic sequences which can show high variability. In this method, single-copy housekeeping genes are sequenced, and the combination of different gene alleles creates a sequence type and defines a strain [98]. MLST has recently been compared to PFGE for strain typing of 93 isolates of *M. abscessus* complex from outbreaks in Brazil [98]. In this study, multiple housekeeping genes were sequenced, and each isolate was assigned a sequence type (ST) from the combination of obtained alleles. PFGE patterns were compared with the MLST. Thirty-three STs and 49 unique PFGE patterns were identified among the 93 isolates. Eight STs were specific for *M. abscessus* subsp. *abscessus*, 21 specific to *M. abscessus* subsp. *bolletii*, and 4 were specific to 6 isolates with unidentified subspecies. MLST showed 100% typeability and grouped monoclonal outbreak isolates congruently with their PFGE pattern. However, MLST was found to be less discriminatory than PFGE, especially for *M. abscessus* subsp. *abscessus*, and some strains were detected that share MLST profiles in patients but had no epidemiological relationship [98]. Investigators have suggested that the discriminatory power might be improved by sequencing more genes [98].

A major advantage of MLST is that the organisms do not require cultivation so that the results are generated more rapidly than PFGE and specialized equipment as for PFGE is not needed. Generally, a laboratory with sequencing capabilities is able to perform the testing.

Enterobacterial Repetitive Intergenic Consensus (ERIC) Sequencing Typing (ERIC PCR)

Enterobacterial repetitive consensus sequences are repetitive elements usually found in gram-negative bacteria distributed along the bacterial chromosome at intergenic flanking (open reading frame) regions of polycistronic operons. Mycobacterial strain typing is accomplished by ERIC PCR because despite the absence of ERIC repeats (approximately 126-bp) in the available *Mycobacterium* genomes, non-specific amplification with appropriate primers can occur in the absence of ERIC sequences. ERIC PCR was recently used to evaluate isolates of *M. abscessus* in outbreaks in Brazil [99, 100]. The method had a higher discriminatory power for isolates of *M. abscessus* that had previously shown smear patterns with PFGE. However, the method was not as discriminatory when evaluating isolates of *M. fortuitum* [99, 100]. The variability of ERIC PCR sequences has been applied in the estimation of genetic diversity of mycobacterial species including *M. tuberculosis*, *M. gordonae*, *M. intracellulare*, *M. szulgai*, *M. chelonae*, *M. abscessus*, and *M. fortuitum* [17, 100].

Whole Genome Sequencing (WGS)

The genomic sequencing of *M. tuberculosis* ushered in new and exciting changes in the epidemiology of mycobacteria [101]. Ongoing WGS has also been used to compare patient and environmental isolates [102] in the recent outbreaks of disseminated *M. chimaera* infections following open-heart surgery and exposure to contaminated water in the heater-cooler units of heart-lung machines [102]. Although it has not yet been utilized for this application, WGS for in vitro susceptibility of the NTM involved in the outbreaks might be a potential method to assess susceptibility in the context of biofilm formation when conventional AST is problematic. At a local level, WGS, compared with traditional strain typing methods such as VNTR, and PFGE may also provide higher resolution and deeper knowledge about individual strains and information about large-scale outbreaks [17]. WGS may also allow for prediction of additional cases that might go undiagnosed without WGS.

Several studies from 2011 to 2012 by more than one investigator have revealed genomic characteristics of specific strains of *M. abscessus* subsp. *abscessus* and *M. abscessus* subsp. *massiliense* [103–107]. Interestingly, the strains had >4500 coding regions. Core regions contained more than 4 million-bp with a GC content of 64% and accessory regions in at least one strain >450,000-bp with a CG content of 61%. The core genome of one strain of *M. abscessus* subsp. *abscessus* revealed >4500 coding sequences and >40 RNAs. There were also >400 putative genes distributed in the subsystem of cofactors, vitamins, prosthetic groups, and pigments and >250 genes in the carbohydrate subsystem. Additionally, in the accessory genome, there were >500 coding sequences including 10 putative genes with subsystem of phages, prophages, transposable elements, and plasmids [104]. Similarly, one strain of *M. abscessus* subsp. *massiliense* revealed approximately 100 genes that encoded virulence factors [107]. Annotation of another strain of *M. abscessus* subsp. *massiliense* predicted >400 genes for amino acids and derivatives and >30 genes for virulence, intracellular existence, disease, defense, antibiotic resistance, and toxic compounds [105].

Information derived from sequencing of these regions described above should elucidate aspects of species diversity and strain-specific properties including survival in the environment [104], resistance to antibiotics and disinfectants [108], and genetic determinants related to pathogenicity of the strain/species [106].

In a 2016 publication by Tortoli et al., WGS was used to support the separation of three subspecies of the *M. abscessus* complex: *M. abscessus* subsp. *abscessus*, *M. abscessus* subsp. *massiliense*, and *M. abscessus* subsp. *bolletii*. Most importantly, differentiation at the subspecies level helps to resolve the controversy of the taxonomic position of this major group of RGM, thus emphasizing another application of WGS [109].

Antimicrobial Susceptibility Testing

Principles of Antimicrobial Susceptibility Testing (AST) Following the 2003 CLSI initial recommendations for laboratory AST [110], the 2007 diagnosis and treatment guidelines for NTM were set by the joint publication of the American Thoracic Society (ATS) and the Infectious Disease Society of America (IDSA) and generally apply best to a few commonly encountered NTM species or complexes including

Mycobacterium avium complex, *M. kansasii*, *M. marinum*, and some rapidly growing mycobacteria (RGM). Insufficient data is available regarding other less common NTM species [111]. Table 4 shows treatment recommendations for some of the most often encountered NTM based upon their in vitro susceptibility patterns.

The initial CLSI guidelines were in major part based on clinical data, organism population distribution, and experience of experts in NTM [110]. A major caveat is that laboratories that perform antimicrobial susceptibility of NTM should implement quality control and quality assurance systems designed to recognize potential problems in cultures of NTM, including contamination, mixed cultures, unusual growth characteristics, and results discrepant from other isolates of the species.

In 2011, the initial laboratory testing guidelines were revised, and the broth microdilution method was recommended as the gold standard for AST of NTM [113]. Specific recommendations for the test method (i.e., broth microdilution), including breakpoints for most antimicrobials used in the treatment of NTM infections, were included along with new information regarding the importance of the erythromycin ribosomal resistance methylase (*erm*) gene as applied to RGM, especially isolates of *M. abscessus*.

Evaluation and Interpretation of Isolates for AST The CLSI has also continued to emphasize the need to ascertain the clinical significance of the isolates, the clinical relevance of the isolate as related to the pathogenicity of the NTM species and host factors including immunologic status, and the clinical findings such as unexplained fever, granulomas, inflammatory lesions, and site of infection [14]. Because of the presence of NTM in the environment (soil and water), determination of the clinical significance of the isolate may not always be clearly indicated. A classic example is the isolation of the *M. mucogenicum* group (a group of RGM frequently encountered in tap water), from a respiratory sample which usually is indicative of environmental contamination, whereas the same organism recovered from blood cultures or a central venous catheter site is usually associated with mycobacterial sepsis. The finding of any NTM in usually sterile body sites such as the blood, tissue, cerebrospinal fluid, pleural fluid, brain, etc. is almost always clinically significant. Likewise, the isolation of NTM in a immunologically suppressed host, especially on corticosteroids or tumor necrosis factor (TNF), increases the possibility of the clinical relevance of the organism, including typically nonpathogenic species [114].

Generally, recovery of species such as *M. gordonae*, *M. terrae* complex, *M. chelonae*, and as described previously, *M. mucogenicum* group from respiratory cultures is rarely clinically significant as they usually represent environmental contamination (all are present in tap water). Some newer species, including *M. botniense*, *M. chlorophenolicum*, *M. aromaticivorans*, *M. hodleri*, *M. murale*, *M. pallens*, *M. rufum*, *M. rutilum*, *M. litorale*, *M. arabiense*, *M. sediminis, and M. paragordonae*, have only been recovered from environmental samples and as yet are not recognized as human pathogens. Other species, including *M. cookii*, *M. fredericksbergense*, and *M. psychrotolerans*, have rarely been recovered from human samples, and thus, their pathogenicity is also questionable.

Table 4 Antimicrobials used for treatment of commonly encountered species of nontuberculous mycobacteria

Species	Growth type	Type(s) of infection and/or disease	Treatment options (based on MICs)
M. abscessus subsp. *abscessus*	Rapid	Chronic pulmonary infection, localized posttraumatic wound infection, catheter infection, disseminated cutaneous infections, eye infections	Linezolid (~50%), clarithromycin-azithromycin[a] (~20%), amikacin, tigecycline, cefoxitin, imipenem[b] (~50%)
M. abscessus subsp. massiliense	Rapid	Chronic pulmonary disease, localized posttraumatic wound infection, postsurgical wound infection, catheter infection	Clarithromycin-azithromycin[c], linezolid (~50%), amikacin, tigecycline, cefoxitin, imipenem[b] (~50%)
M. chelonae	Rapid	Disseminated cutaneous infection, localized posttraumatic wound infection, sinusitis, eye infection (not considered a pulmonary pathogen)	Clarithromycin-azithromycin, linezolid, moxifloxacin (~25%), ciprofloxacin (~20%), doxycycline (~20%), tobramycin, linezolid, amikacin (~50%), imipenem[b] (~60%), tigecycline
M. fortuitum	Rapid	Localized posttraumatic wound infection, catheter infection, surgical wound infection, cardiac surgery, augmentation mammaplasty; rarely a pulmonary pathogen (except in cases of achalasia, lipoid pneumoniae)	Ciprofloxacin, levofloxacin, moxifloxacin, trimethoprim-sulfamethoxazole, linezolid, doxycycline (~50%), clarithromycin-azithromycin[a] (~20%), imipenem, tigecycline, linezolid, amikacin, cefoxitin (~50%)
M. neoaurum-M. bacteremicum	Rapid (pigmented)	Catheter sepsis/bacteremia, localized posttraumatic wound infection, postsurgical wound infection	Ciprofloxacin, levofloxacin, moxifloxacin, doxycycline, linezolid, trimethoprim-sulfamethoxazole, clarithromycin-azithromycin[d], amikacin, tobramycin, linezolid, imipenem, tigecycline, cefoxitin
M. marinum	Intermediate	Localized posttraumatic wound infection, tenosynovitis	Clarithromycin-azithromycin, rifampin-rifabutin, ciprofloxacin (~50%), trimethoprim-sulfamethoxazole, moxifloxacin, linezolid, doxycycline (~50%), amikacin, linezolid
M. avium complex (11 accepted species)	Slow	Chronic pulmonary infection (including cystic fibrosis), disseminated infection (usually associated with AIDS), lymphadenitis, localized cutaneous infection with tenosynovitis	Clarithromycin-azithromycin[e], rifampin-rifabutin, ethambutol, moxifloxacin (<50%), ciprofloxacin (<25%), amikacin, streptomycin, linezolid (<50%)

Table 4 (continued)

Species	Growth type	Type(s) of infection and/or disease	Treatment options (based on MICs)
M. kansasii	Slow	Chronic pulmonary infection, disseminated infection in AIDS	Clarithromycin-azithromycin, rifampin-rifabutin[f], trimethoprim-sulfamethoxazole, ethambutol, isoniazid, moxifloxacin, ciprofloxacin, linezolid, amikacin
M. malmoense[g]	Slow	Chronic pulmonary infection	Clarithromycin-azithromycin, rifabutin, rifampin
M. simiae[h]	Slow	Chronic pulmonary infection	Clarithromycin-azithromycin, moxifloxacin (~60%), trimethoprim-sulfamethoxazole, amikacin
M. xenopi[i]	Slow	Chronic pulmonary infection, joint and soft tissue infection	Clarithromycin-azithromycin, rifampin-rifabutin, ethambutol, moxifloxacin, amikacin, streptomycin

This table is modified from Ref. [112]

[a]Many strains of the *M. fortuitum* group and *M. abscessus* subsp. *abscessus* contain functional *erm* genes, so extended incubation shows clarithromycin MICs to be resistant, while with a routine 3-day incubation, the MICs may appear to be susceptible [57]

[b]Susceptibility testing with imipenem with the *M. abscessus-M. chelonae* group is known to be problematic (lack of reproducibility)

[c]Isolates of *M. abscessus* subsp. *massiliense* contain a truncated (nonfunctional) *erm* gene; thus, macrolide MICs remain susceptible even with extended incubation. Isolates of *M. abscessus* subsp. *bolletii* are rare in the USA but contain a functional *erm* gene and are resistant to macrolides

[d]There is a bimodal distribution of isolates that are resistant/susceptible to macrolides with extended incubation. Testing for functional *erm* genes has not been performed in large numbers of this species

[e]Clarithromycin is recommended as the class agent for the testing of the newer macrolides because clarithromycin and azithromycin share cross-resistance and susceptibility. The macrolides are the only antimicrobials for which the clinical response can be correlated with the in vitro results in isolates of *Mycobacterium avium* complex (MAC). *Testing of the first-line antituberculous agents with MAC isolates is not recommended*

[f]Previously untreated strains of *M. kansasii* should be tested in vitro only for rifampin and clarithromycin. Isolates with susceptibility to rifampin will be susceptible to rifabutin

[g]Susceptibility testing may require adjustment of pH of the media and/or extended incubation

[h]The antimicrobial treatment of *M. simiae* can be difficult, and the clinical response may not correlate with in vitro susceptibility

[i]The optimal antimicrobial treatment for *M. xenopi* has not been established, but most experts recommend a combination of macrolide, rifampin, and ethambutol with moxifloxacin. Poor correlation of in vitro results with clinical response may relate to the difficulty in testing for this species by current standardized methods. Some strains have been reported to have variable antimicrobial susceptibilities, and testing may need to be performed at 45 °C for optimal growth in broth

As relates to host site of infection, the major RGM respiratory pathogen is *M. abscessus*. Other RGM such as the *M. fortuitum* group are rarely respiratory pathogens except in cases of lipoid pneumonia or achalasia [5], and *M. chelonae*, although sometimes recovered from respiratory samples, has not been proven to be a respiratory pathogen. Slowly growing NTM respiratory pathogens most commonly include *M. avium* and *M. intracellulare* (the *M. avium* complex) and *M. kansasii, M. xenopi, and M. simiae*, while RGM such as *M. chelonae*, *M. fortuitum* group, and slowly growing NTM including *M. ulcerans*, *M. marinum*, *M. haemophilum*, newer species including *M. arupense*, *M. virginiense*, and *M. heraklionense* (both in the *M. terrae* complex), are most often encountered in skin, soft tissue, and bone infections [111, 115].

For respiratory samples, the number of positive cultures and the number of colonies present are also important in the determination of clinical significance of the NTM. Isolates recovered from multiple samples and in large quantities are almost always clinically significant, whereas those in small numbers in a single sample are less likely to be clinically relevant [14].

If mycobacterial cultures remain positive following 6 months of appropriate antimicrobial treatment, repeat AST is warranted. Monitoring the development of mutational drug resistance is achieved by repeat AST and, if available and indicated, gene sequencing to molecularly detect the presence or absence of specific mutations (e.g., adenine 2058 or 2059 in the 23S rRNA gene for macrolide non-inducible resistance and amikacin mutation in the 16S rRNA gene) that cause the NTM to develop resistance against specific antimicrobials [116–119].

Although AST for clinically significant isolates is recommended, performing AST on nonsignificant clinical isolates is a waste of laboratory time and patient and laboratory resources, and most importantly, results can be misleading and even detrimental for patient care [111]. Clear communication of accurate laboratory data, including the number of cultures positive, the degree (semi-quantitation) of positivity, and the results of AFB smears between the laboratory and the clinicians, is of vital importance, although ultimately careful evaluation of the host factors and the clinical setting is the responsibility of the clinician.

AST attempts to predict the clinical efficacy of specific antimicrobials against isolates of NTM. However, although indicated for treatment regimens, in vitro susceptibility of the antituberculous agents (isoniazid, rifampin, ethambutol, rifabutin, and streptomycin) for some species such as *M. avium* complex does not predict clinical response [120]. Moreover, just as performance of AST on nonsignificant clinical isolates may be detrimental to patient outcomes, providing susceptibility results for these drugs may also lead to incorrect treatment regimens which can also lead to irreversible genetic mutation, which would eliminate the usage of specific antimicrobials. Thus, both the CLSI and the ATS/IDSA have recommended that these agents not be reported against isolates of MAC [111, 113].

Agar Disk Diffusion

The modified Kirby-Bauer agar disk diffusion method of susceptibility testing has been previously described and recommended by several authors for RGM [6, 121]. Although the agar disk diffusion method is highly standardized for other (non-mycobacterial) rapidly growing organisms, it has several serious limitations and is no longer recommended for testing RGM because of technical problems associated with it.

Agar Disk Elution

The agar disk elution method utilizes a system in which commercial antibiotic disks are eluted into a molten agar medium to produce specific concentrations of antibiotics within each well [113, 122]. The methodology for preparing and reading the plates is very similar to the standard susceptibility method used for testing *M. tuberculosis*. This method has been recommended by the CLSI only for the testing of fastidious species such as *M. haemophilum* [113]. Potential problems with this method include difficulty in controlling inoculum and difficulty in reading the endpoint of some antimicrobials.

Epsilon Tests (E-Test)

The E-test (AB Biodisk, Piscataway, NJ) combines the use of a predefined exponential gradient strip impregnated with antimicrobial concentrations that covers 15 twofold concentration to achieve an MIC in agar [123]. This method was introduced in the 1990s [124] for susceptibility testing of the RGM and for some slowly growing NTM including *M. kansasii* and *M. avium* complex [124–126]. In 2000, a multicenter study revealed poor correlation of the E-test method compared to the accepted gold standard broth microdilution [127]. Thus, to date, the CLSI has not recommended this method for AST of the RGM [127]. No multicenter studies with other NTM have been performed.

Broth Microdilution

The only method for AST of NTM that has been recommended to date by the CLSI is the broth microdilution method [113, 128]. This method uses twofold dilutions prepared in cation-adjusted Mueller Hinton broth. Although the MIC is set up as twofold dilutions, the MIC of the isolate does not represent an absolute value. For example, if the MIC is reported as 32 μg/mL, the "true MIC" would fall between the lowest concentration (16 μg/mL) that inhibits the growth (the reported MIC result) and the next lowest concentration (32 μg/mL). This means that the generally acceptable MIC values are within one twofold dilution of the actual endpoint (i.e., inhibition of growth), and thus standards have been proposed to help insure quality results of the test [113]. Appropriate inoculum is critical to the accuracy of the interpretation of MICs. A too heavy inoculum may result in false resistance, whereas too light inoculum may yield falsely susceptible MIC readings due to inadequate growth of the organism in broth culture.

For AST of RGM, the CLSI recommended battery of antimicrobials includes amikacin, cefoxitin, ciprofloxacin, clarithromycin, doxycycline/minocycline, imipenem/meropenem, linezolid, moxifloxacin, tobramycin (reported only for isolates of *M. chelonae* but can be used to taxonomically differentiate isolates of *M. immunogenum* from *M. chelonae*), and trimethoprim-sulfamethoxazole. Details of the procedure may be found in the latest version of the CLSI document [113]. A modification to include tigecycline to the battery has been proposed but not yet accepted. There are currently no breakpoints for this agent with RGM [113]. Therefore, until these are addressed, the proposal to the CLSI recommends reporting tigecycline MICs without interpretation [113, 128–130].

Clarithromycin is the macrolide of choice for AST and has been selected as the "class agent" to be tested as high concentrations of azithromycin are technically difficult to achieve. Initial clarithromycin MICs should be examined after 72 h incubation [113. 131]. If susceptible, the MICs should be reincubated up to 14 days in order to determine inducible macrolide resistance due to the presence of a ribosomal methylase (erythromycin ribosomal resistance methylase, *erm* gene) [51–57, 113].

Most isolates of *M. abscessus* subsp. *abscessus* and *M. abscessus* subsp. *bolletii* harbor a functional *erm* gene [57]. In contrast, RGM with no functional *erm* genes includes *M. chelonae*, *M. mucogenicum* complex (*M. mucogenicum*, *M. phocaicum*, *M. aubagnense*), *M. immunogenum*, *M. abscessus* subsp. *massiliense*, *M. peregrinum*, and *M. senegalense* [51, 54, 57, 58]. The *erm* gene has been previously discussed in the section on identification of RGM. For the aforementioned isolates that have been identified using genetic sequencing, 14-day incubation to determine inducible macrolide resistance is not necessary, and this proposal has been made to the CLSI but not yet accepted.

Antimicrobial Susceptibility Testing of Slowly Growing Nontuberculous Mycobacteria

Mycobacterium avium Complex *(MAC)*

For testing of isolates of MAC, it is important to select transparent colonies if present as older, pigmented strains usually represent a more susceptible phenotype. MAC is the most common NTM respiratory isolate seen in laboratories worldwide. Originally the complex was composed of two species: *M. avium* and *M. intracellulare* [132]. Since that time, molecular studies have revealed the presence of at least nine other species which should be included in the MAC. Along with the above two species, these include *M. colombiense*, *M. chimaera*, *M. vulneris*, *M. marseillense*, *M. paraintracellulare*, *M. timonense*, *M. arosiense*, *M. bouchedurhonense*, *and M. yongonense* [1, 132, 133]. Other than *M. chimaera*, the other eight species of MAC (generally designated "MAC-X") have not been recognized as pulmonary pathogens [133].

AST of MAC has primarily involved testing of clarithromycin (again chosen because of the inability to achieve high concentrations in azithromycin) [134–136]. Testing of first-line antituberculous agents should not be performed since there is no correlation of in vitro results with clinical response when testing these agents. Recently, studies have shown that in vitro MICs of amikacin also correlate with clinical response, and thus a proposal has been made to the CLSI to include testing of amikacin along with clarithromycin for the AST of MAC [113, 116, 134–136]. Other agents such as moxifloxacin and linezolid may have limited utility in treatment of MAC, and until studies are performed to show lack of correlation of in vitro MICs with clinical response, these agents may also be tested [113, 137], but they are not to be considered as substitute treatment options for the primary antituberculous agents (i.e., rifampin/rifabutin, ethambutol) which should be used in combination with a macrolide (clarithromycin or azithromycin) with or without amikacin for initial treatment of MAC.

Since macrolides are more active *in vitro* in mildly alkaline conditions, a pH of 7.3–7.4 has been recommended for the test media. Cation-adjusted Mueller Hinton broth has been recommended since the pH of this media falls within this range [113]. Previously, a commercial macrodilution broth method was available which used a pH of 6.8 adjusted to 7.3–7.4, but this method is no longer available. Although the CLSI recommendation of Mueller Hinton broth rather than Middlebrook 7H9 is controversial, a 2003 multicenter study [138] showed that reproducibility of AST was related more to experience of the laboratory personnel with the method rather than the specific methodology used [138]. It is also known that MICs in agar-based Middlebrook 7H10 medium are higher for some antimicrobials including amikacin and doxycycline than those measured in Mueller Hinton media and previous unpublished studies have shown the same is true with Middlebrook broth MICs also [139]. Because using different media would require the reestablishment of separate breakpoints for each media type, the CLSI continues to recommend the use of cation-adjusted Mueller Hinton broth for AST of NTM. Incubation of isolates of MAC should be performed for 7–14 days at 35–37° C. For laboratories that lack experience with NTM AST, the CLSI recommends referral of isolates for AST to an experienced reference laboratory [111, 114].

In general, all wild-type (i.e., untreated) strains of MAC are macrolide susceptible, and in vitro clarithromycin susceptibility also applies to azithromycin although testing of azithromycin is not recommended for reasons previously discussed [140, 141]. Macrolide resistance in MAC has been defined as a clarithromycin MIC ≥32 μg/mL [113]. Intermediate MICs (16 μg/mL) are rare and may suggest a mixed population of MAC organisms or signal impending macrolide resistance [113]. Thus, MICs of 16 μg/mL should be confirmed with repeat culture, and patients with MAC isolates with an MIC of 16 μg/mL should be carefully monitored for the possibility of emerging macrolide resistance. Macrolide resistance should also be carefully assessed, and sequencing of the 23S rRNA gene to detect a mutation in the adenine at position 2058 or 2059 can be used to confirm a clarithromycin MIC ≥32 μg/mL [112, 119, 141, 142].

Mycobacterium kansasii

The CLSI recommends AST of *M. kansasii* be performed using the broth microdilution method, similar to that used for MAC. *M. kansasii* conventional treatment includes ethambutol, rifampin, and isoniazid (INH) or a newer regimen of clarithromycin substituted for INH [143]. Rifabutin should be used instead of rifampin for patients with HIV infection on protease inhibitors. Isolates susceptible to rifampin are cross-susceptible to rifabutin, so no additional MIC testing is necessary. For untreated isolates susceptible to rifampin, AST to other agents is not usually required since MICs usually fall within a narrow range with the exception of INH and ethambutol; the CLSI has no breakpoints recommended for INH, and a proposal has been made not to test ethambutol for isolates of slowly growing NTM due to technical difficulties associated with testing the drug. MICs of INH in isolates of *M. kansasii* range from 0.5 to 5 µg/mL so that the standard critical concentration of 0.2 µg/mL usually shows resistance and the 1.0 µg/mL concentration often yields variable results, making testing and determination of breakpoints for INH technically difficult. Thus, the CLSI has excluded this agent from the list of antimicrobials recommended for testing [113].

Treatment failure with isolates of *M. kansasii* is unusual but almost always associated with rifampin resistance and occasionally in vitro resistance to one or more companion agents including clarithromycin as well [144–146]. Therefore, the ATS/IDSA and CLSI have recommended AST on isolates from patients who fail or have poor response to appropriate therapy for *M. kansasii* [14, 111, 113]. The current CLSI recommendation is to test initial isolates of *M. kansasii* to rifampin and clarithromycin only. However, if the isolate is found to be rifampin resistant (defined as MIC >1 µg/mL), a secondary battery of antimicrobials including amikacin, ciprofloxacin, moxifloxacin, linezolid, rifabutin, doxycycline/minocycline, and trimethoprim sulfamethoxazole should be tested. Although streptomycin may be used in clinical practice, there are no accepted CLSI breakpoints for this agent [113].

Mycobacterium marinum

M. marinum is classified as having intermediate growth rate but is traditionally included with the slowly growing NTM since testing performed is the same as previously discussed except that due to its optimal growth at lower temperatures, an incubation for 7 days at 30 °C is recommended for AST [113, 147].

The same antimicrobials and test methodology (i.e., broth microdilution) that have been recommended for rifampin-resistant *M. kansasii* are recommended for testing of isolates of *M. marinum* [113]. Previously, the CLSI did not recommend routine testing of *M. marinum* due to its consistently pan-susceptible pattern [113]. However, recently a number of doxycycline/minocycline and ciprofloxacin-resistant isolates have been seen, and a proposal has been made to the CLSI to recommend susceptibility testing of isolates of *M. marinum*, especially if patients have inferior response to initial treatment.

Miscellaneous Slowly Growing NTM

As previously indicated among the slowly growing NTM, the ATS/IDSA and CLSI recommendation best apply to MAC, *M. kansasii*, and *M. marinum* [111, 113]. However, other NTM species may be human pathogens, and AST of these isolates should be considered if the isolate is deemed clinically significant. For example, isolates from sterile body fluids, tissues, or multiple respiratory sites or sputum samples should have AST performed.

Species such as the *M. terrae* complex (e.g., *M. arupense*, *M. heraklionense*, *M. virginiense*) [115], *M. simiae*, *M. szulgai*, *M. celatum*, and *M. xenopi* [111, 148–150] can cause disease, and although AST data is lacking for most of these species, these species should be tested using the criteria and recommendations as set forth for isolates of rifampin-resistant *M. kansasii* [113]. The latter species, *M. xenopi*, may require extended incubation longer than the routine 7 days and temperature (42–45 °C) adjustments [113, 150].

Fastidious Slowly Growing NTM

AST data is also lacking, and there are no standard recommended methods of AST for several other pathogenic slowly growing NTM including *M. genavense*, *M. malmoense*, *M. ulcerans*, and *M. haemophilum* [113, 151–155]. The latter species, *M. haemophilum*, as previously stated requires supplementation with iron or hemin, extended incubation (2–3 weeks), and lower incubation temperature (28–30 °C) for optimal growth [113]. The previously discussed disk elution method is suitable for isolates of this species. Isolates of *M. genavense* and *M. malmoense* require a pH alteration (5.5–6.0) of media due to their acidophilic nature and extended incubation (more than 6 weeks); *M. ulcerans* also requires lengthy incubation (4–6 weeks), and optimal growth is achieved at 30 °C. Ji et al., however, acknowledged that more studies are necessary to evaluate the usefulness of the in vitro data and establish breakpoints for interpretation of MICs against *M. ulcerans* [153]. Laboratories who encounter these organisms must establish in-house validation for AST or send them to a qualified reference laboratory [113].

A 2006 study of the in vitro AST of 29 isolates of *M. ulcerans* on 10% OADC-enriched 7H11 agar with antimicrobials added to the media and incubated at 30 °C showed that none of the MIC_{90}s of rifampin, streptomycin, amikacin, moxifloxacin, and linezolid (chosen because of their in vivo activity in experiments with mice) were unreasonably increased after 60 days incubation compared to quality control with the *M. tuberculosis* H37rv. MICs for four agents except rifampin (MIC = 2.0 μg/mL) were within the acceptable susceptible ranges for NTM, and although there is no acceptable range established for streptomycin with NTM, the MIC90 was 0.5 μg/mL for 29 isolates of *M. ulcerans* [153].

Antibiotic susceptibility testing for NTM is also discussed in detail in chapter "Drug Susceptibility Testing of Nontuberculous Mycobacteria".

Whole Genome Sequencing (WGS) to Assess Mechanisms of Antimicrobial Resistance

Currently there are no commercial systems recommended for the detection of genetic mechanisms of resistance in NTM. However, as previously mentioned, there are some species-specific assays including those based on sequencing of the gene and comparison to untreated type strains of that species. These sequencing algorithms include the genetic analysis for inducible macrolide resistance due to the *erm*(41) gene in the *M. abscessus* complex [51]; the 23S rRNA gene mutation in the adenine 2058 or 2059 position for non-inducible (mutational) macrolide resistance primarily in the *M. abscessus* complex and MAC [117, 140], *rpo*B gene mutation for rifampin susceptibility in *M. kansasii* [145], and the 16S rRNA gene mutation to assess amikacin resistance [112, 116, 139, 156].

A 2014 comparative genomic and proteomic analysis of 44 pathogenic (P) (including *M. tuberculosis*), opportunistic (OP), and nonpathogenic (NP) mycobacterial species elucidated the presence of novel alternative pathways with possible roles in metabolism, host-pathogen interaction, virulence, and intracellular survival which could lead to detection of new interventions against mycobacterial disease between *M. indicus pranii* (an ancestor of MAC) (NP), *M. tuberculosis* (P), and *M. intracellulare* (OP) [157]. The clinical significance of this finding is important because *M. indicus pranii* is a species with strong immunomodulatory properties, currently used in the treatment of leprosy, thus placing it as an emerging pathogen [157].

Stinear and colleagues investigated the close genetic relationship of *M. marinum* and *M. ulcerans* using WGS at a global level [158]. A recent major global application of WGS with NTM occurred in the recent cystic fibrosis *M. abscessus* subsp. *massiliense* outbreaks in the UK [159] and in Seattle, WA [160], in comparison with the outbreak strain from Brazil [108, 161–164], Malaysia [103–106], Korea [165], and France [107]. The findings of the comparisons revealed high genetic relatedness for the two CF studies as compared with the Brazilian soft tissue outbreaks. However, the soft tissue strain from Brazil did not show mutational resistance to macrolides or amikacin unlike the strains from the CF outbreaks [166].

Using WGS and phylogenomic analysis of specific areas (e.g., single nucleotide polymorphisms and/or genomic islands) of the genome may enable the more accurate detection of molecular antimicrobial resistance markers and large-scale genetic deletions in NTM and provide insight into the susceptibility and virulence mechanisms of NTM [161]. As previously mentioned, WGS may be applicable to determine susceptibility of NTM in biofilms [16]. Moreover, WGS already has been instrumental in elucidating the epidemiological significance of the data obtained by mapping specific regions of target genes from every level including global (population), local (community), individual (patient), down to specific pathogen (strain) [17].

Antimicrobial Susceptibility Testing of Novel (Non-CLSI Recommended) Agents

Although the CLSI has approved specific batteries of antimicrobial agents to be tested against RGM and slowly growing mycobacteria, requests are often made to laboratories to perform AST on additional agents that have not yet been addressed by the CLSI.

Antimicrobial agents including clofazimine, bedaquiline, and tedizolid have been shown to have in vitro activity against NTM recently [120, 167, 168]. The caveat for testing and reporting these antimicrobials (and any new antimicrobials) includes an initial agreement with the pharmaceutical manufacturing companies (e.g., bedaquiline (Janssen Inc.) and tedizolid (Merck Pharmaceuticals)). Once the agreement to test the isolates is executed, the requesting party should contact the manufacturer's suggested reference laboratory for submission protocol. MICs should be reported without interpretation until breakpoints have been addressed by the CLSI. The reference laboratory report should include a disclaimer remark that testing of the new agents has currently not been addressed by the CLSI and that no MIC breakpoints are available for interpretation of the MIC.

Major Caveats for Laboratory Identification and Antimicrobial Susceptibility Testing (AST) of Nontuberculous Mycobacteria (NTM)

- Definitive species identification of NTM should be based on molecular or proteomic testing.
- Differentiation of *Mycobacterium chelonae* from *M. abscessus* is mandatory in order to determine accurate treatment regimens. Furthermore, subspecies identification of *M. abscessus* is necessary in order to determine accurate treatment regimens. Identification of *M. abscessus/M. chelonae* complex is no longer acceptable.
- The recommended AST method is a broth-based (preferably broth microdilution system). No agar AST has been standardized.
- After 4–5 days incubation, caution should be exercised when interpreting the validity of AST results, especially imipenem and tetracyclines (doxycycline, minocycline, and tigecycline). Generally, the only agents which show stability after extended incubation are the aminoglycosides and clarithromycin. Thus, it

may be advisable, in cases in which isolates do not grow sufficiently after 4 days, to consider molecular testing for mutational resistance (16S rRNA gene) for amikacin and inducible (*erm* gene) resistance to clarithromycin.

- Incubation times for AST of rapidly growing mycobacteria should not exceed 5 days. If isolates do not grow adequately in broth after 5 days incubation, repeat testing and/or referral to a reference laboratory experienced in NTM should be sought.
- Testing of carbapenems against some isolates of *M. abscessus* may be problematic. However, imipenem remains the carbapenem with the highest in vitro activity when compared to meropenem or ertapenem.
- Sulfonamide resistance in species that are typically susceptible (i.e., *M. fortuitum* group, *M. mucogenicum* group) should be investigated, and repeat testing or referral to a laboratory experienced in NTM AST should be sought.
- Quinolone resistance in species that are typically susceptible (i.e., *M. fortuitum* group) should be investigated. The finding of quinolone resistance in these species often is associated with previous long-term therapy with quinolones.
- Aminoglycoside resistance in isolates of *M. abscessus* is often associated with prior treatment with amikacin and/or tobramycin. Mutational resistance is associated with a point mutation in the 16S rRNA gene.
- Macrolide resistance in isolates of *M. abscessus* at 3 days is usually associated with a point mutation in the 23S rRNA gene. This resistance is in contrast to the inducible macrolide resistance conferred by the *erm* gene.
- First-line antituberculous agents *should not be tested* with isolates of MAC. There is no correlation between clinical response and in vitro MICs. However, macrolide-containing regimens including rifampin or rifabutin and ethambutol, with or without amikacin or streptomycin *should be implemented* without regard to MICs for the antituberculous agents.
- The only antimicrobials which should be routinely tested with isolates of *Mycobacterium avium* complex are clarithromycin and amikacin. Clarithromycin is the class drug for the macrolides as azithromycin cannot be tested in vitro due to technical problems in higher concentrations. Thus, the only standardized macrolide breakpoints apply to clarithromycin. Isolates that are susceptible to clarithromycin are also susceptible to azithromycin and vice versa.
- There have been no studies to show correlation of clinical response to MICs for moxifloxacin or linezolid in MAC. Thus, these agents may be tested against isolates of MAC. However, the clinical efficacy of these agents *has not been established*, and these agents are *not recommended* as substitutions for any of the first-line antituberculous agents *which should be used* in combination with a macrolide (with or without amikacin or streptomycin) as stated above.

Acknowledgments The author wishes to thank Richard J. Wallace, Jr. for his expert review of the chapter and Joanne Woodring for her excellent clerical skills.

This chapter is dedicated to my beloved husband, Clyde Elliott, and my dear mother Clifford Brown, who passed away during the preparation of this chapter. They provided constant support and encouragement to me throughout my years in the laboratory.

References

1. Tortoli E. Microbiological features and clinical relevance of new species of the genus *Mycobacterium*. Clin Microbiol Rev. 2014;27:727–52.
2. Tortoli E. Impact of genotypic studies on mycobacterial taxonomy: the new mycobacteria of the 1990s. Clin Microbiol Rev. 2003;16:319–54.
3. Turenne CY, Tschetter L, Wolfe J, Kabani A. Necessity of quality-controlled 16S rRNA gene sequence databases: identifying nontuberculous *Mycobacterium* species. J Clin Microbiol. 2001;39:3637–48.
4. Vestal AL. Procedures for the isolation and identification of mycobacteria, vol. 1995. In: US Department of Health E, and Welfare, editor. Washington, DC: United States Government Printint Office; 1969. p. 1–118.
5. Brown-Elliott BA, Wallace RJ Jr. Clinical and taxonomic status of pathogenic nonpigmented or late-pigmenting rapidly growing mycobacteria. Clin Microbiol Rev. 2002;15:716–46.
6. Wallace RJ Jr, Swenson JM, Silcox VA, Good RC. Disk diffusion testing with polymyxin and amikacin for differentiation of *Mycobacterium fortuitum* and *Mycobacterium chelonei*. J Clin Microbiol. 1982;16:1003–6.
7. Wallace RJ Jr, Wiss K, Bushby MB, Hollowell DC. *In vitro* activity of trimethoprim and sulfamethoxazole against the nontuberculous mycobacteria. Rev Infect Dis. 1982;4:326–31.
8. Wallace RJ Jr, Brown BA, Onyi GO. Susceptibilities of *Mycobacterium fortuitum* biovar. *fortuitum* and the two subgroups of *Mycobacterium chelonae* to imipenem, cefmetazole, cefoxitin, and amoxicillin-clavulanic acid. Antimicrob Agents Chemother. 1991;35:773–5.
9. Wallace RJ Jr, Brown BA, Onyi G. Skin, soft tissue, and bone infections due to *Mycobacterium chelonae* subspecies *chelonae* – importance of prior corticosteroid therapy, frequency of disseminated infections, and resistance to oral antimicrobials other than clarithromycin. J Infect Dis. 1992;166:405–12.
10. Wilson RW, Steingrube VA, Böttger EC, Springer B, Brown-Elliott BA, Vincent V, Jost KC Jr, Zhang Y, Garcia MJ, Chiu SH, Onyi GO, Rossmoore H, Nash DR, Wallace RJ Jr. *Mycobacterium immunogenum* sp. nov., a novel species related to *Mycobacterium abscessus* and associated with clinical disease, pseudo-outbreaks, and contaminated metalworking fluids: an international cooperative study on mycobacterial taxonomy. Int J Syst Evol Microbiol. 2001;51:1751–64.
11. Wallace RJJ, Brown-Elliott BA, Brown J, Steigerwalt AG, Hall L, Woods G, Cloud J, Mann L, Wilson R, Crist C, Jost KC Jr, Byrer DE, Tang J, Cooper J, Stamenova E, Campbell B, Wolfe J, Turenne C. Polyphasic characterization reveals that the human pathogen *Mycobacterium peregrinum* type II belongs to the bovine pathogen species *Mycobacterium senegalense*. J Clin Microbiol. 2005;43:5925–35.
12. Wallace RJ Jr, Bedsole G, Sumter G, Sanders CV, Steele LC, Brown BA, Smith J, Graham DR. Activities of ciprofloxacin and ofloxacin against rapidly growing mycobacteria with demonstration of acquired resistance following single-drug therapy. Antimicrob Agents Chemother. 1990;34:65–70.
13. Wallace RJ Jr, Silcox VA, Tsukamura M, Brown BA, Kilburn JO, Butler WR, Onyi G. Clinical significance, biochemical features, and susceptibility patterns of sporadic isolates of the *Mycobacterium chelonae*-like organism. J Clin Microbiol. 1993;31:3231–9.
14. Clinical and Laboratory Standards Institute. Laboratory detection and identification of mycobacteria; approved guidelines. CLSI document M48-A. 2008.
15. Brown-Elliott BA, Wallace RJ Jr. Enhancement of conventional phenotypic methods with molecular-based methods for the more definitive identification of nontuberculous mycobacteria. Clin Microbiol Newsl. 2012;34:109–15.
16. Soini H, Musser JM. Molecular diagnosis of mycobacteria. Clin Chem. 2001;47:809–14.
17. Jagielski T, Minias A, van Ingen J, Rastogi N, Brzostek A, Żaczek A, Dziadek J. Methodological and clinical aspects of the molecular epidemiology of *Mycobacterium tuberculosis* and other mycobacteria. Clin Microbiol Rev. 2016;29:239–90.

18. Tortoli E, Pecorari M, Fabio G, Messinò M, Fabio A. Commercial DNA probes for mycobacteria incorrectly identify a number of less frequently encountered species. J Clin Microbiol. 2010;48:307–10.
19. Cook VJ, Turenne CY, Wolfe J, Pauls R, Kabani A. Conventional methods versus 16S ribosomal DNA sequencing for identification of nontuberculous mycobacteria: cost analysis. J Clin Microbiol. 2003;41:1010–5.
20. Reisner BS, Gatson AM, Woods GL. Use of Gen-Probe AccuProbes to identify *Mycobacterium avium* complex, *Mycobacterium tuberculosis* complex, *Mycobacterium kansasii*, and *Mycobacterium gordonae* directly from BACTEC TB broth cultures. J Clin Microbiol. 1994;32:2995–8.
21. LeBrun L, Espinasse F, Poveda JD, Vincent-Levy-Frebault V. Evaluation of nonradioactive DNA probes for identification of mycobacteria. J Clin Microbiol. 1992;30:2476–8.
22. Goto M, Oka S, Okuzumi K, Kimura S, Shimada K. Evaluation of acridinium-ester-labeled DNA probes for identification of *Mycobacterium tuberculosis* and *Mycobacterium avium-Mycobacterium intracellulare* complex in culture. J Clin Microbiol. 1991;29:2473–6.
23. Brown-Elliott BA, Wallace RJ Jr. *Mycobacterium*: clinical and laboratory characteristics of rapidly growing mycobacteria. In: Manual of clinical microbiology, vol. 1. 11th ed. Washington, DC: ASM Press; 2015.
24. de Zwaan R, van Ingen J, van Soolingen D. Utility of *rpoB* gene sequencing for identification of nontuberculous mycobacteria in the Netherlands. J Clin Microbiol. 2014;52:2544–51.
25. Blauwendraat C, Dixon GLJ, Hartley JC, Foweraker J, Harris KA. The use of a two-gene sequencing approach to accurately distinguish between the species within the *Mycobacterium abscessus* complex and *Mycobacterium chelonae*. Eur J Clin Microbiol Infect Dis. 2012;31:1847–53.
26. Tortoli E, Mariottini A, Mazzarelli G. Evaluation of INNO-LiPA MYCOBACTERIA v2: improved reverse hybridization multiple DNA probe assay for mycobacterial identification. J Clin Microbiol. 2003;41:4418–20.
27. Simner PJ, Stenger S, Richter E, Brown-Elliott BA, Wallace RJ Jr, Wengenack NL. *Mycobacterium*: clinical and laboratory characteristics of slowly growing mycobacteria. In: Manual of clinical microbiology, vol. 1. 11th ed. Washington, DC: ASM Press; 2015.
28. Richter E, Rüsch-Gerdes S, Hillemann D. Evaluation of the GenoType *Mycobacterium* assay for identification of mycobacterial species from cultures. J Clin Microbiol. 2006;44:1769–75.
29. Russo C, Tortoli E, Menichella D. Evaluation of the new GenoType mycobacterium assay for identification of mycobacterial disease. J Clin Microbiol. 2006;44:334–9.
30. Quezel-Guerraz NM, Arriaza MM, Avila JA, Sanchez-Yebra R, Martinez-Lirola MJ. Evaluation of the Speed-oligo(R) Mycobacteria assay for identification of *Mycobacterium* spp. from fresh liquid and solid cultures of human clinical samples. Diagn Microbiol Infect Dis. 2010;68:123–31.
31. Lara-Oya A, Mendoza-Lopez P, Rodriguez-Granger J, Fernandez-Sanchez AM, Bermudez-Ruiz MP, Toro-Peinado I, Palop-Bornas B, Navarro-Mari JM, Martinez-Lirola MJ. Evaluation of the speed-oligo direct *Mycobacterium tuberculosis* assay for molecular detection of mycobacteria in clinical respiratory specimens. J Clin Microbiol. 2013;51:77–82.
32. Hofmann-Thiel S, Turaev L, Alnour T, Drath L, Mullerova M, Hoffmann H. Multi-centre evaluation of the speed-oligo Mycobacteria assay for differentiation of *Mycobacterium* spp. in clinical isolates. BMC Infect Dis. 2011;11:353–9.
33. Telenti A, Marchesi F, Balz M, Bally F, Böttger EC, Bodmer T. Rapid identification of mycobacteria to the species level by polymerase chain reaction and restriction enzyme analysis. J Clin Microbiol. 1993;31:175–8.
34. Steingrube VA, Gibson JL, Brown BA, Zhang Y, Wilson RW, Rajagopalan M, Wallace RJ Jr. PCR amplification and restriction endonuclease analysis of a 65-kilodalton heat shock protein gene sequence for taxonomic separation of rapidly growing mycobacteria [ERRATUM 1995;33:1686]. J Clin Microbiol. 1995;33:149–53.
35. Devallois A, Goh KS, Rastogi N. Rapid identification of mycobacteria to species level by PCR-restriction fragment length polymorphism analysis of the *hsp*65 gene and proposition of an algorithm to differentiate 34 mycobacterial species. J Clin Microbiol. 1997;35:2969–73.

36. Dai J, Chen Y, Lauzardo M. Web-accessible database of *hsp65* sequences from *Mycobacterium* reference strains. J Clin Microbiol. 2011;49:2296–303.
37. Macheras E, Roux A-L, Bastian S, Leão SC, Palaci M, Silvadon-Tardy V, Gutierrez C, Richter E, Rüsch-Gerdes S, Pfyffer G, Bodmer T, Cambau E, Gaillard J-L, Heym B. Multilocus sequence analysis and *rpo B* sequencing of *Mycobacterium abscessus* (sensu Lato) strains. J Clin Microbiol. 2011;49:491–9.
38. Hall L, Doerr KA, Wohlfiel SL, Roberts GD. Evaluation of the MicroSeq System for identification of mycobacteria by 16S ribosomal DNA sequencing and its integration into a routine clinical mycobacteriology laboratory. J Clin Microbiol. 2003;41:1447–53.
39. Kirschner P, Springer B, Vogel U, Meier A, Wrede A, Kiekenbeck M, Bange FC, Böttger EC. Genotypic identification of mycobacteria by nucleic acid sequence determination: report of a 2 year experience in a clinical laboratory. J Clin Microbiol. 1993;31:2882–9.
40. Patel JB, Leonard DGB, Pan X, Musser JM, Berman RE, Nachamkin I. Sequence-based identification of *Mycobacterium* species using the MicroSeq 500 16S rDNA bacterial identification system. J Clin Microbiol. 2000;38:246–51.
41. Clinical and Laboratory Standards Institute. Interpretive criteria for identification of bacteria and fungi by DNA target sequencing: approved guideline. CLSI document MM18-A. 2008.
42. Brown BA, Springer B, Steingrube VA, Wilson RW, Pfyffer GE, Garcia MJ, Menendez MC, Rodriguez-Salgado B, Jost KC Jr, Chiu SH, Onyi GO, Bottger EC, Wallace RJ Jr. *Mycobacterium wolinskyi* sp. nov. and *Mycobacterium goodii* sp. nov., two new rapidly growing species related to *Mycobacterium smegmatis* and associated with human wound infections: a cooperative study from the International Working Group on Mycobacterial Taxonomy. Int J Syst Bacteriol. 1999;49:1493–511.
43. Tortoli E. Phylogeny of the genus Mycobacterium: many doubts, few certainties. Infect Genet Evol. 2012;12:827–31.
44. Adékambi T, Colson P, Drancourt M. *rpoB*-based identification of nonpigmented and late pigmented rapidly growing mycobacteria. J Clin Microbiol. 2003;41:5699–708.
45. Ben Salah I, Adekambi T, Raoult D, Drancourt M. *rpoB* sequence-based identification of *Mycobacterium avium* complex species. Microbiology. 2008;154:3715–23.
46. Kim B-J, Lee SH, Lyu MA, Kim SJ, Bai GH, Chae GT, Kim EC, Cha CY, Kook YH. Identification of mycobacterial species by comparative sequence analysis of the RNA polymerase gene (*rpoB*). J Clin Microbiol. 1999;37:1714–20.
47. Lee H, Bang H-E, Bai G-H, Cho S-N. Novel polymorphic region of the *rpoB* gene containing *Mycobacterium* species-specific sequences and its use in identification of mycobacteria. J Clin Microbiol. 2003;41:2213–8.
48. Tortoli E. Standard operating procedure for optimal identification of mycobacteria using 16S rRNA gene sequences. Stand Genomic Sci. 2010;3:145–52.
49. Ringuet H, Akoua-Koffi C, Honore S, Varnerot A, Vincent V, Berche P, Gaillard JL, Pierre-Audigier C. *hsp*65 sequencing for identification of rapidly growing mycobacteria. J Clin Microbiol. 1999;37:852–7.
50. Turenne CY, Semret M, Cousins DV, Collins DM, Behr MA. Sequencing of *hsp*65 distinguishes among subsets of the *Mycobacterium avium* complex. J Clin Microbiol. 2006;44:433–40.
51. Nash KA, Brown-Elliott BA, Wallace RJ Jr. A novel gene, erm(41), confers inducible macrolide resistance to clinical isolates of *Mycobacterium abscessus* but is absent from *Mycobacterium chelonae*. Antimicrob Agents Chemother. 2009;53:1367–76.
52. Kim H-Y, Kim B-J, Kook Y, Yun Y-J, Shin JH, Kook YH. *Mycobacterium massiliense* is differentiated from *Mycobacterium abscessus* and *Mycobacterium bolletii* by erythromycin ribosome methyltransferase gene (*erm*) and clarithromycin susceptibility patterns. Microbiol Immunol. 2010;54:347–53.
53. Koh WJ, Jeon K, Lee NY, Kim B-J, Kook Y-H, Lee S-H, Park Y-K, Kim CK, Shin SJ, Huitt GA, Daley CL, Kwon OJ. Clinical significance of differentiation of *Mycobacterium massiliense* from *Mycobacterium abscessus*. Am J Respir Crit Care Med. 2011;183:405–10.
54. Nash KA, Andini N, Zhang Y, Brown-Elliott BA, Wallace RJ Jr. Intrinsic macrolide resistance in rapidly growing mycobacteria. Antimicrob Agents Chemother. 2006;50:3476–8.

55. Nash KA. Intrinsic macrolide resistance in *Mycobacterium smegmatis* is conferred by a novel *erm* gene, *erm*(38). Antimicrob Agents Chemother. 2003;47:3053–60.
56. Nash DR, Wallace RJ Jr, Steingrube VA, Udou T, Steele LC, Forrester GD. Characterization of beta-lactamases in *Mycobacterium fortuitum* including a role in beta-lactam resistance and evidence of partial inducibility. Am Rev Respir Dis. 1986;134:1276–82.
57. Brown-Elliott BA, Vasireddy S, Vasireddy R, Iakhiaeva E, Howard ST, Nash KA, Parodi N, Strong A, Gee M, Smith T, Wallace RJ Jr. Utility of sequencing the *erm*(41) gene in isolates of *Mycobacterium abscessus* subsp. *abscessus* with low and intermediate clarithromycin MICs. J Clin Microbiol. 2015;53:1211–5; ERRATUM J Clin Microbiol 1254:1172, April 2016.
58. Brown-Elliott BA, Hanson K, Vasireddy S, Iakhiaeva E, Nash KA, Vasireddy R, Parodi N, Smith T, Gee M, Strong A, Baker A, Cohen S, Muir H, Slechta ES, Wallace RJ Jr. Absence of a functional *erm* gene in isolates of *Mycobacterium immunogenum* and the *Mycobacterium mucogenicum* group, based on *in vitro* clarithromycin susceptibility. J Clin Microbiol. 2015;53:875–8.
59. Adékambi T, Drancourt M. Dissection of phylogenetic relationships among nineteen rapidly growing mycobacterium species by 16S rRNA, *hsp*65, *sod*A, *rec*A, and *rpoB* gene sequencing. Int J Syst Evol Microbiol. 2004;54:2095–105.
60. Park H, Jang H, Kim J, Chung B, Chang CL, Park SK, Song S. Detection and identification of mycobacteria by amplification of the internal transcribed spacer regions with genus- and species-specific PCR primers. J Clin Microbiol. 2000;38:4080–5.
61. Macheras E, Roux A-L, Ripoll F, Sivadon-Tardy V, Gutierrez C, Gaillard J-L, Heym B. Inaccuracy of single-target sequencing for discriminating species of the *Mycobacterium abscessus* group. J Clin Microbiol. 2009;47:2596–600.
62. Adékambi T, Raoult D, D M. *Mycobacterium barrassiae* sp. nov., a *Mycobacterium moriokaense* group species associated with chronic pneumonia. J Clin Microbiol. 2006;44:3493–8.
63. Roth A, Reischl U, Streubel A, Naumann L, Kroppenstedt RM, Habicht M, Fischer M, Mauch H. Novel diagnostic algorithm for identification of mycobacteria using genus-specific amplification of the 16S-23S rRNA gene spacer and restriction endonucleases. J Clin Microbiol. 2000;38:1094–104.
64. Mohamed AM, Kuyper DJ, Iwen PC, Ali HH, Bastola DR, Hinrichs SH. Computational approach involving use of the internal transcribed spacer 1 region for identification of *Mycobacterium* species. J Clin Microbiol. 2005;43:3811–2817.
65. Frothingham R, Wilson KH. Molecular phylogeny of the *Mycobacterium avium* complex demonstrates clinically meaningful divisions. J Infect Dis. 1994;169:305–12.
66. Frothingham R, Wilson KH. Sequence-based differentiation of strains in the *Mycobacterium avium* complex. J Bacteriol. 1993;175:2818–25.
67. Kim H-Y, Kook Y, Yun Y-J, Park CG, Lee NY, Shim TS, Kim B-J, Kook Y-H. Proportions of *Mycobacterium massiliense* and *Mycobacterium bolletii* in Korean *Mycobacterium chelonae-Mycobacterium abscessus* group isolates. J Clin Microbiol. 2008;46:3384–90.
68. Bao JR, Master RN, Schwab DA, Clark RB. Identification of acid-fast bacilli using pyrosequencing analysis. Diagn Microbiol Infect Dis. 2010;67:234–8.
69. Heller LC, Jones M, Widen RH. Comparison of DNA pyrosequencing with alternative methods for identification of mycobacteria. J Clin Microbiol. 2008;46:2092–4.
70. Tuohy MJ, Hall GS, Sholtis M, Procop GW. Pyrosequencing as a tool for the identification of common isolates of *Mycobacterium* sp. Diagn Microbiol Infect Dis. 2005;51:245–50.
71. Arnold C, Barrett A, Cross L, Magee JG. The use of *rpoB* sequence analysis in the differentiation of *Mycobacterium abscessus* and *Mycobacterium chelonae*: a critical judgement in cystic fibrosis? Clin Microbiol Infect. 2012;18:E131–3.
72. Buckwalter SP, Olson SL, Connelly BJ, Lucas BC, Rodning AA, Walchak RC, Deml SM, Wohlfiel SL, Wengenack NL. Evaluation of matrix-assisted laser desorption ionization-time of flight mass spectrometry for identification of *Mycobacterium* species, *Nocardia* species, and other aerobic actinomycetes. J Clin Microbiol. 2016;54:376–84.

73. Saleeb PG, Drake SK, Murray PR, Zelazny AM. Identification of mycobacteria in solid-culture media by matrix-assisted laser desorption ionization-time of flight mass spectrometry. J Clin Microbiol. 2011;49:1790–4.
74. Rodríguez-Sánchez B, Ruiz-Serrano MJ, Ruiz A, Timke M, Kostrzewa M, Bouza E. Evaluation of MALDI biotyper mycobacterial library v3.0 for identification of nontuberculous mycobacteria. J Clin Microbiol. 2016;54:1144–7.
75. Hettick JM, Kashon ML, Slaven JE, Ma Y, Simpson JP, Siegel PD, Mazurek GN, Weissman DN. Discrimination of intact mycobacteria at the strain level: a combined MALDI-TOF MS and biostatistical analysis. Proteomics. 2006;6:6416–25.
76. Lefmann M, Honsich C, Böcker S, Storm N, von Wintzingerode F, Schlötelburg C, Moter A, van den Boom D, Göbel UB. Novel mass spectrometry-based tool for genotypic identification of mycobacteria. J Clin Microbiol. 2004;42:339–46.
77. Lotz A, Gerroni A, Beretti J-L, Dauphin B, Carbonnelle E, Guet-Revillet H, Veziris N, Heym B, Jarlier V, Gaillard J-L, Pierre-Audigier C, Frapy E, Berche P, Nassif X, Bille E. Rapid identification of mycobacterial whole cells in solid and liquid culture media by matrix-assisted laser desorption ionization-time of flight mass spectrometry. J Clin Microbiol. 2010;48:4481–6.
78. Tan N, Sampath R, Abu Saleh OM, Tweet MS, Jevremovic D, Alniemi S, Wengenack NL, Sampathkumar P, Badley AD. Disseminated *Mycobacterium chimaera* infection after cardiothoracic surgery. Open Forum Infect Dis. 2016;3:1–3.
79. Tenover FC, Arbeit RD, Goering RV, Mickelsen PA, Murray BE, Persing DH, Swaminathan B. Interpreting chromosomal DNA restriction patterns produced by pulsed-field gel electrophoresis: criteria for bacterial strain typing. J Clin Microbiol. 1995;33:2233–9.
80. Wallace RJ Jr, Zhang Y, Brown-Elliott BA, Yakrus MA, Wilson RW, Mann L, Couch L, Girard WM, Griffith DE. Repeat positive cultures in *Mycobacterium intracellulare* lung disease after macrolide therapy represent new infections in patients with nodular bronchiectasis. J Infect Dis. 2002;186:266–73.
81. Zhang Y, Yakrus MA, Graviss EA, Williams-Bouyer N, Turenne C, Kabani A, Wallace RJ Jr. Pulsed-field gel electrophoresis study of *Mycobacterium abscessus* isolates previously affected by DNA degradation. J Clin Microbiol. 2004;42:5582–7.
82. Howard ST, Newman KL, McNulty S, Brown-Elliott BA, Vasireddy R, Bridge L, Wallace RJ Jr. Insertion site and distribution of a genomic island conferring DNA phosphorothioation in the *Mycobacterium abscessus* complex. Microbiology. 2013;159:2323–32.
83. Hector JSR, Pang Y, Mazurek GH, Zhang Y, Brown BA, Wallace RJ Jr. Large restriction fragment patterns of genomic *Mycobacterium fortuitum* DNA as strain-specific markers and their use in epidemiologic investigation of four nosocomial outbreaks. J Clin Microbiol. 1992;30:1250–5.
84. Lai KK, Brown BA, Westerling JA, Fontecchio SA, Zhang Y, Wallace RJ Jr. Long-term laboratory contamination by *Mycobacterium abscessus* resulting in two pseudo-outbreaks: recognition with use of random amplified polymorphic DNA (RAPD) polymerase chain reaction. Clin Infect Dis. 1998;27:169–75.
85. Brown-Elliott BA, Wallace RJ Jr. Nontuberculous mycobacteria. In: Mayhall CG, editor. Hospital epidemiology and infection control. 4th ed. Philadelphia: Lippincott Williams & Wilkins; 2012. p. 593–608.
86. Griffith DE, Brown-Elliott BA, Langsjoen B, Zhang Y, Pan X, Girard W, Nelson K, Caccitolo J, Alvarez J, Shepherd S, Wilson R, Graviss EA, Wallace RJ Jr. Clinical and molecular analysis of macrolide resistance in *Mycobacterium avium* complex lung disease. Am J Respir Crit Care Med. 2006;174:928–34.
87. Wallace RJ Jr, Zhang Y, Brown BA, Dawson D, Murphy DT, Wilson R, Griffith DE. Polyclonal *Mycobacterium avium* complex infections in patients with nodular bronchiectasis. Am J Respir Crit Care Med. 1998;158:1235–44.
88. Zelazny AM, Root JM, Shea YR, Colombo RE, Shamputa IC, Stock F, Conlan SS, McNulty S, Brown-Elliott BA, Wallace RJ Jr, Olivier KN, Holland SM, Sampaio EP. Cohort study of

molecular identification and typing of *Mycobacterium abscessus, Mycobacterium massiliense* and *Mycobacterium bolletii.* J Clin Microbiol. 2009;47:1985–95.

89. Cangelosi GA, Freeman RJ, Lewis KN, Livingston-Rosanoff D, Shah KS, Milan SJ, Goldberg SV. Evaluation of a high-throughput repetitive-sequence-based PCR system for DNA fingerprinting of *Mycobacterium tuberculosis* and *Mycobacterium avium* complex strains. J Clin Microbiol. 2004;42:2685–93.
90. Zhang Y, Rajagopalan M, Brown BA, Wallace RJ Jr. Randomly amplified polymorphic DNA PCR for comparison of *Mycobacterium abscessus* strains from nosocomial outbreaks. J Clin Microbiol. 1997;35:3132–9.
91. Sax H, Bloemberg G, Hasse B, Sommerstein R, Kohler P, Achermann Y, Rössle M, Falk V, Kuster SP, Böttger EC, Weber R. Prolonged outbreak of *Mycobacterium chimaera* infection after open-chest heart surgery. Clin Infect Dis. 2015;61:67–75.
92. Sommerstein R, Rüegg C, Kohler P, Bloemberg G, Kuster SP, Sax H. Transmission of *Mycobacterium chimaera* from heater-cooler units during cardiac surgery despite an ultraclean air ventilation system. Emerg Infect Dis. 2016;22:1008–14.
93. Inagaki T, Nishimori K, Yagi T, Ichikawa K, Moriyama M, Nakagawa T, Shibayama T, Uchiya K-I, Nikai T, Ogawa K. Comparison of a variable-number tandem-repeat (VNTR) method for typing *Mycobacterium avium* with mycobacterial interspersed repetitive-unit-VNTR and IS1245 restriction fragment length polymorphism typing. J Clin Microbiol. 2009;47:2156–64.
94. Wallace RJ Jr, Iakhiaeva E, Williams M, Brown-Elliott BA, Vasireddy S, Vasireddy R, Lande L, Peterson D, Sawicki J, Kwait R, Tichenor W, Turenne C, Falkinham JO III. Absence of *Mycobacterium intracellulare* and the presence of *Mycobacterium chimaera* in household water and biofilm samples of patients in the U.S. with *Mycobacterium avium* complex respiratory disease. J Clin Microbiol. 2013;51:1747–52.
95. Iakhiaeva E, Howard S, Brown-Elliott BA, McNulty S, Falkinham JO III, Newman K, Williams M, Kwait R, Lande L, Vasireddy R, Turenne C, Wallace RJ Jr. Variable number tandem-repeat (VNTR) analysis of respiratory and household water biofilm isolates of "*Mycobacterium avium* subspecies *hominissuis*" with establishment of a PCR database. J Clin Microbiol. 2016;54:891–901.
96. Iakhiaeva E, McNulty S, Brown-Elliott BA, Falkinham JO III., Williams MD, Vasireddy R, Wilson RW, Turenne C, Wallace RJ Jr. Mycobacterial interspersed repetitive-unit-variable-number tandem-repeat (MIRU-VNTR) genotyping of *Mycobacterium intracellulare* for strain comparison with establishment of a PCR database. J Clin Microbiol. 2013;51:409–16.
97. Wong YL, Ong CS, Ngeow YF. Molecular typing of *Mycobacterium abscessus* based on tandem-repeat polymorphism. J Clin Microbiol. 2012;50:3084–8.
98. Machado GE, Matsumoto CK, Chimara E, da Silva Duarte F, de Freitas D, Palaci M, Hadad DJ, Batista KV, Lopes LML, Ramos JP, Campos CE, Caldas PC, Heym B, Leão SC. Multilocus sequence typing scheme versus pulsed-field gel electrophoresis for typing *Mycobacterium abscessus* isolates. J Clin Microbiol. 2014;52:2881–91.
99. Sampaio JL, Chimara E, Ferrazoli L, da Silva Telles MA, Del Guercio VM, Jericó ZV, Miyashiro K, Fortaleza CM, Padoveze MC, Leão SC. Application of four molecular typing methods for analysis of *Mycobacterium fortuitum* group strains causing post-mammaplasty infections. Clin Microbiol Infect. 2006;12:142–9.
100. Sampaio JL, Viana-Niero C, de Freitas D, Höfling-Lima AL, Leão SC. Enterobacterial repetitive intergenic consensus PCR is a useful tool for typing *Mycobacterium chelonae* and *Mycobacterium abscessus* isolates. Diagn Microbiol Infect Dis. 2006;55:107–18.
101. Cole ST, Brosch R, Parkhill J, Garnier T, Churcher C, Harris D, Gordon SV, Eiglmeier K, Gas S, Barry CE, Tekaia F, Badcock K, Basham D, Brown D, Chillingworth T, Connor R, Davies R, Devlin K, Feltwell T, Gentles S, Hamlin N, Holroyd S, Hornsby T, Jagels K, Krogh A, McLean J, Moule S, Murphy L, Olivier K, Osborne J, Quail MA, Rajandream MA, Rogers J, Rutter S, Seeger K, Skelton J, Squares R, Squares S, Sulston JE, Taylor K, Whitehead

S, Barrell BG. Deciphering the biology of *Mycobacterium tuberculosis* from the complete genome sequence. Nature. 1998;393:537–44.
102. Kohler P, Kuster SP, Bloemberg G, Schulthess B, Frank M, Tanner FC, Rössle M, Böni C, Falk V, Wilhelm MJ, Sommerstein R, Achermann Y, Ten Oever J, Debast SB, Wolfhagen MJHM, Bravo Bruinsma GJB, Vos MC, Bogers A, Serr A, Beyersdorf F, Sax H, Böttger EC, Weber R, van Ingen J, Wagner D, Hasse B. Healthcare-associated prosthetic heart valve, aortic vascular graft, and disseminated *Mycobacterium chimaera* infections subsequent to open heart surgery. Eur Heart J. 2015;36:2745–53.
103. Choo SW, Wong YL, Tan JL, Ong CS, Wong GJ, Ng KP, Ngeow YF. Annotated genome sequence of *Mycobacterium massiliense* strain M154, belonging to the recently created taxon *Mycobacterium abscessus* subsp. *bolletii* comb. nov. J Bacteriol. 2012;194:4778.
104. Ngeow YF, Wee WY, Wong YL, Tan JL, Ongi CS, Ng KP, Choo SW. Genomic analysis of *Mycobacterium abscessus* strain M139, which has an ambiguous subspecies taxonomic position. J Bacteriol. 2012;194:6002–3.
105. Ngeow YF, Wong YL, Lokanathan N, Wong GJ, Ong CS, Ng KP, Choo SW. Genomic analysis of *Mycobacterium massiliense* strain M115, an isolate from human sputum. J Bacteriol. 2012;194:4786.
106. Ngeow YF, Wong YL, Tan JL, Arumugam R, Wong GJ, Ong CS, Ng KP, Choo SW. Genome sequence of *Mycobacterium massiliense* M18, isolated from a lymph node biopsy specimen. J Bacteriol. 2012;194:4125.
107. Tettelin H, Sampaio EP, Daugherty SC, Hine E, Riley DR, Sadzewicz L, Sengamalay N, Shefchek K, Su Q, Tallon LJ, Conville P, Olivier KN, Holland SM, Fraser CM, Zelazny AM. Genomic insights into the emerging human pathogen *Mycobacterium massiliense*. J Bacteriol. 2012;194:5450.
108. Chan J, Halachev M, Yates E, Smith G, Pallen M. Whole-genome sequence of the emerging pathogen *Mycobacterium abscessus* strain 47J26. J Bacteriol. 2012;194:549.
109. Tortoli E, Kohl TA, Brown-Elliott BA, Trovato A, Cardoso Leao S, Garcia MJ, Vasireddy S, Turenne CY, Griffith DE, Philley JV, Balden R, Campana S, Cariani L, Colombo C, Taccetti G, Teri A, Niemann S, Wallace RJ Jr, Cirillo DM. Emended description of *Mycobacterium abscessus, Mycobacterium abscessus* subsp. *abscessus* and *Mycobacterium abscessus* subsp. *bolletii* and designation of *Mycobacterium abscessus* subsp. *massiliense* subsp. comb. nov. Int J Syst Evol Microbiol. 2016;66:4471–9.
110. Woods GL, Brown-Elliott BA, Desmond EP, Hall GS, Heifets L, Pfyffer GE, Ridderhof JC, Wallace RJ Jr., Warren NG, Witebsky FG. Susceptibility testing of mycobacteria, norcardiae, and other aerobic actinomycetes; approved standard. NCCLS document M24-A. 2003.
111. Griffith DE, Aksamit T, Brown-Elliott BA, Catanzaro A, Daley C, Gordin F, Holland SM, Horsburgh R, Huitt G, Iademarco MF, Iseman M, Olivier K, Ruoss S, von Reyn CF, Wallace RJ Jr, Winthrop K. An official ATS/IDSA statement: diagnosis, treatment and prevention of nontuberculous mycobacterial diseases. Am J Respir Crit Care Med. 2007;175:367–416.
112. Brown-Elliott BA, Nash KA, Wallace RJ Jr. Antimicrobial susceptibility testing, drug resistance mechanisms, and therapy of infections with nontuberculous mycobacteria. Clin Microbiol Rev. 2012;25:545–82.
113. Clinical and Laboratory Standards Institute. Susceptibility testing of mycobacteria, nocardiae, and other aerobic actinomycetes: approved standard—second edition. CLSI document M24-A2. 2011.
114. Forbes BA, Banaiee N, Beavis KG, Brown-Elliott BA, Della Latta P, Elliott LB, Hall GS, Hanna B, Perkins MD, Siddiqi SH, Wallace RJ Jr., Warren NG. Laboratory detection and identification of mycobacteria; approved guideline. CLSI document M48-A. 2008.
115. Vasireddy R, Vasireddy S, Brown-Elliott BA, Wengenack NL, Eke UA, Benwill JL, Turenne C, Wallace RJ Jr. *Mycobacterium arupense, Mycobacterium heraklionense,* and a newly proposed species, "*Mycobacterium virginiense*" sp. nov., but not *Mycobacterium nonchromogenicum*, as species of the *Mycobacterium terrae* complex causing tenosynovitis and osteomyelitis. J Clin Microbiol. 2016;54:1340–51.

116. Brown-Elliott BA, Iakhiaeva E, Griffith DE, Woods GL, Stout JE, Wolfe CR, Turenne CY, Wallace RJ Jr. *In vitro* activity of amikacin against isolates of *Mycobacterium avium* complex with proposed MIC breakpoints and finding of a 16S rRNA gene mutation in treated isolates. J Clin Microbiol. 2013;51:3389–94. ERRATUM J Clin Microbiol 3352:1311, 2014.
117. Wallace RJ Jr, Meier A, Brown BA, Zhang Y, Sander P, Onyi GO, Bottger EC. Genetic basis for clarithromycin resistance among isolates of *Mycobacterium chelonae* and *Mycobacterium abscessus*. Antimicrob Agents Chemother. 1996;40:1676–81.
118. Wallace RJ Jr, Hull SI, Bobey DG, Price KE, Swenson JM, Steele L, Christensen L. Mutational resistance as the mechanism of acquired drug resistance to aminoglycosides and antibacterial agents in *Mycobacterium chelonae*: Evidence based on plasmid analysis, mutational frequencies, and aminoglycoside modifying enzyme assays. Am Rev Respir Dis. 1985;132:409–16.
119. Nash KA, Inderlied CB. Genetic basis of macrolide resistance in *Mycobacterium avium* isolated from patients with disseminated disease. Antimicrob Agents Chemother. 1995;39:2625–30.
120. van Ingen J, Egelund EF, Levin A, Totten SE, Boeree MJ, Mouton JW, Aarnoutse RE, Heifets LB, Peloquin CA, Daley CL. The pharmacokinetics and pharmacodynamics of pulmonary *Mycobacterium avium* complex disease treatment. Am J Respir Crit Care Med. 2012;186:559–65.
121. Wallace RJ Jr, Swenson JM, Silcox VA. The rapidly growing mycobacteria: characterization and susceptibility testing. Antimicrob Newsl. 1985;2:85–92.
122. Stone MS, Wallace RJ Jr, Swenson JM, Thornsberry C, Christensen LA. Agar disk elution method for susceptibility testing of *Mycobacterium marinum* and *Mycobacterium fortuitum* complex to sulfonamides and antibiotics. Antimicrob Agents Chemother. 1983;24:486–93.
123. Biehle JR, Cavalieri SJ, Saubolle MA, Getsinger LJ. Evaluation of Etest for susceptibility testing of rapidly growing mycobacteria. J Clin Microbiol. 1995;33:1760–4.
124. Fabry W, Schmid EN, Ansorg R. Comparison of the E test and a proportion dilution method for susceptibility testing of *Mycobacterium kansasii*. Chemotherapy. 1995;41:247–52.
125. Fabry W, Schmid EN, Ansorg R. Comparison of the E test and a proportion dilution method for susceptibility testing of *Mycobacterium avium* complex. J Med Microbiol. 1996;44:227–30.
126. Jarboe E, Stone BL, Burman WJ, Wallace RJ Jr, Brown BA, Reves RR, Wilson ML. Evaluation of a disk diffusion method for determining susceptibility of *Mycobacterium avium* complex to clarithromycin. Diagn Microbiol Infect Dis. 1998;30:197–203.
127. Woods GL, Bergmann JS, Witebsky FG, Fahle GA, Boulet B, Plaunt M, Brown BA, Wallace RJ Jr, Wanger A. Multisite reproducibility of Etest for susceptibility testing of *Mycobacterium abscessus, Mycobacterium chelonae,* and *Mycobacterium fortuitum*. J Clin Microbiol. 2000;38:656–61.
128. Woods GL, Bergmann JS, Witebsky FG, Fahle GA, Wanger A, Boulet B, Plaunt M, Brown BA, Wallace RJ Jr. Multisite reproducibility of results obtained by the broth microdilution method for susceptibility testing of *Mycobacterium abscessus, Mycobacterium chelonae,* and *Mycobacterium fortuitum*. J Clin Microbiol. 1999;37:1676–82.
129. Fernandez-Roblas R, Martin-de-Hijas NZ, Fernandez-Martinez AI, et al. *In vitro* activities of tigecycline and 10 other antimicrobials against nonpigmented rapidly growing mycobacteria. Antimicrob Agents Chemother. 2008;52:4184–6.
130. Brown BA, Wallace RJ Jr, Onyi GO. Activities of the glycylcyclines N, N-dimethylglycylamido-minocycline and N, N-dimethylglycylamido-6-demethyl-6-deoxytetracycline against *Nocardia* spp. and tetracycline-resistant isolates of rapidly growing mycobacteria. Antimicrob Agents Chemother. 1996;40:874–8.
131. Brown BA, Wallace RJ Jr, Onyi GO, De Rosas V, Wallace RJ III. Activities of four macrolides, including clarithromycin, against *Mycobacterium fortuitum*, *Mycobacterium chelonae*, and *M. chelonae*-like organisms. Antimicrob Agents Chemother. 1992;36:180–4.
132. Turenne CY, Wallace RJ Jr, Behr MA. *Mycobacterium avium* in the postgenomic era. Clin Microbiol Rev. 2007;20:205–29.

133. van Ingen J, Turenne C, Tortoli E, Wallace RJ Jr, Brown-Elliott BA. A Definition of the *Mycobacterium avium* complex for taxonomic and clinical purposes. IJSEM. 2018. In Press.
134. Babady NE, Hall L, Abbenyi AT, Eisberner JJ, Brown-Elliott BA, Pratt CJ, McGlasson MC, Beierle KD, Wohlfiel SL, Deml SM, Wallace RJ Jr, Wengenack NL. Evaluation of *Mycobacterium avium* complex clarithromycin susceptibility testing using SLOMYCO sensititre panels and JustOne strips. J Clin Microbiol. 2010;48:1749–52.
135. Brown BA, Wallace RJ Jr, Onyi GO. Activities of clarithromycin against eight slowly growing species of nontuberculous mycobacteria, determined by using a broth microdilution MIC system. Antimicrob Agents Chemother. 1992;36:1987–90.
136. Eisenberg E, Barza M. Azithromycin and clarithromycin. Curr Clin Top Infect Dis Chest. 1994;14:52–79.
137. Brown-Elliott BA, Crist CJ, Mann LB, Wilson RW, Wallace RJ Jr. *In vitro* activity of linezolid against slowly growing nontuberculous mycobacteria. Antimicrob Agents Chemother. 2003;47:1736–8.
138. Woods GL, Williams-Bouyer N, Wallace RJ Jr, Brown-Elliott BA, Witebsky FG, Conville PS, Plaunt M, Hall G, Aralar P, Inderlied C. Multisite reproducibility of results obtained by two broth dilution methods for susceptibility testing of *Mycobacterium avium* complex. J Clin Microbiol. 2003;41:627–31.
139. van Ingen J, Boeree MJ, van Soolingen D, Mouton JW. Resistance mechanisms and drug susceptibility testing of nontuberculous mycobacteria. Drug Resist Updat. 2012;15:149–61.
140. Meier A, Heifets L, Wallace RJ Jr, Zhang Y, Brown BA, Sander P, Böttger EC. Molecular mechanisms of clarithromycin resistance in *Mycobacterium avium*: observation of multiple 23S rDNA mutations in a clonal population. J Infect Dis. 1996;174:354–60.
141. Meier A, Kirschner P, Springer B, Steingrube VA, Brown BA, Wallace RJ Jr, Böttger EC. Identification of mutations in 23S rRNA gene of clarithromycin-resistant *Mycobacterium intracellulare*. Antimicrob Agents Chemother. 1994;38:381–4.
142. Nash KA, Inderlied CB. Rapid detection of mutations associated with macrolide resistance in *Mycobacterium avium* complex. Antimicrob Agents Chemother. 1996;40:1748–50.
143. Griffith DE, Brown-Elliott BA, Wallace RJ Jr. Thrice-weekly clarithromycin-containing regimen for treatment of *Mycobacterium kansasii* lung disease: results of a preliminary study. Clin Infect Dis. 2003;37:1178–82.
144. Burman WJ, Stone BL, Brown BA, Wallace RJ Jr, Böttger EC. AIDS-related *Mycobacterium kansasii* infection with initial resistance to clarithromycin. Diagn Microbiol Infect Dis. 1998;31:369–71.
145. Klein JL, Brown TJ, French GL. Rifampin resistance in *Mycobacterium kansasii* is associated with *rpo*B mutations. Antimicrob Agents Chemother. 2001;45:3056–8.
146. Wallace RJ Jr, Dunbar D, Brown BA, Onyi G, Dunlap R, Ahn CH, Murphy DT. Rifampin-resistant *Mycobacterium kansasii*. Clin Infect Dis. 1994;18:736–43.
147. Jernigan JA, Farr BM. Incubation period and sources for cutaneous *Mycobacterium marinum* infection: case report and review of the literature. Clin Infect Dis. 2000;31:439–43.
148. Tortoli E, Piersimoni C, Bacosi D, Bartoloni A, Betti F, Bono L, Burrini C, De Sio G, Lacchini C, Mantella A, Orsi PG, Penati V, Simonetti MT, Böttger EC. Isolation of the newly described species *Mycobacterium celatum* from AIDS patients. J Clin Microbiol. 1995;33:137–40.
149. Tortoli E, Piersimoni C, Kirschner P, Bartoloni A, Burrini C, Lacchini C, Mantella A, Muzzi G, Passerini-Tosi C, Penati V, Scarparo C, Simonetti MT, Böttger EC. Characterization of mycobacterial isolates phylogenetically related to, but different from *Mycobacterium simiae*. J Clin Microbiol. 1997;35:697–702.
150. MacSwiggan DA, Collins CH. The isolation of *M. kansasii* and *M. xenopi* from water systems. Tubercle. 1974;55:291–7.
151. Buchholz UT, McNeill MM, Keyes LE, Good RC. *Mycobacterium malmoense* infections in the United States, January 1993 through June 1995. Clin Infect Dis. 1998;27:551–8.
152. Heginbothom ML, Lindholm-Levy PJ, Heifets LB. Susceptibilities of *Mycobacterium malmoense* determined at the growth optimum pH (pH 6.0). Int J Tuberc Lung Dis. 1998;2:430–4.

153. Ji B, Lefrançois S, Robert J, Chauffour A, Truffot C, Jarlier V. *In vitro* and *in vivo* activities of rifampin, streptomycin, amikacin, moxifloxacin, R207910, linezolid, and PA-824 against *Mycobacterium ulcerans*. Antimicrob Agents Chemother. 2006;50:1921–6.
154. Vadney FS, Hawkins JE. Evaluation of a simple method for growing *Mycobacterium haemophilum*. J Clin Microbiol. 1985;28:884–5.
155. McBride ME, Rudolph AH, Tschen JA, Cernoch P, Davis J, Brown BA, Wallace RJ Jr. Diagnostic and therapeutic considerations for cutaneous *Mycobacterium haemophilum* infections. Arch Dermatol. 1991;127:276–7.
156. Prammananan T, Sander P, Brown BA, Frischkorn K, Onyi GO, Zhang Y, Böttger EC, Wallace RJ Jr. A single 16S ribosomal RNA substitution is responsible for resistance to amikacin and other 2-deoxystreptamine aminoglycosides in *Mycobacterium abscessus* and *Mycobacterium chelonae*. J Infect Dis. 1998;177:1573–81.
157. Rahman SA, Singh Y, Kohli S, Ahmad J, Ehtesham NZ, Tyagi AK, Hasnain SE. Comparative analyses of nonpathogenic, opportunistic, and totally pathogenic mycobacteria reveal genomic and biochemical variabilities and highlight the survival attributes of Mycobacterium tuberculosis. mBio. 2014;5:e02020–14; ERRATUM mBio 02015;02026(02021):e02343–02014.
158. Stinear TP, Seemann T, Pidot S, Frigui W, Reysset G, Garnier T, Meurice G, Simon D, Bouchier C, Ma L, Tichit M, Porter JL, Ryan L, Johnson PDR, Davies JK, Jenkin GA, Small PLC, Jones LM, Tekaia F, Laval F, Daffé M, Parkhill J, Cole ST. Reductive evolution and niche adaptation inferred from the genome of *Mycobacterium ulcerans*, the causative agent of Buruli ulcer. Genome Res. 2007;17:192–300.
159. Bryant JM, Grogono DM, Greaves D, Foweraker J, Roddick I, Inns T, Reacher M, Haworth CS, Curran MD, Harris SR, Peacock SJ, Parkhill J, Floto RA. Whole-genome sequencing to identify transmission of *Mycobacterium abscessus* between patients with cystic fibrosis: a retrospective cohort study. Lancet. 2013;381:1551–60.
160. Aitken ML, Limaye A, Pottinger P, Whimbey E, Goss GH, Tonelli MR, Cangelosi GA, Ashworth M, Olivier KN, Brown-Elliott BA, Wallace RJ Jr. Respiratory outbreak of *Mycobacterium abscessus* subspecies *massiliense* in a lung transplant and cystic fibrosis center. Am J Respir Crit Care Med. 2012;185:231–3.
161. Davidson RM, Hasan N, Reynolds PR, Totten S, Garcia B, Levin A, Ramamoorthy P, Heifets L, Daley CL, Strong M. Genome sequencing of *Mycobacterium abscessus* isolates from patients in the United States and comparisons to globally diverse clinical strains. J Clin Microbiol. 2014;52:3573–82.
162. Duarte RS, Silva Lourenço MC, de Souza Fonseca L, Leão SC, Amorim EDLT, ILL R, Coelho FS, Viana-Niero C, Gomes KM, da Silva MG, de Oliveira Lorena NS, Pitombo MC, Ferreira RMC, de Oliveira Garcia MH, de Oliveira GP, Lupi O, Vilaça BR, Serradas LR, Chebato A, Marques EA, Teixeira LM, Dalcolmo M, Senna SG, Sampaio JLM. Epidemic of postsurgical infections caused by *Mycobacterium massiliense*. J Clin Microbiol. 2009;47:2149–55.
163. Leão SC, Matsumoto CK, Carneiro A, Ramos RT, Nogueira CL, Lima JD Jr, Lima KV, Lopes ML, Schneider H, Azevedo VA, Da Costa da Silva A. The detection and sequencing of a broad-host-range conjugative IncP-Ibeta plasmid in an epidemic strain of *Mycobacterium abscessus* subsp *bolletii*. PLoS One. 2013;8:e60746.
164. Raiol T, Ribeiro GM, Maranhão AQ, Bocca AL, Silva-Pereira I, Junqueira-Kipnis AP, Brigido MM, Kipnis A. Complete genome sequence of *Mycobacterium massiliense*. J Bacteriol. 2012;194:5455.
165. Kim B-J, Kim B-R, Hong S-H, Seok S-H, Kook Y-H. Complete genome sequence of *Mycobacterium massiliense* clinical strain Asan 505945, belonging to the type II genotype. Genome Announc. 2013;1:e00429–00413.
166. Tettelin H, Davidson RM, Agrawal S, Aitken ML, Shallom S, Hasan NA, Strong M, de Moura VCN, De Groote MA, Duarte RS, Hine E, Parankush S, Su Q, Daugherty SC, Fraser CM, Brown-Elliott BA, Wallace RJ Jr, Holland SM, Sampaio EP, Olivier KN, Jackson M, Zelazny AM. High-level relatedness among *Mycobacterium abscessus* subsp. *massiliense* strains from widely separated outbreaks. Emerg Infect Dis. 2014;20:364–71.

167. Brown-Elliott BA, Philley JV, Griffith DE, Thakkar F, Wallace RJ Jr. 2017. *In vitro* susceptibility testing of bedaquiline against *Mycobacterium avium* complex. Antimicrob Agents Chemother. In Press.
168. Brown-Elliott BA, Philley JV, Griffith DE, Wallace RJ Jr. Comparison of in vitro susceptibility testing of tedizolid and linezolid against isolates of nontuberculous mycobacteria, 1st ASM-Microbe Meeting, 2016, Boston, MA.

Drug Susceptibility Testing of Nontuberculous Mycobacteria

Jakko van Ingen

Introduction

Most NTM species have a wide array of mechanism that lend them natural resistance to most classes of antibiotics and poor susceptibility, compared to common gram-positive and gram-negative bacteria, even to agents used in treatment of NTM disease. Both the natural and acquired mutational resistance are important determinants of treatment outcomes [1]. Still, the exact role of drug susceptibility testing (DST) in the design of treatment regimens has not been settled, as the correlation between in vitro activity and in vivo outcomes of treatment has not been elucidated for many antimycobacterial drugs [2].

This chapter describes the most important mechanisms of resistance in NTM, the current DST methodologies, current guidelines, and the correlation between in vitro and in vivo efficacy of tested drugs. This latter section includes an overview of current knowledge of pharmacokinetics and pharmacodynamics of antimycobacterial drugs relevant to NTM disease management.

Mechanisms of Antibiotic Resistance in NTM

DST measures the result of a highly complex interplay between natural resistance, inducible resistance, and mutational resistance acquired during suboptimal drug exposure and selection. The role of these three determinants of drug susceptibility differs for the various drugs used to treat NTM disease. Knowledge of their relative importance is essential for the selection and optimization of drug treatment regimens. A graphic overview of the various determinants of resistance is presented in

J. van Ingen
Department of Medical Microbiology, Radboud University Medical Center, Nijmegen, The Netherlands

D. E. Griffith (ed.), *Nontuberculous Mycobacterial Disease*, Respiratory Medicine,
https://doi.org/10.1007/978-3-319-93473-0_3

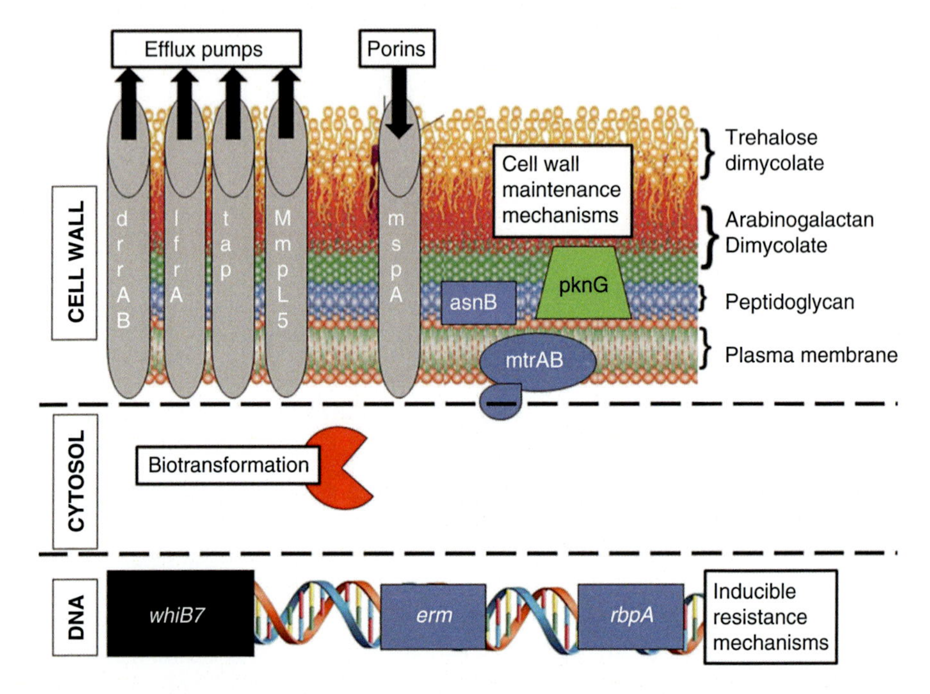

Fig. 1 Simplified overview of the mycobacterial cell wall architecture and resistance determinants (Note: Natural drug resistance in mycobacteria is conferred by their highly lipophilic cell wall and the various mechanisms that control the cell wall content, a low number of porins, broad range of efflux pumps, active biotransformation by cytosolic enzymes, and inducible resistance mechanisms under centralized command (see text for details))

Fig. 1. All mechanisms of resistance and associated genes or proteins are recorded per drug class in Table 1.

Natural Resistance: The Role of the Cell Wall

Natural resistance to antimycobacterial drugs is conferred by a variety of mechanisms that interfere with uptake of the drug by the mycobacterium, enable its biotransformation in the cell, or decrease the affinity with the drug target. The first physical barrier is the mycobacterial cell wall. Natural drug resistance in mycobacteria is related, in large part, to mechanisms that affect the content, hydrophobicity, and thereby permeability of that cell wall. The lipid-rich cell wall of mycobacteria forms an important barrier to the penetration of antimicrobial compounds [3].

Several genes and systems involved in cell wall maintenance are important to maintain the multidrug-resistant phenotype; these include protein kinase G, *fbpA* (encoding the so-called antigen 85 complex), and *asnB* in *M. smegmatis*, the species that serves as a model organism for the genus *Mycobacterium*, the *mtrAB* two-component system in *M. smegmatis* and *M. avium*, *kasB* in *M. marinum*, and

Table 1 Antimycobacterial drugs and mechanisms of resistance in nontuberculous mycobacteria

Drug	Cell wall	Biotransformation	Inducible target protection	Efflux pumps	Acquired mutations in target gene
Rifampicin	*PknG*, *kasB*, *Maa2520*, *pks12*, *asnB*, *fbpA*, *mtrAB*, *mspA* porin	ADP ribosyltransferase	*RbpA*	*efpA*	*rpoB*
Ethambutol	*PknG*				
Quinolones	*kasB*, *Maa2520*, *pks12*, *mtrAB*	Acetylation, nitrosation		*lfrA*, *efpA*, *pstB* gene	*gyrA*, *gyrB*
Macrolides	*PknG*, *asnB*, *kasB*, *Maa2520*, *pks12*, *fbpA*, *mtrAB*, *mspA* porin		*erm*	MAV_1406, MAV_3306	23S rRNA
Aminoglycosides		Aminoglycoside-modifying enzymes		*tetV*, *tap*, P55	16S rRNA
Linezolid					23S rRNA
β-Lactams	*Maa2520*, *pks12*, *mtrAB*, *PknG*, *fbpA*, *mspA* porin	*blaS*, *blaE*, BlaMab β-lactamases		*lfrA*	
Bedaquiline				MmpL5	*MmpL5*, *atpE*
Tetracyclines				*tetV*, *tap*, P55	

Note: See text for relevant references

Maa2520 and *pks12* of *M. avium*. Disruption of these genes generally reduces to hydrophobicity of the mycobacterial cell wall and increases susceptibility to lipophilic antibiotics including the rifamycins, macrolides, ciprofloxacin, vancomycin, and β-lactam antibiotics including imipenem [4–10].

For the clinically important NTM species *M. avium* and *M. abscessus*, distinct colony variants are known that result from differences in the cell wall content. Cell wall glycopeptidolipid content is low in the rough, invasive *M. abscessus* phenotype, but high in the noninvasive, colonizing smooth phenotype associated with biofilm formation [11]. For *M. avium*, smooth transparent and smooth opaque colony types, as well as rare rough types are discerned, though the link between particular phenotypes and virulence is less strong than in *M. abscessus* [12]. The smooth opaque colony type is more susceptible to ciprofloxacin, clarithromycin, and penicillin [13]; this increased susceptibility is regulated by the *mtrAB* two-component system [4]. The differences in drug susceptibility between these variants may in part result from a switch to a stationary metabolic phase of the bacteria during biofilm formation [11, 14]; the metabolic state of mycobacteria, too, is an important determinant of their drug susceptibility [15].

Natural Resistance: Drug Transport Across Porins and Efflux Pumps

Transport of molecules across the membrane into the mycobacterial cell is partly controlled by porins, channel proteins that cross the outer membrane. Mycobacteria utilize porins for nutrient acquisition, but their number is significantly lower than in gram-negative bacteria [2]; only the porin *mspA* in *M. smegmatis* has been extensively studied, and its activity determines susceptibility to small hydrophilic antibiotic molecules including norfloxacin, chloramphenicol, and β-lactam antibiotics but also the hydrophobic vancomycin, erythromycin, and rifampicin [16, 17].

Whereas porins can restrict entry of molecules into the cell, efflux pumps are utilized to evacuate potentially harmful substances out of the mycobacterial cell. In recent years, the role of efflux pumps has received significant attention, including assessments of their suitability as a target for adjunctive therapies [18]. Hollow fiber model experiments have established that antibiotic stress-induced overexpression of efflux pumps in *Mycobacterium avium* is a crucial first step, increasing the resistance to macrolides and permitting survival and later development of mutational resistance [19].

The P55 is an efflux pump that is likely present in all *Mycobacterium* species that permits efflux of at least tetracycline and aminoglycosides [20]. The best characterized efflux pumps in nontuberculous mycobacteria are the *tap* [21], *tetV* [22] *lfrA* [23] and *efpA* [24] efflux pumps that confer tetracycline, aminoglycoside, β-lactam, fluoroquinolone, rifamycin and isoniazid resistance to *M. fortuitum* and *M. smegmatis*. More recently, the MmpL5/MmpS5 efflux pump system has been shown to be important in the development of resistance to clofazimine and bedaquiline in both *M. tuberculosis* and *Mycobacterium avium* complex (MAC) bacteria during treatment [25, 26].

One member of the ABC transporter superfamily has been found in nontuberculous mycobacteria, the phosphate transporter encoded by the *pstB* gene of *M. smegmatis*. This pump is important for fluoroquinolone efflux, and its overexpression confers resistance [27]; there are likely more members of this superfamily of transporters in NTM, but this remains to be investigated.

Part of this extensive armament of mycobacteria is controlled by a single putative transcriptional activator, whiB7, also dubbed the "resistome." WhiB7 is induced by antibiotics and controls the expression of at least *erm* and the *tap* efflux pump. Given the presence of the whiB7 system in all *Streptomyces* and *Mycobacterium* species sequenced to date, it is likely an ancestral trait of a presumed soil-dwelling ancestor [28].

Natural Resistance: Biotransformation in the Intracellular Environment

Biotransformation of the antimicrobial compounds by mycobacteria has been described for β-lactams, quinolones, aminoglycosides, and rifampicin. The β-lactam antibiotics, more specifically imipenem and cefoxitin, are used only in the treatment of infections by rapidly growing mycobacteria, such as *M. abscessus*, *M. chelonae*, and *M. fortuitum* [1]. Their use is limited because of potent β-lactamases – mostly cephalosporinases – present in all mycobacteria [29, 30]. It was recently discovered that the β-lactamases produced by *M. abscessus* can be inhibited by avibactam, which is already in clinical use in combination with ceftazidime [31]. This opens up new opportunity to use β-lactam antibiotics to improve treatment outcome in *M. abscessus* disease.

Acetylation and nitrosation of both norfloxacin and ciprofloxacin have been noted in various rapidly growing *Mycobacterium* species; the acetylation and nitrosation create molecules that have 2–1000 times less antimycobacterial activity [32]. Aminoglycoside susceptibility is influenced by three distinct classes of aminoglycoside-modifying enzymes: aminoglycoside O-nucleotidyltransferases, aminoglycoside O-phosphotransferases, and aminoglycoside N-acetyltransferases. The latter two have been identified in the genomes of *Mycobacterium* species: a phosphotransferase conveys streptomycin resistance in *M. fortuitum*, and 12 homologues have been identified in the genome of *M. abscessus* [33, 34]. Distinct N-acetyltransferases have been identified in the genomes of the *M. tuberculosis* complex (*M. tuberculosis*, *M. bovis*), *M. kansasii*, and MAC [35], as well as the rapid growers *M. fortuitum*, *M. smegmatis*, and, again, *M. abscessus* [33, 34]. These enzymes, in part, determine the susceptibility specific to amikacin (or tobramycin, for *M. chelonae*) and resistance to gentamicin in NTM. Their homology with similar enzymes in other genera may imply that they have been acquired by lateral gene transfer [33, 34]. Last, chromosomally encoded rifampicin ADP-ribosyltransferase (Arr) proteins modify and thereby inactivate rifampicin in select *Mycobacterium* spp.; this mechanism has been studied in *M. smegmatis* but is present in many genera and likely in multiple *Mycobacterium* species [36].

Natural Resistance: The Role and Control of Inducible Target Binding Disruption

The best known inducible resistance mechanism in mycobacteria is the group of erythromycin resistance methylase (*erm*) genes that confer macrolide resistance through methylation of the 23S ribosomal RNA which impairs binding of the macrolides to the ribosomes [37]. These methylases are present in several, but not all, clinically important rapidly growing nontuberculous mycobacteria; their distribution is presented in Table 2 [37, 38]. This inducible macrolide resistance mechanism impacts on treatment outcome of macrolide-based regimens in disease caused by *M. abscessus* (see below).

A less known inducible mechanism of mycobacteria, the RNA polymerase-binding protein A (RbpA), increases tolerance to rifampicin in *M. tuberculosis* and *M. smegmatis*; this protein binds to the RNA polymerase, where it hampers binding of rifampicin [39]. Its distribution and clinical relevance in other slowly growing nontuberculous mycobacteria (SGM) that are treated with rifampicin-based regimens remain unknown.

Table 2 Distribution and functionality of erm genes in rapidly growing mycobacteria

Species	*erm* gene
M. abscessus subsp. *abscessus*	Functional[a]
M. abscessus subsp. *bolletii*	Functional
M. abscessus subsp. *massiliense*	Nonfunctional
M. fortuitum	Functional
M. chelonae	Absent
M. immunogenum	Absent
M. mucogenicum group	Absent
M. peregrinum	Absent
M. senegalense	Absent

[a]A minority of strains have mutations in the gene rendering it nonfunctional

Acquired Resistance Through Genomic Mutations

The increasing number of patients treated for NTM diseases have led to increasing numbers of reports documenting acquired mutational resistance to most key anti-mycobacterial drugs. Since the macrolide antibiotics play a key role in treatment of NTM disease, mutational resistance to this group has received most attention. Mutational resistance to macrolides in *Mycobacterium avium* complex (MAC) disease can be prevented by the use of multidrug regimens that also include rifampicin and ethambutol; macrolide monotherapy or regimens including only quinolones and macrolides are risk factors for the development of macrolide resistance in MAC bacteria [40]. Mutations in codon 2058 or 2059 of the 23S ribosomal RNA gene (*rrl*) have been associated with high-level macrolide resistance in both the *M. avium* complex species and the rapid growers of the *M. abscessus* group [41, 42]; in a case series of patients with macrolide-resistant MAC disease, 96% of patients had isolates with mutations in these two codons [40]. The *rrl* is also the target of linezolid. In experimental settings, mutations in codons inside as well as outside the peptidyl transferase center, the target of linezolid, have decreased susceptibility to linezolid in *M. smegmatis* [43].

In patients treated for MAC disease with amikacin-containing regimens, mutational resistance to amikacin based on mutations in codon 1408 of the 16S ribosomal RNA (*rrs*) gene has been documented [44, 45]. Mutational resistance to aminoglycosides in *M. abscessus* is seen particularly in cystic fibrosis patients and patients with otomastoiditis, who receive long-term (topical) aminoglycoside therapy. Similar mutations in codon 1408 of the 16S ribosomal RNA (*rrs*) gene are responsible for high-level aminoglycoside resistance in both *M. abscessus* and *M. chelonae* after therapy as well as in vitro selection [45, 46].

Rifampicin is the key component of treatment regimens for disease causes by *M. kansasii*. Acquired rifampicin resistance with mutations in codons 513, 526, and 531 of the *rpoB* gene has been observed in *M. kansasii*. These mutations are identical to those observed in rifampicin-resistant *M. tuberculosis* complex isolates [47].

Recently, acquired resistance to bedaquiline was recorded in patients receiving bedaquiline-containing regimens for MAC pulmonary disease; this resulted from mutations in the regulator gene of the MmpS5/MmpL5 efflux system. These mutations are known to lead to cross-resistance to clofazimine [26].

Acquired Resistance Through Acquisition of Plasmids

In contrast to general bacteriology, drug resistance in NTM is thought not to be related to spread of plasmids with antibiotic resistance-determining gene variants. NTM do harbor plasmids, but these are species specific and there is no evidence of spread from one species to another. Whole genome sequencing efforts during the recent outbreak of severe *M. chimaera* disease transmitted by contaminated heater-cooler units in cardiac surgery have revealed that *M. chimaera* isolates yield up to five plasmids and that these plasmids can vary between isolates, in their gene content [48].

For *M. marinum*, it is known that its plasmid harbors genes involved in mercury resistance [49]. One study did show that *M. abscessus* strains in Brazil harbored the pMAB01 plasmid, which encodes for several aminoglycoside-converting enzymes and a dihydropteroate synthase type 1 gene involved in susceptibility to sulfonamides. Strains harboring the plasmid were less susceptible to kanamycin [50], a drug not used for NTM infections. Even if newly acquired plasmids yield gene variants that reduce susceptibility to particular antibiotics, it is questionable whether these will have an effect, since the impermeable cell wall and broad repertoire efflux pumps of NTM already pose a major barrier to the activity of many drug classes. This effect was also seen in the *M. abscessus* strains in Brazil, where the dihydropteroate synthase type 1 gene in the pMAB01 plasmid did not confer sulfonamide resistance as the species is already sulfonamide-resistant [50].

Current Drug Susceptibility Testing Methodologies

Methods: Historical Perspective

In 1963, George Canetti and colleagues published the first consensus statement on drug susceptibility testing of mycobacteria [51]. Three procedures based on dilution of the antimycobacterial drugs isoniazid, PAS, and streptomycin in Löwenstein-Jensen medium were described: the absolute concentration method, the resistance ratio method, and the proportion method. Along with proposals for standardization of test methodology, decision rules for interpretation of the results were also provided. Advice on testing of other antimycobacterial drugs (kanamycin, cycloserine, viomycin, thioacetazone, ethionamide) was provided, but all statements refer to testing *M. tuberculosis* complex bacteria only, not NTM [51]. At the time, susceptibility testing of NTM mainly served identification purposes, i.e., to exclude that a strain represented *M. tuberculosis* [52]. The Löwenstein-Jensen medium was soon replaced by the 7H10 medium developed by Middlebrook and Cohn, on which more strains grow and growth is faster [53]. Over the following decades, a variety of methods for drug susceptibility testing of clinical NTM isolates has been tried. For RGM, test methods used in general bacteriology were adapted, whereas methods designed for *M. tuberculosis* were also applied to SGM. While several methods have been tried in research settings, very few have been used extensively in the clinical setting. Only methods in current or past clinical use are detailed here.

Methods: Absolute Concentration, Resistance Ratio, and Proportion Methods

In the absolute concentration method, the minimum inhibitory concentrations (MICs) are determined by incubating standardized inocula of mycobacteria on media containing various concentrations of the drug to be tested, including the critical concentration. Bacterial growth that exceeds that of a 1:100 dilution of the inoculum on drug-free medium at and above the critical concentration is interpreted as resistance [51]. While this method is still in use for *M. tuberculosis* and has been widely used for DST of NTM, cutoff points for resistance have not been properly defined for NTM [54]. Moreover, none have been clinically validated [2, 54].

The resistance ratio method is methodologically similar to the absolute concentration method, but the MIC is identified and divided by that of the *M. tuberculosis* H37Rv reference strain, to come to a ratio. Low ratios (<2) are interpreted as susceptible, ratios >8 as resistant. This method is more relevant for *M. tuberculosis* than for NTM. Nonetheless, this method was applied to NTM in the treatment trial of *M. avium* complex, *M. xenopi*, and *M. malmoense* pulmonary disease by the British Thoracic Society [55].

The proportion method, as its name suggests, estimates of the proportion of bacteria in the inoculum that is resistant to the drug at the tested concentration. Drug-containing media as well as drug-free media are inoculated with two dilutions of the initial inoculum. If the number of colonies on drug-containing media is >1% of that on drug-free media, the isolate is considered resistant to the drug at the tested concentration [51].

Methods: Disk Diffusion/Disk Elution

In the late 1970s, a series of techniques from general bacteriology were adapted to use for NTM. In disk diffusion a disk with an established quantity of the antimicrobial drug is placed on solid medium inoculated with the test strain. The size of the zone of growth inhibition is measured and compared to established breakpoints (if available). Disk diffusion on solid Mueller-Hinton medium yielded inhibition zones which diameters showed a near linear correlation with MIC determined by agar dilution with Mueller-Hinton medium [56]. Its use of commercially available products in routine use in bacteriology laboratories was a benefit, but the technique proved difficult to adapt to slow-growing organisms: even isolates of *M. marinum*, which is characterized by a pace of growth intermediate between rapid and slow growers, grow too slowly to be tested according to unpublished observations by Richard Wallace Jr. [57]. To support growth of all relevant NTM species, oleic acid, albumin, dextrose, and catalase (OADC) had to be added to solid Mueller-Hinton medium [57].

In disk elution, antibiotic-containing disks are added to liquid OADC supplement first, to which melted medium is then added; contents are mixed by stirring prior to solidification of the medium, so disks are in rather than on the medium, and the drug is mixed in rather than diffused through the medium. Disk elution is used to add fixed quantities of the antibiotic to the medium, and the results are read as growth or no growth at that (breakpoint) concentration, not by reading inhibition zone diameter as in disk diffusion. To overcome the problems observed in disk diffusion, disk elution was applied to *M. fortuitum* and *M. marinum*, where its results were comparable to that of broth microdilution in liquid Mueller-Hinton medium [57].

Methods: Broth Macrodilution

Akin to the proportion method, MICs can be determined by inoculating vials with liquid media and the antibiotics to be tested in their required concentrations. A 1:100 diluted inoculum is added to a drug-free control vial, so the actual MIC can be determined, being the lowest drug concentration that yields less growth than the drug-free vial (thus the lowest concentration that kills >99% of the bacteria in the inoculum) [51].

These methods were developed in the late 1970s to determine drug susceptibility of *M. tuberculosis*. The first commercially available radiometric broth macrodilution method (BacTec460, Becton Dickinson, Sparks, Md) used measurement of $^{14}CO_2$ produced during the metabolism of ^{14}C-incorporated palmitic acid and was adapted to testing SGM [58–60]. It proved less applicable to RGM, because results were difficult to interpret clinically [61, 62]. Although it was a gold standard for long, the radiometric BacTec460 system is no longer available. Its successor, the Mycobacterial Growth Indicator Tube (MGIT; BD Biosciences, Sparks, MD) system, has already become the gold standard for primary culture as well as DST of *M. tuberculosis* [63, 64]. But despite 20 years of clinical use, this platform has not been excessively tested for NTM DST. Initial studies revealed largely concordant results between the MGIT and BacTec460 methods, except for MICs of ethambutol for the *M. avium* complex [65]. Thereafter, only sporadic reports of its use exist [66], and proper protocols and breakpoints have not been established. A major advantage of the MGIT system is that this broth macrodilution methodology is already widely available in mycobacteriology laboratories.

Methods: Broth Microdilution

Broth microdilution allows the measurement of exact MICs by inoculating small (usually 100 μl) volumes of broth with a standardized inoculum of 5×10^5 CFU, typically in a 96-well plate format. Growth density is measured optically and compared to growth in drug-free control vials, to determine MICs; these systems are now available commercially (Sensititre™, Trek Diagnostics, San Diego, USA) and as in-house assays. Broth microdilution made what was to become a lasting impact in general bacteriology after the report by Ericsson and Sherris in 1971, which effectively rendered it the gold standard [67]. A decade later, in 1982, the first report on the use of broth (Mueller-Hinton) microdilution for drug susceptibility testing of RGM was published [68]. The initial study revealed that not all RGM would grow in the cation-adjusted Mueller-Hinton broth (CAMHB) medium. This issue could be overcome by supplementing the media with oleic acid, albumin, dextrose, and catalase (OADC). A larger follow-up study confirmed the suitability of this method for RGM [69]. Thereafter, the microdilution was set up for use with slowly growing nontuberculous mycobacteria (SGM), applying the Middlebrook 7H9 medium [70]. Microdilution MICs for rifampicin, ethambutol, and streptomycin proved lower than in agar, leading to discrepant interpretations for streptomycin [70]. In one multisite reproducibility study, DST for MAC by broth microdilution in both Middlebrook 7H9 and Mueller-Hinton medium were compared: end point readings proved easier in 7H9 medium leading to more reproducible results than in Mueller-Hinton medium [71]; nonetheless, the current CLSI document recommends the use of CAMHB [63].

Methods: E-Tests

Introduced in the late 1990s, the Epsilon-tests, better known as E-tests, are plastic strips calibrated with a continuous logarithmic MIC scale that covers 15 twofold dilutions of the test drug. These strips are pressed onto a solid medium plate that has been swabbed with a suspension of the mycobacteria with a preset inoculum. E-tests were and soon thereafter applied to mycobacteria [72]. Most studies that have assessed this methodology have used Mueller-Hinton blood plates, although for *M. marinum* Middlebrook 7H11 solid medium was preferred as Mueller-Hinton blood media did not support growth of all strains of *M. marinum* [72]. For the slowly growing NTM *M. kansasii*, *M. marinum*, and *M. avium* complex (MAC) bacteria as well as for the rapidly growing *M. chelonae* and *M. fortuitum*, MICs determined by E-tests were similar to those measured by the proportion method or absolute concentration method on Löwenstein-Jensen or Middlebrook 7H10 medium for most drugs [73–76]. For *M. marinum*, MICs for clarithromycin proved two- to threefold lower by E-test and for ethambutol >3-fold lower, although this was most prominent at very low MICs and did not change their interpretation, based on extrapolated cutoffs of *M. tuberculosis* and those proposed in earlier studies of agar dilution methods [76]. After the initial enthusiasm for these E-tests, multisite reproducibility studies revealed that for RGM, reproducibility of the E-tests was inferior to that of broth microdilution particularly for susceptibility testing to amikacin, imipenem, and ciprofloxacin, three key drugs [77]. It is important to realize that E-tests were calibrated for MIC readings after 18–24 h of incubation, which is impossible even for the most rapidly growing NTM.

Methods: Molecular Methods

In-house methods based on sequencing of the target gene and comparisons with wild-type strains of the same species have been developed, mostly for MAC, *M. kansasii*, and *M. abscessus* [40–47]. The few published studies have focused on molecular detection of resistance mechanisms for macrolides, rifamycins, and aminoglycosides.

Owing to the important role of the macrolides in treatment of NTM disease and the fact that MAC and *M. abscessus* group organisms are the most frequent causative agents, molecular analysis of macrolide susceptibility in MAC and *M. abscessus* based on sequence analysis of the 23S rRNA gene was studied first [40–42]. *rpoB* gene mutation analysis has been applied for assessment of rifamycin susceptibility in *M. kansasii*; the mutations associated with rifampicin resistance overlap largely with those known in *M. tuberculosis* [47]. To detect aminoglycoside resistance in MAC and *M. abscessus*, 16S rRNA gene sequence analysis can be used [43–46]; most resistant isolates harbor mutations in codon 1408 of the 16S rRNA gene.

In recent years, several studies have focused on sophistication of existing methods for molecular detection of the *erm* gene, responsible for inducible macrolide resistance in RGM and most notorious in *M. abscessus* [37, 38]. One commercial line probe assay has now been developed for the detection of resistance-conferring mutations in 16S and 23S rDNA genes (i.e., to diagnose aminoglycoside and macrolide resistance), as well as detection of (mutations in) the *erm* gene (GenoType NTM-DR, Hain Lifescience, Nehren, Germany). This assay can be used for MAC and *M. abscessus* isolates. In a study at the French national reference laboratory, the line probe test results were concordant with phenotypic susceptibility testing by broth microdilution in 96/102 (94.1%) isolates, with 4 clarithromycin-resistant and 2 amikacin-resistant isolates not harboring mutations [78].

Having such techniques available is important, particularly as a tool to confirm exceptional phenotypic susceptibility profiles with clinical impact (i.e., macrolide or amikacin resistance in all species, rifampicin resistance in *M. kansasii*).

With time, all the above single-target sequencing approaches and line probe assays will likely be superseded by whole genome sequencing, which is already under investigation for this specific purpose.

Methods: Pitfalls of Current Susceptibility Testing Practices

When evaluating new or existing antibiotics for their antimycobacterial activity, it is attractive to do so in existing and well-validated platforms, such as broth microdilution. But it is essential to understand that many factors determine MICs, including many factors unrelated to the bacteria that are being tested. The key aspect to consider is the medium that is used. The choice of medium determines the rate of growth of NTM and thus the timing of reading the results. More importantly, the medium constituents can influence the activity of antibiotics. The best known factor is pH. Middlebrook media have been designed to have a slightly acidic pH (6.8) to better mimic the intracellular space. However, a low pH is known to affect the activity of some antibiotics. The macrolides are the most important example in the NTM field. Their activity is markedly reduced in acidic environments [79]. Hence, testing in acidic media will yield higher MICs. The CLSI guidelines accommodated for this by using different (higher) breakpoints to determine resistance when testing in Middlebrook or 12B medium, as compared to those for tests done in CAMHB with its pH of 7.3 [63]. Similar decreases of activity at lower pH have been recorded for ciprofloxacin, clofazimine, and ethambutol [80]. This same phenomenon has also been noted for bedaquiline, which is most active at neutral pH and loses activity in both acidic and alkaline environments [81].

Apart from pH, the chemical composition of the medium can affect the activity of some antibiotics, including important antimycobacterial drugs. A well-known example is tigecycline; its activity decreases with increasing manganese content of the test medium [82]. To optimize tigecycline susceptibility testing, manganese concentrations should be controlled and as low as possible. Trimethoprim/ sulfamethoxazole susceptibility tests should not be performed in media that contain thymidine, which allows mycobacteria to bypass the folic acid biosynthesis pathway, the target of trimethoprim/ sulfamethoxazole. The use of agar-based media largely circumvents this problem, although potential presence of trace elements dictates the use of 80% growth reduction as the cutoff for MIC determination [83]. Last, testing susceptibility to cycloserine is only possible in media devoid of pyruvate, as pyruvate inactivates cycloserine [84].

The other key aspect, particularly important when testing slow-growing organisms yet often overlooked, is the stability of the antibiotic to be tested. Chemical stability of the antibiotic has a major impact on MIC measurements. Doxycycline is known to degrade rapidly, i.e., in 14 days, in Mueller-Hinton medium. Test materials thus need to be prepared fresh [85]. Similarly, reduction to <50% activity was noted within 1 week for trimethoprim and minocycline and within 2 weeks for kanamycin, amikacin, trimethoprim/sulfamethoxazole, and rifampicin in Middlebrook 7H10 medium [86, 87].

Drug stability is especially problematic in β-lactam antibiotics, particularly for penicillins and carbapenems. The clinically important imipenem degrades very rapidly in the media used for broth macrodilution DST, and daily addition of the antibiotic may be necessary to obtain useful test results [88]. Broth microdilution plates must be read no later than after 3 days of incubation [63]. Although it has been stated that high imipenem MICs (>8 μg/ml) merit retesting [89], this is questionable as this does not resolve the stability issue; the practice and clinical merit of carbapenem susceptibility testing may be questionable until better methods become available.

Current Guidelines and Recommendations

General Considerations

Drug susceptibility testing is only useful if the following conditions are all met:

1. There is an infection/disease that needs antimicrobial therapy.
2. Effective antimicrobial drugs are available to the patient.
3. There is a known suitable platform to perform the test for the relevant drugs.
4. The activity of the drugs in vitro is related to their effect in vivo.
5. The in vitro activities of the drugs vary (i.e., susceptibility varies or resistance can emerge).

Although the first two conditions are generally met in NTM disease, the latter three are not necessarily met for drugs used to treat NTM disease or not yet fully understood.

The current CLSI recommendations are built on a series of comparative studies of the various methods and experience of a small number of reference laboratories. For many of the drugs recommended to be tested, breakpoints to consider isolates “resistant” stem from laboratory observations, i.e., are derived from wild-type MIC distributions. This is an acceptable strategy but the results may not be clinically relevant [90]. Many breakpoints in the current guideline have not been clinically validated and are not yet supported by preclinical or clinical studies of the pharmacokinetics (i.e., distribution of the drug in the body over time, after administration) and pharmacodynamics (relationship between pharmacokinetics of the drug and its effect size) of the drugs. Such studies, preferably within clinical trials, are important to set clinically relevant breakpoints for resistance (see below).

When Should Drug Susceptibility Testing Be Performed?

It is recommended to perform DST at least for isolates obtained at the start of treatment and in case of treatment failure, i.e., persistence of culture positivity after 3 months of treatment for disseminated disease or 6 months of treatment for pulmonary NTM disease. Testing at the start of treatment is particularly relevant for patients who have received courses of treatment or prophylaxis with antibiotics also relevant to treatment of their NTM disease.

Only for *M. marinum* routine DST at the start of treatment is not recommended, because this species is susceptible to most classes of drugs and treatment outcomes are generally very good. The use of DST is restricted to cases of treatment failure, where acquired resistance – only documented for tetracyclines – may be detected [63].

Which Drugs Should Be Tested and by Which Method?

In 2011, the National Committee on Clinical Laboratory Standards (NCCLS, now Clinical Laboratory Standards Institute – CLSI) published the second version of its document M24-A [63]. The current CLSI recommendations for NTM are summarized in Table 3.

It is important to note that for MAC bacteria, it is not recommended to determine MICs for the classic antituberculosis drugs rifampicin and ethambutol, despite the fact that these are part of guideline-recommended treatment regimens. This is because their MICs do not predict the outcome of multidrug treatment that features these two antibiotics (see below).

In the absence of guidelines by its European counterpart, the European Committee on Antimicrobial Susceptibility Testing (EUCAST), the CLSI approved methods can for now be regarded as the current gold standard. These guidelines advise to use broth microdilution using cation-adjusted Mueller-Hinton medium for the DST of all NTM [63].

Only for MAC, the CLSI document proposed to use the radiometric BACTEC460 method with Middlebrook 7H12B medium, which is now out of production, leaving broth microdilution using 5% OADC-enriched cation-adjusted Mueller-Hinton medium as the sole preferred method now, as for other SGM. Microdilution DST plates for SGM are read after 7 days of incubation at 35° to 37° Celsius; if growth is poor, reading is to be repeated after 10 and 14 days [63].

Results for RGM are read after 3 days of incubation at 30° Celsius, except for macrolides, which should also be read after 14 days of incubation, unless the isolate is certain to belong to a species such as *M. abscessus* subsp *massiliense*, that has no (functional) *erm* gene for inducible resistance (Table 2).

Because of the growth characteristics of *M. marinum*, microdilution trays for DST for this species should be incubated at 28° to 30° Celsius for 7 days.

Table 3 Most frequently isolated NTM and recommended DST practices

Growth rate	Species	Main sites of infection	CLSI recommended DST platform	Alternative(s)	Key drugs to be tested
Slow	*M. avium* complex (*M. avium*, *M. intracellulare*, minor species)	Pulmonary, lymph node	Broth microdilution in MH	Not established	CLA, MOX, LNZ, (AMI)
	M. kansasii	Pulmonary	Broth microdilution in MH	Macrodilution, agar proportion	RIF, CLA (+RIB, MOX, LNZ, AMI, STR, INH, EMB, CIP, SXT)
	M. xenopi	Pulmonary	Broth microdilution in MH [a]	Not established	(RIF, CLA, RIB, MOX, LNZ, AMI, STR, INH, EMB, CIP, SXT)
	M. malmoense (North-Western Europe)	Pulmonary	Broth microdilution in MH	Not established	(RIF, CLA, RIB, MOX, LNZ, AMI, STR, INH, EMB, CIP, SXT)
	M. simiae	Pulmonary	Broth microdilution in MH	Not established	(RIF, CLA, RIB, MOX, LNZ, AMI, STR, INH, EMB, CIP, SXT)
Intermediate	*M. marinum*	Skin	Broth microdilution in MH	No recommendation	RIF, EMB, CLA, DOX, MIN, SXT
Rapid	*M. abscessus*	Pulmonary, skin	Broth microdilution in MH	Not established	CLA, AMI, FOX, IMI, LNZ, CIP, MOX, DOX, TIG, SXT
	M. chelonae	Skin, soft tissues	Broth microdilution in MH	Not established	CLA, AMI, TOB, FOX, IMI, LNZ, CIP, MOX, DOX, SXT
	M. fortuitum	Skin, soft tissues, pulmonary	Broth microdilution in MH	Not established	CLA, AMI, TOB, FOX, IMI, LNZ, CIP, MOX, DOX, SXT

Data summarized from Ref. [63]; for drugs in brackets, breakpoints have not been formally set
MH cation-adjusted Mueller-Hinton broth with OADC supplement, CLA clarithromycin, MOX moxifloxacin, LNZ linezolid, AMI amikacin, RIF rifampicin, RIB rifabutin, EMB ethambutol, STR streptomycin, CIP ciprofloxacin, SXT co-trimoxazole, INH isoniazid, DOX doxycycline, MIN minocycline, TIG tigecycline, FOX cefoxitin, IMI imipenem, TOB tobramycin
[a]*M. xenopi* grows poorly in this medium

In Vitro-In Vivo Correlations

There are very limited amounts of good-quality data on in vitro-in vivo correlations in NTM disease treatment. This has fed the myth that there are no such correlations and that DST of NTM thus has no clinical utility. Any discussion of in vitro-in vivo correlations starts with the cautionary note that such correlations can only be established for antibiotics whose role is to actively kill the infecting NTM. For drugs with an adjunctive role, i.e., drugs only in regimens for their synergy with, or other potentiation of, key antimycobacterial drugs, such correlations may be very difficult to prove. In the case of NTM, this may pertain to rifampicin and ethambutol against MAC or clofazimine against MAC and *M. abscessus*.

Immediately after NTM were found to be causative agents of human lung disease, discrepancies between in vitro drug susceptibility and in vivo outcomes of drug treatment were observed, i.e., the single-drug MICs of classic antituberculosis drugs including rifampicin, isoniazid, and ethambutol could not predict the ultimate outcome of treatment with combination regimens that featured these drugs [1, 91]. This same conclusion was also drawn in the clinical trial of rifampicin-ethambutol regimens with or without additional isoniazid, performed by the British Thoracic Society [55]. As in all arms of this trial, long-term follow-up showed that only 31% of MAC pulmonary disease patients were alive and considered cured [92]. This, coupled with the observation that their MAC isolates were uniformly resistant to isoniazid (100%) and mostly resistant to rifampicin (86% of all isolates) and ethambutol (68%) [92], could also lead to the conclusion that these drugs were inactive in vitro as well as in vivo and thus have good in vitro-in vivo correlations. One small study in Sweden has shed interesting light on this issue, too. In five patients who failed on rifampicin-ethambutol regimens (one received additional clarithromycin, three received additional amikacin), synergistic activity of rifampicin and ethambutol that was measured in baseline isolates was lost in isolates obtained at treatment failure [93]. So, even if single-drug MICs did not change and did not predict failure [55, 92, 93], loss of synergy may be related to treatment failure of rifampicin-ethambutol-based regimens. This observation needs to be confirmed in larger-scale studies.

For the key antimycobacterial drugs, the macrolides and aminoglycosides, there is clear evidence of an association between in vitro susceptibility and outcome of treatment. The best evidence for relationships between MICs and outcomes of treatment has been gathered for MAC. This evidence comes mostly from trials of HIV-associated disseminated disease.

Trials of monotherapy with rifampicin, ethambutol, clofazimine, or clarithromycin for disseminated *M. avium* disease in HIV-infected patients established that only drug susceptibility testing results for clarithromycin predicted outcome of treatment with this drug. Chaisson and co-workers showed that follow-up isolates of patients with *M. avium* bacteremia who experienced failure of clarithromycin monotherapy had clarithromycin MICs >32 μg/ml, while their pretreatment MICs had been ≤4 μg/ml in broth macrodilution [36]. Such clear relationships between in vitro drug susceptibility and culture conversion rates or symptomatic improvement during monotherapy with the respective drugs could not be proven for rifampicin, ethambutol, and clofazimine; the minimum inhibitory concentrations to these drugs did not predict outcome of monotherapy with these drugs [94, 95].

The macrolide antibiotic clarithromycin was also the first drug for which a clear relationship between in vitro activity and in vivo efficacy could be demonstrated in patients with pulmonary MAC disease: case series from the United States [96] and Japan [97] observed that the outcome of treatment with regimens including clarithromycin, rifampicin, and ethambutol, with adjunctive aminoglycosides and fluoroquinolones, in terms of the percentage of patients who showed long-term conversion to negative cultures, was significantly better in patients in whom baseline and follow-up isolates had low MICs to clarithromycin. For example, in the case series reported from Japan, culture conversion rates were 71.8% overall, but only 25% in patients whose primary isolates were already macrolide resistant [97].

Amikacin is now the most widely used aminoglycoside and an important component of treatment regimens for severe MAC lung disease [1]. Over the past 5 years, retrospective case series and one clinical trial have shown that patients with MAC isolates resistant to amikacin (MIC >64 mg/l and with codon 1408 mutations in the *rrs* gene) fail on treatment regimens that rely on intravenous or inhaled (liposomal) amikacin [44, 45].

For *M. kansasii*, data on in vitro-in vivo correlations stem from large retrospective series. The currently recommended treatment regimen consists of isoniazid, rifampicin, and ethambutol for a 12-month duration [1]. The very few patients that experienced treatment failure of this regimen generally had strains isolated after treatment failure that showed rifampicin resistance, while their primary isolates had been tested susceptible [1, 98–100]. Increases in MICs of isoniazid or ethambutol may also be seen but usually in combination with rifampicin resistance [99]. Hence, rifampicin is the key drug to be tested. In rifampicin-resistant *M. kansasii* isolates, CLSI guidelines recommend additional tests of isoniazid, amikacin, streptomycin, ciprofloxacin, moxifloxacin, clarithromycin, rifabutin, and co-trimoxazole [63]. In vitro-in vivo correlations for these drugs have not been proven, although there is, again retrospective, evidence of the efficacy of sulfonamide-based regimens [99, 100] and macrolide-based regimens [101].

For clinically significant slow growers other than MAC or *M. kansasii*, it has been recommended to test rifampicin and the set of drugs tested for rifampicin-resistant *M. kansasii* (Table 3) [63]. Results of these tests should be interpreted with caution as in vitro activity and in vitro efficacy need not be correlated for any of these drugs. Here, *M. simiae* warrants specific attention; this species is characterized by high levels of drug resistance and a lack of synergistic activity of rifampicin and ethambutol, in vitro [102]. In clinical practice, treatment outcomes in *M. simiae* disease are poor and may reflect the high level of drug resistance [102, 103].

In *M. marinum* infections, results of monotherapy with clarithromycin, doxycycline or minocycline, co-trimoxazole or clarithromycin, or combinations with rifampicin and ethambutol are mostly good, and in vitro susceptibility to these compounds is commonplace [104], suggesting good in vitro-in vivo correlation. Failure of doxycycline treatment with the emergence of doxycycline resistance has been described [105]. DST is warranted in such cases of treatment failure and should cover all abovementioned drugs and a fluoroquinolone [63].

For the RGM, in vitro drug susceptibility test results are used to design treatment regimens. As a result, these often feature two to five drug combinations including a macrolide (depending on whether [inducible] macrolide resistance is measured) with an aminoglycoside, a fluoroquinolone, cefoxitin, imipenem, co-trimoxazole, a tetracycline, or more recently assessed drugs including tigecycline, clofazimine, and linezolid [1, 106]. Yet, there is a very limited evidence base that proves that relationships between in vitro drug susceptibility and treatment outcome exist.

For the prime pathogenic RGM, *M. abscessus*, culture conversion rates in case series applying regimens tailored on basis of DST in 69 patients (48%) and outcomes of fixed regimens including agents to which susceptibility is not expected (ciprofloxacin, doxycycline) in 65 patients (58%) revealed no clear differences [107, 108]; although this suggests that in vitro-in vivo correlations are limited, this conclusion cannot be drawn as switches in treatment were commonplace in both series. Still, there is now evidence from two independent case series that correlate macrolide susceptibility in vitro and outcomes of macrolide-based treatment regimens in vivo. In the first case series from South Korea, culture conversion rates were much better in disease caused by *erm*-gene defective *M. abscessus* group strains (*M. abscessus* subsp *massiliense*) at 88%, versus only 25% in patients with disease caused by *M. abscessus* subsp *abscessus*. A later case series from Japan recorded culture conversion rates of 50% (*M. abscessus* subsp *massiliense*) versus 31% (*M. abscessus* subsp *abscessus*) [109, 110].

Patients who develop *M. abscessus* disease despite (topical) aminoglycoside treatment, such as in ear infections and pulmonary disease in cystic fibrosis patients, often have isolates resistant to amikacin, mostly with codon 1408 mutations in the *rrs* gene. In one trial of treatment regimens that rely on inhaled (liposomal) amikacin, disease caused by such amikacin-resistant isolates could not be cured by liposomal amikacin, suggesting a good in vitro-in vivo correlation [45].

For remaining RGM, one study has reported cure rates of 90% after mostly monotherapy with trimethoprim-sulfamethoxazole for *M. fortuitum* disease and 72% after mostly amikacin with cefoxitin treatment for *M. chelonei* (now separated in *M. chelonae* and the *M. abscessus* group) disease, for patients whose isolates had proven susceptible to the relevant drugs [111].

A New Look at Drug Susceptibility: Pharmacokinetics and Pharmacodynamics

To set clinically meaningful breakpoint concentrations for resistance, both the spread in MICs of wild-type strains and the pharmacokinetics (distribution of the drug in the body over time, after administration of a given dose) of the key drugs in patients need to be known [112]. The level of susceptibility is important, but can

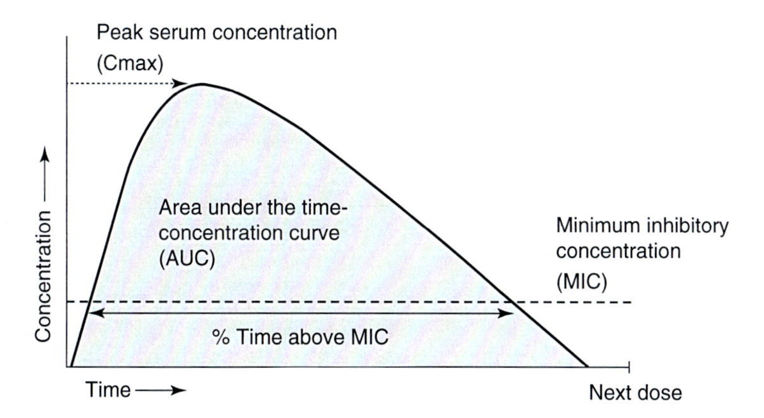

Fig. 2 Graphic representation of the key concepts in pharmacokinetics/pharmacodynamics

that level be overcome by standard doses of the antibiotic under study? To answer that question, we need to know, for each drug, whether its killing effect is driven by the ratio of peak serum concentration (Cmax) over MIC, the ratio of area under the time-concentration curve (AUC) over MIC, or the time (i.e., percentage of the dosing interval) that the serum concentration is above the MIC. Last, if we know the relevant parameter, we need to calculate, based on clinical trial data or preclinical experiments in hollow fiber models or animal models, how high that ratio of Cmax/MIC, AUC/MIC, or percentage of time above MIC must be to achieve maximal killing. This field of pharmacodynamic research, in NTM disease, is in its infancy. The key parameters of pharmacokinetics and pharmacodynamics are visualized in Fig. 2.

In recent years, three studies have explored the pharmacokinetics of key drugs in MAC and *M. abscessus* lung disease. These studies confirmed that important pharmacokinetic interactions occur in treatment, particularly with regimens that combine rifamycins and macrolides, as the regimens for MAC disease generally do. Simultaneous administration of rifampicin lowers macrolide concentrations in blood by 30% (azithromycin) to 60% (clarithromycin) [113–115]. Rifabutin has a similar but weaker effect. As a result, 42% (once daily 500 mg dosing) to 84% (twice daily dosing) of the patients met the applied pharmacodynamic index (serum concentration above MIC for >50% of the dosing interval) of clarithromycin [113]. Expectedly, this percentage is higher in patients treated for *M. abscessus* disease, who do not use rifamycins [114]. These pharmacokinetic data support the current CLSI breakpoints of macrolides for *M. abscessus* (≤2 mg/L) and *M. avium* complex (≤8 mg/L) [63, 113], but the use of time-above-MIC pharmacodynamic index, i.e., considering macrolides to have time-dependent activity, has a limited evidence base. The time-dependent activity of macrolides has been shown in in vitro studies [116], but not in in vivo studies of mycobacterial disease.

The moxifloxacin pharmacokinetics offer little support to the new CLSI breakpoint for moxifloxacin for *M. avium* complex and RGM (≤1 mg/L); with an average AUC of 18.81 mg·h/l in patients with MAC lung disease, only MICs as low as

0.125 mg/L would lead to AUC/MIC ratios >100, associated with bactericidal activity of fluoroquinolones [113]. Here, too, rifampicin lowers concentrations of moxifloxacin and may contribute negatively to moxifloxacin efficacy in MAC lung disease [113]. Thus, even if a strain is tested "susceptible," with an MIC of 0.5 mg/L, moxifloxacin is most likely not an effective drug by itself for the patient infected by this strain.

Serum concentration of rifampicin and ethambutol is generally within ranges considered adequate in tuberculosis treatment; those of rifampicin are actually higher than those observed in TB patients [113–115]. Yet, NTM show MICs to these agents that are 10–20-fold higher than for *M. tuberculosis* [2, 54, 103, 113] and thus pharmacodynamic indices for bactericidal activity of rifampicin against *M. tuberculosis* (fAUC/MIC>24.14) and ethambutol (Cmax/MIC>1.23) as determined in the hollow fiber model were only attained by 18% and 57% of patients with MAC lung disease [113]. If these two antibiotics are in the regimen for their killing capacity, that killing does not occur in the current dosing.

For *M. abscessus*, studies applying the hollow fiber pharmacodynamic model have suggested new breakpoints for resistance for amikacin, moxifloxacin, and tigecycline [117–119]. In this model, live mycobacteria are exposed to antibiotics in the concentrations as they occur over time at the site of infection (lung, bloodstream) with different dosing regimens of the study drug, for several weeks. The number of surviving mycobacteria and the emergence of resistance in them are monitored at regular intervals [117]. This allows correlation of the pharmacokinetics of the different dosing regimens with the killing capacity measured over time and, knowing the MIC of the mycobacteria for the study drug, can help to optimize dosing and to set breakpoints for resistance.

The breakpoints for resistance for amikacin, moxifloxacin, and tigecycline for *M. abscessus*, according to the hollow fiber model experiments, would be 16 mg/l (equal to CLSI), 0.25 mg/l (CLSI breakpoint 2 mg/l), and 0.5 mg/l (no CLSI breakpoint) [117–119]. Particularly for moxifloxacin with its median MIC of 16 mg/l (range 0.13–32 mg/l), this shows only a small minority of isolates can be considered susceptible to moxifloxacin in its current dose. Similar experiments should be performed for all drugs potentially relevant to treatment of *M. abscessus* disease and disease caused by other NTM, to assess their potency and set meaningful breakpoints for resistance.

Summary Statement

Susceptibility testing of NTM, with all its caveats, has become part of the standards of care for NTM disease patients and is helpful in treatment guidance. With broth microdilution, there is a gold standard technique that is flexible, easy to integrate in existing infrastructure, and proven to be well reproducible within and between laboratories. Its rollout and implementation of quality control measures require more attention.

Its main weakness lies in its (perceived) lack of in vitro-in vivo correlations for some drugs. For macrolides, amikacin, fluoroquinolones, and probably tigecycline, these correlations are good. For others they still need to be studied. But we first need to learn which drugs are in regimens for their killing capacity and which are in only for synergy and potentiation of other drugs. For the latter grouping, in vitro-in vivo correlations are likely less important, and susceptibility testing is probably not warranted.

The final hurdle to take is the setting of clinically meaningful breakpoints for resistance. Given the complexities of organizing clinical trials for NTM disease and the need for multidrug treatment regimens, preclinical models like the hollow fiber model or well-validated animal models should be applied to come to meaningful breakpoints based on pharmacokinetics/pharmacodynamics science. These can then be tested in focused clinical studies.

Literature

1. Griffith DE, Aksamit T, Brown-Elliot BA, et al. An official ATS/IDSA statement: diagnosis, treatment, and prevention of nontuberculous mycobacterial diseases. Am J Respir Crit Care Med. 2007;175:367–416.
2. van Ingen J, Boeree M, van Soolingen D, Mouton J. Resistance mechanisms and drug susceptibility testing of nontuberculous mycobacteria. *Drug Resist Updat.* 2012;15:149–61.
3. Lambert PA. Cellular impermeability and uptake of biocides and antibiotics in Gram-positive bacteria and mycobacteria. J Appl Microbiol. 2002;92(Suppl):46S–54S.
4. Cangelosi GA, Do JS, Freeman R, Bennett JG, Semret M, Behr MA. The two-component regulatory system mtrAB is required for morphotypic multidrug resistance in Mycobacterium avium. Antimicrob Agents Chemother. 2006;50:461–8.
5. Gao LY, Laval F, Lawson EH, Groger RK, Woodruff A, Morisaki JH, Cox JS, Daffe M, Brown EJ. Requirement for kasB in Mycobacterium mycolic acid biosynthesis, cell wall impermeability and intracellular survival: implications for therapy. Mol Microbiol. 2003;49:1547–63.
6. Nguyen L, Chinnapapagari S, Thompson CJ. FbpA-Dependent biosynthesis of trehalose dimycolate is required for the intrinsic multidrug resistance, cell wall structure, and colonial morphology of Mycobacterium smegmatis. J Bacteriol. 2005;187:6603–11.
7. Nguyen HT, Wolff KA, Cartabuke RH, Ogwang S, Nguyen L. A lipoprotein modulates activity of the MtrAB two-component system to provide intrinsic multidrug resistance, cytokinetic control and cell wall homeostasis in Mycobacterium. Mol Microbiol. 2010;76:348–64.
8. Philalay JS, Palermo CO, Hauge KA, Rustad TR, Cangelosi GA. Genes required for intrinsic multidrug resistance in Mycobacterium avium. Antimicrob Agents Chemother. 2004;48:3412–8.
9. Ren H, Liu J. AsnB is involved in natural resistance of Mycobacterium smegmatis to multiple drugs. Antimicrob Agents Chemother. 2006;50:250–5.
10. Wolff KA, Nguyen HT, Cartabuke RH, Singh A, Ogwang S, Nguyen L. Protein kinase G is required for intrinsic antibiotic resistance in mycobacteria. Antimicrob Agents Chemother. 2009;53:3515–9.
11. Howard ST, Rhoades E, Recht J, Pang X, Alsup A, Kolter R, Lyons CR, Byrd TF. Spontaneous reversion of Mycobacterium abscessus from a smooth to a rough morphotype is associated with reduced expression of glycopeptidolipid and reacquisition of an invasive phenotype. Microbiology. 2006;152:1581–90.
12. Schorey JS, Sweet L. The mycobacterial glycopeptidolipids: structure, function, and their role in pathogenesis. Glycobiology. 2008;18:832–41.

13. Cangelosi GA, Palermo CO, Laurent JP, Hamlin AM, Brabant WH. Colony morphotypes on Congo red agar segregate along species and drug susceptibility lines in the Mycobacterium avium-intracellulare complex. Microbiology. 1999;145(Pt 6):1317–24.
14. Falkinham JO 3rd. Growth in catheter biofilms and antibiotic resistance of *Mycobacterium avium*. J Med Microbiol. 2007;56:250–4.
15. Nguyen L, Thompson CJ. Foundations of antibiotic resistance in bacterial physiology: the mycobacterial paradigm. Trends Microbiol. 2006;14:304–12.
16. Stephan J, Mailaender C, Etienne G, Daffe M, Niederweis M. Multidrug resistance of a porin deletion mutant of Mycobacterium smegmatis. Antimicrob Agents Chemother. 2004;48:4163–70.
17. Danilchanka O, Pavlenok M, Niederweis M. Role of porins for uptake of antibiotics by Mycobacterium smegmatis. Antimicrob Agents Chemother. 2008;52:3127–34.
18. van Ingen J, Ferro BE, Hoefsloot W, Boeree MJ, van Soolingen D. Drug treatment of pulmonary nontuberculous mycobacterial disease in HIV-negative patients: the evidence. Expert Rev Anti Infect Ther. 2013;11(10):1065–77.
19. Schmalstieg AM, Srivastava S, Belkaya S, Deshpande D, Meek C, Leff R, van Oers NSC, Gumbo T. The antibiotic resistance arrow of time: Efflux pump induction is a general first step in the evolution of mycobacterial drug resistance. Antimicrob Agents Chemother. 2012;56:4806–15.
20. Silva PE, Bigi F, Santangelo MP, Romano MI, Martin C, Cataldi A, Ainsa JA. Characterization of P55, a multidrug efflux pump in Mycobacterium bovis and Mycobacterium tuberculosis. Antimicrob Agents Chemother. 2001;45:800–4.
21. Ramon-Garcia S, Martin C, Ainsa JA, De Rossi E. Characterization of tetracycline resistance mediated by the efflux pump Tap from Mycobacterium fortuitum. J Antimicrob Chemother. 2006;57:252–9.
22. De Rossi E, Blokpoel MC, Cantoni R, Branzoni M, Riccardi G, Young DB, De Smet KA, Ciferri O. Molecular cloning and functional analysis of a novel tetracycline resistance determinant, tet(V), from Mycobacterium smegmatis. Antimicrob Agents Chemother. 1998;42:1931–7.
23. Sander P, De Rossi E, Boddinghaus B, Cantoni R, Branzoni M, Bottger EC, Takiff H, Rodriquez R, Lopez G, Riccardi G. Contribution of the multidrug efflux pump LfrA to innate mycobacterial drug resistance. FEMS Microbiol Lett. 2000;193:19–23.
24. Li XZ, Zhang L, Nikaido H. Efflux pump-mediated intrinsic drug resistance in *Mycobacterium smegmatis*. *Antimicrob Agents Chemother*. 2004;48:2415–23.
25. Briffotaux J, Huang W, Wang X, Gicquel B. MmpS5/MmpL5 as an efflux pump in *Mycobacterium* species. Tuberculosis. 2017;107:13–9.
26. Alexander DC, Vasireddy R, Vasireddy S, et al. The emergence of mmpT5 variants during bedaquiline treatment of *Mycobacterium intracellulare* Lung Disease. J Clin Microbiol. 2017;55:574–84.
27. Bhatt K, Banerjee SK, Chakraborti PK. Evidence that phosphate specific transporter is amplified in a fluoroquinolone resistant Mycobacterium smegmatis. Eur J Biochem. 2000;267:4028–32.
28. Morris RP, Nguyen L, Gatfield J, Visconti K, Nguyen K, Schnappinger D, Ehrt S, Liu Y, Heifets L, Pieters J, Schoolnik G, Thompson CJ. Ancestral antibiotic resistance in Mycobacterium tuberculosis. Proc Natl Acad Sci U S A. 2005;102:12200–5.
29. Flores AR, Parsons LM, Pavelka MS Jr. Genetic analysis of the beta-lactamases of Mycobacterium tuberculosis and Mycobacterium smegmatis and susceptibility to beta-lactam antibiotics. Microbiology. 2005;151:521–32.
30. Nash DR, Wallace RJ Jr, Steingrube VA, Udou T, Steele LC, Forrester GD. Characterization of beta-lactamases in Mycobacterium fortuitum including a role in beta-lactam resistance and evidence of partial inducibility. Am Rev Respir Dis. 1986;134:1276–82.
31. Lefebvre AL, Le Moigne V, Bernut A, Veckerlé C, Compain F, Herrmann JL, Kremer L, Arthur M, Mainardi JL. Inhibition of the β-Lactamase BlaMab by avibactam improves the in vitro and in vivo efficacy of imipenem against *Mycobacterium abscessus*. Antimicrob Agents Chemother. 2017;61. https://doi.org/10.1128/AAC.02440-16.

32. Adjei MD, Heinze TM, Deck J, Freeman JP, Williams AJ, Sutherland JB. Acetylation and nitrosation of ciprofloxacin by environmental strains of mycobacteria. Can J Microbiol. 2007;53:144–7.
33. Ramirez MS, Tolmasky ME. Aminoglycoside modifying enzymes. Drug Resist Updat. 2010;13:151–71.
34. Ripoll F, Pasek S, Schenowitz C, Dossat C, Barbe V, Rottman M, Macheras E, Heym B, Herrmann JL, Daffe M, Brosch R, Risler JL, Gaillard JL. Non mycobacterial virulence genes in the genome of the emerging pathogen Mycobacterium abscessus. PLoS One. 2009;4:–e5660.
35. Ho II, Chan CY, Cheng AF. Aminoglycoside resistance in Mycobacterium kansasii, Mycobacterium avium-M. intracellulare, and Mycobacterium fortuitum: are aminoglycoside-modifying enzymes responsible? Antimicrob Agents Chemother. 2000;44:39–42.
36. Baysarowich J, Koteva K, Hughes DW, Ejim L, Griffiths E, Zhang K, Junop M, Wright GD. Rifamycin antibiotic resistance by ADP-ribosylation: structure and diversity of Arr. Proc Natl Acad Sci U S A. 2008;105:4886–91.
37. Nash KA, Brown-Elliott BA, Wallace RJ Jr. A novel gene, erm(41), confers inducible macrolide resistance to clinical isolates of Mycobacterium abscessus but is absent from Mycobacterium chelonae. Antimicrob Agents Chemother. 2009;53:1367–76.
38. Brown-Elliott BA, Hanson K, Vasireddy S, Iakhiaeva E, Nash KA, Vasireddy R, Parodi N, Smith T, Gee M, Strong A, Barker A, Cohen S, Muir H, Slechta ES, Wallace RJ Jr. Absence of a functional erm gene in isolates of Mycobacterium immunogenum and the Mycobacterium mucogenicum group, based on in vitro clarithromycin susceptibility. J Clin Microbiol. 2015;53:875–8.
39. Dey A, Verma AK, Chatterji D. Role of an RNA polymerase interacting protein, MsRbpA, from Mycobacterium smegmatis in phenotypic tolerance to rifampicin. Microbiology. 2010;156:873–83.
40. Griffith DE, Brown-Elliott BA, Langsjoen B, Zhang Y, Pan X, Girard W, Nelson K, Caccitolo J, Alvarez J, Shepherd S, Wilson R, Graviss EA, Wallace RJ Jr. Clinical and molecular analysis of macrolide resistance in Mycobacterium avium complex lung disease. Am J Respir Crit Care Med. 2006;174:928–34.
41. Meier A, Heifets L, Wallace RJ Jr, Zhang Y, Brown BA, Sander P, Bottger EC. Molecular mechanisms of clarithromycin resistance in *Mycobacterium avium*: observation of multiple 23S rDNA mutations in a clonal population. J Infect Dis. 1996;174:354–60.
42. Bastian S, Veziris N, Roux AL, Brossier F, Gaillard JL, Jarlier V, Cambau E. Assessment of clarithromycin susceptibility in strains belonging to the *Mycobacterium abscessus* group by *erm*(41) and *rrl* sequencing. Antimicrob Agents Chemother. 2011;55:775–81.
43. Long KS, Munck C, Andersen TM, Schaub MA, Hobbie SN, Bottger EC, Vester B. Mutations in 23S rRNA at the peptidyl transferase center and their relationship to linezolid binding and cross-resistance. Antimicrob Agents Chemother. 2010;54:4705–13.
44. Brown-Elliott BA, Iakhiaeva E, Griffith DE, Woods GL, Stout JE, Wolfe CR, Turenne CY, Wallace RJ Jr. In vitro activity of amikacin against isolates of *Mycobacterium avium* complex with proposed MIC breakpoints and finding of a 16S rRNA gene mutation in treated isolates. J Clin Microbiol. 2013;51:3389–94.
45. Olivier KN, Griffith DE, Eagle G, et al. Randomized trial of liposomal amikacin for inhalation in nontuberculous mycobacterial lung disease. Am J Respir Crit Care Med. 2017;195(6):814–23.
46. Prammananan T, Sander P, Brown BA, Frischkorn K, Onyi GO, Zhang Y, Bottger EC, Wallace RJ Jr. A single 16S ribosomal RNA substitution is responsible for resistance to amikacin and other 2-deoxystreptamine aminoglycosides in *Mycobacterium abscessus* and *Mycobacterium chelonae*. J Infect Dis. 1998;177:1573–81.
47. Klein JL, Brown TJ, French GL. Rifampin resistance in *Mycobacterium kansasii* is associated with *rpoB* mutations. Antimicrob Agents Chemother. 2001;45:3056–8.
48. van Ingen J, Kohl TA, Kranzer K, Hasse B, Keller PM, Katarzyna Szafrańska A, Hillemann D, Chand M, Schreiber PW, Sommerstein R, Berger C, Genoni M, Rüegg C, Troillet N, Widmer AF, Becker SL, Herrmann M, Eckmanns T, Haller S, Höller C, Debast SB, Wolfhagen MJ,

Hopman J, Kluytmans J, Langelaar M, Notermans DW, Ten Oever J, van den Barselaar P, Vonk ABA, Vos MC, Ahmed N, Brown T, Crook D, Lamagni T, Phin N, Smith EG, Zambon M, Serr A, Götting T, Ebner W, Thürmer A, Utpatel C, Spröer C, Bunk B, Nübel U, Bloemberg GV, Böttger EC, Niemann S, Wagner D, Sax H. Global outbreak of severe Mycobacterium chimaera disease after cardiac surgery: a molecular epidemiological study. Lancet Infect Dis. 2017;17:1033–41.

49. Stinear TP, Seemann T, Harrison PF, Jenkin GA, Davies JK, Johnson PD, Abdellah Z, Arrowsmith C, Chillingworth T, Churcher C, Clarke K, Cronin A, Davis P, Goodhead I, Holroyd N, Jagels K, Lord A, Moule S, Mungall K, Norbertczak H, Quail MA, Rabbinowitsch E, Walker D, White B, Whitehead S, Small PL, Brosch R, Ramakrishnan L, Fischbach MA, Parkhill J, Cole ST. Insights from the complete genome sequence of *Mycobacterium marinum* on the evolution of *Mycobacterium tuberculosis*. Genome Res. 2008;18(5):729–41.
50. Matsumoto CK, Bispo PJ, Santin K, Nogueira CL, Leão SC. Demonstration of plasmid-mediated drug resistance in Mycobacterium abscessus. J Clin Microbiol. 2014;52:1727–9 [Epub ahead of print]
51. Canetti G, Froman S, Grosset J, Hauduroy P, Langerova M, Mahler HT, Meissner G, Mitchison DA, Sula L. Mycobacteria: laboratory methods for testing drug sensitivity and resistance. Bull World Health Organ. 1963;29:565–78.
52. Runyon EH. Anonymous mycobacteria in pulmonary disease. Med Clin North Am. 1959;43:273–90.
53. Middlebrook G, Cohn ML. Bacteriology of tuberculosis: laboratory methods. Am J Public Health Nations Health. 1958;48:844–53.
54. van Ingen J, van der Laan T, Dekhuijzen PNR, Boeree MJ, van Soolingen D. In vitro drug susceptibility of 2275 clinical nontuberculous *Mycobacterium* isolates of 49 species in the Netherlands. Int J Antimicrob Agents. 2010;35:169–73.
55. Research Committee of the British Thoracic Society. First randomised trial of treatments for pulmonary disease caused by *M. avium intracellulare*, *M. malmoense*, and *M. xenopi* in HIV negative patients: rifampicin, ethambutol and isoniazid versus rifampicin and ethambutol. Thorax. 2001;56:167–72.
56. Wallace RJ Jr, Dalovisio JR, Pankey GA. Disk diffusion testing of susceptibility of Mycobacterium fortuitum and Mycobacterium chelonei to antibacterial agents. Antimicrob Agents Chemother. 1979;16:611–4.
57. Stone MS, Wallace RJ Jr, Swenson JM, Thornsberry C, Christensen LA. Agar disk elution method for susceptibility testing of Mycobacterium marinum and Mycobacterium fortuitum complex to sulfonamides and antibiotics. Antimicrob Agents Chemother. 1983;24:486–93.
58. Snider DE Jr, Good RC, Kilburn JO, Laskowski LF Jr, Lusk RH, Marr JJ, Reggiardo Z, Middlebrook G. Rapid drug-susceptibility testing of *Mycobacterium tuberculosis*. Am Rev Respir Dis. 1981;123:402–6.
59. Heifets LB, Iseman MD, Lindholm-Levy PJ. Determination of MICs of conventional and experimental drugs in liquid medium by the radiometric method against *Mycobacterium avium* complex. Drugs Exp Clin Res. 1987;13:529–38.
60. Hoffner SE, Svenson SB, Kallenius G. Synergistic effects of antimycobacterial drug combinations on *Mycobacterium avium* complex determined radiometrically in liquid medium. Eur J Clin Microbiol. 1987;6:530–5.
61. Hansen KT, Clark RB, Sanders WE. Effects of different test conditions on the susceptibility of *Mycobacterium fortuitum* and *Mycobacterium chelonae* to amikacin. J Antimicrob Chemother. 1994;33:483–94.
62. Yew WW, Piddock LJ, Li MS, Lyon D, Chan CY, Cheng AF. In-vitro activity of quinolones and macrolides against mycobacteria. J Antimicrob Chemother. 1994;34:343–51.
63. Clinical Laboratory Standards Institute. Susceptibility testing of mycobacteria, nocardiae, and other aerobic actinomycetes – approved standard. 2nd ed. Wayne, PA: Clinical Standards Institute. CLSI document M24-A2; 2011.

64. Drobniewski F, Rusch-Gerdes S, Hoffner S. Antimicrobial susceptibility testing of *Mycobacterium tuberculosis* (EUCAST document E.DEF 8.1)--report of the Subcommittee on Antimicrobial Susceptibility Testing of *Mycobacterium tuberculosis* of the European Committee for Antimicrobial Susceptibility Testing (EUCAST) of the European Society of Clinical Microbiology and Infectious Diseases (ESCMID). Clin Microbiol Infect. 2007;13:1144–56.
65. Piersimoni C, Nista D, Bornigia S, De Sio G. Evaluation of a new method for rapid drug susceptibility testing of *Mycobacterium avium* complex isolates by using the mycobacteria growth indicator tube. J Clin Microbiol. 1998;36:64–7.
66. Hombach M, Somoskövi A, Hömke R, Ritter C, Böttger EC. Drug susceptibility distributions in slowly growing non-tuberculous mycobacteria using MGIT 960 TB eXiST. Int J Med Microbiol. 2013;303:270–6.
67. Ericsson HM, Sherris JC. Antibiotic sensitivity testing. Report of an international collaborative study. Acta Pathol Microbiol Scand B Microbiol Immunol. 1971;217(Suppl):211.
68. Swenson JM, Thornsberry C, Silcox VA. Rapidly growing mycobacteria: testing of susceptibility to 34 antimicrobial agents by broth microdilution. Antimicrob Agents Chemother. 1982;22:186–92.
69. Swenson JM, Wallace RJ Jr, Silcox VA, Thornsberry C. Antimicrobial susceptibility of five subgroups of *Mycobacterium fortuitum* and *Mycobacterium chelonae*. Antimicrob Agents Chemother. 1985;28:807–11.
70. Wallace RJ Jr, Nash DR, Steele LC, Steingrube V. Susceptibility testing of slowly growing mycobacteria by a microdilution MIC method with 7H9 broth. J Clin Microbiol. 1986;24:976–81.
71. Woods GL, Williams-Bouyer N, Wallace RJ Jr, Brown-Elliott BA, Witebsky FG, Conville PS, Plaunt M, Hall G, Aralar P, Inderlied C. Multisite reproducibility of results obtained by two broth dilution methods for susceptibility testing of *Mycobacterium avium* complex. J Clin Microbiol. 2003;41:627–31.
72. Flynn CM, Kelley CM, Barrett MS, Jones RN. Application of the Etest to the antimicrobial susceptibility testing of *Mycobacterium marinum* clinical isolates. J Clin Microbiol. 1997;35:2083–6.
73. Fabry W, Schmid EN, Ansorg R. Comparison of the E test and a proportion dilution method for susceptibility testing of *Mycobacterium kansasii*. Chemotherapy. 1995;41:247–52.
74. Fabry W, Schmid EN, Ansorg R. Comparison of the E test and a proportion dilution method for susceptibility testing of *Mycobacterium avium* complex. J Med Microbiol. 1996;44: 227–30.
75. Hoffner SE, Klintz L, Olsson-Liljequist B, Bolmstrom A. Evaluation of Etest for rapid susceptibility testing of *Mycobacterium chelonae* and *M. fortuitum*. J Clin Microbiol. 1994;32: 1846–9.
76. Werngren J, Olsson-Liljequist B, Gezelius L, Hoffner SE. Antimicrobial susceptibility of *Mycobacterium marinum* determined by E-test and agar dilution. Scand J Infect Dis. 2001;33:585–8.
77. Woods GL, Bergmann JS, Witebsky FG, Fahle GA, Boulet B, Plaunt M, Brown BA, Wallace RJ Jr, Wanger A. Multisite reproducibility of Etest for susceptibility testing of *Mycobacterium abscessus*, *Mycobacterium chelonae*, and *Mycobacterium fortuitum*. J Clin Microbiol. 2000;38:656–61.
78. Mougari F, Loiseau J, Veziris N, Bernard C, Bercot B, Sougakoff W, Jarlier V, Raskine L, Cambau E. French National Reference Center for Mycobacteria. Evaluation of the new GenoType NTM-DR kit for the molecular detection of antimicrobial resistance in non-tuberculous mycobacteria. J Antimicrob Chemother. 2017;72:1669–77.
79. Truffot-Pernot C, Ji B, Grosset J. Effect of pH on the in vitro potency of clarithromycin against Mycobacterium avium complex. Antimicrob Agents Chemother. 1991;35:1677–8.
80. Heginbothom ML, Lindholm-Levy PJ, Heifets LB. Susceptibilities of Mycobacterium malmoense determined at the growth optimum pH (pH 6.0). Int J Tuberc Lung Dis. 1998;2:430–4.

81. Lounis N, Vranckx L, Gevers T, Kaniga K, Andries K. In vitro culture conditions affecting minimal inhibitory concentration of bedaquiline against M. tuberculosis. Med Mal Infect. 2016;46:220–5.
82. Veenemans J, Mouton JW, Kluytmans JA, Donnely R, Verhulst C, van Keulen PH. Effect of manganese in test media on in vitro susceptibility of Enterobacteriaceae and Acinetobacter baumannii to tigecycline. J Clin Microbiol. 2012;50(9):3077–9.
83. Wallace RJ Jr, Wiss K, Bushby MB, Hollowell DC. In vitro activity of trimethoprim and sulfamethoxazole against the nontuberculous mycobacteria. Rev Infect Dis. 1982;4: 326–31.
84. Tison F, Tacquet A, Guillaume J, Devulder B. Unsuitability of the basic coletsos medium for the measurement of sensitivity to cycloserine. Inactivation of cycloserine by sodium pyruvate. Ann Inst Pasteur Lille. 1963;14:117–24.
85. Wallace RJ Jr, Wiss K. Susceptibility of Mycobacterium marinum to tetracyclines and aminoglycosides. Antimicrob Agents Chemother. 1981;20:610–2.
86. Davis CE Jr, Carpenter JL, Trevino S, Koch J, Ognibene AJ. In vitro susceptibility of Mycobacterium avium complex to antibacterial agents. Diagn Microbiol Infect Dis. 1987;8:149–55.
87. Rynearson TK, Shronts JS, Wolinsky E. Rifampin: in vitro effect on atypical mycobacteria. Am Rev Respir Dis. 1971;104:272–4.
88. Watt B, Edwards JR, Rayner A, Grindey AJ, Harris G. In vitro activity of meropenem and imipenem against mycobacteria: development of a daily antibiotic dosing schedule. Tuber Lung Dis. 1992;73:134–6.
89. Woods GL, Bergmann JS, Witebsky FG, Fahle GA, Wanger A, Boulet B, Plaunt M, Brown BA, Wallace RJ Jr. Multisite reproducibility of results obtained by the broth microdilution method for susceptibility testing of Mycobacterium abscessus, Mycobacterium chelonae, and Mycobacterium fortuitum. J Clin Microbiol. 1999;37:1676–82.
90. Schon T, Chryssanthou E. Minimum inhibitory concentration distributions for Mycobacterium avium complex – towards evidence-based susceptibility breakpoints. Int J Infect Dis. 2017;55:122–4.
91. Goldman KP. Treatment of unclassified mycobacterial infection of the lungs. Thorax. 1968;23:94–9.
92. Research Committee of the British Thoracic Society. Pulmonary disease caused by Mycobacterium avium-intracellulare in HIV-negative patients: five-year follow-up of patients receiving standardised treatment. Int J Tuberc Lung Dis. 2002;6:628–34.
93. Hoffner SE, Heurlin N, Petrini B, Svenson SB, Kallenius G. Mycobacterium avium complex develop resistance to synergistically active drug combinations during infection. Eur Respir J. 1994;7:247–50.
94. Chaisson RE, Benson CA, Dube MP, et al. Clarithromycin therapy for bacteremic *Mycobacterium avium* complex disease. A randomized, double-blind, dose-ranging study in patients with AIDS. AIDS Clinical Trials Group Protocol 157 Study Team. Ann Intern Med. 1994;121:905–11.
95. Sison JP, Yao Y, Kemper CA, Hamilton JR, Brummer E, Stevens DA, Deresinski SC. Treatment of *Mycobacterium avium* complex infection: do the results of in vitro susceptibility tests predict therapeutic outcome in humans? J Infect Dis. 1996;173:677–83.
96. Wallace RJ Jr, Brown BA, Griffith DE, Girard WM, Murphy DT. Clarithromycin regimens for pulmonary *Mycobacterium avium* complex: the first 50 patients. Am J Respir Crit Care Med. 1996;153:1766–72.
97. Tanaka E, Kimoto T, Tsuyuguchi K, et al. Effect of clarithromycin regimen for *Mycobacterium avium* complex pulmonary disease. Am J Respir Crit Care Med. 1999;160:866–72.
98. Ahn CH, Lowell JR, Ahn SS, Ahn SI, Hurst GA. Short-course chemotherapy for pulmonary disease caused by *Mycobacterium kansasii*. Am Rev Respir Dis. 1983;128:1048–50.
99. Ahn CH, Wallace RJ Jr, Steele LC, Murphy DT. Sulfonamide-containing regimens for disease caused by rifampin-resistant *Mycobacterium kansasii*. Am Rev Respir Dis. 1987;135:10–6.

100. Wallace RJ Jr, Dunbar D, Brown BA, Onyi G, Dunlap R, Ahn CH, Murphy DT. Rifampin-resistant *Mycobacterium kansasii*. Clin Infect Dis. 1994;18:736–43.
101. Shitrit D, Baum GL, Priess R, et al. Pulmonary *Mycobacterium kansasii* infection in Israel, 1999–2004: clinical features, drug susceptibility, and outcome. Chest. 2006;129:771–6.
102. van Ingen J, Totten SE, Heifets LB, Boeree MJ, Daley CL. Drug susceptibility testing and pharmacokinetics question current treatment regimens in *Mycobacterium simiae* complex disease. Int J Antimicrob Agents. 2012;39:173–6.
103. van Ingen J, Hoefsloot W, Mouton JW, Boeree MJ, van Soolingen D. Synergistic activity of rifampicin and ethambutol against slow growing nontuberculous mycobacteria is currently of questionable clinical significance. Int J Antimicrob Agents. 2013;42:80–2.
104. Aubry A, Chosidow O, Caumes E, Robert J, Cambau E. Sixty-three cases of *Mycobacterium marinum* infection: clinical features, treatment, and antibiotic susceptibility of causative isolates. Arch Intern Med. 2002;162:1746–52.
105. Ljungberg B, Christensson B, Grubb R. Failure of doxycycline treatment in aquarium-associated *Mycobacterium marinum* infections. Scand J Infect Dis. 1987;19:539–43.
106. Floto RA, Olivier KN, Saiman L, Daley CL, Herrmann J-L, Nick JA, Noone PG, Bilton D, Corris P, Gibson RL, Hempstead SE, Koetz K, Sabadosa KA, Sermet-Gaudelus I, Smyth AR, van Ingen J, Wallace RJ, Winthrop KL, Marshall BC, Haworth CSUS. Cystic Fibrosis Foundation and European Cystic Fibrosis Society consensus recommendations for the management of non-tuberculous mycobacteria in individuals with cystic fibrosis. Thorax. 2016;71(Suppl 1):1–22.
107. Jarand J, Levin A, Zhang L, Huitt G, Mitchell JD, Daley CL. Clinical and microbiologic outcomes in patients receiving treatment for *Mycobacterium abscessus* pulmonary disease. Clin Infect Dis. 2011;52:565–71.
108. Jeon K, Kwon OJ, Lee NY, Kim BJ, Kook YH, Lee SH, Park YK, Kim CK, Koh WJ. Antibiotic treatment of *Mycobacterium abscessus* lung disease: a retrospective analysis of 65 patients. Am J Respir Crit Care Med. 2009;180:896–902.
109. Koh WJ, Jeon K, Lee NY, Kim BJ, Kook YH, Lee SH, Park YK, Kim CK, Shin SJ, Huitt GA, Daley CL, Kwon OJ. Clinical significance of differentiation of *Mycobacterium massiliense* from *Mycobacterium abscessus*. Am J Respir Crit Care Med. 2011;183:405–10.
110. Harada T, Akiyama Y, Kurashima A, et al. Clinical and Microbiological Differences between Mycobacterium abscessus and Mycobacterium massiliense Lung Diseases. J Clin Microbiol. 2012;50:3556–61.
111. Wallace RJ, Swenson JM, Silcox VA, Bulen MG. Treatment of nonpulmonary infections due to *Mycobacterium fortuitum* and *Mycobacterium chelonei* on the basis of in vitro susceptibilities. J Infect Dis. 1985;152:500–14.
112. Mouton JW, Brown DF, Apfalter P, Canton R, Giske CG, Ivanova M, Macgowan AP, Rodloff A, Soussy CJ, Steinbakk M, Kahlmeter G. The role of pharmacokinetics/ pharmacodynamics in setting clinical MIC breakpoints: the EUCAST approach. Clin Microbiol Infect. 2011;18:E37–45.
113. van Ingen J, Egelund EF, Levin A, Totten SE, Boeree MJ, Mouton JW, Aarnoutse R, Heifets LB, Peloquin CA, Daley CL. The pharmacokinetics and pharmacodynamics of pulmonary *Mycobacterium avium* complex disease treatment. Am J Respir Crit Care Med. 2012;186:559–65.
114. Koh WJ, Jeong BH, Jeon K, Lee SY, Shin SJ. Therapeutic drug monitoring in the treatment of *Mycobacterium avium* complex lung disease. Am J Respir Crit Care Med. 2012;186: 797–802.
115. Magis-Escurra C, Alffenaar JW, Hoefnagels I, Dekhuijzen PN, Boeree MJ, van Ingen J, Aarnoutse RE. Pharmacokinetic studies in patients with nontuberculous mycobacterial lung infections. Int J Antimicrob Agents. 2013;42:256–61.
116. Ferro BE, van Ingen J, Wattenberg M, van Soolingen D, Mouton JW. Time-kill kinetics of slowly growing mycobacteria common in pulmonary disease. J Antimicrob Chemother. 2015;70(10):2838–43.

117. Ferro BE, Srivastava S, Deshpande D, Sherman CM, Pasipanodya JG, van Soolingen D, Mouton JW, van Ingen J, Gumbo T. Amikacin pharmacokinetics/pharmacodynamics in a novel hollow fiber *Mycobacterium abscessus* disease model. Antimicrob Agents Chemother. 2016;60(3):1242–8.
118. Ferro BE, Srivastava S, Deshpande D, Pasipanodya JG, van Soolingen D, Mouton JW, van Ingen J, Gumbo T. Moxifloxacin's limited efficacy in the Hollow-Fiber Model of Mycobacterium abscessus disease. Antimicrob Agents Chemother. 2016;60:3779–85. https://doi.org/10.1128/AAC.02821-15.
119. Ferro BE, Srivastava S, Deshpande D, Pasipanodya JG, van Soolingen D, Mouton JW, van Ingen J, Gumbo T. Tigecycline is highly efficacious against mycobacterium abscessus pulmonary disease. Antimicrob Agents Chemother. 2016;60:2895–900. https://doi.org/10.1128/AAC.03112-15.

Vulnerability to Nontuberculous Mycobacterial Lung Disease or Systemic Infection Due to Genetic/Heritable Disorders

Edward D. Chan

Introduction

Nontuberculous mycobacterial (NTM) infections can be broadly classified into three major domains: (i) skin and soft tissue infections, (ii) isolated NTM lung disease (NTM-LD), and (iii) extrapulmonary visceral organ/disseminated infections. Skin and soft tissue infections are almost always the result of accidental/iatrogenic inoculations of otherwise normal hosts. NTM-LD typically occurs in the setting of preexisting structural lung disease, most often emphysema or bronchiectasis, which in turn may be acquired – e.g., localized bronchiectasis from prior unrelated infections, smoking-related emphysema, and pneumoconiosis [1–3] – or the result of genetic/heritable disorders that result in lung architectural and/or immunologic abnormalities. Extrapulmonary visceral organ/disseminated infections generally occur in individuals who are frankly immunocompromised, which may be acquired – e.g., untreated AIDS or the use of tumor necrosis factor-alpha (TNFα) antagonists – or are genetic/heritable in origin. This chapter focuses on the genetic/heritable causes of isolated NTM-LD and of extrapulmonary visceral organ/disseminated infections.

E. D. Chan (✉)
Pulmonary Section, Denver Veterans Affairs Medical Center, Denver, CO, USA

Program in Cell Biology and Department of Academic Affairs, National Jewish Health, Denver, CO, USA

Division of Pulmonary Sciences and Critical Care Medicine, University of Colorado Anschutz Medical Campus, Aurora, CO, USA
e-mail: chane@njhealth.org

D. E. Griffith (ed.), *Nontuberculous Mycobacterial Disease*, Respiratory Medicine,
https://doi.org/10.1007/978-3-319-93473-0_4

Genetic/Heritable Disorders that Predispose to NTM-LD

Cumulative clinical experience indicates that individuals with certain acquired or genetic/heritable disorders that compromise the lung architecture or lung immune system are more vulnerable to NTM-LD (Fig. 1). Some of the more well-accepted acquired disorders – smoking-related emphysema, bronchiectasis as a sequela of prior unrelated infections, pneumoconiosis such as silicosis, chronic aspiration, and use of corticosteroids and other immunosuppressives such as antagonists of TNFα – are diagrammed on the *left side of* Fig. 1. We have also observed that bulky calcified chest adenopathy – due most likely to prior *Histoplasma capsulatum* infection – may impinge on central airways, impairing mucus clearance and predisposing to

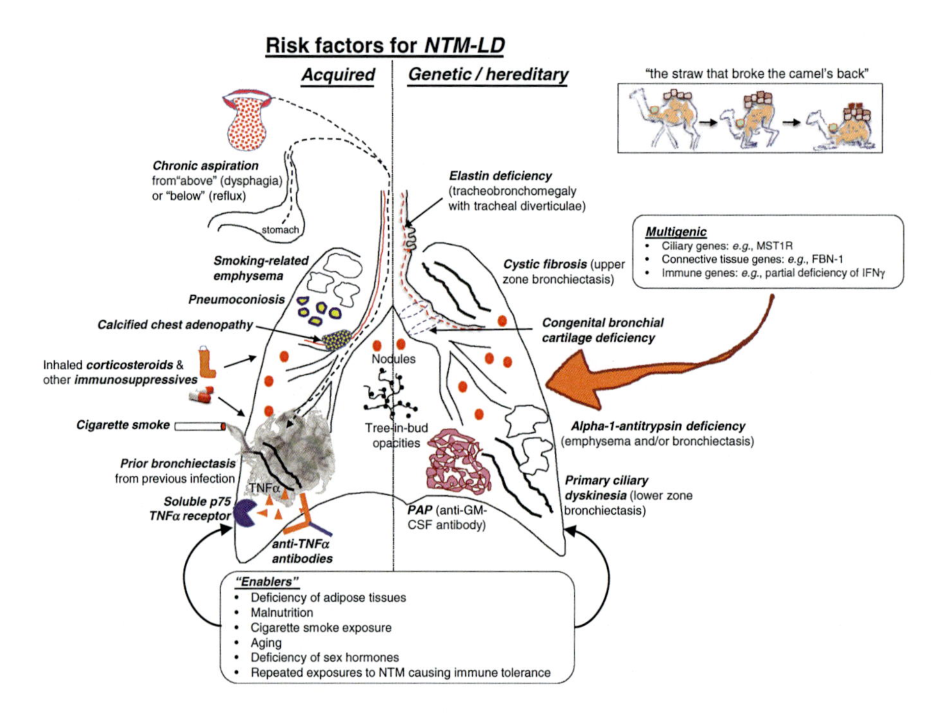

Fig. 1 Diagram of the acquired and genetic/heritable risk factors for NTM-LD. The left side of the lung diagram depicts known acquired risk factors, and the right side illustrates the genetic/heritable risk factors. While congenital bronchial cartilage deficiency is depicted affecting the main stem and lobar bronchi, it is more likely to cause bronchiectasis of the segmental and subsegmental bronchi. Also shown are other "enablers" that may help drive susceptibility to NTM-LD. *Inset (camels)***.** A cartoon to illustrate that accumulation of multiple risk factors – e.g., behaviors of humans and environmental factors that increase exposure to NTM, host genetic susceptibility factors, acquired risk factors, and "enablers" listed – increases one's overall risk for the development of NTM-LD. In this diagram, "bales of straw" are used to depict "the straw that broke the camel's back" aphorism and different colored bales represent the aforementioned risk factors. CFTR cystic fibrosis transmembrane conductance regulator, FBN-1 fibrillin-1, IFNγ interferon-gamma, TNFα tumor necrosis factor-alpha, MST1R macrophage-stimulating 1 receptor

NTM infection in the lung segments normally drained by the affected bronchus [4]. In a case-control study that analyzed environmental and host factors that may be associated with *M. avium* complex lung disease, emphysema, prior hospitalization for pneumonia, thoracic skeletal abnormalities, low body mass index, and corticosteroid and/or immunomodulatory drug usage were each found to be significantly correlated [5]. While there was an increased link between cigarette smoke exposure and NTM-LD, the association did not reach statistical significance [5] although others have found a significant nexus between cigarette smoke exposure and NTM-LD that may be independent of the presence of emphysema [6].

Individuals with genetic/heritable disorders in which bronchiectasis, emphysema, and/or lung immune defects are important sequelae are also predisposed to NTM-LD; these disorders are diagrammed on the *right side of* Fig. 1 and their key clinical signs and diagnostic tests are listed in Table 1; they include cystic fibrosis (CF), discussed in another chapter, primary ciliary dyskinesia (PCD), alpha-1-antitrypsin (AAT) deficiency, congenital bronchial cartilage deficiency (Williams-Campbell syndrome),

Table 1 Clues to the presence of an underlying genetic/heritable cause for isolated NTM-LD

Host risk factor	Clues to presence of host risk factor	Diagnostic test(s)
Cystic fibrosis	Individual (newborn to adult) with bronchiectasis, especially upper zones and the presence of extrapulmonary disorders, e.g., pancreatic insufficiency, sinusitis, etc.	High sweat chloride test Nasal potential CFTR genotyping
Alpha-1-antitrypsin deficiency	Emphysema, bronchiectasis, or both	AAT level and phenotype
Williams-Campbell syndrome	Central bronchiectasis of the segmental and subsegmental airways although more proximal bronchiectasis may occur	Radiographic criteria
Tracheobronchomegaly (Mounier-Kuhn syndrome)	Axial and sagittal (anteroposterior) tracheal diameter >25 mm and >27 mm, respectively, for men; >21 mm and >23 mm, respectively, for women. Right and left mainstem bronchi >21.1 mm and >18.4 mm, respectively, for men; >19.8 mm and >17.4 mm for women. Tracheal and main stem diverticula may be present	Radiographic criteria Evidence of extrapulmonary elastolysis
Hyperimmunoglobulin E (hyper-IgE, Job) syndrome	Recurrent staphylococcal abscesses, sinopulmonary infections resulting in bronchiectasis, bronchopleural fistulae, and pneumatoceles, and eczema due to neutrophil chemotactic defects and defect in T_H17 differentiation. Bronchiectasis is likely the immediate predisposing factor to NTM-LD	Increased IgE (an associated abnormality rather than central in the pathogenesis); genetic testing (STAT3 mutation in autosomal dominant and TYK2, DOCK8, or PGM3 for autosomal recessive hyper-IgE-like disorders)

(continued)

Table 1 (continued)

Host risk factor	Clues to presence of host risk factor	Diagnostic test(s)
Sjogren's syndrome	Dry mouth, dry eyes, parotid gland enlargement, Raynaud's phenomenon, arthralgia, and various multisystem diseases. Pulmonary manifestations include dessicated airways, bronchiolitis, chronic bronchitis, bronchiectasis, and various forms of interstitial lung diseases	Clinical diagnosis supported by serologic tests (ANA, RF, SS-A, and/or SS-B), eye tests to assess tear production (Schirmer) and dry spots on the cornea (Rose Bengal), and salivary gland biopsy
Acquired pulmonary alveolar proteinosis	Alveolar opacification with interlobular septal thickening ("crazy-paving sign")	Anti-GM-CSF antibodies in adults; mutation of SP-B, SP-C, or a subunit of GM-CSF receptor; many causes for secondary PAP
Common variable immunodeficiency	Recurrent respiratory tract infections and bronchiectasis, autoimmune diseases, allergic disease, granulomatous-lymphocytic interstitial lung disease, lymphoid malignancies, and various multisystem disease manifestations	Reduced IgG and IgA and possibly IgM at baseline and reduced IgG response to protein-based vaccine (tetanus and diphtheria) or to polysaccharide vaccine (pneumococcal)
Primary ciliary dyskinesia	Recurrent sinopulmonary infections, infertility	Reduced nasal nitric oxide; abnormal cilia ultrastructure; cilia waveform abnormalities; genotyping of ciliary genes
MPEG1 gene mutation (perforin-2 protein defect)	Bronchiectasis with chronic infections with pyogenic bacteria, NTM, and/or *Aspergillus*	Genotyping

ANA antinuclear antibody, *CFTR* cystic fibrosis transmembrane conductance regulator, *GM-CSF* granulocyte-monocyte colony-stimulating factor, *Ig* immunoglobulin, *PAP* pulmonary alveolar proteinosis, *RF* rheumatoid factor, *SP-B/C* surfactant protein B and C

tracheobronchomegaly (Mounier-Kuhn syndrome), Sjogren's syndrome, pulmonary alveolar proteinosis (PAP), and common variable immunodeficiency (CVID) [7–11]. It is important to emphasize that acquired and genetic/heritable predispositions are not mutually exclusive, and thus it stands to reason that the more risk factors an individual compiles, the greater the likelihood of developing NTM-LD.

PCD is due to defects in the various components of the ciliary ultrastructure such as the microtubules and dynein arms, resulting in ciliary dysfunction, decreased ability to clear airway infections and mucus, and a vicious cycle of airway inflammation, infection, and mucostasis, the denouement of which is bronchiectasis [12]. Bronchiectasis begets more bronchiectasis. PCD is predominantly an autosomal recessive disorder clinically manifested by recurrent sinusitis, otitis media, nasal congestion, bronchiectasis, male > female infertility, and *situs inversus* in ~50% of the cases. It is estimated that 70–90% of adults with PCD have a history of neonatal respiratory distress despite term deliveries. The diagnosis should be suspected in anyone with unexplained bronchiectasis, especially if located in the lower lung zones and accompanied by the aforementioned medical conditions. Supporting diagnostic tests for PCD include nasal nitric oxide (nNO) < 77 nL/min, abnormalities in the ultrastructure of the cilia on transmission electron microscopy, biallelic mutations in one of the known PCD-associated genes, and ciliary waveform abnormalities on high-speed videomicroscopy [13].

AAT deficiency most commonly predisposes to emphysema. But it is less widely appreciated that individuals with frank AAT deficiency may also be vulnerable to bronchiectasis [14]. In a study of 74 subjects with the protease inhibitor (Pi) ZZ phenotype, 70 (95%) had bronchiectatic changes on chest CT scans; although bronchiectasis was generally associated with emphysema, there was a subgroup of patients in whom bronchiectasis was the predominant finding [14]. We previously reported that the presence of AAT anomalies – mostly heterozygous – was more common in patients with NTM-LD compared to the general US population [7]. Moreover, monocyte-derived macrophages (MDM) from PiZZ subjects incubated in autologous plasma – both obtained immediately *after* a session of intravenous AAT augmentation – were better able to control *M. intracellulare* infection than MDM incubated in plasma that were both obtained *before* AAT infusion [15]. Thus, vulnerability of AAT-deficient individuals to NTM-LD may occur as a result of alterations in lung architecture (emphysema and bronchiectasis) as well as impaired macrophage function against NTM.

Congenital bronchial cartilage deficiency syndrome tends to present early in life with bronchiectasis and recurring lung infections [16–18]. Familial cases have been reported although the genetic defect has not been elucidated [19]. Defects in the conducting airway walls typically result in bronchiectasis of the segmental and first few generations of the subsegmental airways (third- to sixth-order bronchi), although more proximal bronchiectasis (lobar and main stem) may also occur [16, 17].

Tracheobronchomegaly may be congenital, acquired, or secondary to other primary disorders – such as Ehlers-Danlos syndrome, generalized elastolysis (cutis laxa), Marfan syndrome (MFS), and Brachmann de Lange syndrome – although the link to secondary causes is not without controversy [20, 21]. Payandeh and coworkers [20] performed an extensive review of the literature and proposed classifying tracheobronchomegaly into six types based on cause: (i) infants who have undergone fetal endoscopic tracheal occlusion procedure, (ii) infants/children after prolonged endotracheal intubation, (iii) a sequela after multiple lung infections, (iv) secondary to lung fibrosis, (v) associated with extrapulmonary elastolysis (e.g., ptosis and drooping of the mucosa of the upper lip below the vermillion border), and (vi) congenital. Pathogenesis of the congenital form is considered to be due to atrophy or absence of elastic fibers and smooth muscle tissues of the large airways, resulting in gross enlargement of the trachea and main bronchi; primary or secondary atrophy of the connective tissue between the cartilage rings weakens the airway walls that may lead to the development of tracheal or bronchial diverticula, potentially serving as static reservoirs for recurrent infections (Fig. 1, *right-hand side*) [16, 20]. Curiously, there is a 4:1 to 8:1 predominance of males in congenital tracheobronchomegaly [20, 22].

PAP is a diffuse lung disease characterized by the accumulation of amorphous, periodic acid-Schiff-positive lipoproteinaceous material in the distal air spaces. The acquired form of PAP is due mostly to the presence of autoantibodies to granulocyte-monocyte colony-stimulating factor (GM-CSF); functional deficiency of this growth factor causes impaired surfactant disposal by lung macrophages, leading to the accumulation of the material in the alveolar spaces and intracellularly in the phagocytes, the latter process further impairing macrophage function and secondarily compromising the activation of adaptive immunity. Infections due to NTM and other opportunistic infections in PAP patients reflect such immune defects [11, 23–29].

The underlying B and T cell defects observed with CVID, also known as acquired hypogammaglobulinema, can lead to recurrent airway infections and bronchiectasis, the latter a prime substrate for subsequent NTM infection. However, to the best of our knowledge, there have been only two reported cases of NTM-LD in the setting of CVID, *M. simiae* in one [30] and *M. intracellulare* in the other [31].

NTM-LD Due to More Newly Identified Genetic/Heritable Predisposing Causes

The occurrence of NTM-LD in individuals without any of the aforementioned, identifiable host risk factor is well recognized [32–34]. Enduring clinical experience has noted that a significant number of such patients possess a lifelong slender body habitus with thoracic cage abnormalities such as pectus excavatum and scoliosis, leading to the notion of an underlying syndrome – perhaps a connective tissue disorder – that predisposes to NTM-LD [6, 33, 35–39].

While an asthenic body habitus may simply be an outward phenotypic manifestation of an underlying disorder that increases vulnerability to NTM lung infections, another, non-mutually exclusive possibility is that reduced body fat – well recognized in NTM-LD subjects – in itself is a risk factor for NTM-LD [6, 35, 37, 40]. Furthermore, an inverse correlation between body mass index and the number of diseased lung segments or NTM-LD-specific mortality would suggest that reduced body weight/fat may also be an adverse prognosticator [41, 42]. One proposed mechanism by which low body fat content may predispose such individuals to NTM-LD is relative deficiency of the fat-derived, satiety hormone leptin since leptin also drives the differentiation of uncommitted T_0 cells toward the T_H1, interferon-gamma (IFNγ)-producing phenotype [43]. This hypothesis is corroborated by the finding that leptin-deficient mice are more susceptible to *M. abscessus* experimental lung infection [44] and that NTM-LD patients have reduced serum leptin levels [39] or a loss in the normal direct relationship between serum leptin concentration and total body fat [6].

A more recent whole exome sequencing study of patients with NTM-LD and their family members indicate that possessing variants of several genes in the immune, connective tissue, ciliary, and CF transmembrane conductance regulator

(CFTR) categories – "multigenic" in nature as opposed to mutation of one dominant gene – may additively increase vulnerability to NTM-LD [45]. Because the presence of variations of immune genes were significantly more common in NTM-LD patients than in unaffected family members, immune gene variants may be the discriminating genetic factor leading to the development of NTM-LD, even with the multigenic paradigm [45]. Becker and colleagues [46] performed whole exome sequencing on 11 NTM-LD subjects with slender body habitus, pectus excavatum, and scoliosis and found 4 (2 being sisters) with heterozygous mutations of the *MACROPHAGE-STIMULATING 1 RECEPTOR* (*MST1R*) gene and in none of 29 NTM-LD patients without pectus excavatum or scoliosis (*see Clinical Vignette 1*, Fig. 2). Significantly, MST1R is a tyrosine kinase receptor important for normal movement of cilia present on cells lining the luminal surface of fallopian tubes and airways [47, 48]. Reduced ciliary movement due to defect in MST1R function is consistent with previous work showing reduced ciliary beat frequency in the nasal epithelium as well as reduced nNO in NTM-LD patients compared to controls; the ciliary beat frequency was increased by NO donors or compounds that increased the concentration of cyclic guanosine monophosphate, a downstream mediator of NO [49]. NO may also have antimicrobial properties, and in a very interesting but preliminary study, inhaled NO administered intermittently at 160 ppm, three to four 30-min inhalations per day for 3–4 weeks, along with antimycobacterial antibiotics, in two cystic fibrosis patients with persistent *M. abscessus* infections significantly reduced NTM load as determined by quantitative PCR of mycobacterial DNA [50]. Recently, Matsuyama and co-workers [51] demonstrated that ex vivo NTM infection of primary human bronchial epithelial cells decreased the expression of genes

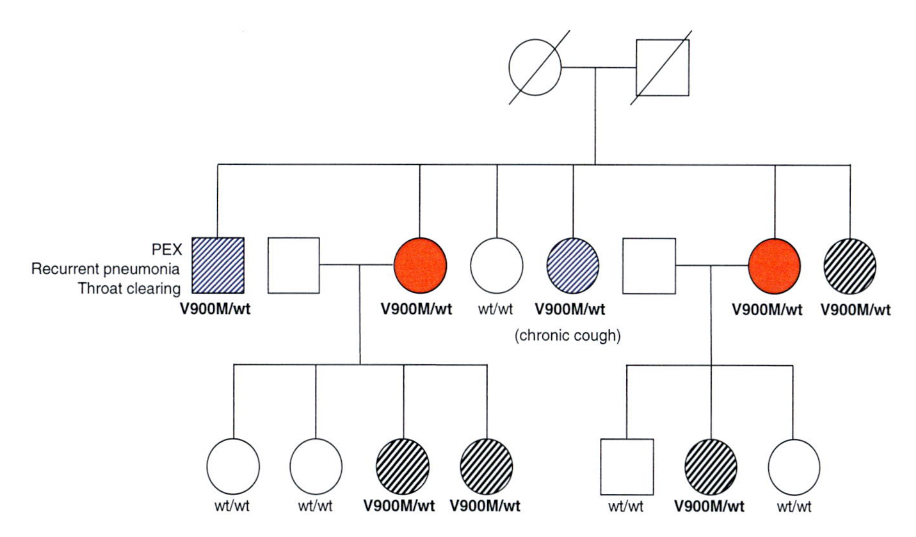

Fig. 2 A genogram of two sisters with pectus excavatum, scoliosis, thin body habitus, and NTM-LD and of their siblings. Of the six surviving siblings tested, five had heterozygous mutation of the *MST1R* gene (V900 M/WT), four of the five with the variant have chronic respiratory disease or symptoms, and two of the five have NTM-LD. (Adapted from Ref. [46])

that encode for ciliary proteins as well as reduced number of ciliated cells; in other words, preexisting ciliary defect predisposes to NTM and that NTM infection itself may, in turn, adversely affect ciliary function [52].

While frank deficiency of IFNγ (e.g., advanced AIDS) or its upstream or downstream signaling molecules (discussed below) predisposes to extrapulmonary visceral/disseminated NTM disease, several studies have found reduced IFNγ production from the peripheral blood mononuclear cells (PBMC) or whole blood from NTM-LD patients [6, 53–57], acknowledging that this is not seen by others [37, 58, 59]. Recently, Cowman and colleagues [60] found decreased IFNγ gene expression in the whole blood of NTM-LD patients compared to control subjects who had bronchiectasis/chronic obstructive pulmonary disease but no NTM infection. While the mechanism(s) for reduced IFNγ are likely to be multifactorial, we found that MST1R may participate to increase IFNγ production by PBMC in addition to its role in maintaining normal ciliary function [46]. The clinical significance of partial reduction in IFNγ levels in blood cells is not known as lung tissue levels of IFNγ were not measured in the aforementioned studies. While a study showed that inhaled IFNγ was not helpful in patients with NTM-LD, the patients were not stratified according to whether their immune cells produced "higher" or "lower" levels of IFNγ [61]; i.e., it is conceivable that only those with relatively low IFNγ levels will benefit from administration of exogenous IFNγ. Some have also reported reduced cellular production of IL-12 [54, 57, 62] or IL-17 [58] in patients with NTM-LD. Whereas a study of PBMC from NTM-LD patients saw an increase in IL-10 with tuberculin purified protein derivative (PPD) or sensitin PPD stimulation [58], others actually saw a decreased IL-10 production in the unstimulated or stimulated (lipopolysaccharide, heat-killed *Staphylococcus epidermidis*, or live *M. intracellulare*) whole blood of NTM-LD patients compared to controls [6]. The finding of such polar differences in the expression of cytokines in these studies is likely related to multiple factors, including genuine differences in immune phenotypes between subjects, differences in the cell types used in the assay (e.g., MDM, PBMC, or whole blood), differences in the type and concentration of stimuli used ex vivo, species and strain differences of NTM that infected patients with differential ability to induce tolerance of immune cells, and differences in external factors such as comorbid medical conditions and administration of antibiotics or corticosteroids that can alter host cytokine expression.

One of the aforementioned 11 patients with NTM-LD who underwent whole exome sequencing also had mutation of the *FIBRILLIN-1* gene [46], an abnormality of which is known to be responsible for MFS, a connective tissue disorder with an array of physical findings that include pectus excavatum, scoliosis, and arachnodactyly. But pectus excavatum and scoliosis have been described in several other connective tissue disorders including Loeys-Dietz syndrome (LDS, due to gain-of-function mutation of transforming growth factor-beta receptors 1/2 – TGFβR1/2) and Shprintzen-Goldberg syndrome (SGS, due to mutation of the Sloan-Kettering Institute (SKI) protein, a downstream inhibitor of TGFβ signaling) [38]. Although all three of these connective tissue disorders are due to monogenic mutations of different genes, each results in increased TGFβ signaling via three separate mechanisms. In light of this, whole blood of NTM-LD patients were found to produce more TGFβ upon ex vivo stimulation with various Toll-like receptor agonists or with *M. intracellulare* as compared to similarly stimulated whole blood from uninfected controls [6]. While TGFβ plays a key role in the physical anomalies associated with these connective tissue disorders, it also inhibits IFNγ gene expression [38] and predisposes to NTM infections [63–65]. Daniels et al. analyzed for the presence of dural ectasia in the lumbar region – enlarged dural sac diameter seen in MFS, LDS, and SGS – in patients with idiopathic bronchiectasis, CF subjects, MFS subjects, and controls and found that the lumbar (L1–L5) dural sac diameter was significantly greater in patients with idiopathic bronchiectasis as compared to controls and to CF subjects, strengthening the supposition of an underlying connective tissue disorder in those with idiopathic bronchiectasis [66]. They also found a strong correlation between dural sac size and NTM-LD as well as dural sac size and long fingers [66]. NTM-LD was also reported in a patient with congenital contractural arachnodactyly, a genetic disorder due to *FIBRILLIN-2* gene mutation and which shares many clinical features with MFS [67]. Finally, while increased TGFβ (and potentially reduced IFNγ) associated with MFS and other connective tissue disorders may compromise host immunity against NTM lung infection, bronchiectasis itself – which independently is a risk factor for NTM-LD – has been linked to MFS, not particularly surprising considering that fibrillin is a connective tissue protein [68–74].

McCormack and colleagues [75] recently described four women with NTM-LD with four different heterozygous variants of the the *MPEG1* gene, which encodes for perforin-2, a pore-forming antibacterial protein. While the combined frequency of the four variants was similar in the NTM-LD and control cohorts, ex vivo and in vitro cell culture studies with innate cells (neutrophils, macrophages) from a patient with the *MPEG1* variant or with mutated *MPEG1* in THP-1 cells found that such gene-altered cells were less capable in controlling *M. avium* and *M. smegmatis* infections, respectively. Chen et al. [76] performed gene linkage analysis on affected and unaffected subjects with familial cases of NTM-LD and identified variants of the TTK protein kinase gene – which normally encodes for a protein kinase involved in repair of damaged DNA – that most strongly cosegregated with the pulmonary NTM families (LOD score >3) as either recessive homozygotes or compound heterozygotes. Gene-level association analysis revealed a gene that encodes for one of the mitogen-activated protein kinases was the most strongly linked to pulmonary NTM disease [76].

If indeed certain genetic/heritable disorders predispose individuals to NTM-LD, it must be explained why some case series of pulmonary NTM patients show a predominance of postmenopausal women. Indeed, in a study of six families with clustering of NTM-LD – 31% of whom had scoliosis – the majority of the affected individuals were women despite the fact that the patterns of transmission were consistent with dominant and recessive modes of inheritance [77]. Reich and Johnson hypothesized that women may be at greater risk because they are less inclined to expectorate forcefully than men (an unproven belief), preventing them from clearing their infections – known by the eponym Lady Windermere syndrome [78].

Another, non-mutually exclusive possibility is that the declining estrogen levels in menopausal women provide a permissive role if one or more additional susceptibility factor(s) are present [35], a hypothesis supported by the finding that estrogen is protective against NTM infection in mice [79]. Thus, an emerging paradigm is that in individuals with genetic/heritable risk factors due to anomalies in one or more "minor" genes, perhaps acquisition of other risk factors with aging (e.g., reduction in body fat, vitamin D level, and/or sex hormones and cumulative exposure to environmental NTM) – labeled as "enablers" in Fig. 1 – results in a "tipping point" where NTM-LD develops, analogous to the aphorism "the straw that broke the camel's back" (Fig. 1, *boxed inset*); this concept may help explain why NTM-LD is significantly more common in the aged population [80].

Genetic/Heritable Disorders that Predispose to Extrapulmonary Visceral Organ and Disseminated NTM Infections

Patients with extrapulmonary visceral organ/disseminated NTM disease have frank immunocompromised states. Acquired disorders that give rise to such profound immunodeficiency include untreated AIDS; use of potent immunosuppressives for inflammatory diseases, cancer, or organ transplantation; and acquired autoantibodies to IFNγ (Fig. 3) [81–84]. One caveat is that there is likely a genetic component to those with anti-IFNγ antibodies as this syndrome appears to be more common in Asians and in those with HLA DRB16:02 or DRB05:02 [81]. However, the presence of extrapulmonary visceral organ/disseminated NTM disease without such acquired risk factors should raise suspicion for genetic/heritable causes. Occurrence of systemic NTM infections in very young individuals – often times infants – should give concern for mutations of genes that are components of the IFNγ-interleukin-12 (IL-12) axes and other immune-related genes that fall under the rubric of Mendelian susceptibility to mycobacterial diseases (MSMD) (Fig. 3) (*see Clinical Vignette 2*) [85–97]. MSDS disorders that have been linked to NTM disease (not just *Mycobacterium bovis* bacillus Calmette-Guerin, BCG), their mode of inheritance, the major clinical presentations, and diagnostic tests are listed in Table 2. Another genetic (X-link) disorder in which reduced IFNγ is a component is seen in individuals with mutations of the NFκB essential modular (NEMO, aka the γ subunit of the IκBα kinase or IKKγ); these affected patients exhibit a wide spectrum of clinical findings that include ectoderm developmental issues (abnormal teeth, sparse hair, anhidrosis), abnormalities of the venous/lymphatic vasculature (limb edema, lymphangiomas), autoimmune/inflammatory conditions, and increased susceptibility to infections with bacteria, mycobacteria (extrapulmonary NTM), viruses, and fungi [91, 93, 98, 99]. Interestingly, mutation or amino acid variation of the IFNγ gene has never been found, likely reflecting the paramount importance of IFNγ for the survival of higher-order species during natural selection [100]. Susceptibility to NTM of individuals with MSDS was experimentally corroborated by the increased

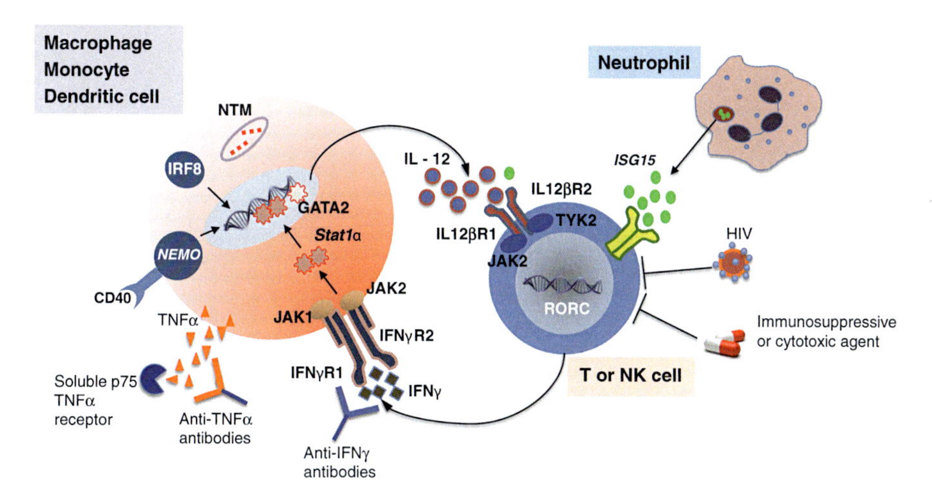

Fig. 3 Cartoon to illustrate the defects in mononuclear phagocytes, T_H1 cells, and neutrophils that predispose individuals to visceral and disseminated mycobacterial infections. Defects in RORC, ISG15, and IRF-8 have not been reported to be associated with NTM but rather disseminated BCG. IFNγR1/2 interferon-gamma receptors 1 and 2, Stat1α signal transducer and activator of transcription 1-alpha, IL-12 interleukin-12, IL-12Rβ1/2 IL-12 receptor (β1 and β2 subunits), TNFα tumor necrosis factor-alpha, TYK2 tyrosine kinase 2, IRF8 interferon regulatory factor-8, NEMO NF-κB essential modulator, JAK2 Janus kinase 2, ISG15 interferon-gamma stimulated gene-15, RORC RAR-related orphan receptor C, HIV human immunodeficiency virus

vulnerability to *Mycobacterium abscessus* in the IFNγ-knockout mice [44]. In individuals who develop extrapulmonary visceral organ/disseminated NTM disease as an adult but without acquired risk factors, the "MonoMAC syndrome" due to mutation of GATA2 transcription factor should be considered (Fig. 3 and Table 2) [101, 102]. Infectious complications typically occur during infancy or childhood in those with gene defects for IFNγR1, IFNγR2, IL-12p40 subunit, IL-12Rβ1 subunit, or Stat1α (Table 2). In contrast, the age range at presentation for MonoMAC syndrome encompasses both children and older adults. One hypothesis for the delayed manifestation of infection into adulthood for the MonoMAC syndrome despite autosomal inheritance is exhaustion of the bone marrow of hematopoietic stem cells with repeated external stressors [103].

Some MSMD disorders – e.g., due to mutations of the genes for *RAR-related orphan receptor C* (*RORC*, a transcription factor essential for differentiation of T cells to the T_H17 phenotype), *ISG15* (expected to reduce IFNγ production by T and NK cells), and interferon regulatory factor-8 (IRF-8, a transcription factor that regulates the differentiation of phagocytes as well as the production of IL-12 and TNFα induced by lipopolysaccharide and IFNγ) – have not been reported to be associated with NTM infections; rather, such affected individuals have been afflicted with disseminated BCG infection [86, 92, 96, 104]. However, the absence of reports of NTM infections in individuals with these rare genetic defects may simply be due to happenstance (the low likelihood of co-occurrence of an uncommon (NTM) disease with a rare disease), the failure to recognize the presence of an underlying genetic

Table 2 Clues to the presence of an underlying genetic/heritable cause for extrapulmonary visceral/disseminated NTM disease

Host gene abnormality (protein) Mode of inheritance	Relative age at presentation	Clues to presence of host risk factor	Diagnostic test(s)
***IFNGR1* mutations** (IFNγR1[a]) AR, PE-, complete[b] AR, PE+, complete AR, PE+, partial[b] AD, PE++, partial	Infants, young children	Disseminated NTM, BCG, and non-typhoidal infections	Surface expression by flow cytometry; functional analysis of IFNγR[c]; gene sequencing
***IFNGR2* mutations** (IFNγR2) AR, PE-, complete AR, PE+, complete AR, PE+, partial AD, PE+, partial	Infants, young children	Disseminated NTM, BCG, and non-typhoidal infections	As above
***IL12B* mutations** (IL-12p40 subunit[d]) AR, PE-, complete	Infants, young children	Disseminated NTM, BCG, and non-typhoidal infections; mucocutaneous candidal infections[e]	Stimulate PBMC with mitogen ± IFNγ, measure IL-12; gene sequencing
***IL12RB1* mutations** (IL-12Rβ1 subunit[d]) AR, PE-, complete AR, PE+, complete AR, PE-, partial-severe	Infants, young children	Disseminated NTM, BCG, and non-typhoidal infections; mucocutaneous candidal infections[e]	Surface expression by flow cytometry; functional testing of IL-12R[f]; gene sequencing
***STAT1* mutation** (Stat1α) AR, PE-, P-, B-, complete AR, PE+, P+, B+, partial AD, PE+, P-, B+, partial AD, PE+, P+, B-, partial	Infants, young children	Disseminated NTM, BCG, and non-typhoidal infections	Functional analysis of Stat1α[g]; gene sequencing
***IKBKG* mutation** (NEMO, IKKγ) X-linked	Male children	Ectoderm developmental abnormalities, venous/lymphatic vasculature abnormalities, autoimmune/inflammatory conditions, and infections with bacteria, extrapulmonary NTM, viruses, and fungi	Functional analysis of NFκB[h]; gene sequencing
***GATA2* mutation** (GATA2) AD	Young child to older adults	Disseminated infection with NTM, fungus, or HPV. May be complicated by lymphedema, PAP, myelodysplasia, acute and chronic myeloid leukemia	Gene sequencing, cytopenias (monocytes, DC, B cells, NK cells); bone marrow shows hypocellularity, fibrosis, multilineage dysplasia, etc.

Table 2 (continued)

Host gene abnormality (protein) Mode of inheritance	Relative age at presentation	Clues to presence of host risk factor	Diagnostic test(s)
Anti-IFNγ autoantibody[i] Associated with DRB16:02 and DRB 05:02	More common in Asian adults	Extrapulmonary visceral/ disseminated infection with NTM, *Salmonella*, fungi, and cytomegalovirus, and varicella-zoster virus reactivation	Anti-IFNγ antibody testing by particle-based technology or ELISA

Adapted from Ref. [93]

AD autosomal dominant, *AR* autosomal recessive, *HPV* human papillomavirus, *IFNγR1* IFNγ receptor subunit 1, *IFNγR2* IFNγ receptor subunit 2, *IL-12Rβ1* IL-12 receptor subunit β1, *NEMO* NFκB essential modulator (= IκBα kinase γ subunit, IKKγ), *P* phosphorylation, *PE* protein expression, *B* DNA binding

[a]Autosomal dominant IFNγR1 deficiency lacks Jak1- and STAT1α binding domains and thus can bind IFNγ but cannot signal downstream. In such cases, there is expression of proteins on the cell membrane, but the defective IFNγR1 accumulates and competes with normally functioning IFNγR1. Most patients with complete autosomal recessive forms do not express IFNγR1 on the cell surface because of stop mutations in the extracellular domain

[b]Complete and partial refers to signaling defect

[c]Stimulation of PBMC with IFNγ and assaying for Stat1α phosphorylation

[d]IL-12 is comprised of two subunits, IL-12p40 (*IL12B* gene) and IL-12p35 (*IL12A* gene). IL-12 receptor is comprised of two subunits, IL-12Rβ1 (*IL12RB1* gene) and IL-12Rβ2 (*IL12RB2* gene)

[e]While mutation is in the IL-12Rβ1 subunit, the susceptibility to mucosal fungal infections is due to defective IL-23 signaling, which is required to stimulate proliferation of T_H17 cells that are required for mucosal antifungal immunity, as IL-12 and IL-23 share the IL-12p40 subunit and IL-12R and IL-23R share the IL-12Rβ1 subunit

[f]Stimulation of PBMC with IL-12 and assaying for Stat4 phosphorylation

[g]Stimulation of PBMC with IFNγ and assay for Stat1α phosphorylation and binding to the *cis*-GAS DNA sequence

[h]Stimulation of PBMC with lipopolysaccharide or relevant cytokine and assay for NFκB binding to its cis-regulatory element

[i]While anti-IFNγ antibody syndrome is considered acquired, it is associated with certain HLA genes

disorder even when such infections are present, and/or the lack of reporting of such cases even if identified; in other words, *the absence of evidence of an association between NTM infection and a specific MSMD disorder does not necessarily mean evidence of absence of such a risk.* While only three cases of NTM infections in chronic granulomatous disease (CGD) patients have been reported in the literature, the unusual and extrapulmonary involvement in these patients suggests that CGD is a predisposing condition for NTM infections: a *M. flavescens* disseminated infection, *M. fortuitum* pneumonia and osteomyelitis, and *M. avium* lung and mediastinal involvement in a 10-month-old infant [105–107]. While this observation would suggest that reactive oxygen species (ROS) is an important host defense factor against NTM infections, we have found that inhibiting ROS with a superoxide dismutase mimetic actually improved macrophage killing of *M. abscessus* by promoting phagosome-lysosome fusion [108, 109].

The Role of Host Immunity in Pathogenesis of NTM-LD

While the previous discussions have focused on defects in host immunity causing NTM-LD, an overexuberant host immune response may also play a part in the pathogenesis of NTM-LD – the proverbial "double-edged sword" paradigm. In response to NTM lung infection, the release of elastase and metalloproteinases by recruited neutrophils can cause damage to the airway epithelium by eroding through mucosal barriers, resulting in NTM-containing microabscesses. Elastase may also cause ciliary dysfunction, mucous gland hyperplasia, and mucus hypersecretion that enhance NTM biofilm formation [16, 110]. Elastase and other proteases also cleave Fcγ receptors and complement receptor 1 from neutrophil surfaces as well as digest immunoglobulins and complement components from mycobacterial surfaces. These activities impair opsonization of mycobacteria and reduce their recognition by neutrophils, leading to decreased phagocytosis and bacterial clearance [16]. Elastase also inhibits efferocytosis, impairing clearance of apoptotic neutrophils [16]. The unphagocytosed, dead neutrophils incite further inflammation and release highly viscous DNA, contributing to the formation of inspissated mucus.

Conclusion

In summary, in determining whether there is an underlying cause for NTM infections, the clinician should first distinguish those with isolated NTM-LD versus those with extrapulmonary visceral/disseminated NTM disease. The reason is that with each of these two major domains of NTM infections, specific acquired and genetic/hereditary risks should be considered in the differential diagnosis of underlying disorders that increases vulnerability to isolated NTM-LD or extrapulmonary visceral/disseminated NTM disease. Acquired disorders should be ruled out first given their greater ease in diagnosis and measures to counter them. If acquired disorders are not apparent, genetic/hereditary risk factors should be entertained. But it is also important to be cognizant that even if an underlying acquired or genetic/hereditary cause is found, other factors play an important role in the pathogenesis of NTM disease, including host factors such as low body fat as well as exposure to environmental sources of NTM coupled to host behaviors that increase exposure and acquisition of NTM infection. Much remains to be elucidated on the role NTM virulence and immune evasive mechanisms – not discussed in this paper – may play in determining the development of bona fide NTM disease.

Clinical Vignette 1

Two sextagenerian sisters were identified to have isolated NTM-LD. Both also have pectus excavatum, scoliosis, and thin body habitus with one having a body mass index of 21 kg/m^2 and another having a 21% body fat [46]. Whole exome sequencing of both as well as four additional surviving siblings revealed heterozygous mutation (V900 M/wt) of the *MACROPHAGE-STIMULATING 1 RECEPTOR* (*MST1R*) gene, which encodes for a tyrosine kinase present on the apical surface of epithelial cells present in the airways and fallopian tubes. MST1R protein functions to maintain normal ciliary beat frequency and to play a role in IFNγ production, the mechanisms of which are not well characterized. Three additional siblings also have the same *MST1R* gene mutation and one is troubled by chronic cough and another has repeated chest infections and severe pectus excavatum. Thus, five of the six siblings have the *MST1R* variant, four of the five with the mutation have chronic respiratory symptoms or disease, and two of the five have NTM-LD.

Clinical Vignette 2

A 7-year-old boy from South Texas presented with cervical lymphadenopathy and radiographic evidence of tibial osteomyelitis. At age 4, he developed cervical lymphadenopathy associated with a positive QuantiFERON® test and received antituberculosis medication with resolution of the lymphadenopathy. The cervical lymphadenopathy recurred, and biopsy of the lymph nodes and a bone biopsy were positive for *M. avium* complex.

A whole blood specimen from the patient was analyzed by the Immunopathogenesis Section at the National Institutes of Health. A pathogenic, heterozygous variant in *IFNGR1* (c.819_822del4) was found, resulting in a stable protein with a truncated intracellular region, deleting the recycling domain and causing an accumulation of dysfunctional IFNγR1 on the cell surface of monocytes. This mutation has been reported previously [111] in patients with disseminated NTM or BCG infections.

Disseminated NTM infection in a child or young adult without a known predisposition is strongly suggestive of an abnormality or defect in the IFNγ-IL-12 axis. This child was treated with azithromycin, rifampin, ethambutol, and isoniazid because of the possibility of NTM and tuberculosis coinfection. He responded to the antibiotics and is being monitored for new infections.

References

1. Kim YM, Kim M, Kim SK, Park K, Jin S-H, Lee US, Kim Y, Chae GT, Lee S-B. Mycobacterial infections in coal workers' pneumoconiosis patients in South Korea. Scand J Infect Dis. 2009;41:656–62.
2. Rosenzweig DY. Pulmonary mycobacterial infections due to *Mycobacterium intracellulare-avium* complex. Clinical features and course in 100 consecutive cases. Chest. 1979;75:115–9.
3. Sonnenberg P, Murray J, Glynn JR, Thomas RG, Godfrey-Faussett P, Shearer S. Risk factors for pulmonary disease due to culture-positive *M. tuberculosis* or non-tuberculous mycobacteria in South African gold miners. Eur Respir J. 2000;15:291–6.
4. Chan ED, Iseman MD. Potential association between calcified thoracic lymphadenopathy due to previous *Histoplasma capsulatum* infection and pulmonary *Mycobacterium avium* complex disease. South Med J. 1999;92:572–6.
5. Dirac MA, Horan KL, Doody DR, Meschke JS, Park DR, Jackson LA, Weiss NS, Winthrop KL, Cangelosi GA. Environment or host?: a case-control study of risk factors for *Mycobacterium avium* complex lung disease. Am J Respir Crit Care Med. 2012;186:684–91.
6. Kartalija M, Ovrutsky AR, Bryan CL, Pott GB, Fantuzzi G, Thomas J, Strand MS, Bai X, Ramamoorthy R, Rothman MS, Nagabhushanam V, McDermott M, Levin AR, Frazer-Abel A, Giclas PC, Korner J, Iseman MD, Shapiro L, Chan ED. Patients with non-tuberculous mycobacterial lung disease exhibit unique body and immune phenotypes. Am J Respir Crit Care Med. 2012;187:197–205.
7. Chan ED, Kaminska AM, Gill W, Chmura K, Feldman NE, Bai X, Floyd CM, Fulton KE, Huitt GA, Strand MJ, Iseman MD, Shapiro L. Alpha-1-antitrypsin (AAT) anomalies are associated with lung disease due to rapidly growing mycobacteria and AAT inhibits *Mycobacterium abscessus* infection of macophages. Scand J Infect Dis. 2007;39:690–6.
8. Noone PG, Leigh MW, Sannuti A, Minnix SL, Carson JL, Hazucha M, Zariwala MA, Knowles MR. Primary ciliary dyskinesia: diagnostic and phenotypic features. Am J Respir Crit Care Med. 2004;169:459–67.
9. Tomii K, Iwata T, Oida K, Kohri Y, Taguchi Y, Nanbu Y, Kubo Y, Yaba Y, Mino M, Kuroda Y. A probable case of adult Williams-Campbell syndrome incidentally detected by an episode of atypical mycobacterial infection. Nihon Kyobu Shikkan Gakkai Zasshi. 1989;27:518–22.
10. Uji M, Matsushita H, Watanabe T, Suzumura T, Yamada M. A case of primary Sjögren's syndrome presenting with middle lobe syndrome complicated by nontuberculous mycobacteriosis. Nihon Kokyuki Gakkai Zasshi. 2008;46:55–9.
11. Witty LA, Tapson VF, Piantadosi CA. Isolation of mycobacteria in patients with pulmonary alveolar proteinosis. Medicine. 1994;73:103–9.
12. Damseh N, Quercia N, Rumman N, Dell SD, Kim RH. Primary ciliary dyskinesia: mechanisms and management. Appl Clin Genet. 2017;10:67–74.
13. Shapiro AJ, Zariwala MA, Ferkol T, Davis SD, Sagel SD, Dell SD, Rosenfeld M, Olivier KN, Milla C, Daniel SJ, Kimple AJ, Manion M, Knowles MR, Leigh MW, Genetic Disorders of Mucociliary Clearance Consortium. Diagnosis, monitoring, and treatment of primary ciliary dyskinesia: PCD foundation consensus recommendations based on state of the art review. Pediatr Pulmonol. 2016;51:115–32.
14. Parr DG, Guest PG, Reynolds JH, Dowson LJ, Stockley RA. Prevalence and impact of bronchiectasis in a1-antitrypsin deficiency. Am J Respir Crit Care Med. 2007;176:1215–21.
15. Chan ED, Bai A, Eichstaedt C, Hamzeh NY, Harbeck R, Honda JR, Frazer-Abel A, Kosmider B, Sandhaus RA, Bai X. Alpha-1-antitrypsin increases autophagosome number and production of host-protective cytokines in *Mycobacterium Intracellulare*-infected macrophages. Am J Respir Crit Care Med. 2017;195:A2906.
16. Chan ED, Iseman MD. Bronchiectasis. In: Broaddus CEJ, King TE, Lazarus SC, Mason R, Murray J, Nadel J, Slutsky AS, editors. Textbook of respiratory medicine. 6th ed: Elsevier Press; 2017. In press.

17. George J, Jain R, Tariq SM. CT bronchoscopy in the diagnosis of Williams-Campbell syndrome. Respirology. 2006;11:117–9.
18. Wayne KS, Taussig LM. Probable familial congenital bronchiectasis due to cartilage deficiency (Williams-Campbell syndrome). Am Rev Respir Dis. 1976;114:15–22.
19. Aldave APN, Saliski DOW. The clinical manifestations, diagnosis and management of Williams-Campbell syndrome. N Am J Med Sci. 2014;6:429–32.
20. Payandeh J, McGillivray B, McCauley G, Wilcox P, Swiston JR, Lehman A. A clinical classification scheme for tracheobronchomegaly (Mounier-Kuhn syndrome). Lung. 2015;193:815–22.
21. Celik B, Bilgin S, Yuksel C. Mounier-Kuhn syndrome: a rare cause of bronchial dilation. Tex Heart Inst J. 2011;38:194–6.
22. Krustins E. Mounier-Kuhn syndrome: a systematic analysis of 128 cases published within last 25 years. Clin Respir J. 2016;10:3–10.
23. Abdul-Rahman JA, Moodley YP, Phillips MJ. Pulmonary alveolar proteinosis associated with psoriasis and complicated by mycobacterial infection: successful treatment with granulocyte-macrophage colony stimulating factor after a partial response to whole lung lavage. Respirology. 2004;9:419–22.
24. Bakhos R, Gattuso P, Arcot C, Reddy VB. Pulmonary alveolar proteinosis: an unusual association with *Mycobacterium avium-intracellulare* infection and lymphocytic interstitial pneumonia. South Med J. 1996;89:801–2.
25. Bedrossian CW, Luna MA, Conklin RH, Miller WC. Alveolar proteinosis as a consequence of immunosuppression. A hypothesis based on clinical and pathologic observations. Hum Pathol. 1980;11(5 Suppl):527–35.
26. Carnovale R, Zornoza J, Goldman AM, Luna M. Pulmonary alveolar proteinosis: its association with hematologic malignancy and lymphoma. Radiology. 1977;122:303–6.
27. Goldschmidt N, Nusair S, Gural A, Amir G, Izhar U, Laxer U. Disseminated *Mycobacterium kansasii* infection with pulmonary alveolar proteinosis in a patient with chronic myelogenous leukemia. Am J Hematol. 2003;74:221–3.
28. Prakash UB, Barham SS, Carpenter HA, Dines DE, Marsh HM. Pulmonary alveolar phospholipoproteinosis: experience with 34 cases and a review. Mayo Clin Proc. 1987;62:499–518.
29. Ramirez J. Pulmonary alveolar proteinosis. Treatment by massive bronchopulmonary lavage. Arch Intern Med. 1967;119:147–56.
30. Arora R, Hagan L, Conger NG. Mycobacterium simiae infection in a patient with common variable immunodeficiency. J Allergy Clin Immunol. 2004;113:S123.
31. Kralickova P, Mala E, Vokurkova D, Krcmova I, Pliskova L, Stepanova V, Bartos V, Koblizek V, Tacheci I, Bures J, Brozik J, Litzman J. Cytomegalovirus disease in patients with common variable immunodeficiency: three case reports. Int Arch Allergy Immunol. 2014;163:69–74.
32. Griffith DE, Girard WM, Wallace RJ. Clinical features of pulmonary disease caused by rapidly growing mycobacteria: an analysis of 154 patients. Am Rev Respir Dis. 1993;147:1271–8.
33. Okumura M, Iwai K, Ogata H, Ueyama M, Kubota M, Aoki M, Kokuto H, Tadokoro E, Uchiyama T, Saotome M, Yoshiyama T, Yoshimori K, Yoshida N, Azuma A, Kudoh S. Clinical factors on cavitary and nodular bronchiectatic types in pulmonary *Mycobacterium avium* complex disease. Intern Med. 2008;47:1465–72.
34. Prince DS, Peterson DD, Steiner RM, Gottlieb JE, Scott R, Israel HL, Figueroa WG, Fish JE. Infection with *Mycobacterium avium* complex in patients without predisposing conditions. N Engl J Med. 1989;321:863–8.
35. Chan ED, Iseman MD. Slender, older women appear to be more susceptible to nontuberculous mycobacterial lung disease. Gend Med. 2010;7:5–18.
36. Iseman MD, Buschman DL, Ackerson LM. Pectus excavatum and scoliosis. Thoracic anomalies associated with pulmonary disease caused by *Mycobacterium avium* complex. Am Rev Respir Dis. 1991;144:914–6.
37. Kim RD, Greenberg DE, Ehrmantraut ME, Guide SV, Ding L, Shea Y, Brown MR, Chernick M, Steagall WK, Glasgow CG, Lin J, Jolley C, Sorbara L, Raffeld M, Hill S, Avila N, Sachdev V, Barnhart LA, Anderson VL, Claypool R, Hilligoss DM, Garofalo M, Fitzgerald A, Anaya-

O'Brien S, Darnell D, DeCastro R, Menning HM, Ricklefs SM, Porcella SF, Olivier KN, Moss J, Holland SM. Pulmonary nontuberculous mycobacterial disease: prospective study of a distinct preexisting syndrome. Am J Respir Crit Care Med. 2008;178:1066–74.
38. Talbert J, Chan ED. The association between body shape and non-tuberculous mycobacterial lung disease. Expert Rev Respir Med. 2013;7:201–4.
39. Tasaka S, Hasegawa N, Nishimura T, Yamasawa W, Kamata H, Shinoda H, Kimizuka Y, Fujiwara H, Hirose H, Ishizaka A. Elevated serum adiponectin level in patients with *Mycobacterium avium-intracellulare* complex pulmonary disease. Respiration. 2010;79:383–7.
40. Wakamatsu K, Nagata N, Maki S, Omori H, Kumazoe H, Ueno K, Matsunaga Y, Hara M, Takakura K, Fukumoto N, Ando N, Morishige M, Akasaki T, Inoshima I, Ise S, Izumi M, Kawasaki M. Patients with MAC lung disease have a low visceral fat area and low nutrient intake. Pulm Med. 2015;2015:218253.
41. Hayashi M, Takayanagi N, Kanauchi T, Miyahara Y, Yanagisawa T, Sugita Y. Prognostic factors of 634 HIV-negative patients with *Mycobacterium avium* complex lung disease. Am J Respir Crit Care Med. 2012;185:575–83.
42. Ikegame S, Maki S, Wakamatsu K, Nagata N, Kumazoe H, Fujita M, Nakanishi Y, Kawasaki M, Kajiki A. Nutritional assessment in patients with pulmonary nontuberculous mycobacteriosis. Intern Med. 2011;50:2541–6.
43. Lord GM, Matarese G, Howard JK, Baker RJ, Bloom SR, Lechler RI. Leptin modulates the T-cell immune response and reverses starvation-induced immunosuppression. Nature. 1998;394:897–901.
44. Ordway D, Henao-Tamayo M, Smith E, Shanley C, Harton M, Troudt JL, Bai X, Basaraba RJ, Orme IM, Chan ED. Animal model of *Mycobacterium abscessus* lung infection. J Leukoc Biol. 2008;83:1502–11.
45. Szymanski EP, Leung JM, Fowler CJ, Haney C, Hsu AP, Chen F, Duggal P, Oler AJ, McCormack R, Podack E, Drummond RA, Lionakis MS, Browne SK, Prevots DR, Knowles M, Cutting G, Liu X, Devine SE, Fraser CM, Tettelin H, Olivier KN, Holland SM. Pulmonary nontuberculous mycobacterial infection. A multisystem, multigenic disease. Am J Respir Crit Care Med. 2015;192:618–28.
46. Becker K, Arts P, Jaeger M, Plantinga T, Gilissen C, van Laarhoven A, van Ingen J, Veltman J, Joosten L, Hoischen A, Netea M, Iseman M, Chan ED, van de Veerdonk F. MST1R mutation as a genetic cause of Lady Windermere syndrome. Eur Respir J. 2017;49:1601478.
47. Sakamoto O, Iwama A, Amitani R, Takehara T, Yamaguchi N, Yamamoto T, Masuyama K, Yamanaka T, Ando M, Suda T. Role of macrophage-stimulating protein and its receptor, RON tyrosine kinase, in ciliary motility. J Clin Invest. 1997;99:701–9.
48. Takano Y, Sakamoto O, Suga M, Suda T, Ando M. Elevated levels of macrophage-stimulating protein in induced sputum of patients with bronchiectasis. Respir Med. 2000;94:784–90.
49. Fowler CJ, Olivier KN, Leung JM, Smith CC, Huth AG, Root H, Kuhns DB, Logun C, Zelazny A, Frein CA, Daub J, Haney C, Shelhamer JH, Bryant CE, Holland SM. Abnormal nasal nitric oxide production, ciliary beat frequency, and Toll-like receptor response in pulmonary nontuberculous mycobacterial disease epithelium. Am J Respir Crit Care Med. 2013;187: 1374–81.
50. Yaacoby-Bianu K, Gur M, Toukan Y, Nir V, Hakim F, Geffen Y, Bentur L. Compassionate nitric oxide adjuvant treatment of persistent *Mycobacterium* infection in cystic fibrosis patients. Pediatr Infect Dis J. 2017. [Epub ahead of print].
51. Matsuyama M, Martins AJ, Shallom S, Kamenyeva O, Kashyap A, Sampaio EP, Kabat J, Olivier KN, Zelazny AM, Tsang JS, Holland SM. Transcriptional response of respiratory epithelium to nontuberculous mycobacteria. Am J Respir Cell Mol Biol. 2018;58:241–52.
52. Honda JR, Bai X, Chan ED. Elucidating the pathogenesis for NTM lung disease: lesson from the six blind men and the elephant (editorial). Am J Respir Cell Mol Biol. 2017. (In press).
53. Greinert U, Schlaak M, Rüsch-Gerdes S, Flad HD, Ernst M. Low *in vitro* production of interferon-gamma and tumor necrosis factor-alpha in HIV-seronegative patients with pulmonary disease caused by nontuberculous mycobacteria. J Clin Immunol. 2000;20:445–52.

54. Kwon YS, Kim EJ, Lee S-H, Suh GY, Chung MP, Kim H, Kwon OJ, Koh W-J. Decreased cytokine production in patients with nontuberculous mycobacterial lung disease. Lung. 2007;185:337–41.
55. Safdar A, Atrmstrong D, Murray HW. A novel defect in interferon-g secretion in patients with refractory nontuberculous pulmonary mycobacteriosis. Ann Intern Med. 2003;138:521.
56. Safdar A, White DA, Stover D, Armstrong D, Murray HW. Profound interferon gamma deficiency in patients with chronic pulmonary nontuberculous mycobacteriosis. Am J Med. 2002;113:756–9.
57. Vankayalapati R, Wizel B, Samten B, Griffith DE, Shams H, Galland MR, Von Reyn CF, Girard WM, Wallace RJ, Barnes PF. Cytokine profiles in immunocompetent persons infected with *Mycobacterium avium* complex. J Infect Dis. 2001;183:478–84.
58. Lim A, Allison C, Price P, Waterer G. Susceptibility to pulmonary disease due to *Mycobacterium avium-intracellulare* complex may reflect low IL-17 and high IL-10 responses rather than Th1 deficiency. Clin Immunol. 2010;137:296–302.
59. de Jong E, Lim A, Waterer G, Price P. Monocyte-derived macrophages do not explain susceptibility to pulmonary non-tuberculous mycobacterial disease. Clin Transl Immunol. 2012;1:e2.
60. Cowman SA, Jacob J, Hansell DM, Kelleher P, Wilson R, Cookson WOC, Moffatt MF, Loebinger MR. Whole blood gene expression in pulmonary non-tuberculous mycobacterial infection. Am J Respir Cell Mol Biol. 2017. [Epub ahead of print].
61. Lam PK, Griffith DE, Aksamit TR, Ruoss SJ, Garay SM, Daley CL, Catanzaro A. Factors related to response to intermittent treatment of *Mycobacterium avium* complex lung disease. Am J Respir Crit Care Med. 2006;173:1283–9.
62. Ryu YJ, Kim EJ, Lee SH, Kim SY, Suh GY, Chung MP, Kim H, Kwon OJ, Koh WJ. Impaired expression of Toll-like receptor 2 in nontuberculous mycobacterial lung disease. Eur Respir J. 2007;30:736–42.
63. Champsi J, Young LS, Bermudez LE. Production of TNF-alpha, IL-6 and TGF-beta, and expression of receptors for TNF-alpha and IL-6, during murine *Mycobacterium avium* infection. Immunology. 1995;84:549–54.
64. Denis M, Ghadirian E. Transforming growth factor beta (TGF-b1) plays a detrimental role in the progression of experimental *Mycobacterium avium* infection; in vivo and in vitro evidence. Microb Pathog. 1991;11:367–72.
65. Judge DP, Dietz HC. Marfan's syndrome. Lancet. 2005;366:1965–76.
66. Daniels ML, Birchard KR, Lowe JR, Patrone MV, Noone PG, Knowles MR. Enlarged dural sac in idiopathic bronchiectasis implicates heritable connective tissue gene variants. Ann Am Thorac Soc. 2016;13:1712–20.
67. Paulson ML, Olivier KN, Holland SM. Pulmonary non-tuberculous mycobacterial infection in congenital contractural arachnodactyly. Int J Tuberc Lung Dis. 2012;16:561–3.
68. Foster ME, Foster DR. Bronchiectasis and Marfan's syndrome. Postgrad Med J. 1980;56: 718–9.
69. Wood JR, Bellamy D, Child AH, Citron KM. Pulmonary disease in patients with Marfan syndrome. Thorax. 1984;39:780–4.
70. Teoh PC. Bronchiectasis and spontaneous pneumothorax in Marfan's syndrome. Chest. 1977;72:672–3.
71. Ras GJ, Van Wyk CJ. Primary ciliary dyskinesia in association with Marfan's syndrome. A case report. S Afr Med J. 1983;64:212–4.
72. Saito H, Iijima K, Dambara T, Shiota J, Hirose S, Uekusa T, Saiki S, Kira S. An autopsy case of Marfan syndrome with bronchiectasis and multiple bullae. Nihon Kyobu Shikkan Gakkai Zasshi. 1992;30:1315–21.
73. Hwang HS, Yi CA, Yoo H, Yang JH, Kim DK, Koh WJ. The prevalence of bronchiectasis in patients with Marfan syndrome. Int J Tuberc Lung Dis. 2014;18:995–7.
74. Desai U, Kalamkar S, Joshi JM. Bronchiectasis in a Marfanoid: diagnosis beyond Marfans. Indian J Chest Dis Allied Sci. 2014;56:105–7.

75. McCormack RM, Szymanski EP, Hsu AP, Perez E, Olivier KN, Fisher E, Goodhew EB, Podack ER. Holland SM. MPEG1/perforin-2 mutations in human pulmonary nontuberculous mycobacterial infections. JCI Insight. 2017;2:89635.
76. Chen F, Szymanski EP, Olivier KN, Liu X, Tettelin H, Holland SM, Duggal P. Whole-exome sequencing identifies the 6q12-q16 linkage region and a candidate gene, TTK, for pulmonary nontuberculous mycobacterial disease. Am J Respir Crit Care Med. 2017;196:1599–604.
77. Colombo RE, Hill SC, Claypool RJ, Holland SM, Olivier KN. Familial clustering of pulmonary nontuberculous mycobacterial disease. Chest. 2010;137:629–34.
78. Reich JM, Johnson RE. *Mycobacterium avium* complex pulmonary disease presenting as an isolated lingular or middle lobe pattern: the Lady Windermere syndrome. Chest. 1992;101:1605–9.
79. Tsuyuguchi K, Suzuki K, Matsumoto H, Tanaka E, Amitani R, Kuze F. Effect of oestrogen on *Mycobacterium avium* complex pulmonary infection in mice. Clin Exp Immunol. 2001;123:428–34.
80. Adjemian J, Olivier KN, Seitz AE, Holland SM, Prevots DR. Prevalence of nontuberculous mycobacterial lung disease in U.S. Medicare beneficiaries. Am J Respir Crit Care Med. 2012;185:881–6.
81. Chi CY, Chu CC, Liu JP, Lin CH, Ho MW, Lo WJ, Lin PC, Chen HJ, Chou CH, Feng JY, Fung CP, Sher YP, Li CY, Wang JH, Ku CL. Anti-IFN-γ autoantibodies in adults with disseminated nontuberculous mycobacterial infections are associated with HLA-DRB1*16:02 and HLA-DQB1*05:02 and the reactivation of latent varicella-zoster virus infection. Blood. 2013;121:1357–66.
82. Doucette K, Fishman JA. Nontuberculous mycobacterial infection in hematopoietic stem cell and solid organ transplant recipients. Clin Infect Dis. 2004;38:1428–39.
83. French AL, Benator DA, Gordin FM. Nontuberculous mycobacterial infections. Med Clin North Am. 1997;81:361–79.
84. Winthrop KL, Baxter R, Liu L, Varley CD, Curtis JR, Baddley JW, McFarland B, Austin D, Radcliffe L, Suhler E, Choi D, Rosenbaum JT, Herrinton LJ. Mycobacterial diseases and antitumour necrosis factor therapy in USA. Ann Rheum Dis. 2013;72:37–42.
85. Altare F, Lammas D, Revy P, Jouanguy E, Döffinger R, Lamhamedi S, Drysdale P, Scheel-Toellner D, Girdlestone J, Darbyshire P, Wadhwa M, Dockrell H, Salmon M, Fischer A, Durandy A, Casanova JL, Kumararatne DS. Inherited interleukin 12 deficiency in a child with bacille Calmette-Guérin and *Salmonella enteritidis* disseminated infection. J Clin Invest. 1998;102:2035–40.
86. Bogunovic D, Byun M, Durfee LA, Abhyankar A, Sanal O, Mansouri D, Salem S, Radovanovic I, Grant AV, Adimi P, Mansouri N, Okada S, Bryant VL, Kong XF, Kreins A, Velez MM, Boisson B, Khalilzadeh S, Ozcelik U, Darazam IA, Schoggins JW, Rice CM, Al-Muhsen S, Behr M, Vogt G, Puel A, Bustamante J, Gros P, Huibregtse JM, Abel L, Boisson-Dupuis S, Casanova JL. Mycobacterial disease and impaired IFN-γ immunity in humans with inherited ISG15 deficiency. Science. 2012;337:1684–8.
87. Bustamante J, Arias AA, Vogt G, Picard C, Galicia LB, Prando C, Grant AV, Marchal CC, Hubeau M, Chapgier A, de Beaucoudrey L, Puel A, Feinberg J, Valinetz E, Jannière L, Besse C, Boland A, Brisseau JM, Blanche S, Lortholary O, Fieschi C, Emile JF, Boisson-Dupuis S, Al-Muhsen S, Woda B, Newburger PE, Condino-Neto A, Dinauer MC, Abel L, Casanova JL. Germline CYBB mutations that selectively affect macrophages in kindreds with X-linked predisposition to tuberculous mycobacterial disease. Nat Immunol. 2011;12:213–21.
88. Döffinger R, Jouanguy E, Dupuis S, Fondanèche MC, Stephan JL, Emile JF, Lamhamedi-Cherradi S, Altare F, Pallier A, Barcenas-Morales G, Meinl E, Krause C, Pestka S, Schreiber RD, Novelli F, Casanova JL. Partial interferon-gamma receptor signaling chain deficiency in a patient with bacille Calmette-Guérin and *Mycobacterium abscessus* infection. J Infect Dis. 2000;181:379–84.
89. Dorman SE, Picard C, Lammas D, Heyne K, van Dissel JT, Baretto R, Rosenzweig SD, Newport M, Levin M, Roesler J, Kumararatne D, Casanova JL, Holland SM. Clinical features of dominant and recessive interferon gamma receptor 1 deficiencies. Lancet. 2004;364:2113–21.

90. Dupuis S, Dargemont C, Fieschi C, Thomassin N, Rosenzweig S, Harris J, Holland SM, Schreiber RD, Casanova JL. Impairment of mycobacterial but not viral immunity by a germline human STAT1 mutation. Science. 2001;293:300–3.
91. Filipe-Santos O, Bustamante J, Haverkamp MH, Vinolo E, Ku CL, Puel A, Frucht DM, Christel K, von Bernuth H, Jouanguy E, Feinberg J, Durandy A, Senechal B, Chapgier A, Vogt G, de Beaucoudrey L, Fieschi C, Picard C, Garfa M, Chemli J, Bejaoui M, Tsolia MN, Kutukculer N, Plebani A, Notarangelo L, Bodemer C, Geissmann F, Israël A, Véron M, Knackstedt M, Barbouche R, Abel L, Magdorf K, Gendrel D, Agou F, Holland SM. Casanova JL. X-linked susceptibility to mycobacteria is caused by mutations in NEMO impairing CD40-dependent IL-12 production. J Exp Med. 2006;203:1745–59.
92. Hambleton S, Salem S, Bustamante J, Bigley V, Boisson-Dupuis S, Azevedo J, Fortin A, Haniffa M, Ceron-Gutierrez L, Bacon CM, Menon G, Trouillet C, McDonald D, Carey P, Ginhoux F, Alsina L, Zumwalt TJ, Kong XF, Kumararatne D, Butler K, Hubeau M, Feinberg J, Al-Muhsen S, Cant A, Abel L, Chaussabel D, Doffinger R, Talesnik E, Grumach A, Duarte A, Abarca K, Moraes-Vasconcelos D, Burk D, Berghuis A, Geissmann F, Collin M, Casanova JL, Gros P. IRF8 mutations and human dendritic-cell immunodeficiency. N Engl J Med. 2011;365:127–38.
93. Haverkamp MH, van de Vosse E, van Dissel JT. Nontuberculous mycobacterial infections in children with inborn errors of the immune system. J Infect. 2014;68:S134–50.
94. Hsu AP, Johnson KD, Falcone EL, Sanalkumar R, Sanchez L, Hickstein DD, Cuellar-Rodriguez J, Lemieux JE, Zerbe CS, Bresnick EH, Holland SM. GATA2 haploinsufficiency caused by mutations in a conserved intronic element leads to MonoMAC syndrome. Blood. 2013;121:3830–7.
95. Minegishi Y, Saito M, Morio T, Watanabe K, Agematsu K, Tsuchiya S, Takada H, Hara T, Kawamura N, Ariga T, Kaneko H, Kondo N, Tsuge I, Yachie A, Sakiyama Y, Iwata T, Bessho F, Ohishi T, Joh K, Imai K, Kogawa K, Shinohara M, Fujieda M, Wakiguchi H, Pasic S, Abinun M, Ochs HD, Renner ED, Jansson A, Belohradsky BH, Metin A, Shimizu N, Mizutani S, Miyawaki T, Nonoyama S, Karasuyama H. Human tyrosine kinase 2 deficiency reveals its requisite roles in multiple cytokine signals involved in innate and acquired immunity. Immunity. 2006;25:745–55.
96. Okada S, Markle JG, Deenick EK, Mele F, Averbuch D, Lagos M, Alzahrani M, Al-Muhsen S, Halwani R, Ma CS, Wong N, Soudais C, Henderson LA, Marzouqa H, Shamma J, Gonzalez M, Martinez-Barricarte R, Okada C, Avery DT, Latorre D, Deswarte C, Jabot-Hanin F, Torrado E, Fountain J, Belkadi A, Itan Y, Boisson B, Migaud M, Arlehamn CS, Sette A, Breton S, McCluskey J, Rossjohn J, de Villartay JP, Moshous D, Hambleton S, Latour S, Arkwright PD, Picard C, Lantz O, Engelhard D, Kobayashi M, Abel L, Cooper AM, Notarangelo LD, Boisson-Dupuis S, Puel A, Sallusto F, Bustamante J, Tangye SG, Casanova JL. Impairment of immunity to *Candida* and *Mycobacterium* in humans with bi-allelic RORC mutations. Science. 2015;349:606–13.
97. Prando C, Samarina A, Bustamante J, Boisson-Dupuis S, Cobat A, Picard C, AlSum Z, Al-Jumaah S, Al-Hajjar S, Frayha H, Alangari A, Al-Mousa H, Mobaireek KF, Ben-Mustapha I, Adimi P, Feinberg J, de Suremain M, Jannière L, Filipe-Santos O, Mansouri N, Stephan JL, Nallusamy R, Kumararatne DS, Bloorsaz MR, Ben-Ali M, Elloumi-Zghal H, Chemli J, Bouguila J, Bejaoui M, Alaki E, AlFawaz TS, Al Idrissi E, ElGhazali G, Pollard AJ, Murugasu B, Wah-Lee B, Halwani R, Al-Zahrani M, Al Shehri MA, Al-Zahrani M, Bin-Hussain I, Mahdaviani SA, Parvaneh N, Abel L, Mansouri D, Barbouche R, Al-Muhsen S, Casanova JL. Inherited IL-12p40 deficiency: genetic, immunologic, and clinical features of 49 patients from 30 kindreds. Medicine (Baltimore). 2013;92:109–22.
98. Wu UI, Holland SM. Host susceptibility to non-tuberculous mycobacterial infections. Lancet Infect Dis. 2015;15:968–80.
99. Hanson EP, Monaco-Shawver L, Solt LA, Madge LA, Banerjee PP, May MJ, Orange JS. Hypomorphic nuclear factor-kappaB essential modulator mutation database and reconstitution system identifies phenotypic and immunologic diversity. J Allergy Clin Immunol. 2008;122:1169–77.

100. Manry J, Laval G, Patin E, Fornarino S, Tichit M, Bouchier C, Barreiro LB, Quintana-Murci L. Evolutionary genetics evidence of an essential, nonredundant role of the IFN-γ pathway in protective immunity. Hum Mutat. 2011;32:633–42.
101. Vinh DC, Patel SY, Uzel G, Anderson VL, Freeman AF, Olivier KN, Spalding C, Hughes S, Pittaluga S, Raffeld M, Sorbara LR, Elloumi HZ, Kuhns DB, Turner ML, Cowen EW, Fink D, Long-Priel D, Hsu AP, Ding L, Paulson ML, Whitney AR, Sampaio EP, Frucht DM, DeLeo FR, Holland SM. Autosomal dominant and sporadic monocytopenia with susceptibility to mycobacteria, fungi, papillomaviruses, and myelodysplasia. Blood. 2010;115:1519–29.
102. Hsu AP, Sampaio EP, Khan J, Calvo KR, Lemieux JE, Patel SY, Frucht DM, Vinh DC, Auth RD, Freeman AF, Olivier KN, Uzel G, Zerbe CS, Spalding C, Pittaluga S, Raffeld M, Kuhns DB, Ding L, Paulson ML, Marciano BE, Gea-Banacloche JC, Orange JS, Cuellar-Rodriguez J, Hickstein DD, Holland SM. Mutations in GATA2 are associated with the autosomal dominant and sporadic monocytopenia and mycobacterial infection (MonoMAC) syndrome. Blood. 2011;118:2653–5.
103. Migliaccio AR, Bieker JJ. GATA2 finds its macrophage niche. Blood. 2011;118:2647–9.
104. Honda JR, Knight V, Chan ED. Pathogenesis and risk factors for nontuberculous mycobacterial lung disease. Clin Chest Med. 2015;36:1–11.
105. Allen DM, Chng HH. Disseminated *Mycobacterium flavescens* in a probable case of chronic granulomatous disease. J Infect. 1993;26:83–6.
106. Chusid MJ, Parrillo JE, Fauci AS. Chronic granulomatous disease. Diagnosis in a 27-year-old man with *Mycobacterium fortuitum*. JAMA. 1975;233:1295–6.
107. Ohga S, Ikeuchi K, Kadoya R, Okada K, Miyazaki C, Suita S, Ueda K. Intrapulmonary *Mycobacterium avium* infection as the first manifestation of chronic granulomatous disease. J Infect. 1997;34:147–50.
108. Oberley-Deegan RE, Rebits BW, Weaver MR, Tollefson AK, Bai X, McGibney M, Ovrutsky AR, Chan ED, Crapo JD. An oxidative environment promotes growth of *Mycobacterium abscessus*. Free Radic Biol Med. 2010;49:1666–73.
109. Oberley-Deegan RE, Lee YM, Morey GE, Cook DM, Chan ED, Crapo JD. The antioxidant mimetic, MnTE-2-PyP, reduces intracellular growth of *Mycobacterium abscessus*. Am J Respir Cell Mol Biol. 2009;41:170–8.
110. Malcolm KC, Nichols EM, Caceres SM, Kret JE, Martiniano SL, Sagel SD, Chan ED, Caverly L, Solomon GM, Reynolds P, Bratton DL, Taylor-Cousar JL, Nichols DP, Saavedra MT, Nick JA. *Mycobacterium abscessus* induces a limited pattern of neutrophil activation that promotes pathogen survival. PLoS One. 2013;8:e57402.
111. Jouanguy E, Lamhamedi-Cherradi S, Lammas D, Dorman SE, Fondanèche MC, Dupuis S, Döffinger R, Altare F, Girdlestone J, Emile JF, Ducoulombier H, Edgar D, Clarke J, Oxelius VA, Brai M, Novelli V, Heyne K, Fischer A, Holland SM, Kumararatne DS, Schreiber RD, Casanova JL. A human IFNGR1 small deletion hotspot associated with dominant susceptibility to mycobacterial infection. Nat Genet. 1999;21:370–8.

Immune Dysfunction and Nontuberculous Mycobacterial Disease

Emily Henkle and Kevin L. Winthrop

Introduction

Nontuberculous mycobacteria (NTM) infection risks are increased for certain immunosuppressed populations. Historically, severely immunocompromised patients with acquired immunodeficiency syndrome (AIDS) and cancer or transplant recipients have been recognized as at risk for disseminated and extrapulmonary NTM disease. More recently population-based studies of immunosuppressive medications have also demonstrated an elevated risk of pulmonary NTM disease in certain individuals. Specific classes of drugs that elevate the risk of NTM infection have also been recently identified including oral and inhaled corticosteroids and biologic therapies that treat rheumatoid arthritis and other autoimmune inflammatory diseases [1, 2].

Similar to disease caused by *Mycobacterium tuberculosis*, the majority of NTM disease in the USA is pulmonary. Approximately 20–25% of NTM disease is extrapulmonary within the general population [3–5]. In 2017 the US Council of State and Territorial Epidemiologists released a standardized extrapulmonary NTM case definition for use in NTM surveillance [6]. Extrapulmonary isolates, from skin or soft tissue, lymph node, urine, or normally sterile sites outside the lungs, are generally considered to represent disease, although this is somewhat species-specific similar to pulmonary isolates. Aside from local inoculation causing skin/soft tissue disease and cervical lymphadenitis among children, extrapulmonary and disseminated NTM are almost exclusively associated with underlying immunosuppression (Fig. 1a). Also similar to tuberculosis, the proportion of extrapulmonary versus pulmonary disease is elevated in those with immunosuppression. At one extreme, 95% of NTM disease in AIDS patients is extrapulmonary and almost exclusively disseminated [7]. On the

E. Henkle (✉) · K. L. Winthrop
OHSU-PSU School of Public Health, Oregon Health and Science University, Portland, OR, USA
e-mail: henkle@ohsu.edu

D. E. Griffith (ed.), *Nontuberculous Mycobacterial Disease*, Respiratory Medicine,
https://doi.org/10.1007/978-3-319-93473-0_5

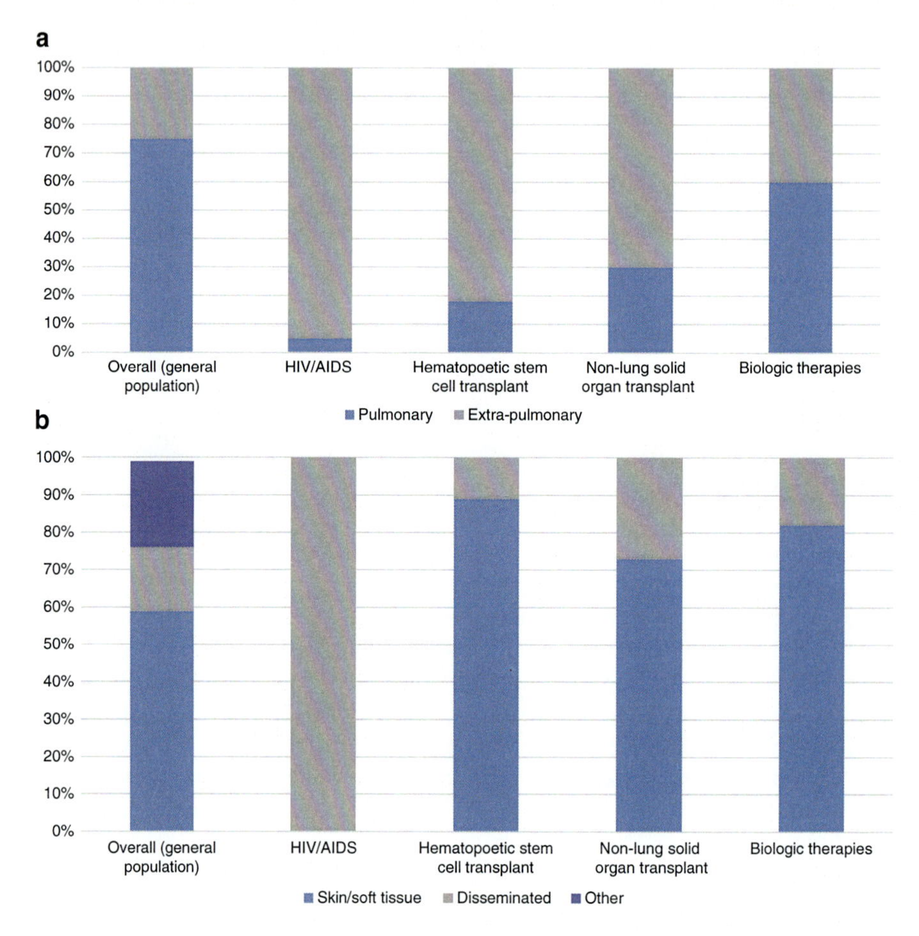

Fig. 1 (**a**) Proportion of nontuberculous mycobacteria cases that are pulmonary vs. extrapulmonary overall and by selected underlying cause of immunosuppression (**b**) Proportion of extrapulmonary disease cases by site of infection, overall and by selected underlying cause of immunosuppression

other hand, in solid organ transplant patients (other than lung) and patients taking biologic therapies for immune-mediated inflammatory disease, the proportion with extrapulmonary and pulmonary disease are approximately similar [8, 9]. The majority of cases associated with these and other types of immunosuppression are skin/soft tissue or catheter-associated infections (Fig. 1b). Although pulmonary isolation is not always associated with meeting ATS/IDSA disease criteria in immunocompetent patients, it is theoretically more likely to cause progressive airway infection and disease in patients with underlying immunosuppression.

While rapid-growing mycobacteria are more frequently found in the extrapulmonary setting (as compared to pulmonary setting), *Mycobacterium avium* complex (MAC) is still the most common cause of pulmonary (80%) and extrapulmonary (50%) NTM disease [5]. Other important slow-growing species include *Mycobacterium kansasii*, the second most common cause of disseminated NTM in AIDS patients

[7]. *M. kansasii* is found in the Southern and Midwestern USA and internationally [10]. *Mycobacterium xenopi* is known to colonize hospital water systems but is rarely identified outside of the Northern USA and Canada [10]. *Mycobacterium marinum* is found in freshwater and salt water and is a rare cause of skin infection that occurs generally in immunocompetent hosts [11]. The most common rapid-growing mycobacteria is *Mycobacterium abscessus,* comprising less than 10% of pulmonary infections [4]. Skin and catheter-related infections and disseminated NTM are more frequently caused by rapid-growing species including *Mycobacterium fortuitum, M. abscessus, Mycobacterium chelonae, Mycobacterium mucogenicum,* and *Mycobacterium neoaurum* [12–14]. Based on case reports, uncommon species such as *Mycobacterium haemophilum* and *Mycobacterium genavense* have been recognized as almost exclusively causing infection in immunocompromised patients such as HIV [15–17].

Mycobacterium tuberculosis has well-established patterns of disease, categorized as primary active disease with the remainder establishing latent or persistent (asymptomatic) infection, which may progress to active disease later, a risk heightened by immunosuppression. However, these concepts of latency and progression do not readily apply to NTM. The timing of exposure or inoculation is a critical element for the development of NTM disease in immunosuppressed patients, given frequent exposure from the environment but infrequent disease. Cell-mediated immunity is thought to play a key role in the immune response to mycobacterial infection, and this is well established for *M. tuberculosis* [18–20]. For NTM, however, it is unclear what role cell-mediated immunity plays in host response [21]. NTM in the lung, and presumably elsewhere, are taken in by macrophages leading to upregulation of interleukin-12, which stimulates T cells or natural killer cells [22, 23]. The stimulated cells in turn upregulate interferon gamma (IFNγ) production, which further upregulates IL-12 and tumor necrosis factor-alpha (TNF-α). TNF-α is responsible for the formulation of granulomatous lesions seen in many NTM infections [24–27]. Further detailed intracellular pathways will be described in the context of genetic mutations (primary immunodeficiency) and pharmacologic disruption (immunosuppressive therapies, including targeted biologic therapy) that increase the risk for NTM disease.

The following sections describe the epidemiology, prevention, diagnosis, and treatment of NTM in different immunosuppressed and immunocompromised patient populations. With NTM isolation, clinicians should seek to determine whether such isolation represents disease and, in the meantime, consider attempts to mitigate any immunosuppressive state that may be present within the patient.

Genetic Immune Dysfunction/Primary Immunodeficiency Disease

Primary immunodeficiency disease (PID) is an immune system dysfunction caused by genetic mutations, which may affect cellular or innate immunity. PID is rare and associated with a small fraction of NTM disease. PID was listed as a comorbid cause of death in 2% of NTM deaths, more common than either hematologic malignancy or HIV/AIDS [28]. Known and theoretical pathways where genetic conditions can disrupt the immune response to mycobacterial infection are discussed further in Chap. 4 [22]. Defects in the IL-12/IFN-ϒ axis, which are necessary for the control of intracellular bacteria such as NTM and *Salmonella*, are associated with an increased risk of NTM disease [29, 30].

Mendelian susceptibility to mycobacterial disease (MSMD) is an extremely rare syndrome caused by a group of monogenic inborn errors in the IL-12/IFN-ϒ pathway that occurs in otherwise healthy individuals [23, 31, 32]. MSMD is caused by ten known genetic defects (see Table 1) and may be autosomal or x-linked (*NEMO* and *CYBB*). Several syndromes that are associated with BCG vaccine infections (BCGosis) but not NTM infections include severe combined immunodeficiency syndrome, complete DiGeorge syndrome, X-linked hyper IgM syndrome, and chronic granulomatous disease [23].

Acquired Immune Dysfunction

There are two main types of acquired immune dysfunction, those caused by infections such as human-acquired immunodeficiency virus (HIV, described in the next section) and those caused by acquired deficiency or autoantibody states. The late-onset abnormalities include anti-IFNγ autoantibodies and *GATA2* deficiency and are associated with disseminated NTM disease, among other outcomes (see Table 2) [32]. The association with anti-IFNγ autoantibodies was noted in 2004, primarily in East Asia, and has been described in several more recent case-control studies in Thailand and Taiwan [33, 34]. Patients present with disseminated disease, lymphadenitis, and skin lesions similar to what is seen in HIV-infected patients, but these cases occur in patients who are previously healthy and are not HIV-infected. *M. abscessus* was associated with the majority of cases acquired anti-IFNγ autoantibodies. In Thailand, over 70% of 70 disseminated NTM cases associated with acquired anti-IFNγ autoantibodies were caused by *M. abscessus*. Rituximab has been used successfully to treat patients with anti-IFNγ autoantibodies [35, 36]. *GATA2* deficiency was initially described as an autosomal dominant "MonoMAC" syndrome, cases of disseminated MAC (and other opportunistic) infections characterized by monocytopenia and associated with myelodysplasia and malignancy [37]. Subsequent work identified 12 distinct *GATA2* mutations as causes of the syndrome [38]. In the initial case series, most (64%) with disseminated mycobacterial disease were caused by MAC, but one to two cases each were associated with *M. kansasii, Mycobacterium scrofulaceum, M. fortuitum,* and *M. abscessus* [37]. *GATA2* deficiency can be reversed by hematopoietic stem cell transplantation [39].

Table 1 Primary and acquired Mendelian susceptibility to mycobacterial disease genetic etiologies and number of patients reported to the Leiden Open Variation Database[a]

	Inheritance	Disease onset	BCG infection	Systemic salmonella infection	Other possible infections	Granuloma formation	Response to antimicrobial therapy	Indication for immunotherapy	Prognosis
Early onset									
IFNGR1/R2									
Complete [16, 24, 25, 30, 31]	AR	Infancy/early childhood	Yes	Yes	Listeriosis, herpes virus, respiratory syncytial virus, parainfluenza virus infections, tuberculosis	No	Very poor	No	Poor
Partial [32–39]	AR	Late childhood	Yes	Yes	Tuberculosis	No report	Favorable	Variable	Good
Partial [24, 40–43]	AD	Late childhood/ adolescence	Yes	Yes	Histoplasmosis, tuberculosis	Yes	Favorable	Yes	Good
*IL*12*B* [43, 44]	AR	Infancy/early childhood	Yes (97%)	Yes (25%)	CMC, disseminated tuberculosis, nocardia, *Klebsiella* spp infection	Yes	Favorable	Yes	Fair
*IL*12*RB*1 [45–48]	AR	Early childhood	Yes (76%)	Yes (43%)	Tuberculosis, CMC (24%), *Klebsiella* spp. infection	Yes	Favorable	Yes	Fair
*STAT*1 LOF									
Complete [49–51]	AR	Infancy (die early without HSCT)	Yes	No	Tuberculosis, fulminant viral infection (mainly herpes)	Yes	Poor	No	Poor
Partial [27, 28, 52–54]	AR	Infancy/early childhood/ adolescence	Yes	Yes (50%)	Severe, curable viral infection (mainly herpes)	No report	Favorable	Yes	Fair

(continued)

Table 1 (continued)

	Inheritance	Disease onset	BCG infection	Systemic salmonella infection	Other possible infections	Granuloma formation	Response to antimicrobial therapy	Indication for immunotherapy	Prognosis
Partial [55–58]	AD	Infancy/early childhood/ adolescence	Yes	No	Tuberculosis	Yes	Favorable	Yes	Good
*IRF*8 [59]	AR	Infancy	Yes	No	CMC	Poorly formed	Poor	No	Poor
*IRF*8 [59]	AD	Late infancy	Yes	No	No report	Yes	Favorable	No	Good
*ISG*15 [60]	AR	Infancy	Yes	Yes	No report	No report	Favorable	Yes	Good
NEMO [61–64]	XR	Early to late childhood	Yes	No	Invasive Hib infection, tuberculosis	Yes	Variable	Yes	Fair
CYBB [15]	XR	Infancy/early childhood	Yes	No	Tuberculosis	Yes	Fair	No	Fair
Late onset									
GATA2 [65–69]	AD	Late childhood/ adulthood	No	No	HPV, CMV, EBV, *Clostridium difficile* histoplasmosis, aspergillosis	Yes	Poor	Yes	Poor
Anti-interferon-γ autoantibodies [70–73]	Acquired	Young adult to elderly	No	Yes	*Salmonella* spp., *Penicillium* spp., *Histoplasma* spp., *Cryptococcus* spp., *Burkholderia pseudomallei*, VZV, CMV infection	Yes	Poor	No	Fair

AE autosomal recessive, *AD* autosomal dominant, *CMC* chronic mucocutaneous candidiasis, *LOF* loss of function, *HSCT* hematopoietic stem cell transplantation, *Hib* Haemophilius influenzae type b, *HPV* human papillomavirus, *CMV* cytomegalovirus, *EBV* Epstein-Barr virus, *VZV* varicella zoster virus

[a]Number of patients as of August 23, 2017: *IL12B* (54), *IL12RB1* (220), *IFNGR1* (142), *IFNGR2* (26), *CYBB* (80); unknown for *STAT1* and *NEMO*

Adapted from Wu UI, Holland SM. Host susceptibility to non-tuberculous mycobacterial infections. The Lancet infectious diseases 2015;15:968–80

Table 2 Biologic agents, mechanisms of action, and date of US Food and Drug Administration-approved indications

Anti-TNF agents	Mechanism (see Fig. 1)	Date of FDA approval	FDA-approved indications	NTM risk level of evidence
Infliximab (Remicade)	Monoclonal antibody against TNF-α	1998	Rheumatoid arthritis, Crohn's disease, ulcerative colitis, psoriasis, ankylosing spondylitis	pop [2],case
Etanercept (Enbrel)	Receptor fusion protein	1998	Rheumatoid arthritis, psoriasis, ankylosing spondylitis, juvenile idiopathic arthritis	pop [2],case
Adalimumab (Humira)	Monoclonal antibody against TNF-α	2008	Rheumatoid arthritis, Crohn's disease, ulcerative colitis, psoriasis, ankylosing spondylitis, juvenile idiopathic arthritis	pop [2],case
Certolizumab (Cimzia)	Monoclonal antibody against TNF-α	2008	Rheumatoid arthritis, Crohn's disease, psoriasis, ankylosing spondylitis	unk
Golimumab (Simponi)	Monoclonal antibody against TNF-α	2009	Rheumatoid arthritis, psoriasis, ankylosing spondylitis, ulcerative colitis	unk
Other agents				
Rituximab (Rituxan)	Monoclonal antibody against CD20 which is found on the surface of B cells	1997	Lymphoma, rheumatoid arthritis, chronic lymphocytic leukemia, granulomatosis with polyangiitis and microscopic polyangiitis	case
Anakinra (Kineret)	IL-1 receptor antagonist	2001	Rheumatoid arthritis	unk
Abatacept (Orencia)	CTLA-4 ligand, selective T-cell costimulation modulator	2005	Rheumatoid arthritis	unk
Ustekinumab (Stelara)	Monoclonal antibody against IL-12 and IL-23	2009	Psoriasis	unk
Tocilizumab (Actemra)	Monoclonal antibody IL-6 receptor	2010	Rheumatoid arthritis, juvenile idiopathic arthritis	case
Belimumab (Benlysta)	Monoclonal antibody against B-cell activating factor	2011	Systemic lupus erythematosus	unk
Tofacitinib (Xeljanz)	Janus kinase inhibitor (JAK1, 3, 2 inhibitor)	2012	Rheumatoid arthritis	case
Secukinumab (Cosentyx)	Monoclonal antibody against IL-17A	2015	Psoriasis	unk

(continued)

Table 2 (continued)

Anti-TNF agents	Mechanism (see Fig. 1)	Date of FDA approval	FDA-approved indications	NTM risk level of evidence
Ixekizumab (Taltz)	Monoclonal antibody against IL-17A	2016	Psoriasis	unk
Guselkumab (Tremfya)	Monoclonal antibody against IL-23	2017	Psoriasis	unk
Sarilumab (Kevzara)	Monoclonal antibody against IL-6 receptor	2017	Rheumatoid arthritis	unk
Brodalumab (Siliq)	Monoclonal antibody against IL-17 receptor A	2017	Psoriasis	unk

Abbreviations: *CTLA-4* Cytotoxic T-Lymphocyte Antigen-4, *FDA* US Food and Drug Administration, *NTM* Nontuberculous Mycobacteria, *IL* Interleukin, *TNF* Tumor Necrosis Factor (pop) population-based studies showing elevated risk; (case) case reports in the literature; (unk) unknown risk

HIV/AIDS

Starting in the early 1980s, coinciding with the developing HIV epidemic, disseminated MAC infections increased in the USA [40]. By the end of the decade disseminated, MAC infection affected up to 24% of all AIDS patients [41]. Disseminated MAC infection is an AIDS-defining illness, primarily associated with CD4+ counts below 50 cells/mm^3, although occurring at levels between 50 and 200 less frequently [40, 41]. In the USA, *M. kansasii* is also associated with disseminated NTM infection although in Oregon during 2007–2012, 95% of disseminated NTM was caused by MAC [7, 42]. With the introduction of highly active antiretroviral therapy (HAART) that suppresses virus and increases CD4+ counts, disseminated MAC has declined [43, 44]. The incidence of disseminated MAC in AIDS patients dramatically declined after the 1997 introduction of highly active antiretroviral therapy (HAART) [43, 44]. Data from Oregon in the modern era of HIV therapy estimated an annual incidence rate of disseminated NTM varying from 0.2 to 0.3/100,000 during 2005–2012 [3, 5]. The median annual incidence of disseminated NTM in HIV-positive patients in Oregon was 110/100,000 HIV person-years during 2007–2012 [42]. In the same study, incidence rates varied inversely with CD4+ counts, ranging from 10/100,000 HIV person-years over 200/cells/mm^3 to 5300/100,000 HIV person-years below 50 cells/mm^3. Several uncommon species such as *M. haemophilum* and *M. genavense* are important causes of disseminated disease in the setting of HIV, though much less common than MAC [15–17].

Pulmonary NTM disease in HIV-infected patients is less well characterized. Without appropriate diagnostic testing including liquid culture media, it is difficult to distinguish NTM from tuberculosis in tuberculosis-endemic countries. Overall in a Southeast Asia study of 1988 HIV-infected patients NTM isolations ($N = 85$) were similar to pulmonary *M. tuberculosis* isolations in frequency ($N = 124$), and the NTM disease prevalence in HIV-infected patients in Thailand and Vietnam was 2% [45]. Among the 85 HIV patients with NTM isolation, 9 were judged to have pulmonary disease and presented with similar disease characteristics as seen in the non-HIV setting. This study suggested that pulmonary NTM disease in the HIV setting is more strongly related to underlying lung disease, as it is in the non-HIV setting, rather than the HIV status of the patients. *M. kansasii* is an important cause of pulmonary NTM disease globally based on studies in South Africa and Southeast Asia [45, 46]. Over half of *M. kansasii* also infections are pulmonary in AIDS patients [7, 10].

Diagnosis, Prevention, and Treatment

Azithromycin or clarithromycin is effective prophylactic therapy, recommended when CD4+ counts dropped below 50 cells/mm^3 [47]. Currently, the recommended treatment regimen is once weekly 1200 mg of azithromycin for HIV-infected patients with CD4+ counts below 50 cells/mm^3 [10]. If HIV disease is well controlled with HAART, a 2002 updated recommendation allowed discontinuation of prophylactic therapy if CD4+ counts rise to 100 cells/mm^3 for at least 3 months [44].

Once a patient develops disseminated MAC, treatment includes antivirals to manage underlying HIV disease and treatment of NTM disease. The preferred recommended treatment for disseminated MAC is clarithromycin (azithromycin as alternative) and ethambutol and may include rifabutin although the benefits of the third drug are less well established [10]. If used, rifabutin should be adjusted for drug-drug interactions with antiretroviral drugs; for macrolide-resistant strains, aminoglycosides and a quinolone should be considered. Disseminated infections are generally treatable, though treatment should continue indefinitely, unless CD4+ cell counts are restored to over 100 cells/mm^3 [44].

In a study of treated patients with disseminated MAC, survival was similar to non-MAC matched controls, while untreated disease shortened the time to death [48]. Mortality was 10% in 19 hospitalized HIV-infected patients reported in a case series from a single hospital [49].

Immune reconstitution inflammatory syndrome (IRIS) occurs in a small subset of patients who initiate HAART while infected with NTM and other opportunistic infections. The period of immune reconstitution may result in host inflammatory responses that cause the presentation of MAC in patients with no prior diagnosis, or unusual presentations including lymphadenitis, musculoskeletal infections or abscesses, and endobronchial mass lesions [50]. Treatment of IRIS includes use of prednisone and rarely anti-TNF-targeted therapies [51, 52].

Immunosuppressive Therapies that Increase NTM Risk

Corticosteroids

Oral corticosteroids suppress inflammatory response and are used to treat chronic conditions such as asthma, COPD, and autoimmune diseases. There is no data to date describing the risk of extrapulmonary NTM infection in patients taking corticosteroids. However, in a single-site case series in Texas, *M. chelonae* disseminated and catheter-related infections were associated with high rates of corticosteroid use [53]. There is some population-based evidence supporting a significantly increased risk of respiratory infection, including pulmonary NTM disease, in patient taking oral corticosteroids. In COPD patients, the risks of pneumonia and tuberculosis are increased fivefold while taking corticosteroids [54], and patients may be at even higher risk of pulmonary NTM. Oral prednisone for at least 1 month prior to pulmonary MAC diagnosis was 8 times more common compared to controls in a case-control study in Oregon and Washington [55]. Brode et al. estimated an increased risk of 60% (aOR 1.6 (1.02–2.52)) for pulmonary NTM disease in older Canadian rheumatoid arthritis patients, comparing high dose vs noncurrent use of oral steroids [56].

To avoid the systemic, nonspecific immunosuppression of oral corticosteroids, inhaled corticosteroids may be used in patients with lung disease to deliver the drug locally with fewer side effects. In Denmark, the use of inhaled corticosteroids was associated with a 24-fold increase in pulmonary NTM in COPD patients [1]. In Japan, asthmatic NTM cases had a longer prior duration and higher dose of inhaled corticosteroid compared to controls in Japan [57]. In both studies with NTM outcomes, there was a clear dose response, with the highest risks at inhaled corticosteroid doses >800 mg fluticasone equivalent. If NTM are isolated, patients should be treated with the minimal necessary doses of oral or inhaled corticosteroids, monitored for symptoms consistent with NTM infection, and treated according to current ATS/IDSA guidelines [10].

Biologic Therapy

Vignette: Patient with Rheumatoid Arthritis

A 62-year-old female with rheumatoid arthritis on etanercept, methotrexate 25 mg/week, and prednisone 5 mg daily and chronic cough of 6-month duration. CT scan reveals bronchiectasis and right middle lobe infiltrate. No cavities. Sputum X 3 reveals *M. avium* on two of three samples. What to do? Consider therapy given underlying immunosuppression and higher risk of progression while using these therapies. Start azithromycin, ethambutol, and rifampin standard multidrug therapy, wean patient off prednisone if possible (strong risk factor for NTM), and discontinue etanercept. Attempt to manage patient on methotrexate and other non-biologic DMARDs as necessary. If rheumatoid arthritis is not manageable with non-biologics, then abatacept is probably the preferred biologic agent based on animal modeling with tuberculosis and very limited clinical trial data [58, 59].

Biologic therapies are used to treat autoimmune diseases such as rheumatoid arthritis, psoriasis, Crohn's disease, ankylosing spondylitis, and others. These medications target inflammatory pathways via monoclonal antibodies or genetically engineered proteins that bind cytokines or receptors to inhibit immune response. Administration is primarily by infusion or injection. As discussed in Chap. 4, there are multiple pathways involved in the immune response to NTM infection. In this section we describe biologic therapies that target these pathways, and the limited population-based evidence of an increased risk of NTM infection as well NTM cases described in the literature or that pose theoretical risks in patients taking these therapies.

The most commonly used biologic therapies are TNF-alpha inhibitors (see Table 2): the anti-TNF-alpha monoclonal antibodies infliximab, adalimumab, golimumab, and certolizumab and soluble receptor fusion protein etanercept [60]. There is a clear increase in risk of NTM infection with the use of anti-TNF therapy. In a US study of new infliximab, etanercept, or adalimumab anti-TNF therapy users with rheumatoid arthritis, rates of NTM disease were increased 5–10 times over unexposed rheumatoid arthritis patients [2]. NTM rates were higher than tuberculosis. In Canada, among seniors ≥67 years with rheumatoid arthritis, NTM cases had a twofold higher odds of current use of anti-TNF medication compared to controls [56]. In the US study, NTM risks varied by TNF-alpha inhibitor drug. The incidence rates were lowest for etanercept, 35 (95% confidence interval 1–69)/100,000 patient-years, and higher for infliximab, 116 (30–203)/100,000, and adalimumab, 122 (3–241)/100,000 [2]. The overall incidence of NTM disease among patients taking TNF antagonists in two South Korean hospital-based reviews was higher, 230/100,000 patients [61, 62].

Other biologic agents used to treat autoimmune diseases include rituximab (anti-CD20 monoclonal antibody), abatacept (T-cell costimulator modulator), tocilizumab (anti-IL-6 monoclonal antibody), ustekinumab (anti-IL-12 monoclonal antibody), anakinra (IL-1 receptor antagonist), and tofacitinib (Janus kinase [JAK] inhibitor, technically a synthetic DMARD) (see Table 2) [60]. There is limited safety data to quantify the risk of NTM infection while taking these drugs. To date, there is one NTM case series that included six patients, two treated with rituximab and four treated with tocilizumab (of whom three were also taking prednisolone +/− tacrolimus and two had prior NTM infection) [63, 64]. Two NTM infections have been reported during clinical trials of tofacitinib to date [65].

Prevention, Diagnosis, and Treatment

Reactivation of latent tuberculosis is a known risk of biologic therapy, and patients should be screened for tuberculosis prior to starting biologics. In addition to the increased risk of NTM from use of biologic therapies, patients with rheumatoid arthritis have an elevated risk of NTM due to associated underlying lung disease including bronchiectasis and interstitial lung disease [2, 13, 62, 66]. Unlike for tuberculosis, screening for NTM is difficult to conceptualize [10]. However, patients with symptoms consistent with pulmonary NTM (i.e., chronic cough with suspected infectious cause) should be considered for CT imaging and sputum culture including acid-fast bacillus testing. Patients with a history of pulmonary NTM should be monitored for recurrence or reinfection. In addition, clinicians should consider NTM with extrapulmonary infections, which accounted for 45% of NTM infections in the setting of TNF blockade reviewed in US Food and Drug Administration MedWatch reports [13]. Overall, just over half of NTM cases in patients taking biologics present as pulmonary disease [2, 13]. Similar to other immunosuppressed patients, around half of cases are caused by MAC, and isolation of rapid growers is common, but data is lacking at the population level [13].

Treatment is species-specific and should follow ATS/IDSA guidelines [10]. Careful consideration of decreasing or stopping biologic therapies should be made [27]. Switching to therapies with a theoretically lower risk, e.g., non-biologics such as methotrexate, should be considered [2]. Two case series from Japan suggest that stopping biologics is generally associated with positive outcomes or non-progression of pulmonary NTM disease [63, 67]. However, some patients were successfully managed on antibiotics while using biologics. High case fatality rates have been reported in pulmonary or extrapulmonary NTM cases taking anti-TNF biologics, ranging from <10% to 15% in US studies with short-term follow-up to 39% with a median of 569 days (range 21–2127) between diagnosis and death [2, 13]. In the Japanese case series, none of the deaths were related to NTM but rather the underlying lung disease [63, 67].

Solid Organ Transplants

Approximately 19,000 kidney, 7800 liver, 3200 heart, and 2300 lung transplants were performed in the USA in 2016 [68]. This represents a record high number of solid organ transplants for the fourth consecutive year. Organ transplantation requires lifelong treatment with immunosuppressive medications to prevent rejection. Maintenance drugs include calcineurin inhibitors (tacrolimus and cyclosporine), mammalian target of rapamycin (mTOR) inhibitor (sirolimus), prednisone, and others that are transplanted organ-dependent (e.g., azathioprine for lung transplants).

Transplant patients have a relatively low risk of NTM infection, though the risk is higher than the general population. The only incidence data to date were reported in 7395 non-lung solid organ transplant patients from a single medical center in South Korea over an 18-year period [69]. The highest incidence of NTM infection was in heart transplant recipients, 115/100,000 patient-years (95% CI 23–336). Lower rates were reported in kidney transplant recipients (28, 95% CI 11–58) and liver transplant recipients (15, 95% CI 3–43). Overall 81% of patients had pulmonary disease, and 59% isolated MAC. This is higher than in the USA, where MAC and *M. abscessus* occur at similar rates, based on reports from 3 institutions where each contributed to 45% of the 49 total NTM infections [70–72]. All other data on NTM infection after transplantation is from case reports and institutional case series. There are no data available to compare differences between transplant centers and limited data that describes risk factors for NTM infection.

Lung transplant patients are more likely to have pulmonary NTM infections. In a 2014 review of 293 solid organ transplant patients with NTM disease, 61% of 100 lung transplant, 26% of 45 heart transplant, 31% of 16 liver transplant, and 17% of 132 kidney transplant patients had pulmonary NTM [9]. Lung transplant patients are at increased risk for NTM infection compared to non-lung transplant (OR 13.6, 95% CI 4.2–43.8, adjusted for age), with a higher risk among bilateral lung transplant recipients [72]. Overall 79% of 19 lung transplant patients had pleuropulmonary infection. Other factors including malnutrition, cytomegalovirus disease, chronic renal insufficiency, acute rejection, cystic fibrosis, receipt of tacrolimus, and receipt of azithiopine were associated with NTM disease in unadjusted analysis. Environmental (water) exposure to NTM has been linked to hospital-associated infections. During an outbreak of 71 *M. abscessus* infections at a US tertiary care hospital, more than half of affected patients were lung transplant recipients [73]. The outbreak was identified after a new hospital wing opened, and the incidence of *M. abscessus* infection increased substantially during 2013–2014. During the period of elevated risk, 79% of infections were pulmonary. Based on the field investigation results, a sterile water protocol was implemented so that high-risk (e.g., lung and heart transplant) patients did not receive or use tap water. The incidence of *M. abscessus* cases among lung transplant patients decreased from 3.0 to 1.0 cases per 10,000 patient-days.

Prevention, Diagnosis, and Treatment

Pre- and posttransplant prophylaxis for infectious agents typically does not include antibiotics targeting NTM. The exception is lung transplant patients, in certain situations. In part due to their underlying lung disease, lung transplant patients are at higher risk of pulmonary NTM colonization and disease pretransplant. Currently most transplant centers do not treat colonized patients, but ATS/IDSA guidelines recommend treatment for pulmonary NTM disease [10, 70]. Globally, many screening centers do not have written policies for NTM screening pre-lung transplant, but screening CF patients is a common practice and recommended by the US Cystic Fibrosis Foundation and European Cystic Fibrosis Society [74, 75]. Pretransplantation culture results from 145 patients at a single institution identified 2 (1.4%) with NTM disease. One continued to isolate despite therapy, and the other isolated MAC but was not treated. Prior pulmonary *M. abscessus* infection in patients with cystic fibrosis is associated with mycobacterial infection posttransplant. There are reports of successful local control and clearance if disease recurs [76]. As a result, the US Cystic Fibrosis Foundation and European Cystic Fibrosis Society consensus recommendations [74] state that (1) the presence of positive NTM cultures should not preclude transplant consideration, if the center has experience with NTM disease; (2) patients with NTM disease should be treated prior to transplant listing; and (3) the presence of *M.* abscessus or *M.* avium complex after therapy is not an absolute contraindication to lung transplant. The results of the previously described *M.*

Table 3 Outcomes after NTM infection in patients who received solid organ transplants

Transplanted organ	N	Resolved	Improvement	Relapse or long-term	Non-NTM death	NTM deaths	NTM death notes	Refs.
Kidney	94[a]	41 (44%)	Not reported	30 (32%)	1(1%)	3(3%)	2 disseminated, 1 pulmonary	[77]
Heart	34	11 (32%)	Not reported	Not reported	9 (26%)	0		[77]
Lung	37[b]	16 (43%)	8 (22%)	10 (27%)		2 (5%)	2 disseminated	[70, 71, 77]

[a]19 (25%) outcome not reported
[b]1 unknown outcome

abscessus outbreak also confirm that tap water avoidance is one method to decrease the risk of NTM in transplant patients.

Treatment is complicated by drug interactions and the limited ability to reduce immunosuppression to improve immune function for effective clearance or control of NTM disease. There is a risk of organ graft-versus-host disease, demonstrated in case studies of patients who decreased immunosuppressive medications [9, 77]. Drug interactions between rifamycin, macrolide, and aminoglycoside antibiotics and antirejection calcineurin inhibitors (cyclosporine, sirolimus) and tacrolimus impact the selection of appropriate antibiotics [77, 78]. Outcomes have been summarized in Table 3 for kidney, heart, and lung transplant patients presented in three reviews. Outcomes vary by transplanted organ, but patients generally respond to treatment and clear infection or see clinical improvement. Just under 1/3 relapse or have long-term infections. Very few NTM-related deaths have been reported. In addition, among 16 transplant patients treated for NTM infection in South Korea, 11 (69%) had a "favorable response," and 5 (31%) had an "unfavorable response," and 5 additional patients with NTM cultures but minimal or no radiographic disease remained untreated [69].

Hematopoietic Stem Cell Transplantation

Increasing from 200 in 1980 to 20,000 in 2010, hematopoietic stem cell transplantation (HSCT) is used primarily to treat hematologic malignancies (i.e., leukemia, multiple myeloma) [79]. HSCT leaves recipients severely immunocompromised over a period of many months to prevent rejection of the transplanted cells. Prior to transplant there is a conditioning phase that involves high-dose chemotherapy or radiation. All patients receive prophylactic antiviral and antifungal therapy, and febrile neutropenia is immediately treated with a broad-spectrum antibiotic. Risk continues after the transplant, during the phase of immune reconstitution.

There is very little data to describe the epidemiology of NTM in HSCT patients. In South Korea, the incidence of NTM infection was 258.7 (95% CI 118.3–491.1) per 100,000 patient-years, higher than solid organ transplant patients [69]. In a 2004 review of NTM-infected HSCT cases reported in the literature, posttransplant graft-versus-host disease was present in 46% of NTM cases, with the NTM risk likely associated with the increase in immunosuppressive medications [77]. The review identified catheter and blood infections as the most common presentations, although pulmonary NTM may account for 20–30%. MAC, *M. abscessus/chelonae* complex, and *M. haemophilum* were the most common species associated with skin and catheter infections. Most *M. haemophilum* infections are associated with skin and catheter infection, but patients receiving bone marrow transplants may also develop pulmonary infections [80–83]. Overall, treatment for such infections is successful especially with the removal of catheters. Treatment for *M. haemophilum* and other rapid-growing bacteria may include surgical removal of the catheter and surrounding tissue and require prolonged antibacterial therapy [10]. Case fatality rates as high as 20% have been reported, based on reviews of 94 and 571 HSCT patients [77, 84].

Solid Tumors and Hematologic Malignancies

Patients with solid tumors and hematologic malignancies have underlying cellular immune dysfunction, chemotherapy-associated immunosuppression, and frequently intravenous catheters that provide access for infections and increase the risk of NTM disease [85–87]. Some of the earliest recognition of disseminated NTM disease in immunosuppressed patients occurred in the early 1980s with the identification of cases associated with hairy cell leukemia [88, 89]. Within a single institution, the incidence of disseminated NTM disease in patients with hairy cell leukemia was 5% (9/186) [90]. Institutional case series have also described NTM disease in other cancer patients. In Taiwan, around 0.3% of 2846 patients with hematologic malignancies developed disseminated NTM disease while 0.9% developed pulmonary NTM disease [85]. Overall in that study, 38% of patients isolated *MAC*, 21% *M. abscessus*, 18% *M. fortuitum*, and 18% *M. kansasii*. In a Texas study limited to 116 cancer patients with rapid-growing mycobacteria isolates, *M. mucogenicum* (39%) was the most common species identified, with 21% isolating *M. fortuitum*, 14% *M. abscessus,* and 12% *M. chelonae*. Additionally, the airway destruction and damage caused by lung tumors is a likely cause of the increased risk of pulmonary NTM disease. Similar to the HSCT setting, the removal of catheters and antibiotic therapy are associated with good treatment response to infection, although in Taiwan 14% of patients died within 30 days of diagnosis (all with disseminated disease) [12, 85].

Conclusion

In conclusion, immunosuppressed and immunocompromised patients are at increased risk for NTM disease. Extrapulmonary infections, particularly disseminated disease, are more common in the immunosuppressed setting and frequently associated with rapid-growing species, although MAC remains the leading cause of NTM disease in immunosuppressed patients. There is a low threshold to start anti-mycobacterial therapy, and interventions should include removal of the source of infection (catheter) or other hardware potentially harboring NTM and modifying immunosuppressive therapies whenever possible (i.e., minimize immunosuppression). For pulmonary NTM disease, in contrast to the immunocompetent hosts where improved lung hygiene and close observation may be appropriate for some patients, species-specific therapy adhering to current ATS/IDSA guidelines should be considered for nearly all immunosuppressed patients who develop pulmonary disease. However, caution is necessary to monitor interactions between therapy and immunosuppressive agents used in transplant patients.

References

1. Andrejak C, Nielsen R, Thomsen VO, Duhaut P, Sorensen HT, Thomsen RW. Chronic respiratory disease, inhaled corticosteroids and risk of non-tuberculous mycobacteriosis. Thorax. 2013;68:256–62.
2. Winthrop KL, Baxter R, Liu L, et al. Mycobacterial diseases and antitumour necrosis factor therapy in USA. Ann Rheum Dis. 2013;72:37–42.
3. Cassidy PM, Hedberg K, Saulson A, McNelly E, Winthrop KL. Nontuberculous mycobacterial disease prevalence and risk factors: a changing epidemiology. Clin Infect Dis. 2009;49:e124–9.
4. Henkle E, Hedberg K, Schafer S, Novosad S, Winthrop KL. Population-based incidence of pulmonary Nontuberculous mycobacterial disease in Oregon 2007 to 2012. Ann Am Thorac Soc. 2015;12:642–7.
5. Henkle E, Hedberg K, Schafer SD, Winthrop KL. Surveillance of Extrapulmonary Nontuberculous mycobacteria infections, Oregon, USA, 2007-2012. Emerg Infect Dis. 2017;23:1627–30.
6. Position Statement 17-ID-07: Standardized Case Definition for Extrapulmonary Nontuberculous Mycobacteria Infections. Council of State and Territorial Epidemiologists. Available at http://c.ymcdn.com/sites/www.cste.org/resource/resmgr/2017PS/2017PSFinal/17-ID-07.pdf, Last accessed 17 Nov 2017.
7. Jones D, Havlir DV. Nontuberculous mycobacteria in the HIV infected patient. Clin Chest Med. 2002;23:665–74.
8. Winthrop KL, Baddley JW, Chen L, et al. Association between the initiation of anti-tumor necrosis factor therapy and the risk of herpes zoster. JAMA. 2013;309:887–95.
9. Knoll BM. Update on nontuberculous mycobacterial infections in solid organ and hematopoietic stem cell transplant recipients. Curr Infect Dis Rep. 2014;16:421.
10. Griffith DE, Aksamit T, Brown-Elliott BA, et al. An official ATS/IDSA statement: diagnosis, treatment, and prevention of nontuberculous mycobacterial diseases. Am J Respir Crit Care Med. 2007;175:367–416.
11. Lewis FM, Marsh BJ, von Reyn CF. Fish tank exposure and cutaneous infections due to Mycobacterium marinum: tuberculin skin testing, treatment, and prevention. Clin Infect Dis. 2003;37:390–7.
12. El Helou G, Hachem R, Viola GM, et al. Management of rapidly growing mycobacterial bacteremia in cancer patients. Clin Infect Dis. 2013;56:843–6.
13. Winthrop KL, Chang E, Yamashita S, Iademarco MF, LoBue PA. Nontuberculous mycobacteria infections and anti-tumor necrosis factor-alpha therapy. Emerg Infect Dis. 2009;15:1556–61.
14. Wentworth AB, Drage LA, Wengenack NL, Wilson JW, Lohse CM. Increased incidence of cutaneous nontuberculous mycobacterial infection, 1980 to 2009: a population-based study. Mayo Clin Proc. 2013;88:38–45.

15. Doherty T, Lynn M, Cavazza A, Sames E, Hughes R. Mycobacterium haemophilum as the initial presentation of a B-cell lymphoma in a liver transplant patient. Case Rep Rheumatol. 2014;2014:742978.
16. Ducharlet K, Murphy C, Tan SJ, et al. Recurrent Mycobacterium haemophilum in a renal transplant recipient. Nephrology. 2014;19(Suppl 1):14–7.
17. Lhuillier E, Brugiere O, Veziris N, et al. Relapsing Mycobacterium genavense infection as a cause of late death in a lung transplant recipient: case report and review of the literature. Exp Clin Transplant. 2012;10:618–20.
18. Turner RD, Chiu C, Churchyard GJ, et al. Tuberculosis infectiousness and host susceptibility. J Infect Dis. 2017;216:S636–S43.
19. Cooper AM. Cell-mediated immune responses in tuberculosis. Annu Rev Immunol. 2009;27:393–422.
20. Stenger S, Modlin RL. T cell mediated immunity to Mycobacterium tuberculosis. Curr Opin Microbiol. 1999;2:89–93.
21. Chan ED, Bai X, Kartalija M, Orme IM, Ordway DJ. Host immune response to rapidly growing mycobacteria, an emerging cause of chronic lung disease. Am J Respir Cell Mol Biol. 2010;43:387–93.
22. Lake MA, Ambrose LR, Lipman MC, Lowe DM. "Why me, why now?" Using clinical immunology and epidemiology to explain who gets nontuberculous mycobacterial infection. BMC Med. 2016;14:54.
23. Wu UI, Holland SM. Host susceptibility to non-tuberculous mycobacterial infections. Lancet Infect Dis. 2015;15:968–80.
24. Ehlers S. Role of tumour necrosis factor (TNF) in host defence against tuberculosis: implications for immunotherapies targeting TNF. Ann Rheum Dis. 2003;62(Suppl 2):ii37–42.
25. Zumla A, James DG. Granulomatous infections: etiology and classification. Clin Infect Dis. 1996;23:146–58.
26. Gardam MA, Keystone EC, Menzies R, et al. Anti-tumour necrosis factor agents and tuberculosis risk: mechanisms of action and clinical management. Lancet Infect Dis. 2003;3:148–55.
27. Winthrop KL, Chiller T. Preventing and treating biologic-associated opportunistic infections. Nat Rev Rheumatol. 2009;5:405–10.
28. Mirsaeidi M, Machado RF, Garcia JG, Schraufnagel DE. Nontuberculous mycobacterial disease mortality in the United States, 1999-2010: a population-based comparative study. PLoS One. 2014;9:e91879.
29. Lee WI, Huang JL, Yeh KW, et al. Immune defects in active mycobacterial diseases in patients with primary immunodeficiency diseases (PIDs). J Formos Med Assoc. 2011;110:750–8.
30. Sexton P, Harrison AC. Susceptibility to nontuberculous mycobacterial lung disease. Eur Respir J. 2008;31:1322–33.
31. Bustamante J, Boisson-Dupuis S, Abel L, Casanova JL. Mendelian susceptibility to mycobacterial disease: genetic, immunological, and clinical features of inborn errors of IFN-gamma immunity. Semin Immunol. 2014;26:454–70.
32. Haverkamp MH, van de Vosse E, van Dissel JT. Nontuberculous mycobacterial infections in children with inborn errors of the immune system. J Infect. 2014;68(Suppl 1):S134–50.
33. Phoompoung P, Ankasekwinai N, Pithukpakorn M, et al. Factors associated with acquired anti IFN- gamma autoantibody in patients with nontuberculous mycobacterial infection. PLoS One. 2017;12:e0176342.
34. Browne SK, Burbelo PD, Chetchotisakd P, et al. Adult-onset immunodeficiency in Thailand and Taiwan. N Engl J Med. 2012;367:725–34.
35. Browne SK, Zaman R, Sampaio EP, et al. Anti-CD20 (rituximab) therapy for anti-IFN-gamma autoantibody-associated nontuberculous mycobacterial infection. Blood. 2012;119:3933–9.
36. Czaja CA, Merkel PA, Chan ED, et al. Rituximab as successful adjunct treatment in a patient with disseminated nontuberculous mycobacterial infection due to acquired anti-interferon-gamma autoantibody. Clin Infect Dis. 2014;58:e115–8.

37. Vinh DC, Patel SY, Uzel G, et al. Autosomal dominant and sporadic monocytopenia with susceptibility to mycobacteria, fungi, papillomaviruses, and myelodysplasia. Blood. 2010;115:1519–29.
38. Hsu AP, Sampaio EP, Khan J, et al. Mutations in GATA2 are associated with the autosomal dominant and sporadic monocytopenia and mycobacterial infection (MonoMAC) syndrome. Blood. 2011;118:2653–5.
39. Hsu AP, McReynolds LJ, Holland SM. GATA2 deficiency. Curr Opin Allergy Clin Immunol. 2015;15:104–9.
40. Horsburgh CR Jr. Mycobacterium avium complex infection in the acquired immunodeficiency syndrome. N Engl J Med. 1991;324:1332–8.
41. Havlik JA Jr, Horsburgh CR Jr, Metchock B, Williams PP, Fann SA, Thompson SE 3rd. Disseminated Mycobacterium avium complex infection: clinical identification and epidemiologic trends. J Infect Dis. 1992;165:577–80.
42. Varley CD, Ku JH, Henkle E, Schafer SD, Winthrop KL. Disseminated Nontuberculous mycobacteria in HIV-infected patients, Oregon, USA, 2007-2012. Emerg Infect Dis. 2017;23:533–5.
43. Kaplan JE, Hanson D, Dworkin MS, et al. Epidemiology of human immunodeficiency virus-associated opportunistic infections in the United States in the era of highly active antiretroviral therapy. Clin Infect Dis. 2000;30(Suppl 1):S5–14.
44. Kaplan JE, Masur H, Holmes KK, Usphs, Infectious Disease Society of A. Guidelines for preventing opportunistic infections among HIV-infected persons--2002. Recommendations of the U.S. Public Health Service and the Infectious Diseases Society of America. MMWR Recomm Rep. 2002;51:1–52.
45. McCarthy KD, Cain KP, Winthrop KL, et al. Nontuberculous mycobacterial disease in patients with HIV in Southeast Asia. Am J Respir Crit Care Med. 2012;185:981–8.
46. Corbett EL, Blumberg L, Churchyard GJ, et al. Nontuberculous mycobacteria: defining disease in a prospective cohort of south African miners. Am J Respir Crit Care Med. 1999;160:15–21.
47. Kaplan JE, Masur H, Holmes KK, et al. USPHS/IDSA guidelines for the prevention of opportunistic infections in persons infected with human immunodeficiency virus: an overview. USPHS/IDSA prevention of opportunistic infections working group. Clin Infect Dis. 1995;21(Suppl 1):S12–31.
48. Horsburgh CR Jr, Havlik JA, Ellis DA, et al. Survival of patients with acquired immune deficiency syndrome and disseminated Mycobacterium avium complex infection with and without antimycobacterial chemotherapy. Am Rev Respir Dis. 1991;144:557–9.
49. Miguez-Burbano MJ, Flores M, Ashkin D, et al. Non-tuberculous mycobacteria disease as a cause of hospitalization in HIV-infected subjects. Int J Infect Dis. 2006;10:47–55.
50. Lawn SD, Bekker LG, Miller RF. Immune reconstitution disease associated with mycobacterial infections in HIV-infected individuals receiving antiretrovirals. Lancet Infect Dis. 2005;5:361–73.
51. Sitapati AM, Kao CL, Cachay ER, Masoumi H, Wallis RS, Mathews WC. Treatment of HIV-related inflammatory cerebral cryptococcoma with adalimumab. Clin Infect Dis. 2010;50:e7–10.
52. Meintjes G, Scriven J, Marais S. Management of the immune reconstitution inflammatory syndrome. Curr HIV/AIDS Rep. 2012;9:238–50.
53. Wallace RJ Jr, Brown BA, Onyi GO. Skin, soft tissue, and bone infections due to Mycobacterium chelonae chelonae: importance of prior corticosteroid therapy, frequency of disseminated infections, and resistance to oral antimicrobials other than clarithromycin. J Infect Dis. 1992;166:405–12.
54. Jick SS, Lieberman ES, Rahman MU, Choi HK. Glucocorticoid use, other associated factors, and the risk of tuberculosis. Arthritis Rheum. 2006;55:19–26.
55. Dirac MA, Horan KL, Doody DR, et al. Environment or host?: A case-control study of risk factors for Mycobacterium avium complex lung disease. Am J Respir Crit Care Med. 2012;186:684–91.

56. Brode SK, Jamieson FB, Ng R, et al. Increased risk of mycobacterial infections associated with anti-rheumatic medications. Thorax. 2015;70:677–82.
57. Hojo M, Iikura M, Hirano S, Sugiyama H, Kobayashi N, Kudo K. Increased risk of nontuberculous mycobacterial infection in asthmatic patients using long-term inhaled corticosteroid therapy. Respirology. 2012;17:185–90.
58. Bigbee CL, Gonchoroff DG, Vratsanos G, Nadler SG, Haggerty HG, Flynn JL. Abatacept treatment does not exacerbate chronic Mycobacterium tuberculosis infection in mice. Arthritis Rheum. 2007;56:2557–65.
59. Schiff M, Keiserman M, Codding C, et al. Efficacy and safety of abatacept or infliximab vs placebo in ATTEST: a phase III, multi-centre, randomised, double-blind, placebo-controlled study in patients with rheumatoid arthritis and an inadequate response to methotrexate. Ann Rheum Dis. 2008;67:1096–103.
60. Novosad SA, Winthrop KL. Beyond tumor necrosis factor inhibition: the expanding pipeline of biologic therapies for inflammatory diseases and their associated infectious sequelae. Clin Infect Dis. 2014;58:1587–98.
61. Lee SK, Kim SY, Kim EY, et al. Mycobacterial infections in patients treated with tumor necrosis factor antagonists in South Korea. Lung. 2013;191:565–71.
62. Yoo JW, Jo KW, Kang BH, et al. Mycobacterial diseases developed during anti-tumour necrosis factor-alpha therapy. Eur Respir J. 2014;44(5):1289–95.
63. Mori S, Tokuda H, Sakai F, et al. Radiological features and therapeutic responses of pulmonary nontuberculous mycobacterial disease in rheumatoid arthritis patients receiving biological agents: a retrospective multicenter study in Japan. Mod Rheumatol. 2012;22:727–37.
64. Lutt JR, Pisculli ML, Weinblatt ME, Deodhar A, Winthrop KL. Severe nontuberculous mycobacterial infection in 2 patients receiving rituximab for refractory myositis. J Rheumatol. 2008;35:1683–5.
65. Cohen SB, Tanaka Y, Mariette X, et al. Long-term safety of tofacitinib for the treatment of rheumatoid arthritis up to 8.5 years: integrated analysis of data from the global clinical trials. Ann Rheum Dis. 2017;76:1253–62.
66. Cortet B, Flipo RM, Remy-Jardin M, et al. Use of high resolution computed tomography of the lungs in patients with rheumatoid arthritis. Ann Rheum Dis. 1995;54:815–9.
67. Yamakawa H, Takayanagi N, Ishiguro T, Kanauchi T, Hoshi T, Sugita Y. Clinical investigation of nontuberculous mycobacterial lung disease in Japanese patients with rheumatoid arthritis receiving biologic therapy. J Rheumatol. 2013;40:1994–2000.
68. U.S. Department of Health & Human Services Organ Procurement and Transplantation Network. Transplant: Organ by Transplant Year (2015–2061). Found at https://optn.transplant.hrsa.gov/data/view-data-reports/build-advanced/. Last accessed 11/6/2017.
69. Yoo JW, Jo KW, Kim SH, et al. Incidence, characteristics, and treatment outcomes of mycobacterial diseases in transplant recipients. Transpl Int. 2016;29:549–58.
70. Huang HC, Weigt SS, Derhovanessian A, et al. Non-tuberculous mycobacterium infection after lung transplantation is associated with increased mortality. J Heart Lung Transplant. 2011;30:790–8.
71. Knoll BM, Kappagoda S, Gill RR, et al. Non-tuberculous mycobacterial infection among lung transplant recipients: a 15-year cohort study. Transpl Infect Dis. 2012;14:452–60.
72. Longworth SA, Vinnard C, Lee I, Sims KD, Barton TD, Blumberg EA. Risk factors for nontuberculous mycobacterial infections in solid organ transplant recipients: a case-control study. Transpl Infect Dis. 2014;16:76–83.
73. Baker AW, Lewis SS, Alexander BD, et al. Two-phase hospital-associated outbreak of Mycobacterium abscessus: investigation and mitigation. Clin Infect Dis. 2017;64:902–11.
74. Floto RA, Olivier KN, Saiman L, et al. US Cystic Fibrosis Foundation and European cystic fibrosis society consensus recommendations for the management of non-tuberculous mycobacteria in individuals with cystic fibrosis: executive summary. Thorax. 2016;71:88–90.
75. Brodhie M. et al. Journal of Heart and Lung Transplantation April 2017 [abstract] NonTuberculous Mycobacteria Infection and Lung Transplantation in Cystic Fibrosis: A Worldwide Survey of Clinical Practice.

76. Lobo LJ, Chang LC, Esther CR Jr, Gilligan PH, Tulu Z, Noone PG. Lung transplant outcomes in cystic fibrosis patients with pre-operative Mycobacterium abscessus respiratory infections. Clin Transpl. 2013;27:523–9.
77. Doucette K, Fishman JA. Nontuberculous mycobacterial infection in hematopoietic stem cell and solid organ transplant recipients. Clin Infect Dis. 2004;38:1428–39.
78. Daley CL. Nontuberculous mycobacterial disease in transplant recipients: early diagnosis and treatment. Curr Opin Organ Transplant. 2009;14:619–24.
79. Pasquini MC, Wang Z. Current use and outcome of hematopoietic stem cell transplantation: CIBMTR Summary Slides. (2013). Available at: http://www.cibmtr.org.
80. Busam KJ, Kiehn TE, Salob SP, Myskowski PL. Histologic reactions to cutaneous infections by Mycobacterium haemophilum. Am J Surg Pathol. 1999;23:1379–85.
81. Kiehn TE, White M, Pursell KJ, et al. A cluster of four cases of Mycobacterium haemophilum infection. Eur J Clin Microbiol Infect Dis. 1993;12:114–8.
82. Straus WL, Ostroff SM, Jernigan DB, et al. Clinical and epidemiologic characteristics of Mycobacterium haemophilum, an emerging pathogen in immunocompromised patients. Ann Intern Med. 1994;120:118–25.
83. White MH, Papadopoulos EB, Small TN, Kiehn TE, Armstrong D. Mycobacterium haemophilum infections in bone marrow transplant recipients. Transplantation. 1995;60:957–60.
84. Weinstock DM, Feinstein MB, Sepkowitz KA, Jakubowski A. High rates of infection and colonization by nontuberculous mycobacteria after allogeneic hematopoietic stem cell transplantation. Bone Marrow Transplant. 2003;31:1015–21.
85. Chen CY, Sheng WH, Lai CC, et al. Mycobacterial infections in adult patients with hematological malignancy. Eur J Clin Microbiol Infect Dis. 2012;31:1059–66.
86. Feld R, Bodey GP, Groschel D. Mycobacteriosis in patients with malignant disease. Arch Intern Med. 1976;136:67–70.
87. Rolston KV, Jones PG, Fainstein V, Bodey GP. Pulmonary disease caused by rapidly growing mycobacteria in patients with cancer. Chest. 1985;87:503–6.
88. Gallo JH, Young GA, Forrest PR, Vincent PC, Jennis F. Disseminated atypical mycobacterial infection in hairy cell leukemia. Pathology. 1983;15:241–5.
89. Weinstein RA, Golomb HM, Grumet G, Gelmann E, Schechter GP. Hairy cell leukemia: association with disseminated atypical mycobacterial infection. Cancer. 1981;48:380–3.
90. Bennett C, Vardiman J, Golomb H. Disseminated atypical mycobacterial infection in patients with hairy cell leukemia. Am J Med. 1986;80:891–6.

Environmental Niches for NTM and Their Impact on NTM Disease

Leah Lande

The prevalence of lung infection from nontuberculous mycobacteria (NTM) has been increasing throughout the world in recent years [1, 2]. This increase in disease prevalence is likely due to a combination of improved microbiological detection techniques, evolving host factors, and an increase in human exposure to NTM in the environment. NTM are ubiquitous in the environment and have been found in multiple different natural and man-made water sources and soils (Table 1). Individuals are likely contracting infection from inhalation of water and soil or dust aerosols and, in select cases, from ingestion of drinking water with subsequent reflux and

Table 1 Environments in which NTM have been isolated [3–9]

Natural water sources	Man-made water sources	Aerosols	Non-water sources
Streams	Drinking water pipelines	Showers	Dusts from natural soils
		Hot tubs	
Rivers	Water tanks – Hot and cold	Humidifiers	
Lakes		Indoor swimming pools	
Seawater	Hot tubs and indoor pools		Dusts from potting soils and peat moss
	Residential plumbing/faucets/showerheads	Operating rooms (heater-cooler units)	
	Hospital plumbing/faucets		
	Ice machines and commercial ice		House dust
	Bottled water		

L. Lande
Division of Pulmonary and Critical Care Medicine, Lankenau Medical Center, Wynnewood, PA, USA

Lankenau Institute for Medical Research, Wynnewood, PA, USA
e-mail: LandeL@MLHS.ORG

D. E. Griffith (ed.), *Nontuberculous Mycobacterial Disease*, Respiratory Medicine,
https://doi.org/10.1007/978-3-319-93473-0_6

pulmonary aspiration. Given the existence of multiple potential exposure sources for each individual, as well as variable periods of time between exposure, onset of symptoms, and diagnosis, it is very difficult to ascertain which specific exposure is the cause of disease in individual patients. Ongoing efforts to try to delineate environmental exposure sources are vital to disease prevention in high-risk individuals and improved disease control in NTM patients.

Environments of concern for NTM transmission include those that have proven capacity for NTM growth, those to whom humans are exposed on a regular basis, and those capable of aerosolization of particles into respirable size. These include drinking water pipelines [10–12], sink faucets [13–15], showers [16, 17], hot tubs [18], humidifiers [19], CPAP machines [20], indoor swimming pools [21], natural and commercially purchased garden soils [22, 23], and house dust [24, 25].

NTM in Water and Plumbing Biofilms

NTM flourish in municipal water systems because they are resistant to most disinfectants, including chlorine and chloramine, and they can survive and proliferate with low levels of nutrients due to their slow growth [19]. In addition, they are associated with water turbidity and higher organic carbon concentration [26]. NTM attach to particulate matter within water and to plumbing biofilms, colonizing the inner surfaces of water pipes and faucets [27, 28]. The survival of NTM in water distribution systems is dependent upon a complex interaction between the water source, transit time, stagnation, pipe surface, nutrient levels, temperature, turbidity, and disinfectant type and residual [29]. In a pilot distribution system, *M. avium* failed to grow at organic carbon concentrations of <50 mg/L, suggesting that reduction in organic matter content might serve to reduce NTM numbers [30].

The distance from the water treatment plant may be an important factor in NTM numbers. Free chlorine concentrations gradually decrease as water travels down the distribution system. Experiments have shown that free chlorine concentrations of 1.0 mg/L eliminated 100,000 colony-forming units (cfu) of the mycobacterial strains tested within 8 h of exposure, whereas a concentration of 0.15 mg/L had virtually no bactericidal effect [31]. Higher mycobacterial numbers have been found in distal distribution samples (average 25,000 fold) than those collected immediately downstream from treatment plants, indicating that mycobacteria actively grow within the distribution system [26]. A study in Massachusetts found that communities were more likely to have patients with *Mycobacterium avium* complex (MAC) isolates if they lived further away from water treatment plants and in more densely populated areas [32].

Pipe composition, size, configuration, and corrosion may also be important in the persistence of NTM. Smaller pipes have a greater surface area for biofilms to form and are usually present at more distal points of the system where complex bends occur [10]. During pressure transients at points of turbulence such as the bends in pipes, sloughing of biofilms can occur. Controlling corrosion on pipes may enable enhanced disinfection by chlorine and better control of biofilm formation [26]. With

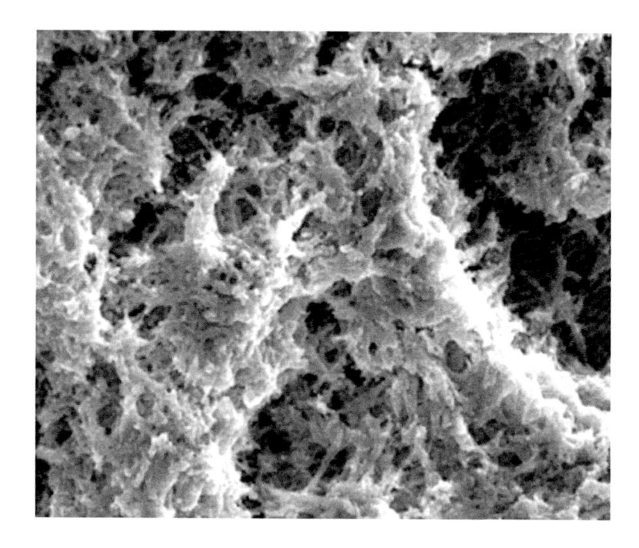

Fig. 1 Scanning electron microscopy of NTM biofilm with presence of extracellular matrix [34]

regard to pipe composition, studies have been conflicting. Some pipe materials, such as iron, have been shown to be more susceptible to *M. avium* biofilm formation [33]. *M. avium* biofilm levels have been shown to be higher on pipes made from iron and galvanized steel than on pipes made from copper or PVC. Furthermore, the biofilm formation was able to be better controlled by free chlorine on noncorroded copper or PVC pipes, with 3–4 logs/cm^2 lower biofilm levels on copper pipe surfaces than on iron pipe surfaces. Monochloramine controlled biofilm formation better on corroded iron pipes, but chlorine controlled biofilm formation better on copper pipes, demonstrating the complex interaction between disinfectant type and pipe surface conditions [30]. In a different study, asbestos cement or PVC pipes seemed to yield the highest growth of mycobacteria [10]. Innovative pipe materials with greater strength and resistance to microbial biofilm attachment should be developed (Fig. 1).

The effects of chemical disinfection on mycobacteria in water are not straightforward. As disinfectants kill off competition from other bacteria, NTM are protected by their lipid-rich outer membrane. This was best shown in a study of hot tubs in Australia, in which the water bacterial load excluding NTM was 1 cfu/mL and the NTM concentration was as high as $4 \times 10^3 - 4 \times 10^4$ cfu/mL [35].

Showerhead biofilms as well as the pipes leading to the showerhead are highly enriched in NTM [29, 36]. Showerhead biofilms have >100-fold more NTM organisms above background water contents, with *M. avium* being the most prevalent species [17]. The genotypes of *M. avium* found in showerheads have been shown to be identical to those found in the respiratory tract of patients [37, 38]. Unfortunately, drying out or flushing of showerheads is ineffective in reducing NTM numbers. Allowing showerheads or sections of shower tubing to dry out in between uses does not decrease the concentration of NTM in the biofilm, once a day flushing does not provide sufficient shearing force to remove biofilm or prevent its formation, and temporal water usage has no effect on showerhead NTM colonization [39]. The impact of regular flushing of water systems with high velocities with elevated disinfectant levels, with a biofilm disrupting compound such as nitrous oxide, or with superheated water is unknown.

Most filters do not prevent the passage of NTM. A pore size of 0.2 micrometer (μm) or smaller is required to prevent passage of NTM. Although effective in removing chlorine, metals, and organic compounds that can impart bad tastes to water, granular-activated charcoal or carbon (GAC) filters do not filter out NTM as their pore size is too big. In fact, GAC filters can promote the growth of NTM, as the mycobacteria attach and grow on the carbon-bound organics and metals that are trapped by these filters [40, 41].

Role of Free-Living Amoebae in the Ecology of NTM

Free-living amoebae are protozoa ubiquitously found in water systems. They mainly feed on bacteria by phagocytosis, and have also been found to phagocytose mycobacteria, with prolonged persistence of the NTM within the amoeba without any detrimental effects to their own survival [42]. NTM have been shown to coexist within amoeba in water systems and may use amoebae as a vehicle for protection and even for replication [43]. In a year-round study of a drinking water network, 87.6% of recovered amoebal cultures carried high numbers of NTM (Fig. 2) [44].

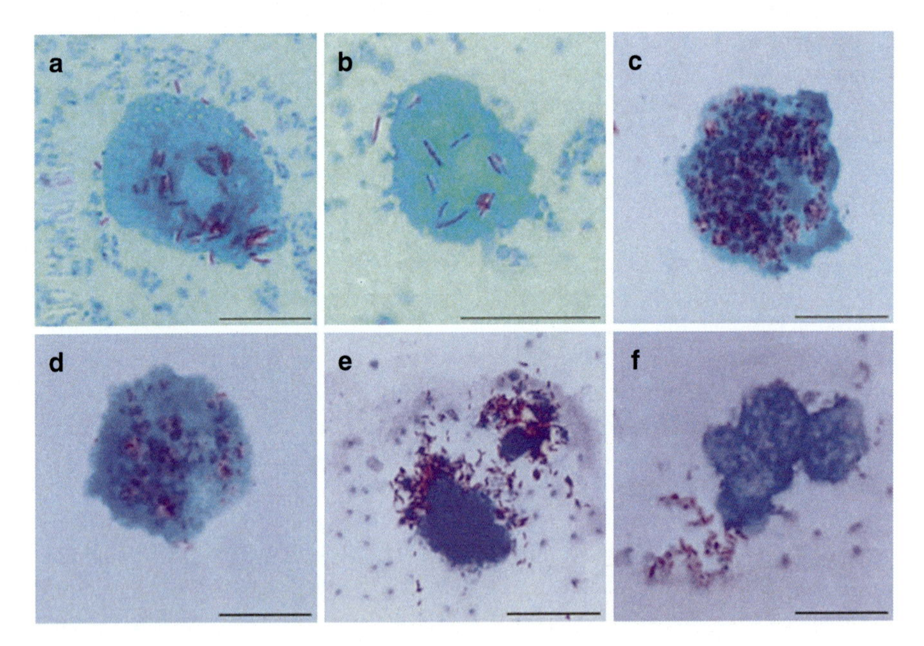

Fig. 2 Acid-fast staining revealing mycobacteria associated within amoebal isolates [44]. *Acanthamoeba* spp. (**a**, **b**), *Protacanthamoeba bohemica* (**c**, **d**), *Vermamoeba vermiformis* (**e**, **f**). (Scale bar represents 10 μm)

Acanthamoeba cysts have been shown to resist inactivation by chlorine and other treatments used to sanitize drinking water [45]. Encysted amoeba may protect NTM from being killed during various disinfection processes [46]. It has been suggested that the ability of the NTM to grow in human macrophages is a consequence of "training" in protozoa and amoebae. Furthermore, *M. avium* cells grown in amoebae have been shown to be more virulent [47].

NTM and Temperature

NTM are relatively heat resistant, with optimal growth between 28 and 37 °C (82–98.6 °F) [26]. Due to their preference for warm water, NTM have been detected in the air or water from 72% of hot tub or therapy pool sites tested [6]. This heat resistance of NTM relative to other drinking water microbes very likely contributes to the high numbers of NTM in apartment and condominium buildings or hospitals with recirculating hot water systems [48, 49]. The heat killing of mycobacteria is dependent upon absolute temperature, the time spent at a given temperature, cell density, sample volume, the pipe material upon which NTM biofilm is attached, and the level of nutrients within the material to which the NTM is attached [30]. In addition, different NTM have different heat susceptibility. For example, colony counts of *M. avium* are reduced at 52–55 °C, whereas *M. chelonae* and *M. xenopi* require temperatures of greater than 60 °C to reliably decrease colony counts [50]. In a study of milk pasteurization, it was shown that *M. avium*, *M. intracellulare*, and *M. kansasii* all survived exposure to a temperature of 63 °C for 30 min [51]. The amount of time spent at these high temperatures is a key variable in the ability of heat to reduce NTM numbers. Since most commercial hot water systems have temperatures set with the goal of killing *Legionella*, the measures that are commonly in place in residential buildings and healthcare institutions are inadequate for killing many of the pathogenic NTM species, which have been shown to be more heat resistant than *Legionella*. Within a given water supply system, it is difficult to achieve a uniform temperature throughout the system, especially given the formation of dense biofilms on the inner surfaces of pipes and faucets. Sebakova et al. investigated the presence of NTM in the hot water systems of four different hospital settings. They found that in those systems that underwent thermal disinfection with temperatures of greater than 50 °C, no NTM were found. In those hot water systems that had no disinfection system, 70% of samples contained NTM [52]. Zwadyk et al. showed that heating of mycobacteria in 100 °C boiling water or in a forced air oven for a minimum of 5 min kills mycobacteria [53]. Although boiling of water is not suitable for disinfection of home or commercial water systems, this can be a reliable method for individual disinfection of drinking water.

NTM have also been detected in cooling towers. Seventy-five percent of samples from 9 cooling towers across the United States contained NTM [54]; 56% of samples from 53 cooling towers in an urban area in Barcelona, Spain, contained

NTM [55]; and 100% of samples from 12 cooling systems in Finland contained NTM, with concentrations up to 7.3×10^4 cfu/L of water [56]. It is unclear at this point whether NTM can be acquired through inhalation of aerosols from air conditioning units.

NTM in Soil and Dust

NTM have been isolated from house dusts as well as from both natural and commercially available soils [25, 57]. Since dusts and soils are easily suspended into the air, the inhalation of NTM from these sources may be a significant source of infection in some individuals [24]. Aerosolized dusts from peat or potting soil contain high numbers of NTM, up to one million organisms per gram of soil, and soil exposure and gardening have been correlated with NTM risk [22, 58]. Fujita et al. found identical genotypes between MAC clinical and soil isolates in individuals in Japan who engaged in gardening activities for more than 2 h per week [59]. 120 strains of NTM were isolated from vacuum cleaner dust collected in Queensland, Australia, including 50 strains of *M. intracellulare*, 44% of which were recognized as serotypes capable of causing human disease [24, 25].

Aerosolization of NTM

NTM can be readily aerosolized, as they concentrate at the air-water interface due to their hydrophobic surface [60]. In an experimental model, hydrophobic NTM cells were shown to adhere to air bubbles rising in a water column, and, at the water surface, the bubbles burst, ejecting droplets to heights of 10 cm. *M. avium* densities were 1000- to 10,000-fold higher in the ejected jet droplets than in the water [61, 62]. In a study of the environment of an indoor heated therapy pool by Angenent et al., it was found that NTM preferentially partition into aerosols, with higher concentrations of *M. avium* found in the air above the pool than in the pool biofilms or water [3]. Wendt et al. investigated rainwater, natural river waters, and their aerosols in the southeastern United States. They recovered MAC species in the aerosols above the river water via an Anderson air sampler, with particle sizes of 0.7–3.3 μm in diameter, a size small enough to reach the lower respiratory tract [63]. The aerodynamic particle size of NTM is between 0.5 and 5 μm in diameter. Penetration into the respiratory tract increases with decreasing particle size, with particles greater than 5 μm lodging in the upper respiratory tract, particles between 1.5 and 5 μm reaching the lower respiratory tract, and particles less than 1.5 μm reaching the alveoli (Fig. 3) [61, 63, 64].

The aerosols to which most individuals are exposed on a daily basis are those generated during showering. Zhou et al. investigated the size of droplets produced

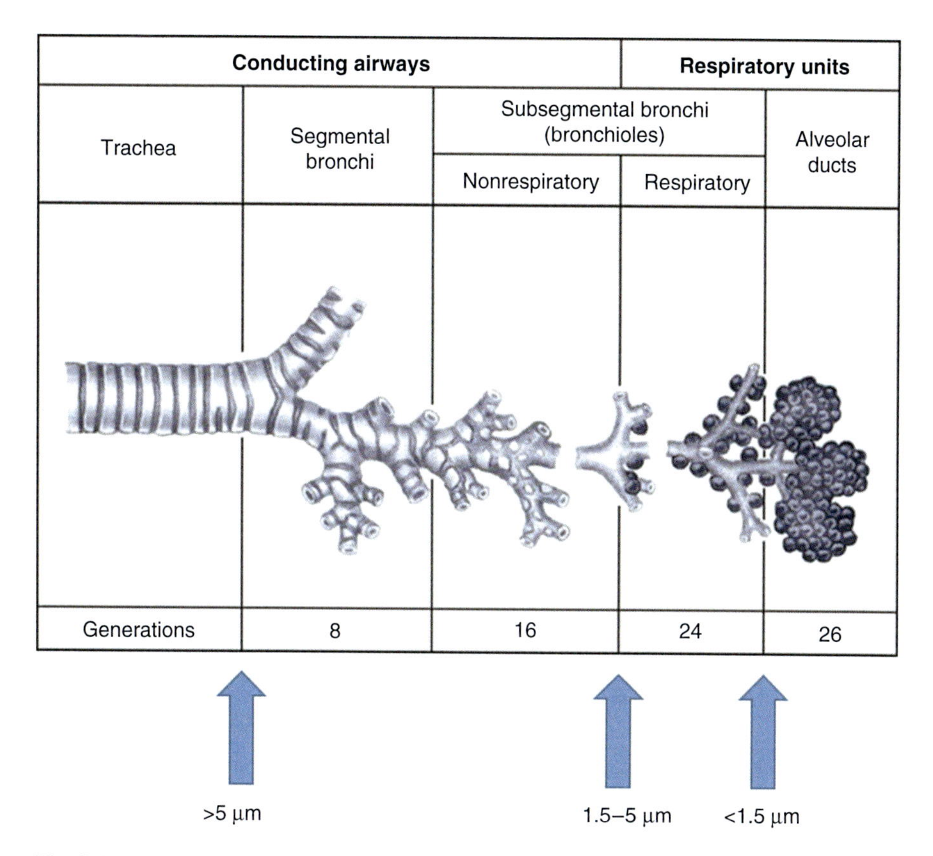

Fig. 3 Penetration into the respiratory tract increases with decreasing particle size

inside the shower and found that for hot water, the mass median diameter of the droplets was 6.3–7.5 µm. The estimated deposition of water droplets within the respiratory tract for hot water aerosols was dependent on water flow rate and breathing method. During mouth breathing, an estimated 50% of particles deposited in the extrathoracic regions and 6–10% in the alveoli. During nose breathing, approximately 86% of particles deposited in the extrathoracic regions and 0.9% in the alveoli [65]. In a separate study by Thomson et al., NTM was cultured from 50% of 20 shower aerosols tested at the level that would have lower respiratory tract penetration, and these shower aerosols contained NTM species that matched patient isolates [13].

Another potential venue for aerosolization of NTM into the respiratory tract is during dental work. Dental unit waterlines support the growth of a dense microbial population that includes NTM. NTM may be aerosolized by dental equipment, with the potential for inhalation or aspiration of NTM during dental procedures [66, 67]. In a study of 21 dental units in 10 offices in Germany, there were a mean NTM concentration of 365 cfu/mL in the water samples and a mean NTM density of 1165 cfu/cm^2 in the biofilm samples [68].

Different Types of NTM May Have Different Preferred Environmental Niches

Though *M. avium* and *M. chimaera* have been definitively found in water, *M. intracellulare* has not been found in water [7, 69], but has been linked to soil, household dust, potting mix, and peat moss [22, 70, 71]. *M. chimaera* has been the organism responsible for the recent contamination of heater-cooler units used during cardiac surgery [72, 73]. Furthermore, the distribution of NTM species that are isolated from clinical samples differs strongly by region [74]. Hoefsloot et al. conducted a worldwide study in which they collected pulmonary NTM culture results from laboratories in different regions of the world to gain further insight into the geographical distribution of different NTM species [75]. They found a wide variation in the incidence of different types of NTM in the different regions. For example, *M. xenopi* was found to predominate in Hungary, where it comprised 49% of all NTM. *M. kansasii* predominated in Poland and Slovakia. Rapidly growing mycobacteria comprised 10–20% of all NTM isolates worldwide, but 50% of all NTM pulmonary isolates in Taiwan. Among MAC species, *M. avium* was most common in North and South America and Europe, but *M. intracellulare* was most common in South Africa and Australia. The reasons for these differences are unclear. They may be related to species-specific favorable conditions with regard to soil composition, water source, temperature, humidity, activities and industries that result in aerosolization, and other environmental factors. In a separate study performed in the United States, high-risk geographic areas for MAC infection were found to have greater proportions of area surface water, higher levels of evapotranspiration (the potential of the atmosphere to absorb water), and higher copper and sodium but lower manganese levels in the soil [2].

It is likely that the predominant NTM type in the environment correlates with the prevalence of specific NTM disease in that region. In addition, a susceptible individual's habits and hobbies may significantly influence their risk of NTM infection and the type of NTM species that they acquire. For example, in cystic fibrosis patients, indoor swimming pool use [76] and high soil exposure [77] were both found to have an elevated odds ratio for NTM infection of 5.9.

Reducing Exposure to NTM

The recurrence rate of NTM pulmonary infection has been reported to be as high as 25–45% [78–80], with a significant percentage of recurrences occurring due to reinfection with a new genotype from the environment [81, 82]. Management of NTM disease should therefore include educating patients regarding modifications that can be instituted in their home environment and personal habits to reduce environmental exposure to NTM. The impact of instituting specific methods to reduce exposure to environmental NTM on disease outcomes has not been studied. Listed below are interventions that can be suggested to individuals at risk of acquiring NTM infection. These are based upon what is known at the present time regarding the persistence of NTM in the environment.

Sink Faucets and Showerheads

The installation of specialized filters with pore sizes less than 0.2 micrometers on faucets and showerheads will prevent the passage of NTM. These filters, however, can easily clog and must be changed every 30 days. Granular activated carbon water filters should be avoided, as they can promote the growth of NTM without preventing their passage [40, 41]. Refrigerator taps and ice machines should be avoided as high numbers of NTM have been isolated from these. Showerheads should be disinfected regularly by either immersion in boiling water or in undiluted bleach for 30 min and then rinsed. The inner surfaces of the showerhead and the pipe to which the showerhead attaches should be manually cleaned to try to remove as much of the biofilm as possible. Showerheads comprised of metal may result in less biofilm attachment than plastic showerheads, and those with large diameter holes that produce a stream as opposed to a fine mist likely result in less aerosolization of NTM organisms. Showers should be ventilated as much as possible, and time spent in the shower should be minimized. After periods of nonuse, faucets and showerheads should be kept on for a few minutes to flush out water that has been left stagnant in the pipes.

Hot Water Heaters

NTM have been found in high numbers in hot water tanks, particularly those set at temperatures under 110 ° F [7]. Raising hot water heater temperatures to 120 ° F or higher has been shown to reduce the growth of *M. avium* and *M. chelonae* in hot water heaters [7]. In addition, hot water tanks should be drained regularly as the highest population of NTM collects at the sediment at the bottom of the tank.

Behavioral and Lifestyle Modifications

Patients should be advised to avoid the use of humidifiers, which are capable of generating aerosols with high numbers of NTM. Humidifiers attached to central heating units have also been shown to harbor NTM and may be a source of inhalational exposure to these organisms through household air [38]. If a humidifier must be used for a positive airway pressure machine, either previously boiled water or sterile water should be used [20]. Indoor pool aerosols have been shown to contain

NTM [3], and in an epidemiologic study of patients with cystic fibrosis, swimming in indoor pools was associated with NTM infection [76]. Patients should be counseled to avoid the use of hot tubs or steam rooms and to consider avoiding the use of indoor pools [18, 21]. Gardening and soil exposure have also been correlated with NTM risk [22]. To decrease the generation of aerosolized dust, potting soil should be moistened prior to handling. Care should be taken during activities such as house dusting, sweeping, and emptying of vacuum cleaners [24, 25]. If activities must be performed that will involve aerosolization of water, dust, or soil, wearing an N95 respirator mask (0.1–0.3 micrometer) may be considered.

Conclusion

Attention to sources and exposures to NTM in the environment is important in order to help prevent new infection in susceptible individuals, to prevent worsening of infection in patients who are culture positive with minimal disease but have proven to be susceptible, to achieve better control of disease in established patients, and to prevent reinfection after completion of treatment courses. It is reasonable for susceptible individuals to adopt measures to try to minimize exposure to NTM in the environment. These include procedures to reduce NTM growth in household water sources and the adoption of changes in individual behaviors and habits to minimize exposure to aerosolized water and soil. Water utilities and owners of apartment buildings, hospitals, swimming pools, and spas must also be involved and must engage the expertise of microbiologists, plumbers, and environmental scientists and engineers to tackle this challenging problem. More research is needed in this area, as manipulating the water environment through changes in disinfection procedures or temperatures may have effects that are poorly understood at this point, such as changing the microbial balance, which could result in the selection of more hardy and potentially more virulent organisms. Species of NTM differ not only in their preferred habitat, but different NTM species or different strains within the same species have variable pathogenicity. Instituting changes in our environment could have the potential to shift the balance of NTM species colonizing our water systems and biofilms with unknown consequences.

References

1. Strollo SE, Adjemian J, Adjemian MK, Prevots DR. The burden of pulmonary nontuberculous mycobacterial disease in the United States. Ann Am Thorac Soc. 2015;12(10):1458–64.
2. Adjemian J, Olivier KN, Seitz AE, Holland SM, Prevots DR. Prevalence of nontuberculous mycobacterial lung disease in U.S. Medicare beneficiaries. Am J Respir Crit Care Med. 2012;185(8):881–6.
3. Angenent L, Kelley S, Amand A, et al. Molecular identification of potential pathogens in water and air of a hospital therapy pool. PNAS. 2005;102(13):4860–5.

4. Falkinham JO. Environmental sources of nontuberculous mycobacteria. Clin Chest Med. 2015;36:35–41.
5. George KL, Parker BC, Gruft H. Epidemiology of infection by nontuberculous mycobacteria. Growth and survival in natural waters. Am Rev Respir Dis. 1980;122(1):89–94.
6. Glazer CS, Martyny JW, Lee B. Nontuberculous mycobacteria in aerosol droplets and bulk water samples from therapy pools and hot tubs. J Occup Environ Hyg. 2007;4(11): 831–40.
7. Lande L, Kwait R, Williams M, et al. Hot water heaters are serving as incubators for nontuberculous mycobacteria in the home environment. Am J Respir Crit Care Med. 2015;191: A5270.
8. Thomson R, Tolson C, Sidjabat H. *Mycobacterium abscessus* isolated from municipal water – a potential source of human infection. BMC Infect Dis. 2013;13:241.
9. Halstrom S, Price P, Thomson R. Review: environmental mycobacteria as a cause of human infection. Int J Mycobacteriol. 2015;4:81–91.
10. Thomson R, Carter R, Tolson C, Huygens F, Hargreaves M. Factors associated with the isolation of nontuberculous mycobacteria (NTM) from a large municipal water system in Brisbane, Australia. BMC Microbiol. 2013;13:89.
11. Torvinen E, Suomalainen S, Lehtola M, Miettinen I, Zacheus O, Paulin L, et al. Mycobacteria in water and loose deposits of drinking water distribution systems in Finland. Appl Environ Microbiol. 2004;70:1973–81.
12. Covert T, Rodgers M, Reyes A, Stelma G Jr. Occurrence of nontuberculous mycobacteria in environmental samples. Appl Environ Microbiol. 1999;65(6):2492–6.
13. Thomson R, Tolson C, Carter R, Coulter C, Huygens F, Hargreaves M. Isolation of NTM from household water and shower aerosols in patients with NTM pulmonary disease. J Clin Microbiol. 2013;51:3006–11.
14. Slosarek M, Kubin M, Jaresova M. Water-borne household infections due to *Mycobacterium xenopi*. Central Eur J Publ Hlth. 1993;1(2):78–80.
15. Donohue MJ, Mistry JH, Donohue JM, O'Connell K, King D, Byran J, et al. Increased frequency of nontuberculous mycobacteria detection at potable water taps within the United States. Environ Sci Technol. 2015;49(10):6127–33.
16. Falkinham J III, Iseman M, de Haas P, van Soolingen D. *Mycobacterium avium* in a shower linked to pulmonary disease. J Water Health. 2008;6:209–13.
17. Feazel L, Baumgartner L, Peterson K, Frank D, Harris J, Pace N. Opportunistic pathogens enriched in showerhead biofilms. PNAS. 2009;106(38):16393–9.
18. Mangione EJ, Huitt G, Lenaway D, Beebe J, Bailey A, Figoski M, et al. Nontuberculous mycobacterial disease following hot tub exposure. Emerg Infect Dis. 2001;7(6):1039.
19. Falkinham JO III. Ecology of nontuberculous mycobacteria – where do human infections come from? Semin Respir Crit Care Med. 2013;34(1):95–102.
20. Assi MA, Beg JC, Marshall WF, et al. Mycobacterium gordonae pulmonary disease associated with a continuous positive airway pressure device. Transpl Infect Dis. 2007;9(3):249–52.
21. Iivanainen E, Northrup J, Arbeit RD, Ristola M, Katila M-L, Von Reyn CF. Isolation of mycobacteria from indoor swimming pools in Finland. APMIS. 1999;107:193–200.
22. De Groote MA, Pace NR, Fulton K, Falkinham JO. Relationships between *Mycobacterium* isolates from patients with pulmonary mycobacterial infection and potting soils. Appl Environ Microbiol. 2006;72:7602–6.
23. Iivanainen E, Martikainen P, Raisanen M, Katila M. Mycobacteria in coniferous forest soils. FEMS Microbiol Ecol. 1997;23(4):325–32.
24. Dawson D. Potential pathogens among strains of mycobacteria isolated from house-dusts. Med J Aust. 1971;1:679–81.
25. Torvinen E, Torkko P, Rintala ANH. Real-time PCR detection of environmental mycobacteria in house dust. J Microbiol Methods. 2010;82:78–84.
26. Falkinham JO III, Norton CD, LeChevallier MW. Factors influencing numbers of Mycobacterium avium, Mycobacterium intracellulare, and other mycobacteria in drinking water distribution systems. Appl Environ Microbiol. 2001;67(3):1225–31.

27. Bendinger B, Rijnaarts HHM, Altendorf K, et al. Physicochemical cell surface and adhesive properties of Coryneform Bacteria related to the presence and chain length of Mycolic acids. Appl Environ Microbiol. 1993;59(11):3973–7.
28. Stelmack PL, Gray MR, Pickard MA. Bacterial adhesion to soil contaminants in the presence of surfactants. Appl Environ Microbiol. 1999;65(1):163–8.
29. Mullis SN, Falkinham JO III. Adherence and biofilm formation of Mycobacterium avium, Mycobacterium intracellulare and Mycobacterium abscessus to household plumbing materials. J Appl Microbiol. 2013;115(3):908–14.
30. Norton CD, LeChevallier MW, Falkinham JO III. Survival of Mycobacterium avium in a model distribution system. Water Res. 2004;38:1457–66.
31. Pelletier PA, du Moulin GC, Stottmeier KD. Mycobacteria in public water supplies: comparative resistance to chlorine. Microbiol Sci. 1988;5(5):147–8.
32. du Moulin GC, Sherman IH, Hoaglin DC, et al. Mycobacterium avium complex, an emerging pathogen in Massachusetts. J Clin Microbiol. 1985;22(1):9–12.
33. Norton CD, LeChevallier M. A pilot study of bacteriological population changes through potable water treatment and distribution. Appl Environ Microbiol. 2000;66(1):268–76.
34. Sousa S, Bandeira M, Carvalho PA, Duarte A, Jordao L. Nontuberculous mycobacteria pathogenesis and biofilm assembly. Int J Mycobacteriol. 2015;4:36–43.
35. Lumb R, Stapledon R, Scroop A, Bond P, et al. Investigation of spa pools associated with lung disorders caused by Mycobacterium avium complex in immunocompetent adults. Appl Environ Microbiol. 2004;70:4906–10.
36. Joseph O, Falkinham JO III. Nontuberculous mycobacteria from household plumbing of patients with nontuberculous mycobacteria disease. Emerg Infect Dis. 2011;17(3):419–24.
37. Nishiuchi Y, Maekura R, Kitada S, et al. The recovery of Mycobacterium avium-intracellulare complex (MAC) from the residential bathrooms of patients with pulmonary MAC. Clin Infect Dis. 2007;45(3):347–51.
38. Lande L, Peterson D, Sawicki J, et al. Municipal water supply as a major source for Mycobacterium Avium pulmonary disease: a comparison of household and respiratory isolates. Am J Respir Crit Care Med. 2013;187:A5100.
39. Whiley H, Giglio S, Bentham R. Opportunistic pathogens Mycobacterium Avium complex (MAC) and Legionella spp. Colonise Model Shower. Pathogens. 2015;4(3):590–8.
40. Falkinham JO III. Surrounded by mycobacteria: nontuberculous mycobacteria in the human environment. J Appl Microbiol. 2009;107(2):356–67.
41. Rodgers MR, Blackstone BJ, Reyes AL, Covert TC. Colonisation of point of use water filters by silver resistant non-tuberculous mycobacteria. J Clin Pathol. 1999;52(8):629.
42. Strahl ED, Gillaspy GE, Falkinham JO III. Fluorescent acid fast microscopy for measuring phagocytosis of Mycobacterium avium, Mycobacterium intracellulare, and Mycobacterium scrofulaceum by Tetrahymena pyriformis and their intracellular growth. Appl Environ Microbiol. 2001;67:4432–9.
43. Ovrutsky AR, Chan ED, Kartalija M, et al. Cooccurrence of free-living amoebae and nontuberculous mycobacteria in hospital water networks, and preferential growth of Mycobacterium avium in Acanthamoeba lenticulata. Appl Environ Microbiol. 2013;79(10):3185–92.
44. Delafont V, Mougari F, Cambau E, et al. First evidence of amoebae-mycobacteria association in drinking water network. Environ Sci Technol. 2014;48(20):11872–82.
45. Coulon C, Collignon A, McDonnell G, Thomas V. Resistance of Acanthamoeba cysts to disinfection treatments used in health care settings. J Clin Microbiol. 2010;48(8):2689–97.
46. Thomas V, Bouchez T, Nicolas V, et al. Amoebae in domestic water systems: resistance to disinfection treatments and implication in Legionella persistence. J Appl Microbiol. 2004;97(5):950–63.
47. Cirillo JD, Falkow S, Tompkins LS, Bermudez LE. Interaction of Mycobacterium avium with environmental amoebae enhances virulence. Infect Immun. 1997;65(9):3759–67.
48. duMoulin GC, Stottmeier KD, Pelletier PA, et al. Concentration of Mycobacterium avium by hospital hot water systems. JAMA. 1988;260:1599–601.

49. Tichenor WS, Thurlow J, McNulty S, Brown-Elliott BA, Wallace RJ Jr, Falkinham JO III. Nontuberculous mycobacteria in household plumbing as possible cause of chronic rhinosinusitis. Emerg Infect Dis. 2012;18(10):1612–7.
50. Schulze-Röbbecke R, Buchholtz K. Heat susceptibility of aquatic mycobacteria. Appl Environ Microbiol. 1992;58(6):1869–73.
51. Grant IR, Ball HJ, Rowe MT. Thermal inactivation of several Mycobacterium spp. in milk by pasteurization. Lett Appl Microbiol. 1996;22(3):253–6.
52. Sebakova H, Kozisek F, Mudra R, et al. Incidence of nontuberculous mycobacteria in four hot water systems using various types of disinfection. Can J Microbiol. 2008;54(11):891–8.
53. Zwadyk P Jr, Down JA, Myers N, et al. Rendering of mycobacteria safe for molecular diagnostic studies and development of a lysis method for strand displacement amplification and PCR. J Clin Microbiol. 1994 Sep;32(9):2140–6.
54. Black WC, Berk SG. Cooling towers – a potential environmental source of slow-growing mycobacterial species. AIHA J (Fairfax, VA). 2003;64(2):238–42.
55. Adrados B, Julián E, Codony F, Torrents E, Luquin M, Morató J. Prevalence and concentration of non-tuberculous mycobacteria in cooling towers by means of quantitative PCR: a prospective study. Curr Microbiol. 2011;62(1):313–9.
56. Torvinen E, Suomalainen S, Paulin L, Kusnetsov J. Mycobacteria in Finnish cooling tower waters. APMIS. 2014;122(4):353–8.
57. Ichiyama S, Shimokata K, Tsukamura M. The isolation of *Mycobacterium avium* complex from soil, water and dusts. Microbiol Immunol. 1998;32:733–9.
58. Reed C, von Reyn CF, Chamblee S, Ellerbrock TV, Johnson JW, Marsh BJ, Johnson LS, Trenschel RJ, Horsburgh CR Jr. Environmental risk factors for infection with Mycobacterium avium complex. Am J Epidemiol. 2006;164(1):32–40.
59. Fujita K, Ito Y, Hirai T. Genetic relatedness of Mycobacterium avium-intracellulare complex isolates from patients with pulmonary MAC disease and their residential soils. Clin Microbiol Infect. 2013;19(6):537–41.
60. Falkinham JO III. Mycobacterial aerosols and respiratory disease. Emerg Infect Dis. 2003;9(7):763–7.
61. Parker BC, Ford MA, Gruft H, Falkinham JO III. Epidemiology of infection by nontuberculous mycobacteria. IV. Preferential aerosolization of *Mycobacterium intracellulare* from natural water. Am Rev Respir Dis. 1983;128:652–6.
62. Gruft H, Katz J, Blanchard DC. Postulated source of Mycobacterium intracellulare (Battey) infection. Am J Epidemiol. 1975;102:311–8.
63. Wendt S, George K, Parker B. Epidemiology of infection by nontuberculous mycobacteria III. Isolation of potentially pathogenic mycobacteria from aerosols. Am Rev Respir Dis. 1980;122(2):259–63.
64. Wells WF. Airborne contagion and air hygiene. Cambridge, MA: Harvard University Press; 1955.
65. Zhou Y, Benson JM, Irvin C, et al. Particle size distribution and inhalation dose of shower water under selected operating conditions. Inhal Toxicol. 2007 Apr;19(4):333–42.
66. Porteous NB, Redding SW, Jorgensen JH. Isolation of non-tuberculosis mycobacteria in treated dental unit waterlines. Oral Surg Oral Med Oral Pathol Oral Radiol Endod. 2004;98(1):40–4.
67. Dutil S, Veillette M, Mériaux A, et al. Aerosolization of mycobacteria and legionellae during dental treatment: low exposure despite dental unit contamination. Environ Microbiol. 2007;9(11):2836–43.
68. Schulze-Röbbecke R, Feldmann C, Fischeder R, et al. Dental units: an environmental study of sources of potentially pathogenic mycobacteria. Tuber Lung Dis. 1995;76(4):318–23.
69. Jr Wallace RJ, Iakhiaeva E, Williams MD, et al. Absence of Mycobacterium intracellulare and presence of Mycobacterium chimaera in household water and biofilm samples of patients in the United States with Mycobacterium avium complex respiratory disease. J Clin Microbiol. 2013;51(6):1747–52.

70. Cayer MP, Veillette M, Pageau P, et al. Identification of mycobacteria in peat moss processing plants: application of molecular biology approaches. Can J Microbiol. 2007;53(1):92–9.
71. Reznikov M, Leggo JH, Dawson DJ. Investigation by seroagglutination of strains of the Mycobacterium intracellulare-M. scrofulaceum group from house dusts and sputum in Southeastern Queensland. Am Rev Respir Dis. 1971;104(6):951–3.
72. Perkins KM, Lawsin A, Hasan NA, et al. Notes from the field: Mycobacterium chimaera contamination of heater-cooler devices used in cardiac surgery – United States. MMWR Morb Mortal Wkly Rep. 2016;65(40):1117–8.
73. Sommerstein R, Rüegg C, Kohler P. Transmission of Mycobacterium chimaera from heater-cooler units during cardiac surgery despite an ultraclean air ventilation system. Emerg Infect Dis. 2016;22(6):1008–13.
74. Marras TK, Daley CL. Epidemiology of human pulmonary infection with nontuberculous mycobacteria. Clin Chest Med. 2002;23:553–6.
75. Hoefsloot W, van Ingen J, Andrejak C. The geographic diversity of nontuberculous mycobacteria isolated from pulmonary samples: an NTM-NET collaborative study. Eur Respir J. 2013;42(6):1604–13.
76. Prevots DR, Adjemian J, Fernandez AG, et al. Environmental risks for nontuberculous mycobacteria. Individual exposures and climatic factors in the cystic fibrosis population. Ann Am Thorac Soc. 2014;11:1032–8.
77. Maekawa K, Ito Y, Hirai T, et al. Environmental risk factors for pulmonary Mycobacterium avium-intracellulare complex disease. Chest. 2011;140:723–9.
78. Lee BY, Kim S, Hong YK, et al. Risk factors for recurrence after successful treatment of Mycobacterium avium complex lung disease. Antimicrob Agents Chemother. 2015;59:2972–7.
79. Boyle DP, Zembower TR, Reddy S, Qi C. Comparison of clinical features, virulence, and relapse among Mycobacterium avium complex species. Am J Respir Crit Care Med. 2015;191:1310–7.
80. Field SK, Fisher D, Cowie RL. Mycobacterium avium complex pulmonary disease in patients without HIV infection. Chest. 2004;126:566–81.
81. Wallace RJ Jr, Zhang Y, Brown-Elliott BA, et al. Repeat positive cultures in Mycobacterium intracellulare lung disease after macrolide therapy represent new infections in patients with nodular bronchiectasis. J Infect Dis. 2002;186:266–73.
82. Boyle DP, Zembower TR, Qi C. Relapse versus reinfection of Mycobacterium avium complex pulmonary disease. Patient characteristics and macrolide susceptibility. Ann Am Thorac Soc. 2016;13:1956–61.

Epidemiology of Nontuberculous Mycobacterial Pulmonary Disease (NTM PD) in the USA

Shelby Daniel-Wayman, Jennifer Adjemian, and D. Rebecca Prevots

Introduction

Nontuberculous mycobacteria (NTM) as a cause of human pulmonary disease (NTM PD) was first described in 1954 [1] when patients with pulmonary disease at tuberculosis (TB) sanitoria were found to have atypical mycobacteria which were not *Mycobacterium tuberculosis*. Since that time, more than 180 species of mycobacteria have been identified [2], and efforts to characterize the prevalence, incidence, cost, and risk factors associated with these mycobacteria have increased. In this chapter, we review the available information regarding these epidemiologic features and highlight current gaps in the knowledge.

Prevalence and Incidence Estimates

Unlike disease from *M. tuberculosis*, NTM associated with human disease is generally not reportable in the USA, with the exception of several states [3]. Even when it is reportable, the condition under surveillance varies greatly, from any isolate of NTM to NTM pulmonary and/or extrapulmonary disease [3, 4]. Thus, estimates of

S. Daniel-Wayman · D. Rebecca Prevots (✉)
Epidemiology Unit, Laboratory of Clinical Immunology and Microbiology, Division of Intramural Research, National Institute of Allergy and Infectious Diseases, National Institutes of Health, Bethesda, MD, USA
e-mail: rprevots@niaid.nih.gov

J. Adjemian
Epidemiology Unit, Laboratory of Clinical Immunology and Microbiology, Division of Intramural Research, National Institute of Allergy and Infectious Diseases, National Institutes of Health, Bethesda, MD, USA

United States Public Health Service, Commissioned Corps, Rockville, MD, USA

D. E. Griffith (ed.), *Nontuberculous Mycobacterial Disease*, Respiratory Medicine,
https://doi.org/10.1007/978-3-319-93473-0_7

NTM PD incidence and prevalence are not routinely available and must be obtained through a variety of special surveillance efforts, studies, or surveys, each with its own advantages and limitations. Because NTM PD is a relatively rare condition, large population-based datasets from various healthcare systems, with microbiology data linked to patient data, have been particularly important to describing the burden of NTM. Two studies were conducted in large health maintenance organizations (HMO), one at five HMOs in different parts of the country and one in the state of Hawaii, a high prevalence state with a large patient population of Asian/Pacific Islander ancestry. These studies have been key to understanding the epidemiology of NTM in the USA, as they included both mycobacteriology laboratory data and other clinical and demographic features. They provide consistent findings, namely, that *Mycobacterium avium* complex (MAC) comprised approximately 80% of all disease and that NTM was more common in older adults, with an increased prevalence among women relative to men (1.1 to 1.9-fold higher). Both studies also found a 2 to 2.9-fold higher prevalence of NTM compared with TB during the same time period [5, 6]. In the study of five healthcare systems, annual prevalence ranged from 1.4 to 6.6 per 100,000 persons across the five regions, which included a large HMO in Southern California as well as managed healthcare systems in Colorado, the Seattle area, and one region of Pennsylvania. From 1994 to 2006, NTM was increasing 2.6% per year among men and 2.9% per year among women (Table 1) [5].

The second HMO study used a large, linked dataset from beneficiaries enrolled in Kaiser Permanente Hawaii (KPH) from 2005 through 2013 and estimated prevalence and trends by species and race/ethnicity [6]. Overall, the period prevalence was 122 per 100,000 persons for this time period (Table 1). The annual prevalence increased from 9 to 19 per 100,000 persons over the study period, and when evaluated by species, the increase was found only for MAC-associated NTM PD (Table 1). Of greatest interest was the interaction of species and race/ethnicity, with the highest period prevalence observed among persons identified as Japanese, Korean, Chinese, or Vietnamese, with a disease prevalence of 160–183 per 100,000 persons and an infection prevalence of 293–336 per 100,000, compared to a disease prevalence of only 89 per 100,000 persons and an infection prevalence of 156 per 100,000 persons for persons identified as white (Fig. 1). Persons who identified as Native Hawaiians or Pacific Islanders had a much lower NTM PD prevalence of only 23 per 100,000 persons and an infection prevalence of 50 per 100,000 persons [6].

Population-based estimates of NTM PD were also derived for three counties in North Carolina, from which researchers obtained clinical and demographic data for all samples tested from 2006 to 2010. In this study, the average annual prevalence was 9.4 per 100,000 persons and was similar among men and women [7]. However, significant differences were observed by sex and race, whereby among women, prevalence was highest among white women, at 10.2 per 100,000 persons, while among men, prevalence was highest for black men, at approximately 7 per 100,000 persons [7]. Although estimates were not adjusted for age, observed rates by age showed a nearly twofold increased prevalence at nearly 40 per 100,000 persons among white individuals aged >60 years compared with nearly 20 per 100,000 persons among black individuals in the same age group [7].

Table 1 NTM prevalence and incidence by population studied and type of measure

Study	Population studied	Annual prevalence	Incidence	Period prevalence\ cumulative incidence	Trends
NTM HMO [5]	4 HMOs in different areas of the USA (Southern California, Seattle, Colorado, Pennsylvania)	1.4–6.6 across sites	Not assessed	By age (2004–2006): <60 yrs.: 1.7 per 100,000 60–69 yrs.: 15 per 100,000 70–79 yrs.: 30 per 100,000 > 80 yrs.: 57 per 100,000	Increasing
NTM HI [6]	KPH beneficiaries	Isolation: 20 cases/100,000 persons in 2005 to 44 cases/100,000 persons in 2013	Not assessed	Isolation: 122 cases/100,000 persons (2005–2013)	Increasing among MAC
North Carolina, 3 counties [7]	All samples tested from 3 labs in North Carolina	Avg annual prevalence of pulmonary isolations: 9.4/100,000 Interaction of race and sex	Not assessed		Not assessed
Oregon [8]	Surveillance data/lab integrated Portland Tri-County region 2005–2006	5.6/100,000 population	Not assessed	Disease: 8.6/100,000 (2005–2006)	Not assessed
Oregon [9]	Oregon residents 2007–2012: all isolates	5.9/100,000 (2010–2011 only)	4.8–5.6 per 100,000 over time period; age-adjusted average annual incidence = 3.8/100,000	Not assessed	Non-significant increase from 4.8 to 5.6, mostly observed with MAC; 2.2% annual increase
Five states (NTM reporting) [10]	Residents of Missouri, Mississippi, Maryland, Ohio, and Wisconsin, 2008–2013	Isolation: 8.7–13.9 per 100,000 persons over the study period	Not assessed	75.8 per 100,00 (2008–2013)	9.9% increase per year

(continued)

Table 1 (continued)

Study	Population studied	Annual prevalence	Incidence	Period prevalence\ cumulative incidence	Trends
CF [11]	CF patients 10 years or older at 21 US centers			13% (3 sputum samples)	Not assessed
CF registry [12]	CF patients 12 years or older in the CF patient registry (2010–2011)	Not assessed	Not assessed	14%	Not assessed
CMS [13]	Adults aged 65 years or older (1997–2007)	20–47 cases/100,000	Not assessed	112 cases/100,000 (1997–2007)	8.2% increase per year
US vital statistics – mortality data [14]	US death records, 1999–2014	N/A	2.3 deaths per 1000,000 person-years	N/A	Increasing among HIV-negative individuals

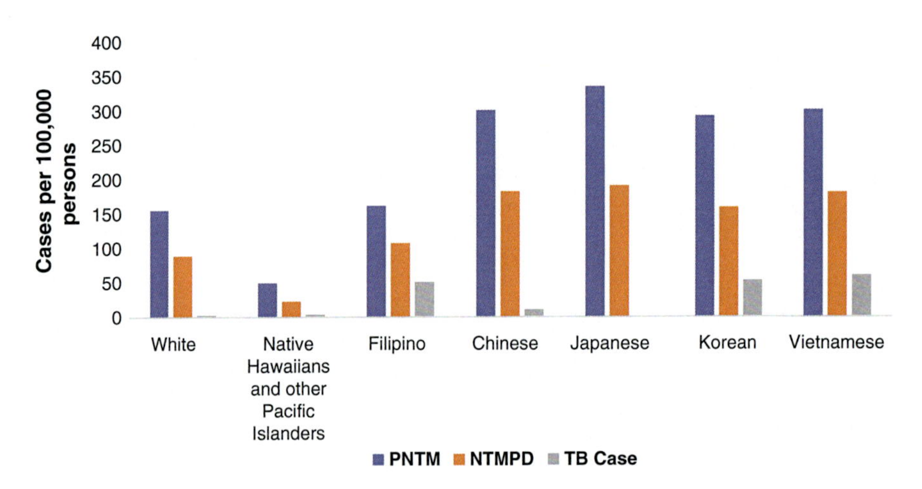

Fig. 1 Overall period prevalence of pulmonary nontuberculous mycobacteria isolation, nontuberculous mycobacterial pulmonary disease, and tuberculosis, by race/ethnicity, among a cohort of Kaiser Permanente Hawaii patients, Hawaii, 2005–2013. (From Adjemian et al. [6])

Similarly, in the state of Oregon, near complete identification of all NTM isolates was achieved by mandating reporting of NTM for a 2-year period, 2005–2006. This reporting period allowed estimation of the prevalence of cases meeting American Thoracic Society (ATS) microbiologic criteria for the NTM PD case definition [15], among patients in the Portland Tri-County metropolitan region, as well as description of the specificity of this criteria for identifying those patients who met the full ATS

NTM disease criteria. Considering only those with evaluable records and meeting clinical, radiographic, and microbiologic disease criteria, the prevalence of NTM disease in this region was found to be 8.6 per 100,000 persons over the 2-year period (Table 1), but this is an underestimate, given that 30% of patients lacked an evaluable record. Of those meeting the microbiologic criteria, 87% also fulfilled the full case definition, 88% of whom had disease caused by MAC [8].

More recently, the same researchers collected data from all microbiologic laboratories serving the state from 2007 to 2012. Using the previous data from 2005 to 2006 to exclude prevalent cases, they found annual incidence rates that ranged from 4.8 per 100,000 persons in 2007 to 5.6 per 100,000 persons in 2012 (Table 1) [16]. This moderate increase was consistent with other studies. Incidence varied by age and sex, with incidence higher among women overall, but among persons aged <60 years, incidence was higher among men compared with women [16]. The close correspondence of the incidence rates with the prevalence estimated from this and other studies suggests that, because most patients only have a single pulmonary specimen tested for NTM, the observed prevalence represents incidence—i.e., the first diagnosis of disease.

In Canada, microbiology data from the province of Ontario has been used to estimate population-based prevalence and trends for that region, comprising 38% of the country's total population [17]. The provincial laboratory processes approximately 95% of the provinces nontuberculous mycobacteria isolates [18]. Using data from this laboratory, the estimated period prevalence of NTM PD, as identified by the ATS microbiologic criteria, was 41.3 per 100,000 in 2006–2010 (Table 1) [18]. As in the USA, the majority of disease is due to MAC, which had a period prevalence of 26.5 per 100,000 from 2006 to 2010. The annual prevalence of NTM PD for Ontario increased from 4.9 per 100,000 in 1998 to 9.8 per 100,000 in 2010, while the prevalence of NTM isolation rose from 11.4 per 100,000 to 22.2 per 100,000 [18].

Administrative claims datasets have been critical to obtaining population-based estimates for the older adult population. Medicare data has been used to estimate prevalence and trends from 1997 to 2007, given that the largest burden of disease is in the >65-year age group. This analysis defined patients with NTM as those for whom one claim with the International Classification of Diseases, Ninth Revision (ICD-9) code for pulmonary NTM was submitted. This likely resulted in an underestimation of prevalence, as other studies have found only 27–50% of patients who met microbiologic criteria for NTM PD [15] received the corresponding ICD-9 code [5, 19]. Estimated period prevalence in this population was 112 cases per 100,000 persons, with annual prevalence increasing from 20 to 47 cases per 100,000 persons, an increase of 8.2% per year (Fig. 2, Table 1) [13]. Consistent with previous managed care studies, this analysis also found significant racial/ethnic variation, with a twofold higher prevalence among Asians/Pacific Islanders compared to whites. Significant geographic variation was found as well, ranging from a period prevalence in the West (including Hawaii) of 149 per 100,000 persons to 78 per 100,000 persons in the Midwest [13].

Medicare data have been combined with other estimates to generate overall burden estimates across all age groups. By combining prevalence estimates with age distribution from NTM survey data [20], case numbers by state were estimated [21].

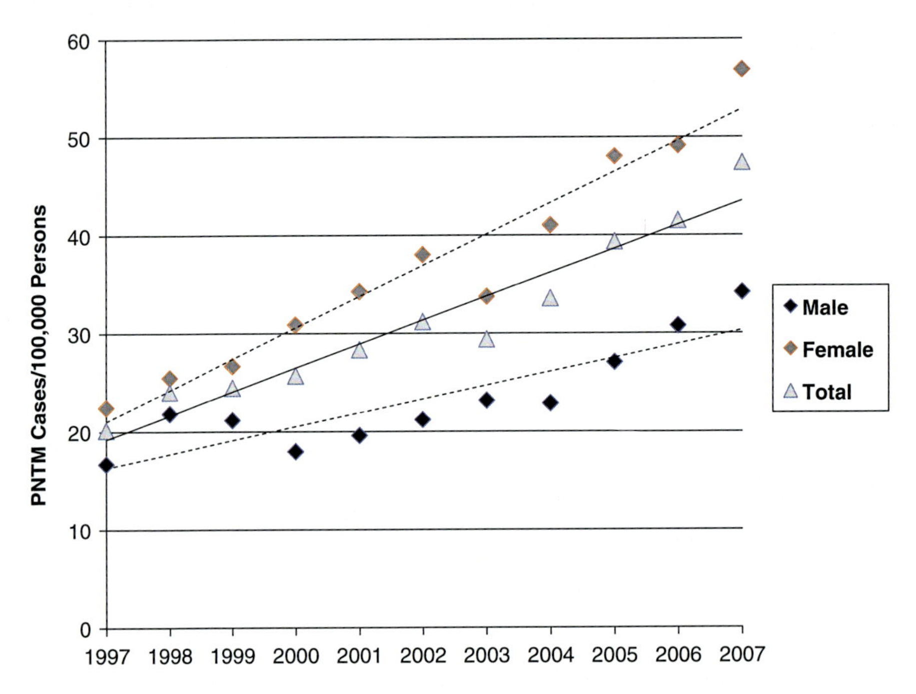

Fig. 2 Annual prevalence of pulmonary nontuberculous mycobacteria cases among a sample of US Medicare Part B enrollees by sex from 1997 to 2007. (Reprinted with permission of the American Thoracic Society. Copyright © 2017 American Thoracic Society. Cite: Adjemian et al. [13]. *The American Journal of Respiratory and Critical Care Medicine* is an official journal of the American Thoracic Society)

Combined with data on treatment cost by type of insurance, this study estimated a total of 86,244 NTM cases in the USA in 2010, at a cost of $815 million. Prescription drug costs were estimated to comprise 76% of this total [21]. This approach is similar to that used by German researchers, who used claims data from public statutory health insurance, although that study found that in that country, hospitalization accounted for the majority (63%) of NTM-associated costs [22].

Regarding surveillance, several states have NTM infection or disease listed as a notifiable condition. The reportable conditions vary, with six states listing any mycobacterial isolate as notifiable, two states having only extrapulmonary as notifiable [3], and three states requiring reporting of NTM disease, with one of these (MS), basing this on laboratory reporting alone (Fig. 3). Regardless of the notifiable condition, analysis of isolation or disease patterns requires additional effort to obtain detailed clinical and demographic information, which is often not provided with the sample. Thus, additional staff time and effort are required to obtain, compile, analyze, and disseminate this information for public health purposes. Systematic data collection will provide ongoing estimates that can be monitored for trends, and potential biases in the system can be evaluated. In the USA, electronic data from five of these states with reporting requirements were used to estimate

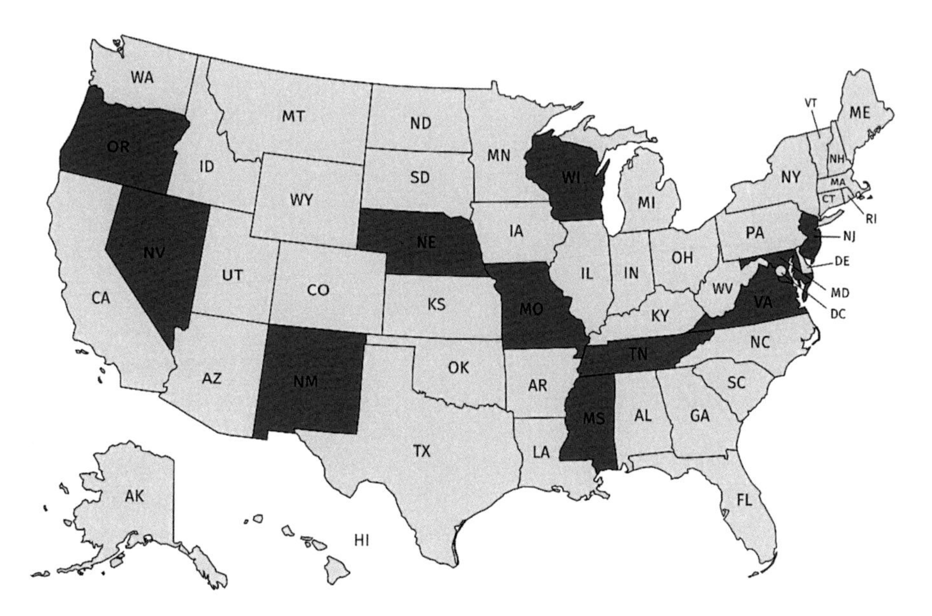

Fig. 3 States requiring reporting of nontuberculous mycobacteria

trends in NTM infection, defined as at least one isolate of NTM (from any body site): for all states combined, the average annual age-adjusted prevalence increased from 8.7 in 2008 to 13.9 in 2013, with a steady increase observed in Ohio and Wisconsin (Table 1) [10]. The overall increase was most marked in the >50 age group, with an annual percent change of 11.8% [10].

Studies in cystic fibrosis (CF) patients have identified a high prevalence in this population. In one cross-sectional study of 21 US CF centers, NTM prevalence among CF patients was estimated at 3% (ranging from 7% to 24% across centers) (Table 1). Prevalence varied substantially by age, ranging from approximately 10% among patients 10–14 years of age to approximately 40% among patients greater than 45 years of age [11]. An analysis of data from the CF Patient Registry from 2010 to 2011 found a similar 2-year period prevalence estimate of 14% [12]. The prevalence estimates from this database showed significant geographic variation, with a prevalence of 5–10% for 11 states and > 20% for 7 states. These studies highlight that while NTM prevalence is substantially higher among CF patients, disease patterns vary by geographic area and age.

Treatment and Mortality Estimates

Survey data have allowed an estimation of other aspects of NTM disease, including treatment adherence. Adherence to ATS treatment guidelines among physicians treating patients for NTM PD was assessed by a survey of US physicians conducted during 2011–2012. Respondents were asked to extract medical record data on the treatment of their last four patients treated for NTM disease [20]. Of MAC patients, only 13% were treated with a regimen consistent with ATS guidelines, and 16% were treated with macrolide monotherapy, which has been shown to increase the risk of macrolide resistance and poor treatment outcomes; 56% were treated with a regimen that did not include a macrolide at all. While pulmonologists were more likely to prescribe the recommended regimen (18%) compared to infectious disease (10%) and family/general practice/internal medicine (9%) physicians, the low levels of adherence to treatment guidelines across specialties raise concerns about physician knowledge regarding NTM treatment.

Vital statistics have been used to estimate prevalence of NTM and NTM-associated morbidity and mortality. Adults aged >65 years identified as pulmonary NTM cases in Medicare data were 40% more likely to die than non-cases, with a higher risk of death among men and among persons identified as black [13]. National mortality data from the National Center for Health Statistics has also been used to estimate the mortality rate from NTM, classified by ICD-10 code, from 1999 to 2014 [14]. The mortality rate during the study period was 2.3 deaths per 1,000,000 person-years, and no significant trend was observed. However, of these NTM-related deaths, the proportion of deaths also associated with human immunodeficiency virus (HIV) decreased from 33% in 1999 to 4% in 2014 and the NTM mortality rate increased among HIV-negative persons [14]. Thus, in an era where anti-retrovirals are now widely used in the HIV-infected population, the burden of NTM has shifted to the HIV-negative population. Mortality was also assessed among Oregon patients with either NTM disease or NTM isolation, and a 5-year age-adjusted mortality rate of 28.7 per 1,000 persons was found among those meeting ATS disease criteria versus 23.4 per 1,000 persons among those not meeting ATS disease criteria [23]. As mortality rates are dependent on both disease incidence and disease duration, the reasons for the increased mortality in some settings remain unclear and are likely to depend on the prevalence of underlying comorbid conditions.

Risk Factors

Risk of NTM disease is determined by a combination of environmental, microbial, and host factors (Table 2). A variety of host factors favor growth of mycobacteria and increase disease risk, including structural, immunologic, and genetic differences. Structural defects correlated with NTM include chronic obstructive pulmonary disease (COPD), which has been identified in 18–38% of patients with NTM [5, 25]. Lung cancer is also associated with NTM lung disease, with one single-center study finding that 25% of lung cancer patients had a positive MAC culture compared to only 3% of patients undergoing bronchoscopy for nonbronchiectatic benign lung disease [30]. Disorders of mucociliary clearance, including CF [11, 12] and pulmonary ciliary dyskinesia [31], are associated with high rates of NTM

Table 2 Risk factors for NTM infection and disease

Risk factor	Population studied	Relative risk, odds ratio, or relative prevalence	Outcome definition
Behavioral exposures			
Indoor swimming [24]	CF patients	5.9 (1.3–26.1)	Incident NTM infection
Spraying plants with a spray bottle [25]	Oregon residents (case-control)	2.7 (1.1–6.7)	NTM disease
High soil exposure (>2 hours per week) [26]	Japanese residents (case-control)	5.9 (1.4–24.7)	MAC disease
Environmental factors			
Soil copper levels, per 1 ppm increase [27]	Medicare beneficiaries	1.2 (1.0–1.4)	NTM disease (ICD-9 code)
Soil sodium levels, per 0.1 ppm increase [27]	Medicare beneficiaries	1.9 (1.2–2.9)	NTM disease (ICD-9 code)
Soil manganese levels, per 100 ppm increase [27]	Medicare beneficiaries	0.7 (0.4–1.0)	NTM disease (ICD-9 code)
Soil pH, per 1 unit increase [28]	Patients of National Jewish Health residing in CO	0.69 (0.51–0.95)	NTM disease due to slowly-growing NTM
Proportion of area as surface water [27]	Medicare beneficiaries	4.6 (1.5–14.6)	NTM disease (ICD-9 code)
Mean daily potential evapotranspiration [27]	Medicare beneficiaries	4.0 (1.6–10.1)	NTM disease (ICD-9 code)
Saturated vapor pressure [12]	CF patients	1.06 (1.02–1.10)	NTM isolation
Average annual vapor pressure	CF patients	Significant R^2	CF center NTM infection prevalence
Host factors			
COPD [25]	Oregon residents (case-control)	10 (1.2–80)	NTM disease
Lung cancer [13]	Medicare beneficiaries	3.4	NTM disease (ICD-9 code)
Anti-TNF-α [29]	Ontario seniors aged ≥67 years with rheumatoid arthritis	2.19 (1.10 to 4.37)	NTM disease (microbiologic criteria)
Oral prednisone [25]	Oregon residents (case-control)	8 (1.6–41.4)	NTM disease
Thoracic skeletal abnormalities [25]	Oregon residents (case-control)	5.4 (1.5–20)	NTM disease
Mitral valve disorder [13]	Medicare beneficiaries	1.4	NTM disease (ICD-9 code)
Low BMI [25]	Oregon residents (case-control)	9.09	NTM disease

disease; low ciliary beat frequency has also been associated with increased NTM disease risk in a case-control study of patients not diagnosed with these other conditions [32]. Immunologic therapy can confer increased risk by inhibiting the immune response to mycobacteria. TNF-α blockers have been repeatedly associated with NTM risk in both cohort studies and case-control studies, as have steroids and other immunosuppressive medications [25, 29, 33].

Evidence for genetic risk factors predisposing persons to NTM comes from family and genetic association studies as well as studies identifying a predisposing morphotype likely connected to genetic risk factors. While structural factors from COPD to CF often predispose patients to disease, NTM disease in patients lacking these disorders was correlated with low body mass index, thoracic skeletal abnormalities, mitral valve prolapse, and connective tissue disorders [25, 34, 35]. NTM disease and these associated traits also appear to cluster in families, suggesting common genetic risk factors [35, 36]. Moreover, a whole exome sequencing study using a candidate gene approach found that patients had more low-frequency variants in genes related to immune function, ciliary movement, and connective tissue, as well as in the gene coding for the cystic fibrosis transmembrane conductance regulator protein (which causes CF when both copies are nonfunctional) compared to both unaffected relatives and control subjects [37].

The geographic variation in risk of NTM infection and disease provides insight into environmental risk factors. Historically, evidence for geographic variation in NTM exposure comes from a *M. intracellulare* skin test sensitization study conducted among navy recruits, which found that exposure was higher in the Southeastern and Southwestern USA [38]. More recently, disease prevalence and clustering has been associated with climatic factors at the population level. Centers for Medicare and Medicaid Services (CMS) claims data were examined at the county level to identify spatial clusters of NTM disease among this population (Fig. 4) [27]. Seven clusters were identified encompassing parts of California, Florida, Hawaii, Louisiana, New York, Oklahoma, Pennsylvania, and Wisconsin. Compared to low-risk counties, these areas had a higher proportion of surface water and higher mean daily potential evapotranspiration, suggesting possible roles for climatic factors in NTM disease risk. Soil factors were also identified as important, with high-risk counties having higher soil levels of copper and sodium and lower levels of manganese [27]. Geographic clustering and climatic factors have also been studied in CF patients, with analysis of data from the CF Patient Registry identifying similar clusters of NTM cases, as well as an association between mean annual saturated vapor pressure and NTM risk (Fig. 4b) [12]. Additionally, a separate study of patients at 21 CF centers found a correlation between NTM prevalence and the average annual atmospheric water vaper content [24]. These associations are consistent with previous microbiologic findings that environmental prevalence of NTM is related to warmer temperature, low dissolved oxygen, high-soluble zinc, low pH, high humic acid, and high fulvic acid [39].

A recent study further illuminating these environmental risks found that risk of NTM PD varies not only by these climactic and soil factors but also by watershed [28]. Using data on non-CF NTM patients treated at National Jewish Health and

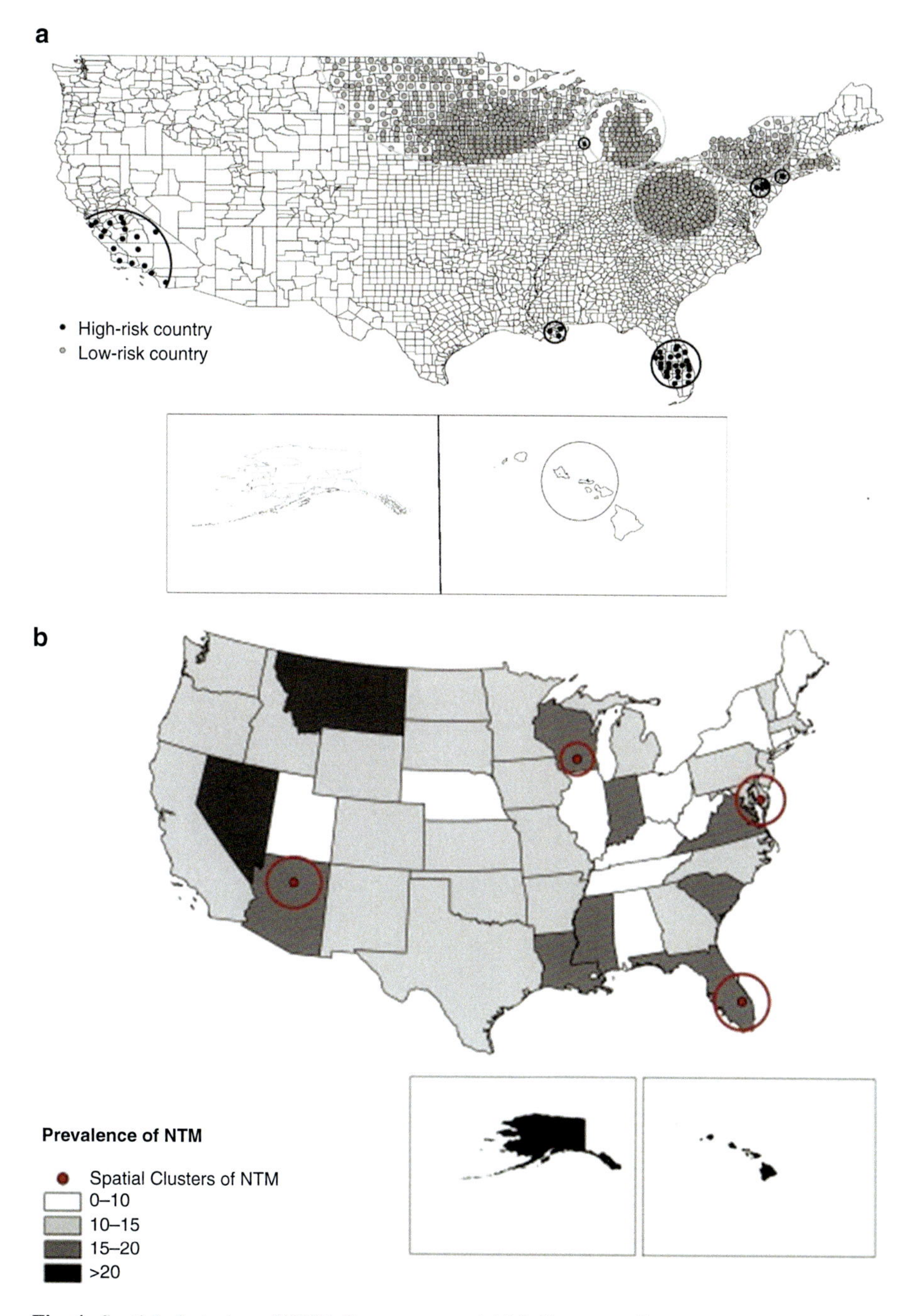

Fig. 4 Spatial clustering of NTM disease among (**a**) Medicare enrollees and (**b**) Cystic fibrosis patients. (Reprinted with permission of the American Thoracic Society. Copyright © 2017 American Thoracic Society. (**a**) Cite: Adjemian et al. [27]; (**b**) Cite: Adjemian et al. [12]. *The American Journal of Respiratory and Critical Care Medicine* is an official journal of the American Thoracic Society.)

resident in Colorado, researchers confirmed that acidic soil increased risk of slow-growing NTM species, while manganese was protective. When considering all NTM species and correcting for socioeconomic and soil factors as well as accessibility to National Jewish Health, they found that three specific watersheds (Blue, Upper South Platte, and Middle South Platte Cherry Creek) were at increased risk of NTM PD [28]. Interestingly, although these watersheds fall on both sides of the continental divide, water is pumped from the Blue watershed to the other two to provide drinking water for Denver and Aurora, indicating that this water may be the mechanism of increased risk for all three [28].

Environmental risk factors for NTM may vary dependent on species, which in turn have different clinical relevance. One study based on microbiologic data from a nationally distributed electronic health record database found that species prevalence varies by region, with the proportion of isolates identified as MAC ranging from 61% in the West South Central region (states included: AR, LA, OK, TX) to 91% in the East South Central region (AL, KY, MS, TN) [40]. The proportion of *M. abscessus/M. chelonae* isolates also varied significantly from 2% in the East South Central region to 18% in the West South Central region. A study among CF patients had similar findings, with the percentage of isolates identified as MAC ranging from only 29% in Louisiana to 100% in Nebraska [12]. These results suggest that the environmental risk factors for infection may vary by species, which likely differ in their environmental niches. Given the high levels of antibiotic resistance and low treatment response rates for *M. abscessus*, the variable prevalence of this species may also lead to geographic variation in NTM PD treatment outcomes.

At the household level, several studies have provided support for household water and water pipe biofilms as a source of NTM exposure. Two studies used repetitive sequence-based PCR to identify genetic matches between variants in environmental samples from patient households and clinical isolates from these same patients [41, 42]. This has led to an investigation of factors affecting NTM recovery from household water, and the discovery that households with water heaters set at temperatures above 55 °C had a lower rate of recovery of NTM [41]. However, significant variability in heat susceptibility of mycobacteria has been demonstrated in a laboratory setting, leading to concerns that alterations to water heater settings might select for more thermo-resistant, and potentially more pathogenic, strains such as *M. xenopi* [43].

Soil and dust have also been identified as potential sources of NTM exposure within the home environment. Aerosols generated from potting soils in households of pulmonary NTM patients were found to contain known pathogenic NTM species, including *M. avium, M. intracellulare*, and *M. kansasii*, some of which matched corresponding patient isolates by pulsed-field gel electrophoresis genotyping [44]. Similar results were obtained from a study examining soil samples from patient households in Japan, which found NTM in potting soil, residential yard soil, and farm soil. Using variable number tandem repeat genotyping, matches between isolates obtained from residential soil samples and directly from patients were identified for six patients with high soil exposure [45].

Behavioral exposures for NTM have been difficult to assess due to the rarity of the disease and the ubiquity of the organism, and the high frequency of household water exposures. However, case-control studies in both high-risk and general populations have identified some risk factors. A nested case-control study in CF patients, a high-risk population, allowed for the assessment of potential exposures that occurred within a 4-month window prior to the patient's first positive NTM culture, compared to control subjects with negative cultures throughout the study period [24]. A wide range of exposures were examined, including showering, residential and drinking water supply, soil exposure through gardening, and nebulizer use, but key factors significantly associated with incident infection included indoor swimming pool use in the previous 4 months and tap water appearing rusty or unclear [24].

In the general population, a case-control population in Oregon found that out of a list of water aerosol-generating activities including showering and jacuzzi use, only spraying plants with a spray bottle was significantly positively correlated with disease [25]. The same study found that dishwashing by hand and swimming were protective against NTM; however, this is thought to be the result of a bias toward these activities in healthier people [25]. None of the soil aerosol-generating activities, including potting plants, gardening, and lawn maintenance, were found to be significant [25]. While neither of these two US-based case-control studies found a relationship between soil exposure and NTM disease, a case-control study of bronchiectasis patients in Japan with and without NTM disease found that case patients were more likely to have high levels of soil exposure ($\geq$2 h per day) [26]. These differing results suggest that the routes of exposure vary by setting.

While NTM is primarily acquired from the environment, recent cases of transmission of *M. abscessus* among CF patients have been detected. Outbreaks of *M. abscessus* subsp. *massiliense* have been identified at multiple CF centers through genetic analyses of patient isolates including whole genome sequencing [46–48]. Whole genome sequencing has also detected a high level of relatedness among outbreak strains from the USA and UK [48, 49]. The small differences in SNPs (<20) between outbreak strains have been suggested as possible proof of recent transmission between continents, although geographical differences in large-scale deletions and insertions raise the possibility of independent selection for more transmissible strains in both countries with local evolution [48, 49]. Additionally, contact tracing has not identified transmission routes between outbreaks, but a recent laboratory study indicated *M. abscessus* may be viable in fomites [50].

Research Gaps

Epidemiological studies performed thus far highlight the increasing prevalence of NTM in the USA and its disproportionate burden on older populations and persons of Asian ancestry, with varying risk in different Asian subpopulations [6]. To further elucidate the risk and burden of NTM disease in the USA, additional research at the population level is needed in several disparate areas, including more detailed species-specific information regarding environmental reservoirs and geographic clustering of disease by species, as well as clinical studies to further identify genes associated with disease susceptibility. Regarding treatment, given that guidelines do not recommend universal treatment following NTM diagnosis due to the high cost, long duration, and significant side effects, further study of factors determining the need for treatment, as well as associated outcomes, is warranted. In addition, poor treatment outcomes may be mediated in great part by antibiotic resistance, but risk of developing antibiotic resistance has not been well studied. For species such as *M. abscessus*, which may have resistance to first-line antibiotics (e.g., macrolides) prior to treatment, studies of the frequency of antibiotic resistance among isolates from treatment-naïve patients would also be useful. Moreover, the high rates of reinfection seen in single-center studies suggest more research into rates of reinfection and risk factors for reinfection may elucidate this process, as well as host susceptibility more generally.

Acknowledgments This research was supported by the Intramural Research Program of the NIAID, NIH.

References

1. Timpe A, Runyon EH. The relationship of atypical acid-fast bacteria to human disease; a preliminary report. J Lab Clin Med. 1954;44(2):202–9.
2. Parte AC. LPSN – list of prokaryotic names with standing in nomenclature. Nucleic Acids Res. 2014;42(Database issue):D613–6. https://doi.org/10.1513/AnnalsATS.201610-802PS.
3. Winthrop KL, Henkle E, Walker A, Cassidy M, Hedberg K, Schafer S. On the Reportability of nontuberculous mycobacterial disease to public health authorities. Ann Am Thorac Soc. 2016;14(3):314–7. https://doi.org/10.1513/AnnalsATS.201610-802PS.
4. Salfinger M, Falkinham JO, Leitman PB. Has the time come to require nontuberculous mycobacteria reporting? NTM-TB Insights. 2016:1–4.
5. Prevots DR, Shaw PA, Strickland D, Jackson LA, Raebel MA, Blosky MA, Montes de Oca R, Shea YR, Seitz AE, Holland SM, Olivier KN. Nontuberculous mycobacterial lung disease prevalence at four integrated health care delivery systems. Am J Respir Crit Care Med. 2010;182:970–6.
6. Adjemian J, Frankland TB, Daida YG, Honda JR, Oliver KN, Zelazny A, Honda S, Prevots DR. Epidemiology of nontuberculous mycobacterial lung disease and tuberculosis, Hawaii, USA. Emerg Infect Dis. 2017;23(3):439–47. https://doi.org/10.3201/eid2303.161827.
7. Smith GS, Ghio AJ, Stout JE, Messier KP, Hudgens EE, Murphy MS, Pfaller SL, Maillard JM, Hilborn ED. Epidemiology of nontuberculous mycobacteria isolations among central North Carolina residents, 2006-2010. J Infect. 2016;72(6):678–86. https://doi.org/10.1016/j.jinf.2016.03.008.
8. Winthrop KL, McNelley E, Kendall B, Marshall-Olson A, Morris C, Cassidy M, Saulson A, Hedberg K. Pulmonary nontuberculous mycobacterial disease prevalence and clinical features: an emerging public health disease. Am J Respir Crit Care Med. 2010;182:977–82.
9. Henkle E, Hedberg K, Schafer SD, Winthrop KL. Surveillance of extrapulmonary nontuberculous mycobacteria infections, Oregon, USA, 2007-2012. Emerg Infect Dis. 2017;23(10):1627–30. https://doi.org/10.3201/eid2310.170845.

10. Donohue MJ, Wymer L. Increasing prevalence rate of nontuberculous mycobacteria infections in five states, 2008-2013. Ann Am Thorac Soc. 2016;13(12):2143–50. https://doi.org/10.1513/AnnalsATS.201605-353OC.
11. Olivier KN, Weber DJ, Wallace RJ Jr, Faiz AR, Lee JH, Zhang Y, Brown-Elliot BA, Handler A, Wilson RW, Schechter MS, Edwards LJ, Chakraborti S, Knowles MR. Nontuberculous mycobacteria. I: multicenter prevalence study in cystic fibrosis. Am J Respir Crit Care Med. 2003;167(6):828–34. https://doi.org/10.1164/rccm.200207-678OC.
12. Adjemian J, Olivier KN, Prevots DR. Nontuberculous mycobacteria among cystic fibrosis patients in the United States: screening practices and environmental risk. Am J Respir Crit Care Med. 2014;190(5):581–6. https://doi.org/10.1164/rccm.201405-0884OC.
13. Adjemian J, Olivier KN, Seitz AE, Holland SM, Prevots DR. Prevalence of nontuberculous mycobacterial lung disease in U.S. medicare beneficiaries. Am J Respir Crit Care Med. 2012;185(8):881–6.
14. Vinnard C, Longworth S, Mezochow A, Patrawalla A, Kreiswirth BN, Hamilton K. Deaths related to nontuberculous mycobacterial infections in the United States, 1999-2014. Ann Am Thorac Soc. 2016;13(11):1951–5. https://doi.org/10.1513/AnnalsATS.201606-474BC.
15. Griffith DE, Aksamit T, Brown-Elliott BA, Catanzaro A, Daley C, Gordin F, Holland SM, Horsburgh R, Huitt G, Iademarco MF, Iseman M, Olivier K, Ruoss S, von Reyn CF, Wallace RJ Jr, Winthrop K, Subcommittee ATSMD, American Thoracic Society, Infectious Disease Society of America. An official ATS/IDSA statement: diagnosis, treatment, and prevention of nontuberculous mycobacterial diseases. Am J Respir Crit Care Med. 2007;175(4):367–416. https://doi.org/10.1164/rccm.200604-571ST.
16. Henkle E, Hedberg K, Schafer S, Novosad S, Winthrop KL. Population-based incidence of pulmonary nontuberculous mycobacterial disease in Oregon 2007 to 2012. Ann Am Thorac Soc. 2015;12(5):642–7. https://doi.org/10.1513/AnnalsATS.201412-559OC.
17. Statistics Canada. Population and dwelling count highlight tables, 2016 census. Ottawa: Statistics Canada; 2017.
18. Marras TK, Mendelson D, Marchand-Austin A, May K, Jamieson FB. Pulmonary nontuberculous mycobacterial disease, Ontario, Canada, 1998-2010. Emerg Infect Dis. 2013;19(11):1889–91.
19. Winthrop KL, Baxter R, Liu L, McFarland B, Austin D, Varley C, Radcliffe L, Suhler E, Choi D, Herrinton LJ. The reliability of diagnostic coding and laboratory data to identify tuberculosis and nontuberculous mycobacterial disease among rheumatoid arthritis patients using anti-tumor necrosis factor therapy. Pharmacoepidemiol Drug Saf. 2011;20(3):229–35. https://doi.org/10.1002/pds.2049.
20. Adjemian J, Prevots DR, Gallagher J, Heap K, Gupta R, Griffith D. Lack of adherence to evidence-based treatment guidelines for nontuberculous mycobacterial lung disease. Ann Am Thorac Soc. 2014;11(1):9–16. https://doi.org/10.1513/AnnalsATS.201304-085OC.
21. Strollo SE, Adjemian J, Adjemian MK, Prevots DR. The burden of pulmonary nontuberculous mycobacterial disease in the United States. Ann Am Thorac Soc. 2015;12(10):1458–64. https://doi.org/10.1513/AnnalsATS.201503-173OC.
22. Diel R, Ringshausen FC, Richter E, Welker L, Schmitz J, Nienhaus A. Microbiological and clinical outcomes of treating non-MAC NTM pulmonary disease: a systematic review and meta-analysis. Chest. 2017;152(1):120–42. https://doi.org/10.1016/j.chest.2017.04.166.
23. Novosad SA, Henkle E, Schafer S, Hedberg K, Ku J, Siegel SA, Choi D, Slatore CG, Winthrop KL. Mortality after respiratory isolation of nontuberculous mycobacteria: a comparison of patients who did and did not meet disease criteria. Ann Am Thorac Soc. 2017;14(7):1112–9. https://doi.org/10.1513/AnnalsATS.201610-800OC.
24. Prevots DR, Adjemian J, Fernandez AG, Knowles MR, Olivier KN. Environmental risks for nontuberculous mycobacteria: individual exposures and climatic factors in the cystic fibrosis population. Ann Am Thorac Soc. 2014;11(7):1032–8. https://doi.org/10.1513/AnnalsATS.201404-184OC.
25. Dirac MA, Horan KL, Doody DR, Meschke JS, Park DR, Jackson LA, Weiss NS, Winthrop KL, Cangelosi GA. Environment or host?: a casecontrol study of risk factors for *Mycobacterium*

avium complex lung disease. Am J Respir Crit Care Med. 2012;186(7):684–91. https://doi.org/10.1164/rccm.201205-0825OC.
26. Maekawa K, Ito Y, Hirai T, Kubo T, Imai S, Tatsumi S, Fujita K, Takakura S, Niimi A, Iinuma Y, Ichiyama S, Togashi K, Mishima M. Environmental risk factors for pulmonary *Mycobacterium avium-intracellulare* complex disease. Chest. 2011;140(3):723–9. https://doi.org/10.1378/chest.10-2315.
27. Adjemian J, Olivier KN, Seitz AE, Falkinham JO 3rd, Holland SM, Prevots DR. Spatial clusters of nontuberculous mycobacterial lung disease in the United States. Am J Respir Crit Care Med. 2012;186(6):553–8. https://doi.org/10.1164/rccm.201205-0913OC.
28. Lipner EM, Knox D, French J, Rudman J, Strong M, Crooks JL. A geospatial epidemiologic analysis of nontuberculous mycobacterial infection: an ecological study in Colorado. Ann Am Thorac Soc. 2017;14(10):1523–32. https://doi.org/10.1513/AnnalsATS.201701-081OC.
29. Brode SK, Jamieson FB, Ng R, Campitelli MA, Kwong JC, Paterson JM, Li P, Marchand-Austin A, Bombardier C, Marras TK. Risk of mycobacterial infections associated with rheumatoid arthritis in Ontario, Canada. Chest. 2014;146(3):563–72. https://doi.org/10.1378/chest.13-2058.
30. Lande L, Peterson DD, Gogoi R, Daum G, Stampler K, Kwait R, Yankowski C, Hauler K, Danley J, Sawicki K, Sawicki J. Association between pulmonary *Mycobacterium avium* complex infection and lung cancer. J Thorac Oncol. 2012;7(9):1345–51. https://doi.org/10.1097/JTO.0b013e31825abd49.
31. Noone PG, Leigh MW, Sannuti A, Minnix SL, Carson JL, Hazucha M, Zariwala MA, Knowles MR. Primary ciliary dyskinesia: diagnostic and phenotypic features. Am J Respir Crit Care Med. 2004;169(4):459–67. https://doi.org/10.1164/rccm.200303-365OC.
32. Fowler CJ, Olivier KN, Leung JM, Smith CC, Huth AG, Root H, Kuhns DB, Logun C, Zelazny A, Frein CA, Daub J, Haney C, Shelhamer JH, Bryant CE, Holland SM. Abnormal nasal nitric oxide production, ciliary beat frequency, and toll-like receptor response in pulmonary nontuberculous mycobacterial disease epithelium. Am J Respir Crit Care Med. 2013;187(12):1374–81. https://doi.org/10.1164/rccm.201212-2197OC.
33. Winthrop KL, Baxter R, Liu L, Varley CD, Curtis JR, Baddley JW, McFarland B, Austin D, Radcliffe L, Suhler E, Choi D, Rosenbaum JT, Herrinton LJ. Mycobacterial diseases and antitumour necrosis factor therapy in USA. Ann Rheum Dis. 2013;72(1):37–42. https://doi.org/10.1136/annrheumdis-2011-200690.
34. Kim RD, Greenberg DE, Ehrmantraut ME, Guide SV, Ding L, Shea Y, Brown MR, Chernick M, Steagall WK, Glasgow CG, Lin JP, Jolley C, Sorbara L, Raffeld M, Hill S, Avila N, Sachdev V, Barnhart LA, Anderson VL, Claypool L, Hilligoss DM, Garofalo M, Fitzgerald A, Anaya-O'Brien S, Darnell D, DeCastro R, Menning HM, Ricklefs SM, Porcella SF, Olivier KN, Moss J, Holland SM. Pulmonary nontuberculous mycobacterial disease: prospective study of a distinct preexisting syndrome. Am J Respir Crit Care Med. 2008;178:1066–74.
35. Leung JM, Fowler C, Smith C, Adjemian J, Frein C, Claypool RJ, Holland SM, Prevots RD, Olivier K. A familial syndrome of pulmonary nontuberculous mycobacteria infections. Am J Respir Crit Care Med. 2013;188(11):1373–6. https://doi.org/10.1164/rccm.201306-1059LE.
36. Colombo RE, Hill SC, Claypool RJ, Holland SM, Olivier KN. Familial clustering of pulmonary nontuberculous mycobacterial disease. Chest. 2010;137(3):629–34. https://doi.org/10.1378/chest.09-1173.
37. Szymanski EP, Leung JM, Fowler CJ, Haney C, Hsu AP, Chen F, Duggal P, Oler AJ, McCormack R, Podack E, Drummond RA, Lionakis MS, Browne SK, Prevots DR, Knowles M, Cutting G, Liu X, Devine SE, Fraser CM, Tettelin H, Olivier KN, Holland SM. Pulmonary nontuberculous mycobacterial infection. A multisystem, multigenic disease. Am J Respir Crit Care Med. 2015;192(5):618–28. https://doi.org/10.1164/rccm.201502-0387OC.
38. Edwards LB, Acquaviva FA, Livesay VT, Cross FW, Palmer CE. Clinical and laboratory studies of tuberculosis and respiratory disease: the navy recruit program. Am Rev Respir Dis. 1969;99(4(2)):1–132.
39. Kirschner RA Jr, Parker BC, Falkinham JO III. Epidemiology of infection by nontuberculous mycobacteria. *Mycobacterium avium, Mycobacterium intracellulare*, and *Mycobacterium scrofulaceum* in acid, brown-water swamps of the southeastern United States and their

association with environmental variables. Am Rev Respir Dis. 1992;145(2 Pt 1):271–5. https://doi.org/10.1164/ajrccm/145.2_Pt_1.271.
40. Spaulding AB, Lai YL, Zelazny AM, Olivier KN, Kadri SS, Prevots DR, Adjemian J. Geographic distribution of nontuberculous mycobacterial species identified among clinical isolates in the United States, 2009-2013. Ann Am Thorac Soc. 2017;14(11):1655–61. https://doi.org/10.1513/AnnalsATS.201611-860OC.
41. Falkinham JO 3rd. Nontuberculous mycobacteria from household plumbing of patients with nontuberculous mycobacteria disease. Emerg Infect Dis. 2011;17(3):419–24. https://doi.org/10.3201/eid1703.101510.
42. Thomson R, Tolson C, Carter R, Coulter C, Huygens F, Hargreaves M. Isolation of nontuberculous mycobacteria (NTM) from household water and shower aerosols in patients with pulmonary disease caused by NTM. J Clin Microbiol. 2013;51(9):3006–11. https://doi.org/10.1128/jcm.00899-13.
43. Schulze-Robbecke R, Buchholtz K. Heat susceptibility of aquatic mycobacteria. Appl Environ Microbiol. 1992;58(6):1869–73.
44. De Groote MA, Pace NR, Fulton K, Falkinham JO 3rd. Relationships between *Mycobacterium isolates* from patients with pulmonary mycobacterial infection and potting soils. Appl Environ Microbiol. 2006;72(12):7602–6. https://doi.org/10.1128/aem.00930-06.
45. Fujita K, Ito Y, Hirai T, Maekawa K, Imai S, Tatsumi S, Niimi A, Iinuma Y, Ichiyama S, Mishima M. Genetic relatedness of *Mycobacterium avium-intracellulare* complex isolates from patients with pulmonary MAC disease and their residential soils. Clinical microbiology and infection: the official publication of the European society of. Clin Microbiol Infect Dis. 2013;19(6):537–41. https://doi.org/10.1111/j.1469-0691.2012.03929.x.
46. Aitken ML, Limaye A, Pottinger P, Whimbey E, Goss CH, Tonelli MR, Cangelosi GA, Olivier KN, Brown-Elliott BA, McNulty S, Wallace RJ Jr. Respiratory outbreak of *Mycobacterium abscessus subspecies massiliense* in a lung transplant and cystic fibrosis center. Am J Respir Crit Care Med. 2012;185:231–2.
47. Bryant JM, Grogono DM, Greaves D, Foweraker J, Roddick I, Inns T, Reacher M, Haworth CS, Curran MD, Harris SR, Peacock SJ, Parkhill J, Floto RA. Whole-genome sequencing to identify transmission of *Mycobacterium abscessus* between patients with cystic fibrosis: a retrospective cohort study. Lancet. 2013;381(9877):1551–60. https://doi.org/10.1016/s0140-6736(13)60632-7.
48. Bryant JM, Grogono DM, Rodriguez-Rincon D, Everall I, Brown KP, Moreno P, Verma D, Hill E, Drijkoningen J, Gilligan P, Esther CR, Noone PG, Giddings O, Bell SC, Thomson R, Wainwright CE, Coulter C, Pandey S, Wood ME, Stockwell RE, Ramsay KA, Sherrard LJ, Kidd TJ, Jabbour N, Johnson GR, Knibbs LD, Morawska L, Sly PD, Jones A, Bilton D, Laurenson I, Ruddy M, Bourke S, Bowler IC, Chapman SJ, Clayton A, Cullen M, Daniels T, Dempsey O, Denton M, Desai M, Drew RJ, Edenborough F, Evans J, Folb J, Humphrey H, Isalska B, Jensen-Fangel S, Jonsson B, Jones AM, Katzenstein TL, Lillebaek T, MacGregor G, Mayell S, Millar M, Modha D, Nash EF, O'Brien C, O'Brien D, Ohri C, Pao CS, Peckham D, Perrin F, Perry A, Pressler T, Prtak L, Qvist T, Robb A, Rodgers H, Schaffer K, Shafi N, van Ingen J, Walshaw M, Watson D, West N, Whitehouse J, Haworth CS, Harris SR, Ordway D, Parkhill J, Floto RA. Emergence and spread of a human-transmissible multidrug-resistant nontuberculous mycobacterium. Science (New York, NY). 2016;354(6313):751–7. https://doi.org/10.1126/science.aaf8156.
49. Tettelin H, Davidson RM, Agrawal S, Aitken ML, Shallom S, Hasan NA, Strong M, de Moura VC, De Groote MA, Duarte RS, Hine E, Parankush S, Su Q, Daugherty SC, Fraser CM, Brown-Elliott BA, Wallace RJ Jr, Holland SM, Sampaio EP, Olivier KN, Jackson M, Zelazny AM. High-level relatedness among *Mycobacterium abscessus* subsp. *massiliense* strains from widely separated outbreaks. Emerg Infect Dis. 2014;20(3):364–71. https://doi.org/10.3201/eid2003.131106.
50. Malcolm KC, Caceres SM, Honda JR, Davidson RM, Epperson LE, Strong M, Chan ED, Nick JA. *Mycobacterium abscessus* displays fitness for fomite transmission. Appl Environ Microbiol. 2017. https://doi.org/10.1128/aem.00562-17.

Global Epidemiology of NTM Disease (Except Northern America)

Dirk Wagner, Marc Lipman, Samantha Cooray, Felix C. Ringshausen, Kozo Morimoto, Won-Jung Koh, and Rachel Thomson

Introduction

In this chapter we provide an overview on the current knowledge on the epidemiology of NTM worldwide. Different sections cover different continents (except Northern America), e.g. Europe (England, Wales, Northern Ireland, Ireland, Scotland and Spain [Section "England, Wales, Northern Ireland, Ireland, Scotland and Spain"] and the remaining Europe [Section "Europe Remain (Europe Except England, Wales, Northern Ireland, Ireland, Scotland and Spain)"], respectively), Asia [Section "Asia"], Oceania [Section "Oceania"], Africa [Section "Africa"] and Middle and Southern America [Section "Central and Southern America"].

D. Wagner (✉)
Division of Infectious Diseases, Department of Internal Medicine II,
Medical Center – University of Freiburg, Faculty of Medicine, Freiburg, Germany
e-mail: Dirk.Wagner@uniklinik-freiburg.de

M. Lipman
UCL Respiratory, University College London & Royal Free London NHS Foundation Trust, London, UK
e-mail: marclipman@nhs.net

S. Cooray
Department of Respiratory Medicine, St Thomas' Hospital, Guys and St Thomas' NHS Foundation Trust, London, UK
e-mail: samantha.cooray@gstt.nhs.uk

F. C. Ringshausen
Department of Respiratory Medicine, Hannover Medical School and German Center for Lung Research, Hannover, Germany
e-mail: Ringshausen.Felix@mh-hannover.de

K. Morimoto
Division of Clinical Research, Fukujuji Hospital, Japan Anti-Tuberculosis Association, Tokyo, Japan
e-mail: morimotok@fukujuji.org

D. E. Griffith (ed.), *Nontuberculous Mycobacterial Disease*, Respiratory Medicine,
https://doi.org/10.1007/978-3-319-93473-0_8

W.-J. Koh
Division of Pulmonary and Critical Care Medicine, Department of Medicine, Samsung Medical Center, Sungkyunkwan University School of Medicine, Seoul, South Korea
e-mail: wjkoh@skku.edu

R. Thomson
Gallipoli Medical Research Institute, University of Queensland, Brisbane, Australia
e-mail: R.Thomson@uq.edu.au

England, Wales, Northern Ireland, Ireland, Scotland and Spain

Marc Lipman and Samantha Cooray

England, Wales and Northern Ireland

In England, Wales and Northern Ireland, hospital laboratories voluntarily report mycobacterial infections to the public health authorities. Although there are a number of factors that limit the conclusions that may be drawn from this (e.g. often the associated clinical information is fairly simple), it provides a reasonable representation of NTM isolation frequency within a large and national population.

Moore et al. reported the frequency of NTM reports received by the public health service between 1995 and 2006 [1]. This demonstrated an overall increase in both the number and rate of reports (from 0.9 per 100,000 population to 2.9 per 100,000 by 2006). The data included both pulmonary and extrapulmonary specimens, though the main increase was noted in lung-derived samples – in particular in people aged 60 years or over.

Forty-three percent of isolates were *Mycobacterium avium* complex (MAC), with other common species being *M. malmoense* (14%) and *M. kansasii* (13%) (Table 1). Four-fifths of isolates were obtained from a pulmonary sample, though this was less often the case in children aged under 15.

The authors noted that less than 1 in 10 of the 4732 reports had useful associated clinical information. Notwithstanding this, it appeared that HIV infection and cystic fibrosis (CF) were generally not contributing to the increased numbers of isolates, though chronic respiratory illness might be partly responsible [1].

More recently the public health service has reported cumulative data from England, Wales and Northern Ireland covering 2007 to 2012 [2]. Here, the authors linked first isolates with individuals and excluded any subsequent positive results. 21,118 people had NTM culture-positive isolates. Sixteen percent of these were *M. gordonae*, and, given its low likelihood of causing disease, results were presented with and without this organism. When excluded, the overall incidence of NTM rose over the study period from 4.8 per 100,000 to 6.3 per 100,000, $p < 0.001$.

Table 1 Studies from England, Wales and Northern Ireland that have published data on isolation of and disease due to NTM

			Various proportional data					Prevalent species				
Study location	Study period	Target population	NTM isolate/culture-positive specimen (%)	Patient with NTM isolate/culture-positive patient (%)	NTM-PD patient[a]/patient with NTM isolate (%)	NTM-PD patient[a]/NTM culture-positive specimens (%)	*Median* or mean age, female %	Proportion of each relevant NTM species among NTM-PD patients	Identification method	Additional information [proportion of prevalent species of isolates]	Annotation	Reference
England, Wales and Northern Ireland (all regions)	1995–2006	Patients, including extrapulmonary disease	NA	NA	8832/10,895 (81%)	NA	NA, 40%	NA		Annual incidence increased from 0.9 to 2.9 MAC incidence increased from 0.4 to 1.2 [MAC 43%, *M. malmoense* 14% *M. kansasii* 13%]	81% had pulmonary disease	[1]
South London, England	2000–2007	Non-HIV-infected patients >18 years	NA	NA	NA	NA	*61*, 32%	NA	Biochemical methods, 16s rDNA PCR	[*M. kansasii* 69%, MAC 12%, *M. xenopi* 12%, *M. malmoense* 7%]	79 NTM-positive isolates, 57 of which had associated clinical data 37% COPD, 70% smokers, 28% alcoholism, 9% bronchiectasis, 5% pulmonary fibrosis Overall mortality 15%	[7]

(continued)

Table 1 (continued)

Study location	Study period	Target population	Various proportional data				*Median* or mean age, female %	Prevalent species	Identification method	Additional information [proportion of prevalent species of isolates]	Annotation	Reference
			NTM isolate/ culture-positive specimen (%)	Patient with NTM isolate/ culture-positive patient (%)	NTM-PD patient[a]/ patient with NTM isolate (%)	NTM-PD patient[a]/ NTM culture-positive specimens (%)		Proportion of each relevant NTM species among NTM-PD patients				
Cambridge, England	2002–2003	Patients with diagnosis of bronchiectasis	10/97 (10.3%)	NA	NA	NA	59, 67%	NA	Biochemical methods	Of 23 total isolates: *M. fortuitum* 26%, MAC 21%, *M. xenopi* 21%, *M. chelonae* 4%, *M. malmoense* 4%, *M. terrae*, 4%, *M. simiae* 4%	10/97 (10.3%) bronchiectasis patients NTM culture positive	[3]
England, Wales and Northern Ireland (all regions)	2007–2012	Patients, including extrapulmonary disease	NA	NA	16,294/21,118 (90.9%)	NA	NA, 42%	MAC 35.6%, *M. gordonae* 16.7%, *M. chelonae* 9.6%, *M. fortuitum* 8.2%, *M. kansasii* 5.9%, *M. xenopi* 5.9%, *M. abscessus* 5%	MALDI-TOF of 16s ribosomal proteins, 16s rDNA comparative sequencing	Annual incidence increased from 5.6 to 7.6 NTM pulmonary disease increased from 4.0 to 6.1 Extrapulmonary disease incidence fell from 0.6 to 0.4 Pulmonary MAC disease incidence increased from 1.3 to 2.2	90.9% had pulmonary disease	[2]

London, England (2 reference laboratories)	2008	Patients, excluding extrapulmonary disease with NTM isolated	NA	NA	NA	NA	NA	*[20% M. abscessus, 19% MAC, 10.6% M. malmoense, 10% M. mucogenicum, 9.4% M. szulgai, 6.7% M. simiae, 5.6% M. terrae, 3.3% M. scrofulaceum, 2.8% M. genavense, 1.1% M. gordonae]*				[13]
United Kingdom	2009	Patients attending CF clinics	300/7122 (4.2%)	NA	NA	NA	NA	NA	Biochemical methods	[*M. abscessus* 62% adults and 68% children, MAC 27% adults and 28% children. Remaining 8% *M. gordonae, kansasii, fortuitum, xenopi, simiae, malmoense, mucogenicum, peregrinum*]	Overall prevalence of 4.2% in the UK (5% in children) Lowest prevalence in Northern Ireland (1.7%) Highest prevalence in SE England (7.5%)	[4]

MAC *Mycobacterium avium* complex, *COPD* chronic obstructive pulmonary disease, *CF* cystic fibrosis
[a]*NTM-PD*: nontuberculous mycobacterial pulmonary disease

Over 90% of the isolates with a known specimen site were from the lung – and the subjects with positive pulmonary samples were older than those with extrapulmonary isolates (mean age 60 years versus 53 years, respectively).

As in the earlier study, the increase in overall incidence reflected a rise in pulmonary isolates, from 3.4 per 100,000 to 5.0 per 100,000. However, extrapulmonary positive cultures actually declined during the same time period (from 0.6 per 100,000 to 0.4 per 100,000) [2]. The data indicated that MAC was the most common NTM cultured from pulmonary samples (present in 35.6% of cases). This was followed by *M. gordonae* (16.7%), *M. chelonae* (9.6%), *M. fortuitum* (8.2%), *M. kansasii* (5.9%), *M. xenopi* (5.9%) and *M. abscessus* (5%).

Whilst most pulmonary isolates increased with increasing age, *M. abscessus* had a biphasic age distribution with peaks between ages 10–19 and 70–79 years. Other rapid-growing NTM did not follow this pattern – and the authors suggested that the younger age group represented patients with CF lung disease and associated *M. abscessus.* Once again MAC was the most frequent organism cultured from extrapulmonary samples (34% of cases). The commonest sites for positive extrapulmonary cultures were blood, lymph nodes and urine [2].

Whilst this study has considerable statistical power due to its large size, it suffers from a lack of clinical data and also the uncertainty as to whether all culture-positive isolates were in fact reported to public health services. In mitigation of this, the authors' use of just the first positive isolate would make it less likely that this was a significant cause of bias. It is also noteworthy that MAC was the commonest overall mycobacterial isolate, and not a more rapid grower (which may have been expected if the increase in NTM over time was a reflection of improved bacteriological techniques). The study also suffers from its inability to speciate NTM accurately. All species that comprise MAC were described as a single group; and *M. abscessus* could not be sub-speciated.

Other reports from England, Wales and Northern Ireland have either focused on specific patient population (CF, the HIV infected or bronchiectatics) [3–6] or have been more local in their geographic setting [7].

These studies suggest that around 5% of CF populations have been diagnosed with pulmonary NTM disease during their clinical care [4]. This appeared to be more likely in patients with indicators of advancing disease, e.g. worse lung physiology, and increased isolation of complex bacteria or fungi [5].

The widespread uptake of antiretroviral therapy in HIV-infected populations has led to dramatic changes in its associated infection epidemiology. The UK is no exception to this, and NTM disease which was typically disseminated, and usually MAC diagnosed on positive blood cultures or another extrapulmonary site, in reports such as that from Yates covering the period 1984–1992, is now much less common [6]. In HIV patients using antiretroviral therapy, NTM disease is usually pulmonary and associated with underlying structural lung damage due to previous infection or emphysema.

The geographic distribution of NTM appears to vary somewhat across England. Whilst Moore [1] showed that the North East of England and London provided the greatest number of reports of positive NTM isolates, an association with predisposing disordered anatomy due to pre-existing lung disease and specific environmental local factors cannot be ignored. A recent report from South London (of 57 patients satisfying the ATS 2007 criteria for pulmonary NTM) noted that the most common organism implicated was not MAC but *M. kansasii*. The authors commented that their patients with this organism had significant pulmonary and systemic symptoms and signs – implying that what they were seeing was not overcalling of a bystander organism [7].

Scotland

To date, few NTM epidemiological studies have been carried out in Scotland. Bollert et al. using Scottish Mycobacteria Reference Laboratory (SMRL) data [8] investigated the prevalence of NTM pulmonary isolates between 1990 and 1993. They had previously noted empirically that a particular geographic area of South Scotland (Lothian, including Edinburgh) appeared to report cases of pulmonary NTM – and in particular *M. malmoense* – at a higher frequency than elsewhere. Overall 248/1005 (24.7%) mycobacterial culture-positive samples were identified as NTM, with 53% of isolates in Lothian being NTM, compared to 18% elsewhere (Table 2). *M. malmoense* was isolated more frequently in Lothian than in regions beyond this area (41/108 (37.9%) compared to 41/140 (29.2%) cases, respectively). Why Lothian should appear to be the site of so much pulmonary NTM and possibly more *M. malmoense*-associated pulmonary NTM disease is unclear and cannot be explained by differences in population size alone or the variation in laboratory experience in NTM identification. The authors described the clinical management of the patients in Lothian with *M. malmoense* (which was one of the aims of their study). They found that 75% of patients received treatment, with one-third dying during follow-up. Information on possible underlying immunosuppression was limited.

A more detailed retrospective study of pulmonary and extrapulmonary NTM isolates from 2000 to 2010 in the SMRL sought to examine trends over time and further explore the clinical significance of pulmonary isolates [using the ATS "microbiological criteria"] [9]. A single positive extrapulmonary culture was taken as significant. *M. gordonae* was considered a likely contaminant/colonizer and so excluded from analysis. Overall there was a mean incidence of 2.43/100000, with no clear trend over time [9]. Sixty-four percent of NTM specimens came from patients over 50, with the exception of adenitis samples; 72% were from children under 5 years of age. Rates were slightly higher in males (2.53) compared to females (2.33), and isolation is more frequent in men (46.5% were in women). The majority of NTM isolates came from pulmonary specimens 933/1370 (68.1%), collected from 557 patients. 86.5% (806/933) were considered clinically significant. MAC was the most common species identified in pulmonary specimens (44.8%), followed by *M. malmoense* 21.7% and *M. abscessus* 13.7% (Table 2). *M. marinum* was the most frequently isolated organism from cutaneous specimens (32.1%), followed by *M. chelonae* 28.3%.

Table 2 Studies from Scotland that have published data on isolation of and disease due to NTM

			Various proportional data					Prevalent species				
Study location	Study period	Target population	NTM isolate/culture-positive specimen (%)	Patient with NTM isolate/culture-positive patient (%)	NTM-PD patient[a]/patient with NTM isolate (%)	NTM-PD patient[a]/NTM culture-positive specimens (%)	*Median* or mean age, female %	Proportion of each relevant NTM species among NTM-PD patients	Identification method	Additional information [proportion of prevalent species of isolates]	Annotation	Reference
Lothian and the rest of Scotland	1990–1993	Patients with culture-positive pulmonary isolates	NA	NA	NA	NA	NA	MAC 32.8%, *M. malmoense* 38%, *M. xenopi* 7.8%, *M. kansasii* 8.5%			Lothian had 2× as much NTM strains as the rest of Scotland. 75% of patients in Lothian positive for M malmoense had clinically significant lung disease	[8]
Scotland (all regions)	2000–2010	Patients, including extrapulmonary disease	931/1370 (68%)	NA	806/931 (86.5%)	NA	NA, 46.5%	MAC 44.8%, *M. malmoense* 21.7%, *M. abscessus* 13.7%, *M. chelonae* 2.6%, *M. fortuitum* 1.4%, *M. kansasii* 3.9%, *M. xenopi* 4.5%, *M. celatum* 1.5%, *M. szulgai* 1%		Mean NTM incidence 2.43	81% of M. *marinum* isolated were from cutaneous specimens. 13% of CF patients had NTM isolated from the lung	[9]

MAC Mycobacterium avium complex, *CF* cystic fibrosis
[a]*NTM-PD*: nontuberculous mycobacterial pulmonary disease

The apparently high frequency of detection of *M. malmoense* has been reported previously from the SMRL, where it comprised 39% of all NTM isolates between 1985 and 1990 [10].

The authors noted that over the study period, 209/557 (37.5%) patients had another episode of infection (defined as an organism identified >12 months from a previous positive specimen). However, in 94% these were the same organism and may in fact have represented relapse since molecular methods were not used to distinguish reinfection from relapse.

Their study could identify patients with cystic fibrosis – and this was strongly associated with *M. abscessus* (68% CF-related NTM episodes compared to 5% in the rest of the population). Of note, females with CF appeared more likely to have *M. abscessus* infection.

In agreement with Bollert et al., this study found the highest rates of NTM in Lothian and Greater Glasgow (32.5 and 3.66 per 100,000, respectively), far greater than the national average rate of 1.34–1.49. The reasons for these differences remain unclear, though they pointed out that there is little guidance for local laboratories to determine which positive samples to send – with some passing on all to the SMRL for speciation and others sending only enough to determine the organism and if this constituted clinical disease.

Ireland

A single population-based study (population >500,000) carried out in South-West Ireland (Counties Cork and Kerry), between 1987 and 2000, investigated rates of NTM compared to MTB and *M. bovis* [11]. Criteria were used to define episodes of disease (ATS 1990 and subsequently 1997 pulmonary NTM criteria; extrapulmonary positive culture). The mean annual incidence of disease-causing isolates was 0.4 per 100,000. This rose gradually between 1995 and 2000 – and was statistically significant ($p < 0.001$), mainly due to an increase in pulmonary NTM. Although this may be explained by better detection methods, the authors note that the increase in NTM had been occurring before more advanced culture methods were introduced.

NTM was identified in 143/960 (18%) of all mycobacterial cultures. 32 (22.3%) of these isolates were thought to be disease-causing – with 17 (53%) from pulmonary samples. The male-to-female ratio of disease-causing isolates was 1:1. MAC was the commonest cause of disease at both pulmonary and extrapulmonary sites. The mean annual incidence of MAC was 0.3 per 100,000. Of 32 isolates, 23 (71.9%) were MAC, 5 (15.6%) *M. malmoense*, 2 (6.3%) *M. abscessus*, 1 (3.1%) *M. kansasii* and 1 (3.1%) *M. marinum* (Table 3).

Another retrospective study in Ireland identified patients with positive NTM isolated from pulmonary specimens through review of historical medical records [12]. The authors excluded people with CF and identified 37 respiratory patients (18 female, 19 male) who had NTM isolated between January 2007 and July 2012. *M. avium* was the most common NTM, present in 60% (22) of patients. Other NTM

Table 3 Studies from Ireland that have published data on isolation of and disease due to NTM

Study location	Study period	Target population	Various proportional data				*Median* or mean age, female %	Prevalent species	Identification method	Additional information [proportion of prevalent species of isolates]	Annotation	Reference
			NTM isolate/ culture-positive specimen (%)	Patient with NTM isolate/ culture-positive patient (%)	NTM-PD patient[a]/ patient with NTM isolate (%)	NTM-PD patient[a]/ NTM culture-positive specimens (%)		Proportion of each relevant NTM species among NTM-PD patients				
South West Ireland (counties Cork and Kerry)	1987–2000	Patients, including extrapulmonary disease	NA	175/960 (18.2%)	NA	NA	40, 50%	NA		Mean annual NTM incidence 0.4 (1987–2000) and 0.63 (1995–2000). Mean annual MAC incidence was 0.3. [MAC 64%, *M. malmoense* 15.6%, *M. abscessus* 6.25%, *M. kansasii* 3.1%, *M. marinum* 3.1%]	53% of NTM cases were pulmonary, 47% were extrapulmonary. 32/175 NTM isolates were considered disease-causing with 143 regarded as contaminants. MAC isolates were from pulmonary specimens	[11]

Ireland (Dublin)	2007–2012	Respiratory patients, excluding CF and extrapulmonary isolates	NA				NA, 48%	*M. avium* 60%, *M. intracellulare* 11%, *M. gordonae* 4%, *M. szulgai* 5%, *M. chelonae* 5%, *M. abscessus* 5%, *M. malmoense* 3%		NTM-positive specimen identified in 37 patients from review of medical records. (18 females, 19 males)	7 male patients with NTM had COPD and 10 females with NTM had bronchiectasis	[12]
Ireland (all counties)	2000	Patients, excluding extrapulmonary disease with NTM isolated	NA	NA	NA	NA	NA	*M. avium* 31%, *M. intracellulare* 12.8%, *M. chelonae* 10.3%, *M. interjectum* 10.3%, *M. malmoense* 9%, *M. kansasii* 7.7%, *M. xenopi* 3.85%, *M. abscessus* 3.85%]				[13]

MAC Mycobacterium avium complex
[a]*NTM-PD*: nontuberculous mycobacterial pulmonary disease

included *M. intracellulare* (11%) (elsewhere often combined with *M. avium* as MAC), *M. gordonae* (4%), *M. szulgai* (5%), *M. chelonae* 5% and *M. abscessus* 5% (Table 3). There was a higher frequency of NTM in men with COPD and women with bronchiectasis.

In a global cross-sectional study of NTM in pulmonary specimens in 2008 [13], Ireland's National Mycobacterial Reference Laboratory reported 78 NTM positive isolates, of which 30.8% (24) were *M. avium*, 12.8% (10) *M. intracellulare*, 10.3% (8) *M. chelonae*, 10.3% (8) *M. interjectum*, 9% (7) *M. malmoense* and 7.8% (6) *M. kansasii* (Table 3). As with other Northern European countries, MAC accounted for the majority of cases, a picture unchanged from the earlier regional studies. The small numbers of other identified NTM species makes it difficult to be precise about variation in their frequency compared to each other and over time. The estimated mean annual incidence rate of pulmonary NTM isolates in 2008 was 1.64 per 100,000 – a marked increase compared to the previous study. However, as noted elsewhere, it is difficult to disentangle the strength of the signal from increased clinical awareness of NTM disease and improved laboratory methods over time.

Spain

Studies of NTM epidemiology in Spain are fairly limited and have focussed on certain at-risk groups such as non-CF bronchiectasis or HIV. An exception is a large isolate-based study that, using data on 11,128 isolates from 41 laboratories, showed more NTM were identified between 1991 and 1996 than in the preceding 20 years [10]. Once again this suffered from a number of potential biases – including issues of improving detection techniques over time.

More recently the 2008 NTM-NET collaborative cross-sectional study identified 805 NTM isolates in pulmonary specimens from patients in Madrid and Barcelona [13]. 39% (313) were MAC, 9.7% *M. gordonae*, 8.9% *M. fortuitum*, 7.7% *M. xenopi*, 3.6% *M. chelonae*, 3.5% *M. abscessus* and 2.1% *M. kansasii* (Table 4).

A multicentre study of 218 adults with non-CF bronchiectasis followed up between 2002 and 2010 for >5 years estimated the NTM prevalence [14]. NTM were isolated from the sputum of 18/218 (8.2%) patients. MAC was again the most common organism (present in 50% of cases). Others were *M. abscessus*, *M. fortuitum*, *M. gordonae*, *M. chelonae* and *M. lentiflavum*. Patients with NTM tended to be older (64 vs. 54.9 years, $p < 0.05$) and have better lung function (FVC >75%), though lower body mass index (?BMI). They were less frequently colonized with *P. aeruginosa* or *H. Influenzae* but more often with *Aspergillus*. In the 18 patients where NTM was isolated, 33% were smear-positive, and 28% met the 2007 ATS criteria for NTM lung disease.

Another study using data collected by the Spanish Ministry of Health's Minimum Basic Data Set of hospital admissions between 1997 and 2010 estimated the incidence of NTM in patients with and without HIV [15]. Inevitably the rate in HIV-infected individuals was 1000-fold higher. However, whilst it decreased over time

Table 4 Studies from Spain that have published data on isolation of and disease due to NTM

Study location	Study period	Target population	Various proportional data				*Median* or mean age, female %	Prevalent species	Identification method	Additional information [proportion of prevalent species of isolates]	Annotation	Reference
			NTM isolate/ culture-positive specimen (%)	Patient with NTM isolate/ culture-positive patient (%)	NTM-PD patient[a]/ patient with NTM isolate (%)	NTM-PD patient[a]/ NTM culture-positive specimens (%)		Proportion of each relevant NTM species among NTM-PD patients				
Spain (all regions – MBDS)	1997–2010	Hospitalized patients screened for HIV	NA	NA	NA	NA	63, 36.8% (HIV negative) 36, 24.1% (HIV positive)			Incidence increased from 2.91 to 3.97 in non-HIV patients. Rate decreased in HIV patients from 2.29 to 0.71	3729 NTM isolates	[15]
Spain (all regions)	2002–2010	Non-CF bronchiectasis patients	NA	18/218 (8.3%)	NA	NA	64, 78%	NA	In situ hybridization and sequencing of 16s rRNA	[MAC 50%, *M. abscessus* 14%, *M. gordonae* 11%, *M. fortuitum* 11%, *M. chelonae* 5%, *M. lentiflavum* 5%]	Prevalence of NTM in study population 8.3%	[14]

(continued)

Table 4 (continued)

Study location	Study period	Target population	Various proportional data					Prevalent species				
			NTM isolate/ culture-positive specimen (%)	Patient with NTM isolate/ culture-positive patient (%)	NTM-PD patient[a]/ patient with NTM isolate (%)	NTM-PD patient[a]/ NTM culture-positive specimens (%)	*Median* or mean age, female %	Proportion of each relevant NTM species among NTM-PD patients	Identification method	Additional information [proportion of prevalent species of isolates]	Annotation	Reference
Ireland (all counties)	2000	Patients, excluding extrapulmonary disease with NTM isolated	NA	NA	NA	NA	NA	*[39% MAC, 9.7% M. gordonae, 8.9% M. fortuitum, 7.7% M. xenopi, 3.6% M. chelonae, 3.5% M. abscessus and 2.1% M. kansasii. Others included M. lentiflavum, mucogenicum, mageritense, scrofulaceum, terrae, simiae, kumamotense, szulgai, peregrinum, interjectum, immunogenum, arupense, elephantis, thermoresistible, asiaticum, parmense and brumae]*				[13]

MAC Mycobacterium avium complex, *CF* cystic fibrosis, *MBDS* Minimum Basic Data Set, Spanish Ministry of Health
[a]*NTM-PD*: nontuberculous mycobacterial pulmonary disease

in HIV-positive patients (from 2.29 to 0.71 per 1000 patient years) presumably due to the increased use of effective antiretroviral therapy, it rose in the HIV-negative population from 0.29 (1997–1999) to 0.4 (2000–2004) per 100,000 patient years. Mortality in the HIV-negative group appeared to increase over time – which may reflect irreversible comorbidities such as chronic respiratory disease, present in over one-third of subjects.

Summary

Published studies from England, Wales, Northern Ireland, Ireland, Scotland and Spain suggest that there has been a consistent increase in reports of NTM. The frequency of positive isolates increases with age – and is generally in pulmonary samples. However the quality and breadth of the data are variable across studies and countries. The most frequently isolated organism is *Mycobacterium avium* complex – generally both from pulmonary and extrapulmonary samples. There appears to be some geographic distribution to isolates, e.g. Scotland has a higher isolation frequency of *M. malmoense*. Local epidemiology in Scotland, however, seems not to be due to different laboratory techniques. The evidence for an association with structural lung disease is reported, though the lack of good clinical data linked to mycobacterial isolates makes it hard to be specific about risk factors such as chronic obstructive pulmonary disease, previous tuberculosis, or the use of confounders such as inhaled steroids which may be used more in people with lung disease.

Europe Remain (Europe Except England, Wales, Northern Ireland, Ireland, Scotland and Spain)

Felix C. Ringshausen

Europe (Remain)

There are numerous recent studies reporting about NTM epidemiology across Europe [16]. However, most of these reports use variable methodology, ranging from single centre [17] to nationally representative [18] and from isolate-based to sentinel-site [19, 20] or population-based approaches [21–24].

In addition, study target populations differ considerably across the available studies. Some studies reported outcomes from positive mycobacterial isolates from respiratory specimens only [25, 26], while others included extrapulmonary samples [27]. One study covered subjects who received treatment for NTM disease only [28], while others focused on hospitalized subjects [17, 22] or subjects with concomitant pulmonary TB [29]. *M. gordonae* was excluded in one study as considered to be non-pathogenic [30], while other studies found this species in up to 12% of subjects with NTM-PD [31]. Some of the available reports provide data on unique NTM species [32–37] or (pseudo-)outbreaks at a single centre (or laboratory) [33, 38], while others focus on populations with increased susceptibility to NTM-PD, such as subjects with cystic fibrosis (CF) or chronic obstructive pulmonary disease (COPD) [33, 39–43], or summarize international multicentre experience [44–46]. Some studies included only HIV-negative subjects [17, 20, 28, 31].

With regard to isolated species, there are considerable differences between and even within studies, which may be attributable to regional geographical rather than country-specific variations. In summary, MAC appears to be the most prevalent species. Hoefsloot and colleagues provided a comprehensive overview of different NTM isolation frequencies between European laboratories in their pivotal "NTM world map" paper [45]. However, these differences in study designs and methodologies often complicate comparison between data and hinder meaningful conclusions to be drawn. The heterogeneity in methodology is essentially caused by the fact that neither NTM isolation nor NTM-PD or clinical significant extrapulmonary NTM infections are notifiable health conditions in most European countries [46].

Overall, data on several aspects of the epidemiology of NTM are available for the following European countries: Austria, Belgium, Croatia, Czech Republic, Denmark, Estonia, Finland, France, Germany, Greece, Hungary, Italy, Luxembourg, the Netherlands, Norway, Poland, Portugal, Slovenia, Slovakia, Spain, Sweden, Switzerland and the United Kingdom of Great Britain and Northern Ireland including Scotland (see Table 5 for selected examples) [22–24, 32, 33, 44–48].

Those studies reporting on trends over time found consistent increases in isolation frequency [36, 49, 50], hospitalization frequency [22] or NTM-PD prevalence [23]. A German population-based study using routine statutory health insurance claim data reported increasing prevalence rates from 2.3 to 3.3 cases per 100,000 population from 2009 to 2014 [23]. Incidence rates of NTM-PD have been reported for a few European studies: in France 0.72 to 0.74 per 100,000 population between 2001 and 2003 [20]; in Greece 0.54 to 0.94 per 100,000 population between 2004 and 2006 [30]; in Denmark, a stable mean annual incidence rate of NTM-PD of 1.08 per 100,000 population between 2003 and 2008 [21]; and in Croatia an annual incidence rate of 0.23 per 100,000 population between 2006 and 2010. Very recently the cumulative incidence rate for NTM-PD was reported as 2.6 per 100,000 insured persons (95% CI 2.2–3.1) in Germany in 2010 and 2011, with an increased mortality rate for patients with NTM-PD compared to a age-, sex- and comorbidity-matched control group (22.4% vs. 6%, $p < 0.001$) [24].

Apparently, European subjects with NTM isolation and NTM-PD differ from North American populations in terms of a slightly younger age and a higher proportion of male subjects [16]. In Europe, COPD appears to be a major risk factor for NTM isolation and disease and is the most prevalent comorbidity [22, 23, 30, 50]. In addition, COPD is associated with increased mortality among patients with NTM disease [21, 24]. In a recent German health economic study, the mean direct

Table 5 Selected studies from Northern, Central, Western, Southern and Southeastern Europe that have published data on isolation of and disease due to NTM

Study location	Study period	Target population	Various proportional data				*Median* or mean age, female %	Prevalent species	Identification method	Additional information [proportion of prevalent species of isolates]	Annotation	Reference
			NTM isolate/culture-positive specimen (%)	Patient with NTM-isolate/culture-positive patient (%)	NTM-PD patient[a]/patient with NTM-isolate (%)	NTM-PD patient[a]/NTM culture-positive specimens (%)		Proportion of each prevalent NTM species among NTM-PD patients				
Croatia	2006–2010	All Croatian residents with NTM isolated from respiratory samples by culture	NA	NA	65/1187 (5%), according to ATS-defined microbiological criteria ("probable NTM-PD")	65/1710 (4%), according to ATS-defined microbiological criteria ("probable NTM-PD")	Median 66 years; 48%	*M. xenopi* 32%, *M. avium* 22%, *M. intracellulare* 12%, *M. gordonae* 8%, *M. kansasii* 6%, *M. fortuitum* 5%, M. abscessus 5%, other 11% ("probable NTM-PD")	GenoType Mycobacterium CM/AS (Hain Lifescience, Germany), supplemented with phenotypic methods	*M. gordonae* 43%, *M. xenopi* 16%, *M. fortuitum* 12%, *M. terrae* 8%, *M. avium* 4%, *M. abscessus* 3%, *M. intracellulare* 2%, *M. kansasii* 1%, other 14%	Annual incidence of probable NTM-PD was 0.23 per 100,000 population. Isolation frequency increased over the study period. Of note, limited clinical data was available. Thus probable NTM-PD was used according to ATS microbiological criteria	[18]

(continued)

Table 5 (continued)

Study location	Study period	Target population	Various proportional data				Median or mean age, female %	Prevalent species	Identification method	Additional information [proportion of prevalent species of isolates]	Annotation	Reference
			NTM isolate/culture-positive specimen (%)	Patient with NTM-isolate/culture-positive patient (%)	NTM-PD patient[a]/patient with NTM-isolate (%)	NTM-PD patient[a]/NTM culture-positive specimens (%)		Proportion of each prevalent NTM species among NTM-PD patients				
Zabok, Croatia	2007–2012	Patients with suspected mycobacterial infection	NA	150/3903 (4%)	0/150 (0%)	0/150 (0%)	Men median 70 years, women median 74 years, 59/150 women (39%)	NA	GenoType Mycobacterium CM/AS (Hain Lifescience, Germany)	*M. gordonae* (90%), *M. fortuitum* (5%), *M. nonchromogenicum* (3%), *M. terrae* and *M. xenopi* (1%, each)	Nosocomial pseudo-outbreak of *Mycobacterium gordonae* associated with a hospital's water supply contamination	[38]

Denmark	1997–2008	All Danish residents aged > = 15 years, with at least one NTM-positive pulmonary bacteriological specimen	NA	NA	573/1282 (45%)	NA	Mean 61 years, 41%	MAC 57%, M. malmoense 8%, *M. xenopi* 8%, *M. gordonae* 5%, other non-RGM 12% (including *M. celatum* 4% and *M. szulgai* 2%), RGM 10% (including *M. abscessus* 7% and *M. fortuitum* 1.3%)	NA		The mean annual age-standardized incidence rate of patients with at least 1 NTM-positive specimen was 2.44 per 100,000 person-years between 1997 and 2008, with an incidence rate per 100,000 person-years of 1.36 for NTM colonization and 1.08 for NTM disease. Patients with NTM colonization and disease had similarly poor prognosis. Negative prognostic factors included high levels of comorbidity, advanced age, male sex and *M. xenopi*	[21]

(continued)

Table 5 (continued)

			Various proportional data					Prevalent species				
Study location	Study period	Target population	NTM isolate/ culture-positive specimen (%)	Patient with NTM-isolate/ culture-positive patient (%)	NTM-PD patient[a]/patient with NTM-isolate (%)	NTM-PD patient[a]/NTM culture-positive specimens (%)	*Median* or mean age, female %	Proportion of each prevalent NTM species among NTM-PD patients	Identification method	Additional information [proportion of prevalent species of isolates]	Annotation	Reference
France	2001–2003	HIV-negative patients with NTM isolates according to ATS-defined bacteriological criteria	NA	NA	NA	NA	Mean age ranged from 70 years among MAC patients to 54 years among subjects with infections due to *M. kansasii*; the proportion of females ranged from 60% among MAC patients to 21% among subjects infected by *M. kansasii* or *M. xenopi*	MAC 48%, *M. xenopi* 25%, *M. kansasii* 13%, *M. abscessus* 10%, other 5%	Identification of isolates was performed using approved methods, including biochemical tests and/or molecular tests such as DNA probes or genomic typing		In the Paris area, M. xenopi was the most frequently isolated species, followed by MAC. Most patients (>50%), except those with M. kansasii, had underlying predisposing factors such as pre-existing pulmonary disease or immune deficiency. The incidence of nontuberculous mycobacteria pulmonary infections in HIV- negative patients was estimated at 0.74, 0.73 and 0.72 cases per 100,000 inhabitants in 2001, 2002 and 2003, respectively	[20]

Marseille, France	2010–2014	Pediatric and adult CF patients	NA	25/354 (7%)	2/6 (33%) with *M. lentiflavum* detection	NA	NA, 56%	NA	Specific real-time PCR assay	*M. abscessus* complex 48%, MAC 32%, *M. lentiflavum* 24%; *M. lentiflavum* was only detected in males (mean age of 22 yrs)	This study describes a *M. lentiflavum* outbreak among CF patients	[33]
Germany	2005–2011	Patients with NTM-PD among >125 million hospitalizations	NA	NA	NA	NA	NA	NA	ICD-10 hospital discharge code A31.0, extracted from official nationwide DRG hospital statistics	No data on species available	Between 2005 and 2011, the overall age-adjusted rate of hospitalizations associated with NTM-PD increased from 0.73 to 1.09 per 100,000 population, with the most pronounced increase among females and subjects with CF. COPD, was the most frequent condition associated with NTM-PD and also showed a significant average annual increase of 5%	[22]

(continued)

Table 5 (continued)

Study location	Study period	Target population	Various proportional data				*Median* or mean age, female %	Prevalent species	Identification method	Additional information [proportion of prevalent species of isolates]	Annotation	Reference
			NTM isolate/ culture-positive specimen (%)	Patient with NTM-isolate/ culture-positive patient (%)	NTM-PD patient[a]/patient with NTM-isolate (%)	NTM-PD patient[a]/NTM culture-positive specimens (%)		Proportion of each prevalent NTM species among NTM-PD patients				
Germany	2009–2014	Patients with prevalent NTM-PD among approximately 4 million insured subjects covered by statutory health insurance	NA	NA	NA	NA	Mean 55–61 years, 43–53%	NA	ICD-10 code A31.0, extracted from routine statutory health insurance claims data	No data on species available	Between 2009 and 2014, prevalence rates increased from 2.3 to 3.3 cases/100,000 population. Prevalence showed a strong association with advanced age and COPD	[23]

Germany	2010–2015	Patients with incident NTM-PD among approximately 5 million insured subjects covered by statutory health insurance	NA	NA	NA	NA	Mean 50 years; 50%	NA	ICD-10 code A31.0, extracted from routine statutory health insurance claims data	No data on species available	The incidence rate for NTM-PD was 2.6 per 100,000 insured persons in 2010 and 2011. Subjects with NTM-PD had a significantly increased risk of death compared to age-, sex- and comorbidity-matched controls, in particular in subjects with coexisting COPD	[24]
Athens, Greece	2007–2013	In- and outpatients at a single centre, with at least one NTM culture-positive respiratory specimen	NA	NA	12/74 (16%), with clinical data available		Median 69 years; 37%	MAC 50%, unidentified 17%, *M. abscessus*, *M. gordonae*, *M. fortuitum*, *M. xenopi*, 8% each	GenoType Mycobacterium CM/AS (Hain Lifescience, Germany)		120/128 (94%) isolates were from respiratory samples; 56/120 (46%) patients fulfilled ATS/IDSA microbiological criteria; NTM incidence was 18.9 and 8.8 per 100,000 in- and outpatients, respectively	[26]

(continued)

Table 5 (continued)

Study location	Study period	Target population	Various proportional data				Median or mean age, female %	Prevalent species	Identification method	Additional information [proportion of prevalent species of isolates]	Annotation	Reference
			NTM isolate/culture-positive specimen (%)	Patient with NTM-isolate/culture-positive patient (%)	NTM-PD patient[a]/patient with NTM-isolate (%)	NTM-PD patient[a]/NTM culture-positive specimens (%)		Proportion of each prevalent NTM species among NTM-PD patients				
Athens, Greece	1990–2013	HIV-negative patients with NTM-PD receiving treatment	467/4241 (11%) in 2005–2013	NA	NA	51/467 in 2005–2013	Mean 59 years; 49%	MAC 63% (*M. avium* 57%), *M. kansasii* 12%, RGM 11% (5 *M. chelonae*, 2 *M. fortuitum*, 1 *M. abscessus*), other SGM 11% (4 *M. malmoense*, 2 *M. xenopi*, 1 *M. celatum* and 1 *M. lentiflavum*)	"Conventional phenotypic tests and molecular characterization" (not specified)		Comorbidity of NTM-PD patients: bronchiectasis 49%, COPD 30%, previous TB 29%, chest wall disease 7%, immuno-suppression 5%.	[28]
Larissa, Greece	2004–2006	Patients with positive mycobacterial cultures	NA	564/682 (82%)	16/156 (10%; after exclusion of *M. gordonae*)	16/389 (4%; after exclusion of *M. gordonae*)	Mean 66 years; 31% females	*M. fortuitum* (13%), *M. avium* (25%), *M. malmoense* (13%)	Genotype CM and MTBC commercial kits (Hain Lifescience, Nehren, Germany)	M. fortuitum (52%), *M. peregrinum* (35%), *M. avium* (7%), *M. chelonae* (5%), *M. malmoense* (2%), *M. scrofulaceum* (1%)	Smoking habits and chronic obstructive pulmonary disease were significant risk factors for NTM disease	[30]

Naples, Italy	2006–2009	Patients with NTM-positive respiratory cultures	NA	NA	16/39 (41%)	16/55 (29%)	Mean 67 years; 31%	*M. intracellulare* 41%, *M. kansasii* 29%, *M. fortuitum* 6% and *M. xenopi* 2%	Molecular identification (Gen-Probe Incorporated, San Diego, CA, USA; INNOLiPA Mycobacterium v2, Innogenetics NV, Ghent, Belgium; and Genotype Mycobacterium CM/AS assay, Hain Lifescience, Nehren, Germany)	*M. intracellulare* 27%, *M. gordonae* 24%, *M. kansasii* 20%, *M. xenopi* 15%, *M. chelonae* 8% and *M. fortuitum* 5%, M. szulgai 4%		[25]
Tuscany, Italy	2004–2014	Patients with suspected mycobacterial infection	NA	147/595 (25%)	NA	NA	NA	NA	Molecular identification: INNOLipa, Innogenetics, Belgium and/or Genotype CM/AS, Hain Lifescience, Germany, and by a multiplex PCR designed to discriminate MAC organisms	*M. avium* 42%, *M. intracellulare* 14%, *M. gordonae* 12%, *M. xenopi* 10%, *M. fortuitum* 7%, *M. kansasii* 5%, *M. celatum* 2%, other species 10% (16 different species in total)	58% of subjects were > 60 years old; 76% of isolated were from respiratory samples	[49]

(continued)

Table 5 (continued)

Study location	Study period	Target population	Various proportional data				*Median* or mean age, female %	Prevalent species	Identification method	Additional information [proportion of prevalent species of isolates]	Annotation	Reference
			NTM isolate/culture-positive specimen (%)	Patient with NTM-isolate/culture-positive patient (%)	NTM-PD patient[a]/patient with NTM-isolate (%)	NTM-PD patient[a]/NTM culture-positive specimens (%)		Proportion of each prevalent NTM species among NTM-PD patients				
Sardinia, Italy	2011–2012	Patients, including suspects with extrapulmonary disease	NA	21/3000 (0.7%)	16/21 (76%)	16/21 (76%)	NA, 43%	*M. avium* 24%, *M. gordonae* 24%, *M. xenopi* 24%, *M. chelonae* 9%, *M. alvei* 5%, *M. marinum* 5%, *M. szulgai* 5% and *M. intracellulare* 5%.	Molecular identification with the hsp65 PCR-RFLP method	24% of patients had an extrapulmonary infection: skin infection (5%), cervical lymphadenitis 14%, gastrointestinal infection 5%		[27]

Nijmegen-Arnhem region, Netherlands	1999–2005	Patients with NTM-positive cultures	NA	NA	53/232 (23%)	NA	Mean 60 years; 38%	*MAC* 49%, *M. kansasii* 23%, *M. szulgai* 8%, *M. xenopi* 6%, *M. celatum* 4%, *M. malmoense* 4%, other 8%	INNOLiPA MYCO-BACTERIA v2 (Innogenetics, Gent, Belgium), Hain GenoType MTBC line probe assay (Hain Lifescience, Nehren, Germany). 16S rDNA gene sequencing (151 bp hypervariable region A)		In 91% of patients NTM were isolated from respiratory specimens; most patients with ATS-defined NTM-PD had pre-existing pulmonary diseases (76%); clinical relevance differed by species, with *M. avium*, *M. malmoense*, *M. kansasii*, *M. xenopi*, *M. szulgai*, *M. celatum* and *M. genavense* possessing the highest clinical relevance. NTM isolation increased over time	[48]

(continued)

Table 5 (continued)

Study location	Study period	Target population	Various proportional data				Median or mean age, female %	Prevalent species	Identification method	Additional information [proportion of prevalent species of isolates]	Annotation	Reference
			NTM isolate/ culture-positive specimen (%)	Patient with NTM-isolate/ culture-positive patient (%)	NTM-PD patient[a]/patient with NTM-isolate (%)	NTM-PD patient[a]/NTM culture-positive specimens (%)		Proportion of each prevalent NTM species among NTM-PD patients				
The Netherlands	1999–2005	Patients with *M. xenopi* isolation	NA	NA	25/49 (51%)	NA	Mean 60 years; 24%	NA	Hain GenoType MTBC line-blot (Hain Lifescience, Nehren, Germany); INNOLiPA MYCO-BACTERIA (Innogenetics, Gent, Belgium); 16S rRNA gene sequence analysis; AccuProbe MTB DNA probe kit (GenProbe, San Diego, CA, USA)		Most patients with ATS-defined NTM-PD had pre-existing pulmonary diseases (84%)	[34]

Lisbon, Portugal	2008–2009	HIV-negative patients with pulmonary infections	58/510 (11%)	58/510 (11%)	58/58 (100%)	58/58 (100%)	Median 55 years, 53%	*MAC* 22% (*M. intracellulare* 16%, *M. avium* 7%), *M. fortuitum* 14%, *M. gordonae* 12%, *M. kansasii* 10%, *M. chelonae* 9%, *M. abscessus* 5%, *M. peregrinum* 5%, *M. triplex* 5%, M. spp.5%, *M. szulgai* 5%, *M. mucogenicum* 3%, M. lentiflavum 2% and *M. simiae* 2%.	Sequencing (Genotype Mycobacterium CM/AS assay, Hain Lifescience, Nehren, Germany)	Convenience sample of 12,491 sputum samples from 5497 patients	In this study, all patients had respiratory symptoms, abnormal chest radiography and two positive cultures from sputum specimens obtained at least 7 days apart	[31]
Porto, Portugal	2008–2012	Hospitalized HIV-negative patients NTM-positive respiratory cultures	NA	NA	36/407 (9%)	36/202 (18%)	Mean 62 years, 39%	*MAC* 94%, *M. kansasii* 3%, and *M. xenopi* 3%.	Sequencing (Genotype Mycobacterium CM/AS assay, Hain Lifescience, Nehren, Germany)	32/36 (89%) patients considered to suffer from NTM-PD met ATS/IDSA 2007 criteria.	The number of isolates increased each year, from 52 in 2008 to 58 in 2009, 74 in 2010, 86 in 2011 and 137 in 2012.	[17]

Abbreviations: *CF* cystic fibrosis, *DRG* diagnosis-related groups, *ICD-10* International Classification of Diseases, 10th revision, *Non-RGM* non-rapid-growing mycobacteria, *RGM* rapid-growing mycobacteria

[a]*NTM-PD*: nontuberculous mycobacterial pulmonary disease

expenditure per NTM-PD patient was almost fourfold that for a matched control (€39,560 vs. €10,007). Hospitalizations were three times higher in the NTM-PD group and accounted for 63% of the total costs [24].

Although NTM-PD is rare in Europe, it is associated with considerable morbidity, mortality and financial burden. Moreover, it should be noted that the epidemiology of NTM needs close and continuous monitoring, in particular taking recent report on nosocomial device-associated *M. chimaera* outbreaks across Europe and beyond into consideration [51–58], which potentially pose patients at risk of infection and add to NTM being a serious public health risk.

Asia

Kozo Morimoto and Won-Jung Koh

Introduction

Monitoring the epidemiology of nontuberculous mycobacteria (NTM) pulmonary disease in Asia is difficult because only a few population-based reports exist. A review article on NTM pulmonary disease in eastern Asia that considered the research conducted through the early 2000s found that *M. abscessus* complex (MABC), such as *M. abscessus* and *M. massiliense*, is the second most common cause of pulmonary infection after *M. avium* complex (MAC), such as *M. avium* and *M. intracellulare*, and that many patients have a previous history of tuberculosis (TB) [59]. In this section, we review the recently published reports to provide epidemiological data of NTM pulmonary disease in all areas within Asia. Several methods have been implemented to analyse the epidemiological climate of this disease, including a tertiary hospital report, an isolation ratio analysis and a population-based analysis, and we include as many of these reports as possible. Figure 1 indicates the countries that have produced articles included in this review.

Each article used various denominators and numerators to analyse the NTM situation in each research target to provide proportional data. For example, acid-fast bacilli (AFB) culture-positive samples, NTM culture-positive cases and NTM culture-positive samples were considered as denominators, and NTM isolates, NTM pulmonary disease cases and other factors were considered as numerators. Moreover, prevalent species proportions were described using denominators, including diagnosed cases, NTM culture-positive patients and NTM culture-positive samples. We summarize the data in the tables. However, each article required several annotations; thus, these tables should be used only as an overview of each category. In the

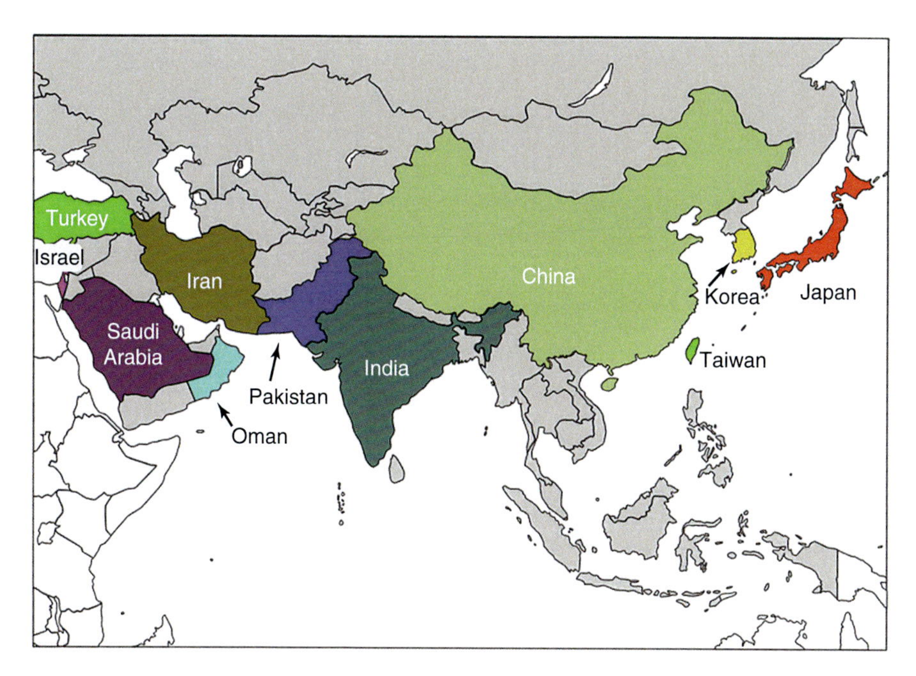

Fig. 1 Asian countries reporting data summarized in this review. Grey indicates countries not analysed here, and white indicates countries outside of Asia

following review, we preferentially described the proportion of NTM to AFB culture-positive specimens that reflect the NTM/TB ratio in each target area. For prevalent species, the diagnosed cases were considered as the denominator, and other data were described if these were not analysed in the article.

Eastern Asia

Japan

The first Japanese epidemiological study dates to the late 1950s, when sensitin studies were conducted in cooperation with the USA [60, 61]. A recent study published in EID clarified the NTM incidence in 2014 following the methodology established by Tsukamura et al. in the 1970s [62–64]. Furthermore, a study based on mortality data and a laboratory-based analysis was proactively conducted recently [65, 66]. Population-based surveys in a southwestern prefecture and certain tertiary hospital surveys have included detailed patient characteristics (Table 6) [67–70].

Table 6 Japan

Study location	Study period	Target population	Various proportional data: NTM isolate/culture positive specimen (%)	Patient with NTM-isolate/culture positive patient (%)	NTM-PD patient[a]/patient with NTM-isolate (%)	NTM-PD patient[a]/NTM culture positive specimens (%)
Nagasaki	2001–2010	Resident	NA	NA	NA	NA
Kyoto	1999–2005	MAC patients	2180/unkown	454/unknown	NA	285/unknown
Kyoto	2000–2013	Patients	6327/10,104 (62.6%) 44.8→81.2%	NA	NA	592[a]/6327
Saitama	1999–2005	MAC patients	NA	NA	NA	634 cases/unknown

MAC Mycobacterium avium complex, *MABC Mycobacterium abscessus* complex, *NB* nodular bronchiectatic type, *FC* fibrocavitary type, *TB* tuberculosis, *COPD* chronic obstructive pulmonary disease
[a]*NTM-PD*: NTM pulmonary disease as defined by ATS criteria

Studies Using the Cutireaction Test

The first epidemiological study was conducted in 1959 using sensitin (PPD-S, PPD-Y and PPD-B) provided by Dr. Edward, who had already conducted a sensitin study in the USA [60]. The surveyed population included 849 cases of pulmonary tuberculosis, 882 healthy children and 573 healthy adults. This study concluded that infections with yellow (*M. kansasii*) and battery bacillus (MAC) were minimal in Japan. The second study was conducted in 1960–1961 using sensitin purified from strains isolated in Japan [61]. The skin reaction test was performed among 5534 patients with TB, 3721 healthy children and adults and an additional 11,561 junior high school students. As a result, the positive rate increased with age among the healthy population, and the average positive rates among junior high school students with non-photochromogen (MAC) and photochromogen (*M. kansasii*) were 5.5% and 2.7%, respectively.

Studies Since 1970

Tsukamura organized a cooperative study group of the national sanatoriums in 1968. The purpose of this group was to investigate the isolation prevalence of each mycobacterium and the disease incidence as well as the geographical differences across the country. The first preliminary data were published in English, and this surveillance continued for 26 years from 1971 to 1997 [63, 64, 71]. All samples were studied at the central institute using para-nitrobenzoic acid (PNB) to reduce selection bias and correctly identify NTM. The fundamental data included the ratio of new NTM cases to all cases of mycobacteriosis. The authors estimated the NTM rate by multiplying by the national TB incidence [63].

In the early 1970s, they reported that the isolated ratio of NTM in all cases of mycobacteria was 6%. The most frequently isolated species were MAC and *M. gordonae*;

however, the NTM diseases were mostly caused by MAC (95%) and *M. kansasii* (3.8%). Geographically, the western part of the country along the Pacific coast region showed a higher disease incidence than the northern part of the country [64]. The incidence of NTM disease was 0.89 per 100,000 in 1971 and remained below 2 until the early 1980s [63]. The incidence exceeded 2 per 100,000 people in 1984 and gradually increased [72]. The final data from this surveillance showed a rate of 3.2 per 100,000 people in 1997. Another study group conducted the survey via questionnaire in 2001 and 2007 following the same method and reported rates of 5.9 and 5.7, respectively (the latter study focused on MAC, *M. kansasii* and MABC) [73]. The study methodology was changed to a questionnaire because the coverage institute had expanded to the general hospital. The most recent survey was conducted in 2014 and was supported by the government. This survey targeted 884 hospitals certified by the Japanese Respiratory Society [62]; it indicated that the incidence rate was 14.7 per 100,000 people. MAC (88.8%) was the major cause of pulmonary disease, as in the 1970s, followed by *M. kansasii* (4.3%) and *M. abscessus* (3.3%). The NTM incidence rate exceeded the TB incidence rate for the first time in this country and is likely the highest in the world (Fig. 2).

This surveillance also revealed an interesting transition in the *M. kansasii* ratio. *M. kansasii* cases were found only in or near Tokyo during the early 1970s [64] but were identified in other areas by the late 1970s. Furthermore, the incidence became prominent during the early 1980s in the Kinki area, where it exceeded incidence in Tokyo by 1984 [72]. Geographical differences between *M. avium* and *M. intracellulare* are also interesting. In the 1990s, a genetic test that differentiates between *M. avium* and *M. intracellulare* became common. Saito et al. first reported geographical differences between these two species [74], and surveys in 2001 and 2014 revealed that the incidence of *M. avium* was higher in the northeast, whereas *M. intracellulare* was higher in the southwest [62, 73].

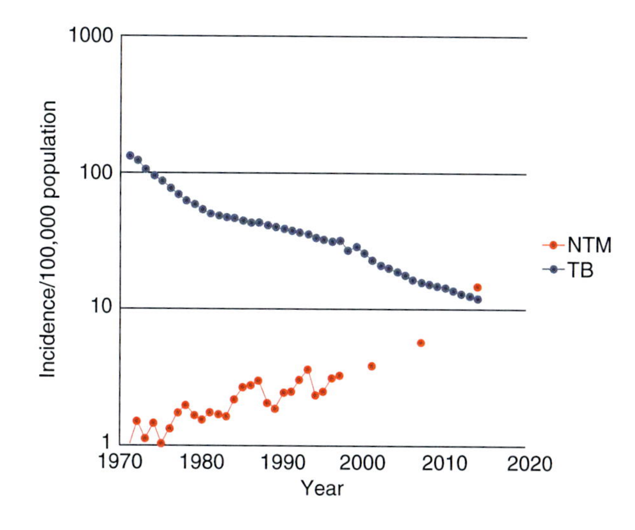

Fig. 2 The incidence of nontuberculous mycobacterial (NTM) pulmonary disease and pulmonary tuberculosis from 1971 to 2014. The epidemiological surveillance of NTM pulmonary disease from 1971 to 1997 was conducted annually by the same research group; another group conducted the surveys in 2001 and 2007

Laboratory-Based Survey

The current and most sophisticated method for the surveillance of NTM disease is a laboratory-based study that uses the microbiological criteria from the 2007 ATS/IDSA statement [75, 76]. In a recent study, researchers analysed more than 110,000 mycobacterial examination results outsourced from the institute (which did not have an in-house laboratory from 2012 to 2013) [66]. NTM was isolated in 27,142 cases, and 27.7% of these cases met the criteria. The two-period prevalence rate was 24/100,000, which was higher in the southwest. Prevalent species were MAC predominant (93.3%), and the incidence in females was approximately twice that in males (1:1.9). Geographically, the data again showed an *M. avium* predominance in the northeast and *M. intracellulare* in the southwest. *M. kansasii* was most prevalent in Kinki, where its incidence surpassed that of Tokyo in the mid-1980s. Interestingly, the prevalence of MABC was highest in Kyushu-Okinawa, which is geographically nearest to Taiwan and South Korea (Fig. 3).

Mortality Data Analysis

Mortality data can reflect the disease status of a country. The first three deaths due to NTM were reported in 1970 and increased over the study period. Females have

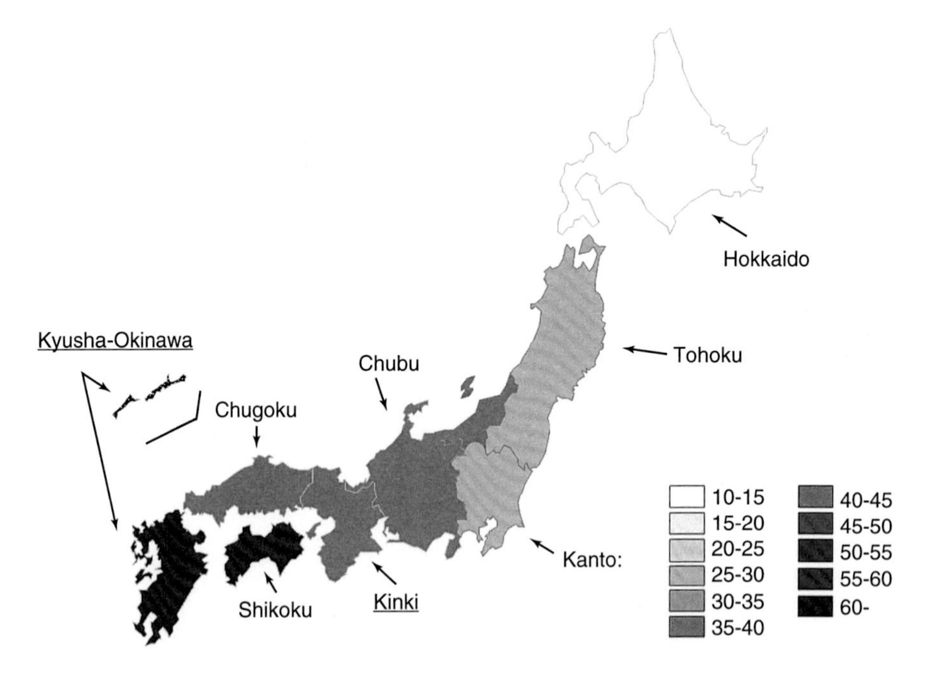

Fig. 3 *M. avium* versus *M. intracellulare* in each region (proportion of *M. intracellulare*). The *M. intracellulare* ratio was highest in the southwest. The *M. avium* ratio was highest in the northeast. The prevalence of *M. kansasii* was highest in Kinki, and *M. abscessus* complex was highest in the Kyushu-Okinawa region (underlined) [66].

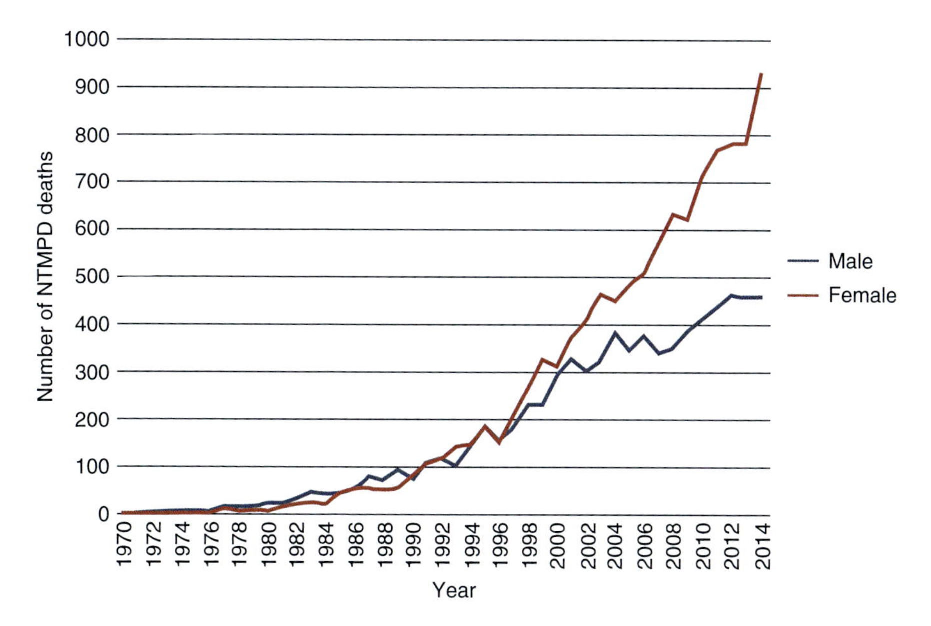

Fig. 4 The number of NTM pulmonary disease deaths from 1970 to 2014. Totals of 460 males and 929 females died in 2014

dominated the mortality rate since 2000 (1121 cases: male, 409; female, 712). The most recent data included 1389 cases (male, 460; female, 929) in 2014 (Fig. 4). The crude mortality and age-adjusted rates in females increased until 2010. The standardized mortality ratios showed a higher ratio in the southwest, especially along the Pacific coast. In this study, the authors estimated the disease prevalence assuming a mortality rate of 1–2%. The NTM disease prevalence was calculated to be 33–65 in 2005 [65].

A Prefecture Study and Tertiary Hospital Studies

Researchers in Nagasaki prefecture, located in the southwest, analysed disease incidence by dividing the prefecture into five administrative areas between 2001 and 2010 using the same methods as the national surveillance values calculated from the TB incidence ratio [67]. These authors found that the incidence increased from 4.6 to 10.1. The patient characteristics of the 601 cases indicated an average age of 71 years and a female predominance (69.7%). The nodular bronchiectatic type was the most common radiological pattern (79.7%), followed by the fibrocavitary type (15.6%). Interestingly, the authors found that the *M. avium* to *M. intracellulare* ratio varied within the prefecture.

A tertiary hospital analysis also provided patient characteristics. In a report from a university hospital in Kyoto, researchers analysed MAC pulmonary disease between 1999 and 2005 [68]. The mean age of the 164 cases was 66 years, and females predominated (56.7%). In a CT analysis of 126 cases, 80 cases (63.5%) showed bronchiectasis, and 30 cases (23.8%) had cavitary lesions. The same researchers analysed laboratory data between 2000 and 2013 [69]. They reported that the proportion of NTM increased from 44.8% (335/792 samples) to 81.2% (688/847 samples) and that the number of newly diagnosed cases increased from 24 in 2000 to 54 in 2013. In an analysis of 634 MAC cases conducted by researchers in Saitama [70], the mean age was 68.9 years, and females predominated (58.5%). The nodular bronchiectatic type was the most common radiological pattern (76%), followed by the fibrocavitary type (16.6%).

Conclusion The incidence of NTM pulmonary disease in Japan increased from 0.89/100,000 in 1971 to 14.7/100,000 in 2014. The number of deaths caused by NTM increased by more than 460-fold during this 40-year period. Laboratory-based and mortality analyses showed that the prevalence and mortality ratios were higher in the western part of the country. MAC was consistently the most prevalent species (approximately 90%) throughout the 40-year period. *M. avium* was more prevalent in the northeast, whereas *M. intracellulare* was most prevalent in the southwest. The prevalence of *M. kansasii* was highest in Kinki and that of MABC was highest in Kyusyu-Okinawa. The incidence of NTM pulmonary disease was highest among middle-aged and elderly (over 60 years of age) patients and women (55–65%). Radiographically, the nodular bronchiectatic type was most common (60–80%).

South Korea

Although no population-based epidemiological reports have been conducted in South Korea, seven reports from tertiary hospitals in Seoul or Goyang (i.e. the northern part of the country; Fig. 5) [77–83] provide detailed information on cases

Fig. 5 South Korea. Goyang city lies north of Seoul. Grey areas are outside of South Korea

and laboratory data including the ratio of NTM isolation among culture-positive specimens (Table 7).

Three institutes reported the proportion of NTM isolates among AFB culture-positive specimens, and two reported increases, from 22.2% to 45.9% between 2002 and 2008 [77] and from 43% to 70% between 2001 and 2011 [78]. The remaining paper reported 78.6% in 2012 [79]. The number of diagnosed cases increased from 4 to 32 between 2007 and 2011 [80]. One institute reported that the transition of patient population-based isolation and disease prevalence increased from 4.6 to 14.8/100,000 patients and from 1.8 to 4.4/100,000 patients, respectively [81].

Middle-aged and elderly woman were the most often affected group (median age, 63 years old; female proportion, 54.5–64.4%) [77, 79–83].

MAC was the most prevalent species (48.2–76.3%) isolated from the diagnosed cases. According to the three reports, the proportion of *M. avium* (28.7–40.9%) was slightly higher than that of *M. intracellulare* (19.5–35.4%) [79, 81, 83]. MABC was the second most prevalent species in all four reports (18.2–33%). The next most prevalent species was *M. fortuitum* (1.8–11% according to three reports) [79, 81, 83].

Conclusion Tertiary centre reports from South Korea showed an increased frequency of isolates and diseases. MAC disease appears to be the cause of disease in middle-aged women. The ratios of *M. avium* and *M. intracellulare* did not significantly differ. MABC was the second most frequent species (10–30%), followed by *M. fortuitum.*

Taiwan

No population-based report has been published from Taiwan; however, six tertiary hospital reports exist [84–89], including one report analysing MAC cases and one report focusing on patients 65 years or older [84, 86]. Three reports performed patient population-based incidence analyses (Table 8) [84, 85, 87].

The ratio of NTM among AFB culture-positive specimens was 35.6–72%, and this ratio was higher in more recent reports [84, 85, 87, 89]. All three patient population-based studies showed an increase: from 9.0 to 21.5/100,000 patients (1997–2002), from 2.1 to 4.4/100,000 patients (2004–2008) and from 1.3 to 8.0/100,000 patients (2000–2008) [84, 85, 87].

Patient age ranged from 60 to 79 years old, and most reports showed a male predominance. The median female percentage was 33.7% (16.7–55%). Two papers found that patients frequently showed a history of TB or chronic obstructive pulmonary disease (COPD) [88, 89]. All but one report found that the most isolated species was MAC (39.1–44.1%), followed by MABC (19.2–44.8%) and *M. fortuitum* (9.5–23.9%). An institute in Southern Taiwan provided an interesting exception, reporting that MABC was the dominant species (approximately 45%), which is contrary to the MAC predominance found in the northern institutes [88] (Fig. 6).

Conclusion Isolation and disease frequency seems to have increased according to Taiwan institutes. MAC is the most prevalent species, followed by MABC and *M.*

Table 7 Korea

Study location	Study period	Target population	Various proportional data				Median or mean age, female %	Prevalent species	Additional information [proportion of prevalent species of isolates]	Annotation	Reference
			NTM isolate/ culture-positive specimen (%)	Patient with NTM isolate/ culture-positive patient (%)	NTM-PD patient[a]/ patient with NTM isolate (%)	NTM-PD patient[a]/NTM culture-positive specimens (%)		Proportion of each prevalent NTM species among NTM-PD patients			
Seoul	2002–2008	Patients	4316/12,760 (33.8%) 22.2 → 45.9%	1801/4012 (44.9%)	651/1801 (36.1%)	651/4316 (15.1%)	*63.3*, 52.4%	MAC (62.9%) MABC (26.7%)	Isolated cases increased from 214 to 302 during the study period		[77]
Seoul	2001–2011	Patients	20,222/32,841 (62%) 43 → 70%	NA	NA	NA	NA	NA	Proportion of MAC increased remarkably [MAC 53%, MABC 25%, *M. fortuitum* 6%, *M. kansasii* 1%]		[78]
Seoul	2012 (July)	Patients	367/467 (78.6%)	NA	NA	111/232 (47.8%)	61.3, 61.3%	MAC 65.8% (*M. avium* 36.9, *M. intracellulare* 28.8%)#, MABC 28.8% (*M. abscessus* 18.9%, *M. massiliense* 9.9%) *M. fortuitum* 1.8%	[*M. avium* 35.7, *M. intracellulare* 23.3%, *M. abscessus* 19.4%, *M. massiliense* ##7.9%, *M. gordonae* 2.2%]	#111 cases analysed ##227 isolates analysed	[79]
Goyang	2007–2011	Patients	NA	NA	NA	90/752 (12%)	64, 64.4%	NA	Case increased from 4 to 32.		[80]
									Third prevalent isolate species was *M. fortuitum*		

Seoul	2006–2010	Patients	NA	1216/26,793 (4.5%)	345/1216 (28.4%)	345/2735 (12.6%)	*63*, 54.5%	MAC 76.3% (*M. avium* 40.9%, *M. intracellulare* 35.4%)	Isolation and disease incidence increased from 4.6 to 14.8 and from 1.8 to 4.4		[81]
								MABC (18.2%)	NB 62.4%, upper lobe cavitary form 21.9%		
								M. fortuitum (2.3%)	42.5% of cases had previous TB history		
Seoul	1993–2006	Patients	NA	NA	NA	NA	[a](63.4, 56.2%)	NA	[*M. avium* 14.5%, *M. intracellulare* 42.4%, *M. fortuitum* 8.9% MABC 16.9%, *M. kansasii* 5.2%]	[a]Isolated patient	[82]
Seoul	2002–2003	Patients	NA	NA	195/794 (24.6%)[a]	195/1548 (12.6%)	[a](63, 54.9%)	MAC 48.2% (*M. avium* 28.7%, *M. intracellulare* 19.5%)	[*M. avium* 15%, *M. intracellulare* 17%, MABC 29%, *M. fortuitum* 17%]	[a]Definite and probable cases	[83]
								MABC 33%			
								M. fortuitum 11%			

MAC Mycobacterium avium complex, *MABC Mycobacterium abscessus* complex, *NB* nodular bronchiectatic type, *TB* tuberculosis
[a]*NTM-PD*: nontuberculous mycobacterial pulmonary disease

Table 8 Taiwan

Study location	Study period	Target population	Various proportional data: NTM isolate/culture-positive specimen (%)	Patient with NTM isolate/culture-positive patient (%)	NTM-PD patient[a]/patient with NTM isolate (%)	NTM-PD patient[a]/NTM culture-positive specimens (%)	*Median* or mean age, female %	Prevalent species: Proportion of each prevalent NTM species among NTM-PD patients	Additional information [proportion of prevalent species of isolates]	Annotation	Reference
Taipei	2004–2008	Patients >65 years old, including extrapulmonary disease	3175/4407 (72%)	NA	326/1633 (19.6%)[a]	326/3175 (10.2%)[a]	NA, 37.4%	MAC 39.1%[aa]	Annual incidence increased from 2.1 to 4.4	[a]90.2% of cases were pulmonary disease	[84]
								MABC 23.1%	[MAC 35.2%, *M. fortuitum* 14.5%, MABC 17.2%]	[aa]294 pulmonary disease cases	
								M. chelonae 13.6%			
Taipei	1997–2003	Patients, including extrapulmonary disease	2650/7442 (35.6%) 26.2 → 41.8%	NA	412/1346 (30.6%)[a]	412/2650 (15.5%)[a]	*64*, 42%	MAC 44.1%[aa]	Disease incidence increased from 9.0 to 21.5 (1997–2002)	[a]59.5% were pulmonary or pleurisy	[85]
								MABC 19.2%	MAC disease incidence increased from 5.5 to 9.5 (1997–2002)	[aa]254 cases of pulmonary disease	
								M. fortuitum 10.6%	[MAC 39%, MABC 18.3%, *M. fortuitum* 12.8%]		
Kaohsiung	2001	MAC patients	NA	NA	12/54 (22.2%)	NA	*73.5*, 16.7%	NA			[86]

Taipei	2000–2008	Patients, including extrapulmonary disease	9204/23,499 (39.2%)	NA	NA	1105/9204	NA	MAC 40.3%[a]	NTM pulmonary disease increased from 1.3 to 8.0	[a]Pulmonary disease	[87]
									[MAC 30%, MABC 17.5%, *M. fortuitum* 13%]		
Kaohsiung	2004–2005	Patients	NA	NA	67/473 (14.2%)	NA	66.6, 29.9%	MABC 44.8%	88.1% had pre-existing lung disease (COPD and TB)		[88]
								M. fortuitum 23.9%			
								MAC 14.9%			
Taipei	2007–2009	Patients	4553/8565 (53.2%)	2149/2890 (74.4%)	481/2149 (22.4%)	481/4553 (10.6%)	60.5–68.8, 34–55%	MAC 39.5%[a]	*M. kansasii* 11.1%[a]	[a]Data of 253 cases	[89]
								M. chelonae-abscessus 30.0%	FC type had prior TB history 45%		
								M. fortuitum 9.5%			

MAC = *Mycobacterium avium* complex, MABC = *Mycobacterium abscessus* complex, FC = fibrocavitary, TB = tuberculosis, COPD = chronic obstructive pulmonary disease
[a]*NTM-PD*: nontuberculous mycobacterial pulmonary disease

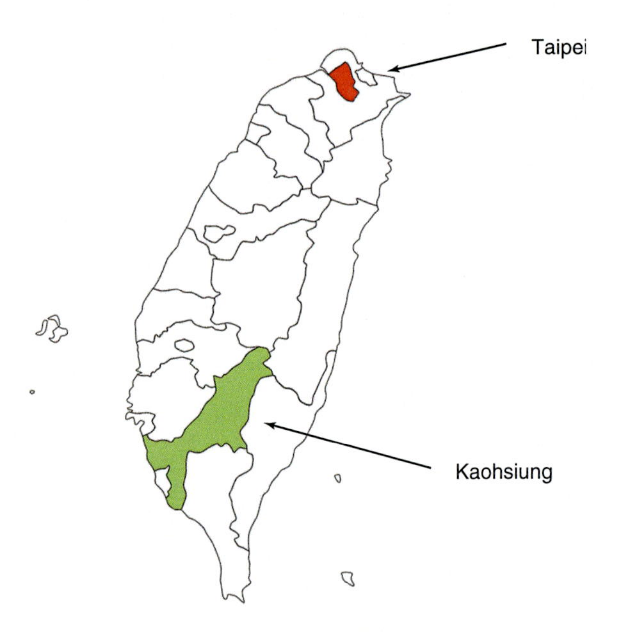

Fig. 6 Taiwan. Four reports from Taipei and two reports from Kaohsiung

fortuitum. The proportion of male and *M. fortuitum* cases was higher in Taiwan than in South Korea. One institute in Southern Taiwan reported that MABC was the most prevalent species, whereas Northern Taiwan institutes found that MAC was the most prevalent species. This difference should be confirmed in the future studies.

China

Several studies based on administrative units have been consecutively conducted in recent years, including in Shanghai and Beijing; however, most reports originate from the eastern coast (Fig. 7) [90–98]. The primary data reported were the proportion of NTM in AFB culture-positive specimens. In addition, a sentinel site survey from Zhejiang province reported population-based data, and a few single-institute studies summarized patient characteristics (Table 9). A systematic review and meta-analysis is separately discussed in this manuscript [99].

Shandong province is located along the east coast at the same latitude as South Korea [90]. The authors analysed 3949 culture-positive mycobacterial data from 2004 to 2009, and the proportion of NTM within the samples was 1.6% (63 samples). As shown in Table 9, most reports from the provinces and cities in eastern China have shown a low ratio of NTM in AFB culture-positive specimens with a gradual increase from north to south (1.6%–5.5%) [90–94, 97, 98]. One report from the south central region also stated that the rate was low (4.0%) [98]. Isolation and

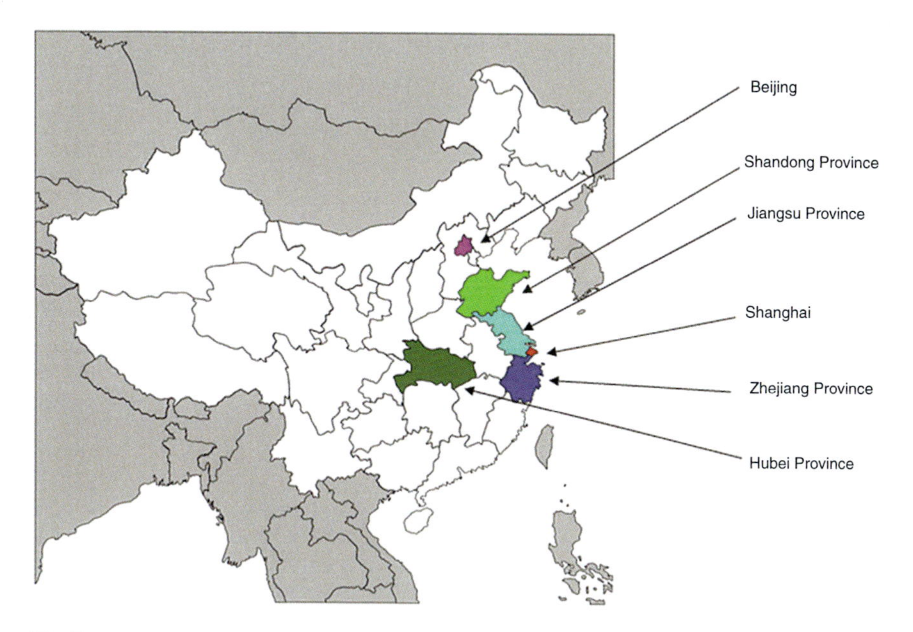

Fig. 7 China. The grey areas are located outside of China

disease prevalence were reported in a sentinel survey in Zhejiang province [96]. These authors reported that the isolation and prevalence rates were 3.1–3.9/100,000 and 0.4–0.5/100,000, respectively.

The sentinel survey and other reports also showed that NTM-isolated patients were predominantly male (53–68.3%) [94, 96–98].

Most isolated species were MAC (35.7–81.2%), except those in Shanghai [95], where *M. kansasii* was predominant (45%). The proportion of *M. intracellulare* in MAC was higher in most provinces (28%–100%), whereas *M. avium* was present in less than 10% of cases. The second and third most prevalent species were MABC (13.3–32.7%) and *M. kansasii* (6.7–25%), respectively, in most provinces. *M. fortuitum* was identified in 4–8% of cases in Shandon, Beijing and Shanghai. As described above, the proportion of *M. kansasii* was higher in Shanghai (45%) and the Zhejiang Province (25%), which is located south of Shanghai [95, 96]. A report from a single institution in Shandong province showed a higher *M. intracellular* ratio among pulmonary disease cases [94].

Conclusion Although the proportion of NTM among AFB culture-positive samples was low (median, 2.8%), the proportion showed a gradual increase from north to south, and some reports have revealed that the ratio is increasing. *M. intracellular* was identified more frequently from NTM culture-positive specimens and NTM pulmonary disease cases, followed by MABC or *M. kansasii*. These individual case data are insufficient. Notably, the report from Shandon clarified that the proportion of NTM was 30.7% among MDR-TB-suspected cases and 4% in TB retreatment cases.

Table 9 China (new)

Study location	Study period	Target population	Various proportional data				*Median* or mean age, female %	Prevalent species		Annotation	Reference
			NTM isolate/culture-positive specimen (%)	Patient with NTM isolate/culture-positive patient (%)	NTM-PD patient[a]/patient with NTM isolate (%)	NTM-PD patient[a]/NTM culture-positive specimens (%)		Proportion of each prevalent NTM species among NTM-PD patients	Additional information [proportion of prevalent species of isolates]		
Beijing	2008–2011	Resident	95/3654 (2.6%)	95/3714 (2.6%)	NA	NA	NA	NA	[*M. avium* 5.3%, *M. intracellulare* 40% MABC 29.5%, *M. fortuitum* 8.4%]		[90]
Shandong Province	2004–2009	Resident	64/3949 (1.6%)	NA	NA	NA	NA	NA	30.7% for MDR-TB-suspected case, 4.0% for retreatment TB cases [*M. intracellulare* 81.2%, *M. kansasii* 7.8%, *M. fortuitum* 4.7%]		[91]

Shandon Province (single institute)	2007–2012	Patients	91/4541 (2%)	NA	NA	NA	NA, 26.4%	*M. intracellulare* (46.1%) *M. chelonae*/*abscessus* (28.6%) *M. kansasii* (12.1%)	NA		[92]
Jiangsu Province	2008	Resident	60/1779 (3.4%)	NA	NA	NA	NA	NA	MABC was present in the southern part *M. kansasii* was found three cities located along the Yangzi River [*M. intracellulare* 68.3%, *M. abscessus-M. immunogenum* 13.3%, *M. kansasii* 6.7%]		[93]
Shanghai	2008–2012	Resident	NA	616/10,407 (5.9%) 3.0 → 8.5%	NA	NA	54, 24.5%	NA	Local residents, previous TB, cavity on X-ray, and sputum smear were risk factors for NTM culture+ [*M. kansasii* 45%, *M. avium* 3.6%, *M. intracellulare* 20.8%, *M. chelonae*/MABC 14.9%]		[94]

(continued)

Study location	Study period	Target population	Various proportional data				*Median* or mean age, female %	Prevalent species	Additional information [proportion of prevalent species of isolates]	Annotation	Reference
			NTM isolate/culture-positive specimen (%)	Patient with NTM isolate/culture-positive patient (%)	NTM-PD patient[a]/patient with NTM isolate (%)	NTM-PD patient[a]/NTM culture-positive specimens (%)		Proportion of each prevalent NTM species among NTM-PD patients			
Shanghai (single institute)	2005–2008	Patients, including extrapulmonary disease	248/4868 (5.1%)	NA	NA	NA	*60.5*, 47%	NA	Proportion of NTM increased 4.3 to 6.4% 60.5 years old, 53% of isolated cases were male [*M. chelonae* 26.7%, *M. fortuitum* 15.4%, *M. kansasii* 14.2%, MAC 13.1%]		[95]
Zhejiang Province	2011–2012	Resident	78/1410 (5.5%)	NA	NA	NA	NA	NA	[*M. avium* 3.8%, *M. intracellulare* 50%, *M. chelonae*/MABC 15.4%, *M. kansasii* 16.7%]		[96]

Zhejiang Province (sentinel)	2011–2013	Resident	NA	100/1831 (5.5%)	11/100 (11%)	NA	[a]58.3, 32.0%	NA	Isolate cases were male predominance (68%), 58.3 years old isolate prevalence 3.5/100,000, disease prevalence 0.36/100,000 [*M. avium* 8%, *M. intracellulare* 55%, *M. kansasii* 25%, *M. chelonae*/MABC 16%]	[a]NTM isolate case	[97]
Hubei Province	2011–2013	Resident	160/3995 (4.0%)	NA	NA	NA	[a]NA, 31.7%	NA	Isolate cases were male predominance (68.3%) 58.4% of case had previous TB treatment history [*M. avium* 3.0%, *M. intracellulare* 32.7%, MABC 32.7%, *M. fortuitum* 7.9%]	[a]Randomly selected 101 isolates	[98]

MAC Mycobacterium avium complex, *MABC Mycobacterium abscessus* complex, *TB* tuberculosis
[a]*NTM-PD*: nontuberculous mycobacterial pulmonary disease

A Systematic Review and Meta-analysis

This article summarizes the results of 105 articles, including 97 Chinese manuscripts, and the results have added new data to the above-described review [99]. A combined analysis revealed that the proportion of NTM among TB-suspected cases was 6.3%. The regional differences showed that the southeastern region (8.6%) presented the highest proportion and that the northeastern region exhibited the lowest proportion (2.7%). The ratio of the composition of slowly growing mycobacteria was significantly higher in the northern part (63.7%) compared with the southern part (53%), whereas that of rapidly growing mycobacteria was higher in the southern (44.1%) part compared with the northern part (21.9%). A regional analysis of the eastern coast showed that the proportion gradually increased from north to south (1.9–16%), and the identified number of species also increased from 6 to 26. Moreover, the proportion of *M. intracellulare* was higher in the north than in the south.

South Asia

Three reports originate from India (Table 10). The median proportion of NTM among AFB culture-positive specimens was 3.9% (2.1–9.9%) [100–102]. Although species frequency was not apparent in the largest report in this country, *M. intracellulare* (28%) and MABC (20%) were the most prevalent species among the diagnosed cases [102]. According to a report from Pakistan, the species identified among patients with pulmonary isolates were *M. fortuitum* (17.4%), MAC (12%) and *M. smegmatis* (12%) [103]. One report showed that the proportion of pre-existing pulmonary disease was higher in these cases.

Conclusion Large studies are lacking in this region. Relatively large-scale data showed that the proportion of NTM among AFB culture-positive specimens is low and that the prevalent species were MAC and rapidly growing mycobacteria. Future reports from this region are highly warranted.

West Asia

To date, all reports from Turkey, Saudi Arabia, Oman, Israel and Iran [104–111] were from single institutes or reference laboratories. Two reviews from Iran are also included here (Table 10).

The data from Turkey, Oman and Iran showed that the NTM ratio among AFB culture-positive specimens was low (median, 3.2%) [104, 108, 109]. The most prevalent species differed by country (MAC, MABC, *M. kansasii* and *M. fortuitum*). One report and two review articles from Iran showed that the most prevalent species

Table 10 South and West Asia

Study location	Study period	Target population	Various proportional data				*Median* or mean age, female %	Prevalent species	Additional information [proportion of prevalent species of isolates]	Annotation	Reference
			NTM isolate/culture-positive specimen (%)	Patient with NTM isolate/culture-positive patient (%)	NTM-PD patient[a]/patient with NTM isolate (%)	NTM-PD patient[a]/NTM culture-positive specimens (%)		Proportion of each prevalent NTM species among NTM-PD patients			
New Delhi, India (northern)	2011–2012	Patients, including extrapulmonary disease	13/131 (9.9%)	NA	7/9 (77.8%)	7/13 (53.8%)	[a]55.1, 0%	*M. xenopi* 2 cases	*M. avium* 1 case	[a]Pulmonary disease	[100]
								M. kansasii 1 case	4 cases had underlying lung conditions		
								M. fortuitum 1 case			
Tamil Nadu, India (southern)	1999–2004	Patients, including extrapulmonary disease	173/4473 (3.9%)	NA	NA	NA	NA	NA	Only 20/173 specimens were from sputum		[101]
									[*M. chelonae* 46%, *M. fortuitum* 41%]		
Mumbai, India (western)	2005–2008	Patients, including extrapulmonary disease	127/6143 (2.1%)	NA	34/103 (33%)	NA	[a]NA, 47%	MAC 29% (*M. avium* 1%, *M. intracellulare* 28%)[aa]	33 (29%) were highly probable cases	[a]Data of the 127 isolated patients	[102]
								M. simiae 23%	Proportion of pre-existing pulmonary disease is high	[aa]Pulmonary disease	
								MABC 20%	[*M. avium* 2.4%, *M. intracellulare* 26.7%, *M. simiae* 22.8%, MABC[aaa] 21.3%]	[aaa]Including extrapulmonary sample	

(continued)

Table 10 (continued)

Study location	Study period	Target population	Various proportional data					Prevalent species			
			NTM isolate/culture-positive specimen (%)	Patient with NTM isolate/culture-positive patient (%)	NTM-PD patient[a]/patient with NTM isolate (%)	NTM-PD patient[a]/NTM culture-positive specimens (%)	*Median* or mean age, female %	Proportion of each prevalent NTM species among NTM-PD patients	Additional information [proportion of prevalent species of isolates]	Annotation	Reference
Karachi, Pakistan (southern)	2010–2011	Patients, including extrapulmonary disease	NA	NA	NA	NA	NA	MAC 4 cases[a]	[*M. fortuitum* 17.4%, MAC 12%, *M. smegmatis* 12%][aa]	[a]Pulmonary disease	[103]
										[aa]92 respiratory specimens	
Izmir, Turkey (western)	2004–2009	Patients, including extrapulmonary disease	208/10,041 (2.1%)	NA	37/77 (40.3%)	31/208 (15%)	[a]53.9, 12.9%	MAC 42% (*M. avium* 10%, *M. intracellulare* 32.2%)[a]	44% of contaminant samples were *M. gordonae*	[a]Pulmonary disease	[104]
								MABC 16.1%			
								M. kansasii 16.1%			
Haifa, Israel (western)	2004–2010	Patients	NA	NA		21/215 (9.8%)	[a](NA, 38.8%)	*M. kansasii* 33.3%	24 (11.2%) were probable	[a]Isolated case	[105]
								MAC 28.6%	The number of *M. xenopi* and *M. simiae isolation* was increasing		
								MABC, *M. xenopi*, *M. fortuitum*, 10%, respectively	[*M. xenopi* 39.5%, *M. simiae* 24.2%]		

Saudi Arabia (all regions)	2009–2010	Patients, including extrapulmonary disease (referral lab)	NA	NA	49/72 (67.1%)	49/95 (51.6%)	[a](43, 44.2%)	MABC 30.6%[aa]	[MABC 30.5%, *M. fortuitum* 29.5%, *M. avium* 5.2%, *M. intracellulare*[aaa] 12.6%]	[a]Isolated case	[106]
								M. fortuitum 28.6%		[aa]Pulmonary disease	
								MAC 16.3% (*M. avium* 6%, *M. intracellulare* 10.2%)		[aaa]Including extrapulmonary sample	
Riyadh, Saudi Arabia (central)	2006/2012	Patients	NA	NA	40/142 (28%)	40/380 (10.5%)	54, 42%	MAC 47.5%	50% had underlying lung disease		[107]
								MABC 25%	NB type 45%, FC type 40%		
								M. kansasii 10%	[MAC 35%, *M. fortuitum* 24%, *M. chelonae*/MABC 17%]		
Muscat, Oman (northern)	2006–2007	Patients, including extrapulmonary disease(referral lab)	46/491 (9%)	NA	8/13 (63.6%)[a]	NA	[a]43, 54%	*M. intracellulare* 2, *M. chimaera* 2, *M. simiae* 1, *M. colombiense* 1, [aa]	38 (80%) of the NTM samples were from respiratory samples	[a]Randomly selected 13 strains from all isolates	[108]
								M. marinum 1, MAC 1	[MAC 9, *M. kansasii*, *M. simiae*, each one][a]	[aa]Eight diagnosed cases	

Table 10 (continued)

			Various proportional data					Prevalent species			
Study location	Study period	Target population	NTM isolate/culture-positive specimen (%)	Patient with NTM isolate/culture-positive patient (%)	NTM-PD patient[a]/patient with NTM isolate (%)	NTM-PD patient[a]/NTM culture-positive specimens (%)	*Median* or mean age, female %	Proportion of each prevalent NTM species among NTM-PD patients	Additional information [proportion of prevalent species of isolates]	Annotation	Reference
Tehran, Iran (northern)	2004–2014	Patients (referral lab)	175/5314 (3.2%)	NA	NA	NA	NA	NA	Proportion of NTM isolation increased from 1.2% to 3.8%		[109]
									Origin of samples was unknown		
									[*M. simiae* 29.1%, *M. chelonae* 16%, *M. kansasii* 14.9%, *M. avium* 4%, *M. intracellulare* 5.7%]		
Iran	1992–2014	Review of 13 studies analysing NTM isolates, including extrapulmonary	NA	NA	NA	NA	NA	NA	Proportion of NTM isolation increased from 8% in 1992 to 18% in 2013		[110]
									[*M. fortuitum* 28.3%, *M. simiae* 21.4%]		

Iran	1990–2014	Review of 19 original articles, including extrapulmonary disease	NA	NA	NA	NA	NA	NA	Proportion of NTM isolation was 10.2%		[111]
									[*M. simiae* 43.3%, *M. intracellulare* 27.3%, *M. fortuitum* 22.7%, MABC 14%]		

MAC Mycobacterium avium complex, *MABC Mycobacterium abscessus* complex, *NB* nodular bronchiectatic type, *FC* fibrocavitary type

[a]*NTM-PD*: nontuberculous mycobacterial pulmonary disease

were *M. simiae* and *M. fortuitum*. Each report/review stated that the proportion of NTM isolation increased throughout the study period [110, 111].

Conclusion The reported proportion of NTM among AFB culture-positive specimens was low. Several similar prevalent species were reported, including MAC, MABC, *M. kansasii* and *M. fortuitum*. However, *M. simiae* was most prevalent in Iran. The data in this region were not consistent or conclusive and should be clarified by future study.

Summary

The epidemiological data of NTM pulmonary disease in Asia, which includes many countries with a high prevalence of TB, is scarce, but most of the reported studies have shown an increasing tendency in recent years. However, in Japan, the incidence of NTM pulmonary disease exceeds that of TB, and Japan is considered to have a high NTM burden because the prevalence rate was estimated to be over 100/100,000 individuals. MAC was the most prevalent species in Asian countries, and the ratio of *M. intracellulare* in MAC is higher than that of *M. avium* in MAC in most countries except the northern part of Japan. The other characteristics in this area were the same as those detailed in a previous review: predominantly found in males, many patients have a previous history of TB, and MABC was the second most prevalent species. However, the nodular bronchiectatic type observed in female patients was dominant in South Korea and Japan, and *M. kansasii* was found to be prevalent in specific areas, such as Kinki and Shanghai.

Asia is undergoing rapid growth, has an ageing population and will present a decreasing incidence of TB and an increasing prevalence of NTM in the next few decades. To obtain a better understanding of NTM ecology and the respective clinical situation, future epidemiological reports from this area are highly warranted.

Oceania

Rachel Thomson

Western Pacific

The Western Pacific region stretches over a vast area from China in the north and west to New Zealand in the south and French Polynesia in the east. The population is approximately 1.8 billion. However, very little data exists on the prevalence of NTM. Like African countries the lack of laboratory infrastructure accounts for little culture or identification of NTM. A call for more attention to this issue was raised by Yew in 2015 [112], given the identified problems with increasing NTM in neighbouring countries such as Australia and New Zealand (Fig. 8).

French investigators retrospectively looked at NTM isolates from TB suspects in French Polynesia between 2008 and 2013 [113]. During this time 67.4% of isolates were NTM, with a startling predominance of rapid-growing isolates. Figure 9 (Map) of 87 isolates (from 83 patients) partial rpoB sequencing identified 95.4%: 42(48.3%) *M. fortuitum* complex, 28(32.2%) *M. abscessus* complex and 8 (9.2%) *M. mucogenicum* complex. In contrast to many countries, only 5.7% were in the *M. avium* complex. The remaining four isolates were identified using 16s rRNA sequencing as *M. cosmeticum*, *M. farcinogenes*, *M. phocaicum* (urine isolate) and *M. acapulcensis*. Using ATS criteria, only 4.8% of patients were considered to have NTM disease; the remainder considered to be colonized. Three of the patients who had disease were infected with *M. porcinum* and the other with *M. senegalense*. *M. porcinum* has been reported in water supplies in tropical and subtropical areas of the USA [114], Iran [115], Brazil [114] and Mexico [116]. Rapid growers are also more

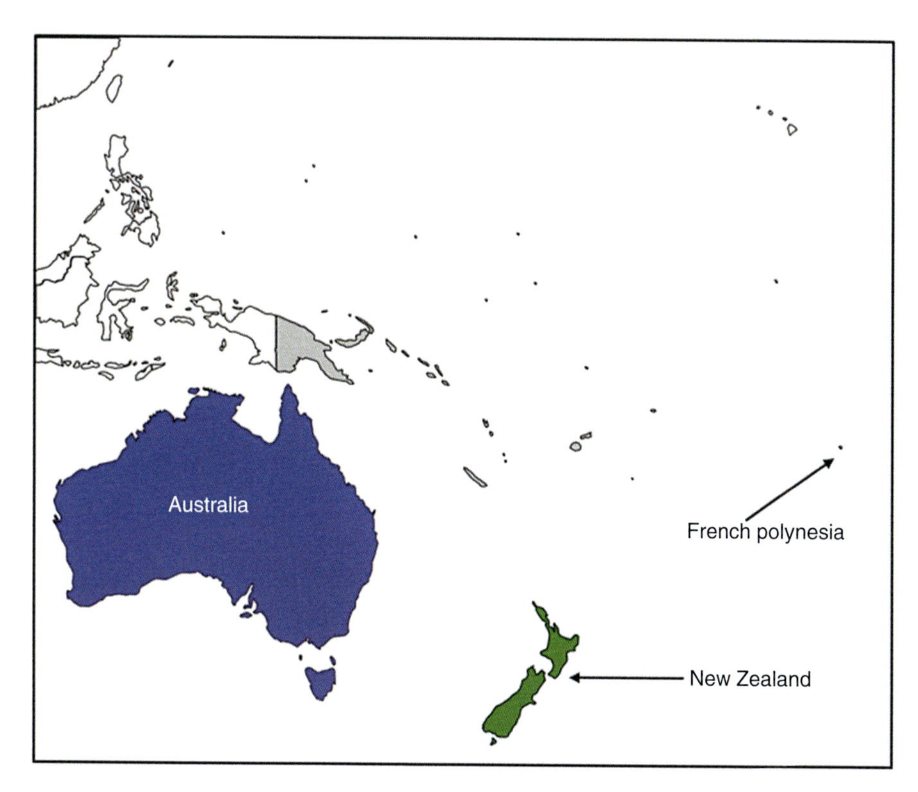

Fig. 8 The map shows countries from Ozeania with data included in the review. Please note that data for Papua New Guinea were added after this map had been drawn

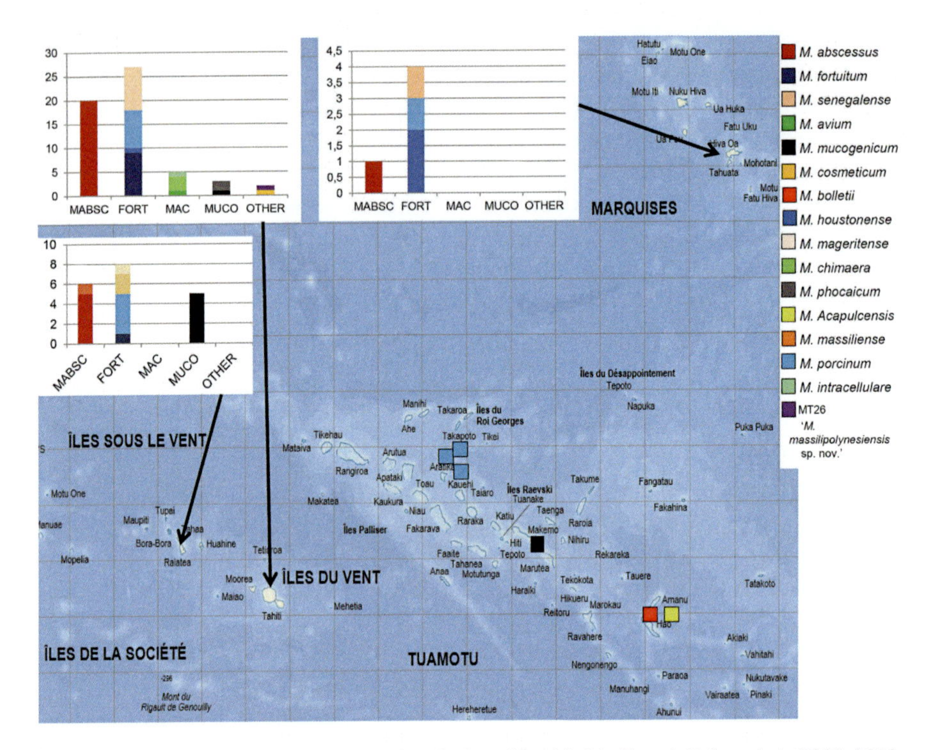

Fig. 9 Geographical distribution of 87 NTM isolates identified in French Polynesia in 2008–2013 MABSC, *Mycobacterium abscessus* complex; FORT, *M. fortuitum* complex; MAC, *M. avium* complex; MUCO, *M. mucogenicum* complex; other NTM [113]

common in Asia [117, 118]. These areas share similar geographic latitude and climate, which may account for the species distribution. Of the *M. abscessus* isolates in this study, 14 occurred in 2013, in 14 patients that underwent bronchoscopy in the same centre, raising the likelihood of a pseudo-outbreak due to tap water contamination. *M. abscessus* and *M. fortuitum* have been isolated from water supplies in northeastern Australia [119–121], also of similar latitude to neighbouring French Polynesia, again suggesting climatic and environmental factors may contribute to the high prevalence of rapid-growing mycobacteria.

Papua New Guinea (PNG)

As with many other TB endemic countries, little information is known about the prevalence of NTM in PNG. A TB case detection study was performed in selected provincial hospitals (Madang, Goroka and Alotau), and 335 sputum samples from TB cases ≥15 years of age were referred to the Queensland Mycobacterium Reference laboratory in Australia for culture and species identification. Of 225 culture-positive samples, NTM were detected in only 4% (9/225). Five samples grew NTM only – *M. fortuitum* (3), *M. terrae* (1) and *M. intracellulare* (1). Four isolates had a mixture of TB and NTM – three MTBC and *M. avium* and one MTBC and *M. intracellulare* [122].

Limited clinical features of the cases were presented, but all had respiratory or systemic symptoms, and interestingly eight of the nine cases were female. Whilst this was a relatively limited study, compared to other studies of TB suspects, the proportion of NTM was relatively low.

Australia and New Zealand

The challenges of an aging population and high prevalence of chronic disease in Western countries contrast with that of developing countries like Africa.

Since the introduction of TB services in Australia, NTM have been notifiable in many Australia states, to avoid confounding of smear-positive cases with TB. This notification process has provided a unique opportunity to study the clinical significance of NTM isolates. As in many parts of the world, an increasing incidence of pulmonary disease and pulmonary isolation of NTM has been demonstrated.

A national survey of mycobacterial reference laboratories in Australia in 2000 revealed a conservative estimate of 1.8 cases of pulmonary disease per 100,000 population [123]. This study only included isolates that had been referred for speciation to a reference laboratory. MAC isolates predominated, accounting for 67.6% of pulmonary and 80% of lymphatic infections. *M. kansasii* accounted for 19.4% of pulmonary disease. *M. fortuitum, M. chelonae, M. abscessus* and *M. marinum* were commonly isolated from soft tissue. *M. ulcerans* was only isolated from north Queensland and Victoria.

Pang reported a threefold increase in *M. kansasii* disease in Western Australia during 1962–1982 and 1982–1987 [124]. A report from the Northern Territory of Australia also documented an increase in incidence and prevalence of NTM disease between 1989 and 1997 of 2.7 to 4.7 per 100,000 population [125]. MAC was the most common species isolated (78%), and the yearly incidence of pulmonary MAC (HIV negative) was 2.1/100,000.

A national report from New Zealand examined the clinical significance of isolates referred to each of New Zealand's mycobacterium reference laboratories [126]. The overall incidence of disease in 2004 was 1.92/100,000 (pulmonary 1.17, lymph node 0.39 and ST 0.24). Of 368 patients with NTM, 21% were considered clinically significant by the treating physicians applying 1997 ATS criteria. This varied according to the site of infection with 15% pulmonary, 100% lymph node and 89% soft tissues isolates considered significant. MAC was the most common species accounting for 83% respiratory, 88% lymph node and 44% soft tissue infections.

More detailed reports have emerged from the state of Queensland, Australia, using data from the notifiable conditions database. The incidence of clinically reported cases of pulmonary disease caused by MAC has increased from 0.63 per 100,000 in 1985 to 2.2/100,000 in 1999. A detailed analysis compared clinical and epidemiological data from 1999 to 2005 and showed a further increase [127]. The total number of isolates in 1999 was 14.8/100,000, and the estimated total in 2005 was 22.1/100,000. Using clinical information, clinician advice and ATS criteria, the authors could determine the proportion who were likely to have clinically significant pulmonary disease was 3.2/100,000 in 2005. The species accounting for most of the change were *M. intracellulare* and *M. abscessus*. The clinical significance of the different species varied from 10% for *M. gordonae* to 52% for *M. kansasii*. Approximately 33% of MAC isolates were considered clinically significant or associated with invasive disease.

Of those cases that met ATS criteria for significant pulmonary disease in 2005, MAC contributed 72.1% of cases (*M. intracellulare* 56.8%, *M. avium* 15.3%) followed by *M. kansasii* (8.1%). Patients were predominantly female and had nodular bronchiectatic disease on radiological imaging [127].

Whilst the incidence of pulmonary disease has been increasing consistently, that of extrapulmonary disease has remained fairly stable [127]. This was also found to be the case in a detailed report of NTM infections in children from a tertiary referral centre in Victoria between 2000 and 2010 [128]. Lymphadenitis (86% submandibular or cervical, 7.5% preauricular) accounted for the majority of infections, and MAC was the predominant species (78.9%). This study observed a seasonal variation in the incidence of NTM disease, with the lowest incidence in autumn and the highest in late winter to spring. It has been postulated that seasonal variation in vitamin D levels may play a role.

The QLD Communicable Diseases Branch now issues annual reports from the notifiable diseases system, and the trend of increasing NTM isolation continues at approximately 17% per year [129]. In 2015 there were 1222 cases of NTM notified giving a rate of 25.9/100,000 population (Fig. 10). The age range of cases was 9 months to 95 years, median 66 years. Again, the 65–69-year age group predominated, but males now represented 49% and females 51% (Fig. 11). Of the 1222 cases, speciation of the isolate was performed on 953 cases (78%). The most common species continues to be *M. intracellulare* (35% of total, 40% of pulmonary isolates), *M. avium* (9% of both total and pulmonary) and *M. abscessus* (9% of total, 7% of pulmonary). *M. fortuitum* was the most common nonpulmonary species accounting for 50% of isolates.

In keeping with the documented worldwide differences in geographical distribution of NTM species, within-state differences were documented according to health service district. Rates per 100,000 in 2015 varied from 15 in the south west, 47.2 in Wide Bay to 72.4 in the central west. These differences were examined in detail on the data from 2001 to 2011 in a geospatial analysis [130]. For selected more common species, a Bayesian spatial conditional autoregressive model was constructed at the postcode level, with covariates including income, land use category, soil, rainfall and temperature variables. Each species showed distinct spatial patterns.

The risk for *M. intracellulare*, *M. avium M. kansasii* and *M. abscessus* demonstrated significant spatial heterogeneity after accounting for the covariates in the models.

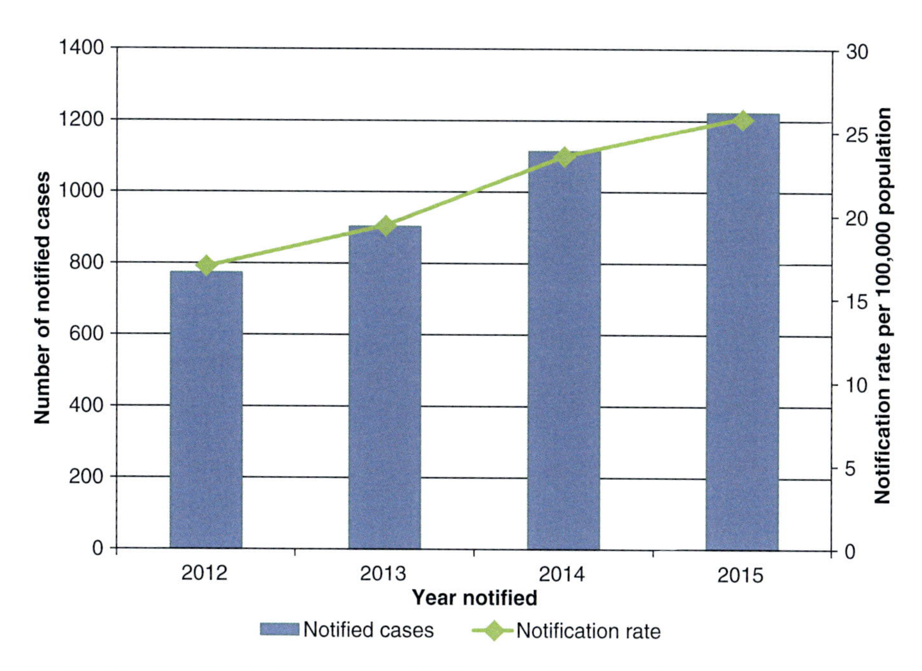

Fig. 10 Number of notified cases and notification rate[a] of nontuberculous mycobacteria and site of disease, Queensland, 2012–2015 [129]. [a]Notification rates calculated using Queensland Estimated Resident Population 2012–2014

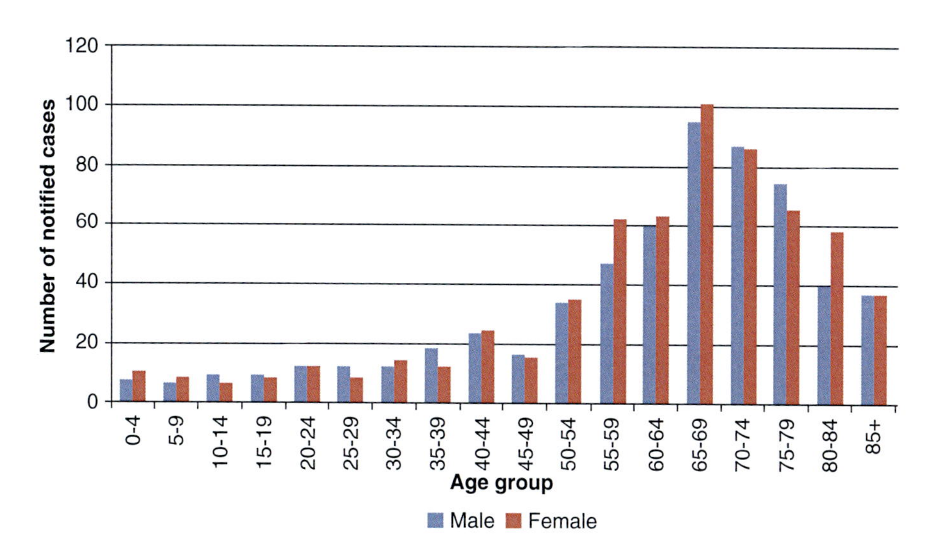

Fig. 11 Number of notified cases of nontuberculous mycobacteria by sex and age group, Queensland, 2015 [129]

Clusters of high relative risk for *M. intracellulare* and *M. kansasii* were found in the Darling Downs region of south central QLD. Geographically the clusters were located in a region overlying the Surat Division of the Great Artesian Basin (Figs. 12 and 13). The western geologic border of the Surat Basin corresponds well with the western edge of the *M. intracellulare* cluster (Fig. 13). This region is well known for its agricultural and mining activities with a number of developed petroleum and coal seam gas wells. In addition, the water supply for many communities in the region is primarily from private and communal bores, aquifers and rainwater tanks.

The cluster for *M. kansasii* was more geographically focused around the town of Roma, Yuleba and Clifford areas. Yuleba is a small town and the site of a major processing facility for silica deposits. Silicosis is a well-documented risk factor for *M. kansasii* disease. Mining activity is common in this area, and there have been several studies linking *M. kansasii* to industrial activities such as gold mining and iron manufacturing in South Africa, Japan and the former Czechoslovakia [131–134]. In addition, Yuleba has had a reticulated supply from bore water since the 1960s. *M. kansasii* has been identified as a cause of opportunistic infections from drinking water distribution systems.

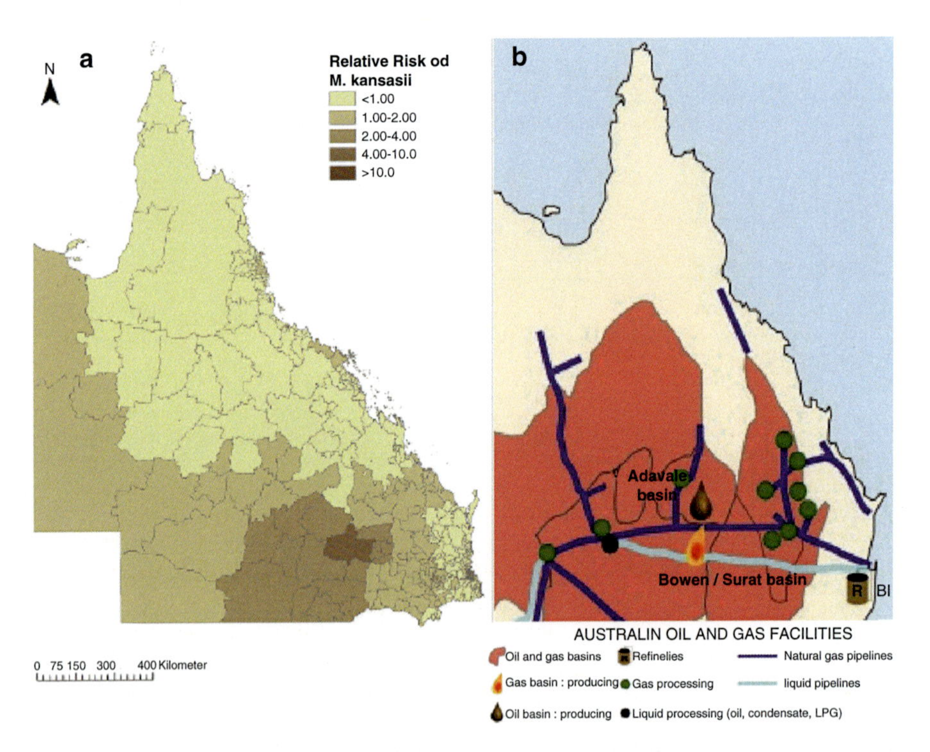

Fig. 12 Map of QLD showing spatial risk of *M. kansasii* disease [130] (**a**). Highest risk overlies the Surat Basin and major site of coal seam gas production [138] (**b**)

For *M. intracellulare* and *M. kansasii* significant covariates were found: soil depth, soil bulk density, earning <$52,000 and earning <$32,000. Soil pH was found to be a significant predictor variable for *M. fortuitum*, and soil nitrogen content a significant predictor for *M. chelonae* [130].

Risk of *M. intracellulare* was estimated to increase by 21% for every meter increase in topsoil depth. Potential reasons may include the poor rooting of vegeta-

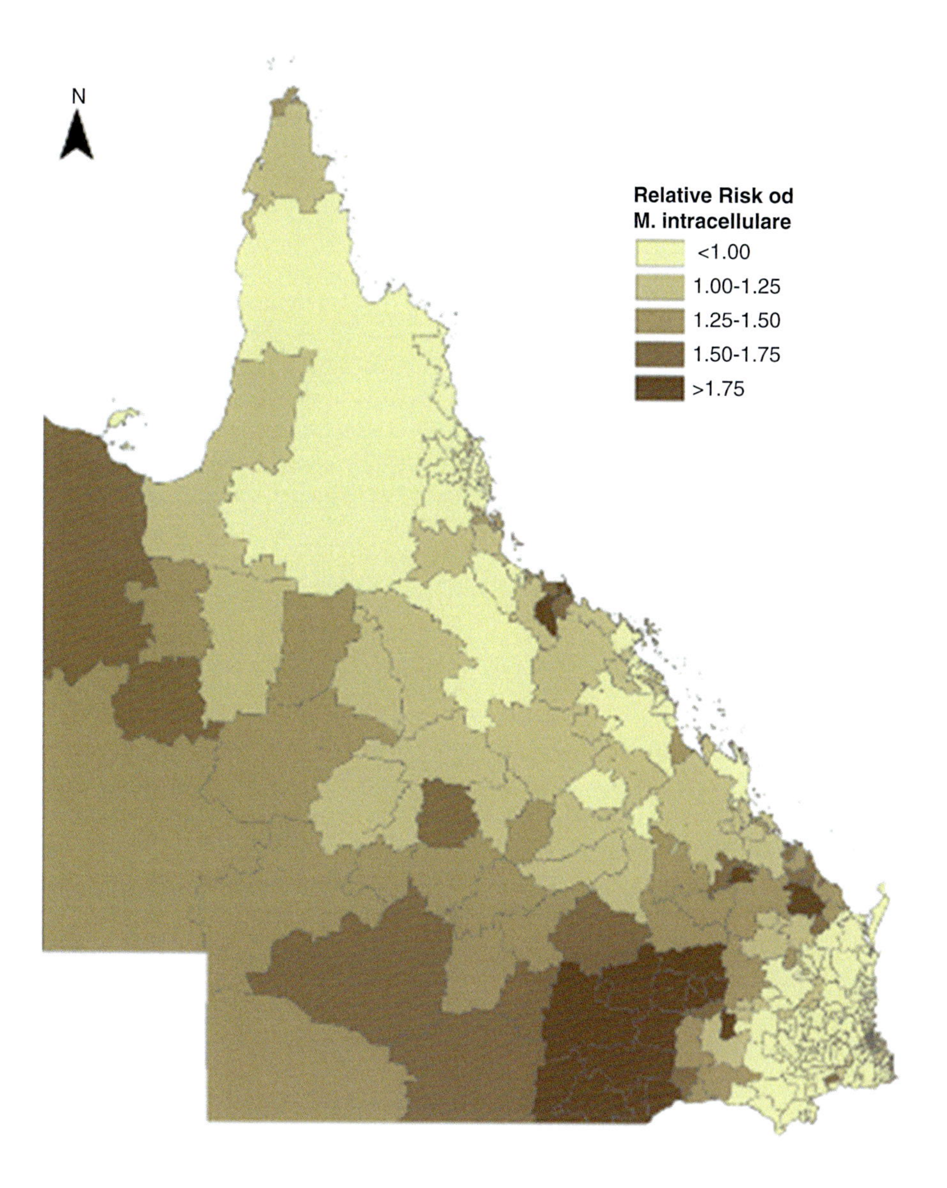

Fig. 13 Map of QLD showing relative spatial risk of *M. intracellulare* [130]

tion in shallow soil, leading to decreased uptake of soil nutrients, leaving a nutrient-rich topsoil environment in which mycobacteria may thrive [135]. A shallow soil depth is usually associated with low plant available water capacity (PAWC) and low crop yields. To improve yields, a process of "deep ripping" is undertaken, which may aerosolize soil particles and increase the potential for inhalation of mycobacteria.

Risk of *M. kansasii* infection was estimated to increase by 79% per mg/m^3 increase of soil bulk density. This variable can be affected by several other soil factors such as permeability, composition and depth. It is also affected by human activity, such as constant or inappropriate tilling and compressive forces from heavy agricultural machinery [136]. These activities similarly may increase exposure to mycobacteria from soil. Increased bulk density impairs root growth, further contributing to nutrient-rich topsoil environments conducive to mycobacterial growth.

There was also a cluster of high relative risk for *M. abscessus* in the Whitsunday region of north QLD (Fig. 14). This region is well known for its tropical climate and pristine marine environments and is a popular tourist destination. Isolates of *M. abscessus* from swimming pools and rainwater tanks have been linked to patient isolates in Brisbane [121]. Water characteristics in a coastal lagoon were correlated with *M. abscessus* by Jacobs et al. [137] Positive correlations with water temperature, nitrogen and phosphorous content and negative correlations with depth and salinity were found. However, in the Chou study soil nitrogen and phosphorus content was not associated with NTM [130].

Summary

NTM are commonly isolated from patients in Australia, and detailed epidemiological analyses confirm an increasing incidence of pulmonary disease, particularly nodular bronchiectatic disease due to *M. intracellulare* in postmenopausal females. Geographic hot spots and environmental predictors of disease have been identified that warrant further investigation.

Information regarding NTM in the Western Pacific region is scant, but like African countries where TB predominates, NTM is likely to confound the diagnosis of TB and contribute to the proportion of presumed drug-resistant cases, where laboratory infrastructure is lacking for culture and identification of NTM.

Africa

Rachel Thomson

Introduction

Many African countries are renowned for high rates of tuberculosis and HIV infection. The rising problem of MDR and XDR TB has led to an increased interest in accurate identification of isolates and the effect of NTM confounding in TB diagnosis [139]. As such NTM reports from many African countries have increased in the

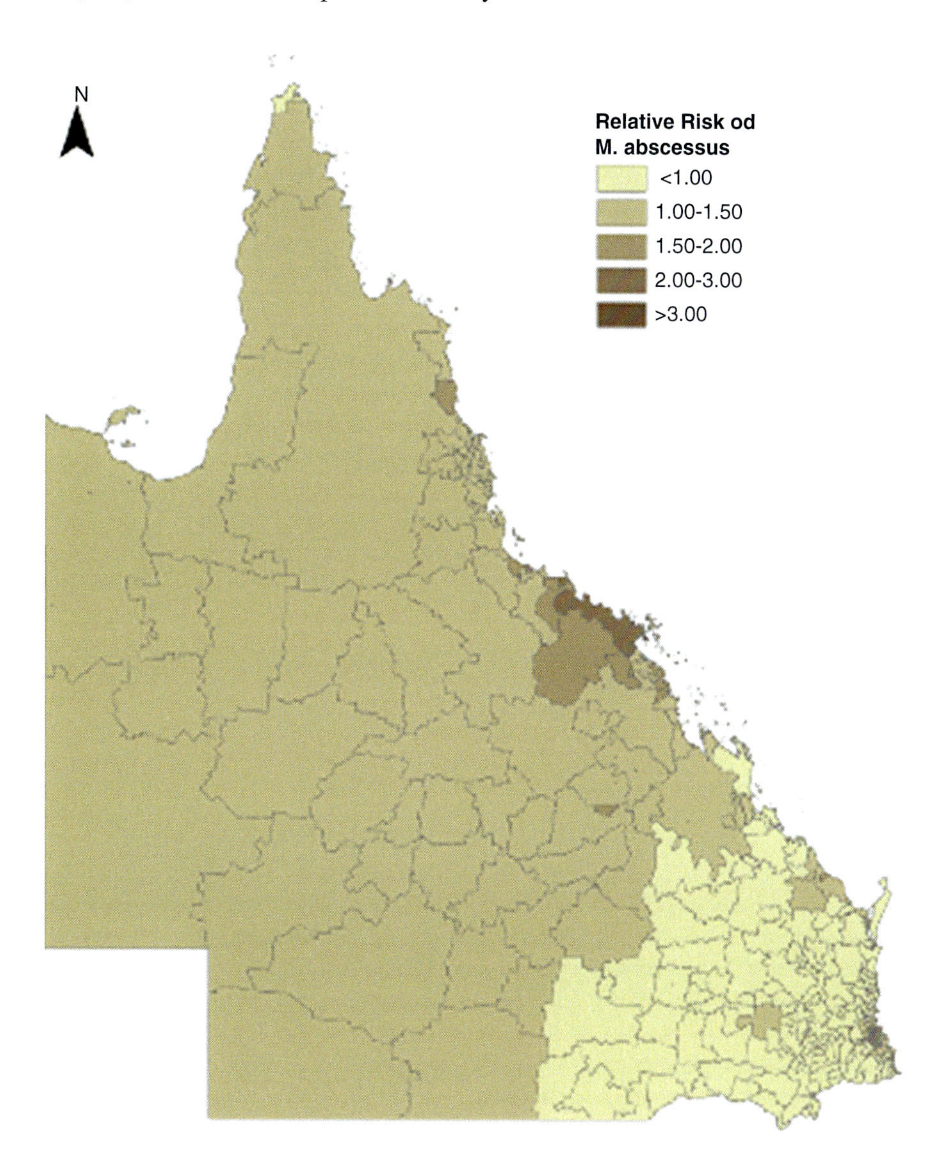

Fig. 14 Map of QLD showing relative spatial risk of *M. abscessus* [130]

last decade. Still, the contribution of NTM to mycobacterial diseases in Africa has only been examined on a small scale. One of the main reasons for this is that laboratory infrastructure has been lacking for culture and identification of NTM. In many resource poor settings with a high prevalence of TB, patients with AFB-positive sputum or those with consistent CXR findings are presumed to have pulmonary TB. In general, they are treated with standard short course TB regimens. Those who fail to respond or develop "chronic TB" are often assumed to have drug-resistant disease. The misclassification of chronic TB cases elicits unnecessary use of second-line drugs, with associated toxicity and expense. The blind management of these cases is also likely to foster the development of further antimicrobial resistance. Recent reports suggest NTM are responsible for varying proportions of such cases. The high prevalence of HIV in African countries has also led to an interest in NTM, given their potential to cause both pulmonary and disseminated disease.

The presence of NTM in the environment has been reported in a number of African locations including soil and water [140]; biofilms in drinking water distribution systems in South Africa [141]; soil, water and faecal matter from cattle and pigs in Uganda [142]; and wildlife species and indigenous cattle in Tanzania [143]. Table 11. In many parts of Africa, there is a high degree of interaction between humans, wildlife and the environment, increasing the potential for animal- and human-based contamination of natural water sources and transmission of NTM. Hence there has been some interest in the species of NTM in the environment and infecting animals.

However, there are no inclusive national population-based studies providing good epidemiological data on NTM. Published reports include case-based series, individual institutional reviews and regional laboratory-based studies. These provide snapshots of the species and geographical distribution of NTM in some African countries. Figure 15 indicates the countries that have produced articles included in this review; Table 12 summarizes the published data.

South and Eastern Africa

In South Africa, NTM have been reported back as far as the 1950s and 1960s [164].

Prevalence rates of NTM colonization of 1400–6700/100,000 persons have been reported [165, 166]. A high incidence of *M. kansasii* infection has been reported in the South African gold-mining workforce, estimated at 320/100,000 [167]. Among this population there is a high prevalence of risk factors for NTM disease, including a high burden of TB, silicosis, HIV infection and occupational exposure to aerosolized water used for dust control within the mines [156, 167–169].

A retrospective case series of HIV-negative miners from 1993 to 1996 revealed *M. kansasii* and *M. scrofulaceum* accounted for 68% and 14% of cases, respectively. There were also 18 (6%) *M. avium-intracellulare (*MAI) isolates and 36 (12%) isolates of NTM species not normally considered pathogenic. More than 80% were smear-positive, and the majority (>75%) had cavitary changes on CXR. The authors

Table 11 Species diversity of rapid-growing (a) and slow-growing (b) NTM reported from African countries

(a) Rapid-growing				
NTM species	Sample source	Host	Country	Reference
M. abscessus	Sputum	Humans	Nigeria, South Africa, Zambia	[139, 144–147, 179]
M. acapulcensis	Sputum, soil, animal swabs and tissue	Humans, animals	Tanzania, South Africa	[140, 148]
M. arupense	Water	Household water	Uganda, South Africa	[140, 142]
M. chelonae	Sputum, animal tissues	Humans, animals	Burkina Faso, Tanzania, Zambia, Uganda, South Africa	[139, 140, 143–145, 149]
M. chitae	Animal tissue	Animals	South Africa	[140]
M. chubuense	Water	Envt	Uganda	[142]
M. confluentis	Soil, animal tissues, sputum	Animals, envt, human	South Africa, Ethiopia	[140]
M. elephantis	Sputum, animal tissue	Humans, animals	Zambia, South Africa	[139, 140, 150]
M. flavescens	Sputum, animal tissue, animal swabs	Humans, animals	Zambia, South Africa	[139, 140, 150]
M. fortuitum	Sputum, soil, water, faeces, animal tissue	Humans, animals, envt	Nigeria, Uganda, Zambia, Tanzania, Mali, Nigeria, South Africa	[139, 140, 142–148, 150–154, 179]
M. gilvum	Sputum, animal tissue	Human, animals	Zambia, Tanzania	[139, 143, 152]
M. goodie	Sputum, water, animal tissue	Humans, mines, animals	Zambia, South Africa	[140, 144, 152]
M. holsaticum	Animal tissue	Animals	South Africa	[140]
M. lacticola	Animal tissue	Animals	South Africa	[140]
M. madagascariense	Water	Envt	South Africa	[140]
M. mageritense	Sputum	Human	Nigeria	[147]
M. monasence	Animal tissue	Animals	South Africa	[140]
M. moriokaense	Sputum, animal tissue, soil, water	Humans, animals, envt	Mali, South Africa	[139, 140, 154]
M. neoaurum	Sputum, animal tissues, soil	Humans, Animals, envt	Zambia, Uganda, South Africa, Tanzania	[140, 143, 150, 152]
M. parafortuitum	Soil, animal tissue	Envt, animals	South Africa	[140]

(continued)

Table 11 (continued)

(a) Rapid-growing

NTM species	Sample source	Host	Country	Reference
M. peregrinum	Sputum, soil, water, faeces, animal tissue	Humans, animals, envt	Nigeria, Uganda, Zambia	[140, 142, 144, 146, 150–152]
M. phlei	Animal tissue	Animals	South Africa	[139]
M. porcinum	Sputum	Humans	Zambia	[152]
M. poriferae	Sputum	Humans	Tanzania	[143]
M. pulveris	Animal tissue	Animals	South Africa	[140]
M. senegalense	Animal Tissue	Animals	South Africa	[139]
M. septicum	Soil, animal tissue	Envt, animals	South Africa	[140]
M. sphagni	Sputum	Humans	Tanzania	[143]
M. thermoresistibile	Soil, animal tissue	Envt, Cattle	South Africa, Tanzania	[140, 143]
M. vaccae	Animal tissue, animal swabs, soil	Animals, envt	South Africa	[139, 140]
M. vanbaalenii	Water, animal swabs & tissue, soil	Envt, animals	Uganda, South Africa	[140, 142]
M. wolinskyi	Animal tissue	Animals	South Africa	[140]

(b) Slow-growing

NTM species	Sample source	Host	Country	Reference
M. asiaticum	Sputum, animal tissue, soil	Humans, Animals, Envt	Zambia, South Africa	[139, 140, 152]
M. avium complex	Sputum, blood, water, soil	Humans, animal shelters, household water	Burkina Faso, Zambia, South Africa, Uganda	[140, 142, 152, 155, 156]
M. aurum	Sputum	Humans	Zambia	[152]
M. austroafricanum	Soil, water	Envt	South Africa	[140]
M. avium	Sputum, blood, water, soil, faeces	Humans, environment, animals	Uganda, Burkina Faso, Malawi, Tanzania, Zambia, Guinea-Bissau, Mali, Mozambique, Nigeria, South Africa	[142, 144, 145, 147–150, 152, 154, 157–161]
M. boucherdurhonense	Sputum	Humans	Zambia, South Africa	[150, 160]
M. celatum	Sputum	Humans	Tanzania	[148]
M. chimaera	Sputum	Humans	Zambia	[150, 179]
M. colombiense/M. stomatepiae	Sputum	Humans	Tanzania	[148]

Table 11 (continued)

(b) Slow-growing				
NTM species	Sample source	Host	Country	Reference
M. conspicuum	Sputum	Humans	Zambia	[152]
M. duvalii	Soil, animal swags and tissue	Envt, Animals	South Africa	[140]
M. engbaekii	Soil, animal tissue, water	Animal shelter, envt	Uganda, South Africa	[140, 142]
M. europaeum	Sputum	Humans	Zambia	[150]
M. farcinogenes	Animal Tissue	Animals	South Africa	[139]
M. fluoroanthenivorans	Water	Envt	South Africa	[140]
M. gastri	Sputum, gastric lavage	Humans (children)	South Africa	[160]
M. genavense	Sputum	Humans	Tanzania	[143]
M. gordonae	Sputum, soil, water, faeces, animal tissue	Humans, animals, envt, mines, household water	Nigeria, Uganda, Zambia, Tanzania, South Africa, Mali, Kenya, Ethiopia	[140, 142–146, 148, 150–152, 154, 159, 160, 162]
M. heckershornense	Sputum	Human	South Africa	[145]
M. heidelbergense	Animal tissue	Animals	South Africa	[139]
M. hiberniae	Soil, animal tissue, sputum	Animal shelter, animals, humans	Uganda, South Africa, Tanzania	[139, 142, 148]
M. indicus pranii	Sputum	Humans	Zambia	[150]
M. interjectum	Sputum, animal tissue	Humans, Animals	Nigeria, Tanzania, South Africa	[139, 140, 145, 146, 148]
M. intermedium/ triplex	Sputum	Humans	Tanzania	[148]
M. intermedium	Sputum, soil, water, animal tissue	Humans, Envt, animals	Tanzania, South Africa	[140, 143, 148]
M. intracellulare	Sputum, water, soil	Humans, envt, cattle, wildlife, household water	Nigeria, Uganda, Burkina Faso, Zambia, Tanzania, Guinea-Bissau, South Africa	[142–146, 148–152, 160, 161, 179]
M. kansasii	Sputum, water	Humans, gold mines	South Africa, Nigeria, Tanzania	[144–146, 148, 156, 159, 160]
M. kubicae	Sputum, water	Humans Household water	Mali, Uganda	[142, 154]
M. kumanotomense	Sputum	Humans	Zambia Tanzania	[148, 150]

(continued)

Table 11 (continued)

(b) Slow-growing				
NTM species	Sample source	Host	Country	Reference
M. lentiflavum	Sputum, animal tissue	Humans, animals, envt	Zambia, Uganda, Kenya, Tanzania, South Africa	[139, 143, 145, 148, 150–152]
M. malmoense	Sputum	Humans	Nigeria, South Africa	[145, 146]
M. marinum	Sputum	Humans	Tanzania	[148]
M. montefiorense	Water, soil	Gold mines, envt	South Africa	[140, 144]
M. nonchromogenicum	Soil, water, faeces, sputum, animal tissues	Animals, envt, household water	Uganda, Zambia, South Africa, Ethiopia, Tanzania	[140, 142, 143, 150]
M. novocastrense	Animal tissue	Animals	South Africa	[140]
M. palustre	Sputum, soil, animal tissue	Humans, envt, animals	Mali, South Africa	[140, 154]
M. paraffinicum	Soil, water	Envt	South Africa	[140]
M. parascrofulaceum	Water	Gold mines	South Africa	[144]
M. saskatchawanense	Sputum	Humans	Guinea-Bissau	[161]
M. scrofulaceum	Sputum	Humans	South Africa, Nigeria, Uganda, Tanzania	[145, 146, 148, 151, 160]
M. senuense	Water	Envt	Uganda, South Africa	[140, 142]
M. setense	Water	Gold mines	South Africa	[144]
M. sherisii	Sputum, blood	Humans	Burkina Faso, Tanzania	[148, 155, 163]
M. simiae	Sputum, blood, water, soil, animal tissue	Humans, household water, envt, animals	Burkina Faso, Malawi, Zambia, Tanzania, Uganda, South Africa, Mali, Mozambique	[140, 142, 143, 148, 149, 152, 154, 157, 158]
M. stomatepiae	Sputum, animal tissue	Humans, Animal	Tanzania, Zambia	[148, 179]
M. szulgai	Sputum, water	Humans, envt	Uganda, South Africa	[140, 151, 160]
M. terrae complex	Sputum	Humans	Tanzania	[148]
M. terrae	Water, soil, animal tissue, sputum	Household water, animals, envt, humans	South Africa, Ethiopia, Zambia, Uganda	[139, 140, 142, 143, 152, 162]
M. triplex	Soil, water	Envt	South Africa	[140]
M. triviale	Soil, water, animal tissue	Envt, animals	South Africa	[140]
M. tusciae	Water	Coal mines	South Africa	[144]
M. xenopi	Sputum	Humans	Nigeria	[146]

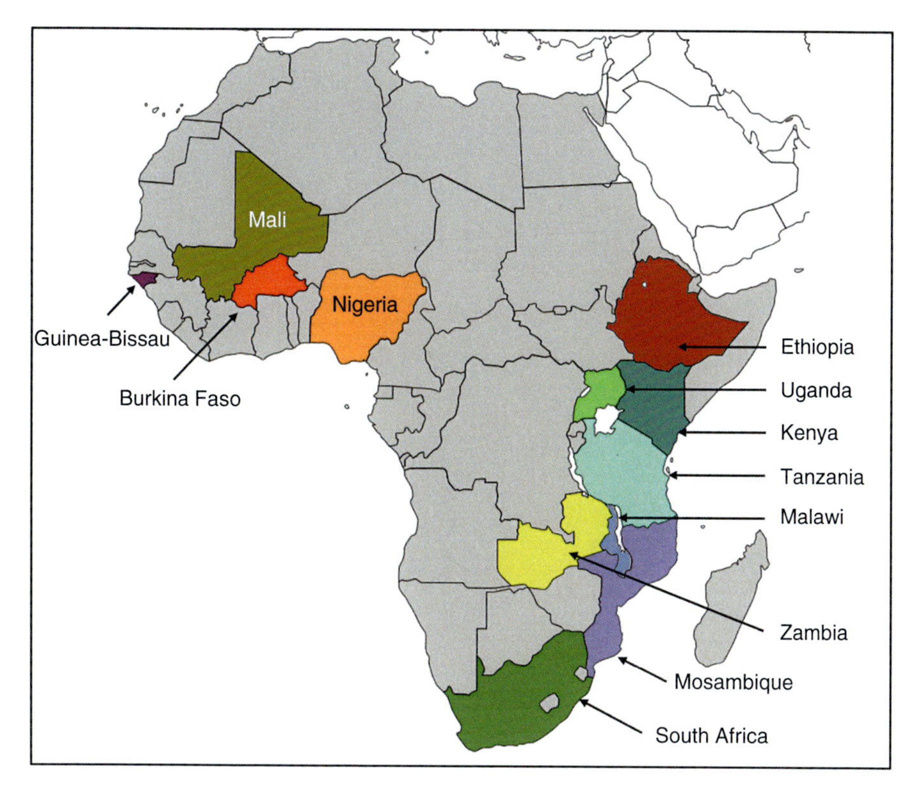

Fig. 15 Countries reporting data summarized in this review. Grey indicates countries without reports, and white indicates countries outside of Africa

defined NTM disease incidence as a positive isolate in conjunction with new cavitation on CXR, and estimated overall rates of 101/100,000 for NTM, 66/100,000 for *M. kansasii* and 12/100,000 for *M. scrofulaceum* [169].

Kwenda et al. [144] investigated the presence of *M. kansasii* in gold and coal mine and associated hostel water supplies and compared the genetic diversity of clinical (obtained 2005–2007) and environmental isolates of *M. kansasii*. Potentially pathogenic mycobacteria from the gold-mining region included *M. kansasii, M. avium, M. fortuitum, M. peregrinum, M. chelonae, M. abscessus, M. parascrofulaceum, M. setense* and *M. montefiorense*, whilst those from the coal-mining region included *M. avium, M. intracellulare, M. peregrinum, M. chelonae* and *M. tusciae. M. kansasii* was not isolated from the coal-mining region. The data did not establish/exclude water as a source of infection.

The introduction of liquid culture, and new speciation methods, is likely to have contributed to the different distribution of NTM species identified by Van Halsema et al. [159] in this population. In an observational study of 232 South African miners recruited at workforce TB screening and routine clinics 2006–2009, *M. gordonae* (60 individuals), *M. kansasii* (50) and *M. avium* complex (MAC: 38) were the commonest species, with notable absence of *M. scrofulaceum* using 16S sequencing.

Table 12 Studies from Africa reporting NTM

Country	Time	Population	Proportional data			Mean/median age, %F	Predominant species of total NTM % total (absolute number)	Identification method		Reference
			Prop NTM/ total pts. screened	Prop NTM/Pos cultures	NTM-PD/total NTM					
Nigeria	2010–2011	TB suspects	69/1603 4.2%	69/444 15%	NA	37 years 43.8%F	*M. intracellulare* 30.4 *M. abscessus* 11.6 *M. fortuitum* 5.8 *M. gordonae* 5.8 *M. scrofulaceum* 4.3 *M. malmoense* 2.9	Liquid culture Hain CM/AS	38% NTM pts. HIV+ (21% TB pts)	[146]
Nigeria	2008–2009	AFB smear + TB suspects	16/137 11.7%	4/97 5.2%	NA	NA	*M. fortuitum I 25 (1 isolate)* *M. fortuitum II/M. mageritense 25* *M. abscessus 25* *M. avium ssp. 25*	Genotype CM	815 HIV+	[147]
Nigeria	2007–2009	TB suspects	7/102 6.9%	7/77 9.1%	100%	NA	*M. fortuitum* 42.9 (3) *M. intracellulare* 28.6 (2) *M. chelonae* 28.6 (2)	NA		[173]
Tanzania	2012–2013	TB suspects	36/372 9.7%	36/121 29.8%	All met micro criteria 80.6% 3/3 sp. pos 19.4% 2/3 sp. pos	40 years 47.3%F	*M. gordonae* 16.7 *M. interjectum* 16.7 *M. intracellulare* 11.1 *M. scrofulaceum* 8.3 *M. avium* 5.5 *M. fortuitum* 5.5	Genotype CM/AS 16s rRNA and hsp65 gene sequencing	30.6% NTM pts. HIV+	[148]
Tanzania	NA	TB suspects	12/472 2.9%	NA	NA	NA	*M. chelonae/abscessus* 25 (3) *M. genavense* 16.7 (2) *M. gilvum* 16.7 (2) *M gordonae* 16.7 (2) *M intermedium* 16.7 (2) *M. poriferae* 16.7 (2) *M. sphagni* 16.7 (2)	16s rRNA gen seq	Comparison with isolates from animals	[143]

Tanzania	2006–2008	Suspected disseminated disease	2/723 Bloodstream infections 0.3%	2/30 blood 6.7% +1 tissue isolate	NA	NA 33.3% F	*M. sherrisii* 2 *M. avium* 1			
Uganda	2009–2011	Children, TB suspects from vaccine programme	26/710 (3.7%) infants 69/1490 (4.6%) adolescents	95/103	NA	NA	*M. fortuitum* 63.5 *M. szulgai* 14.3 *M. gordonae* 9.5 *M. intracellulare* 4.7	Genotype CM/AS		[151]
Burkina Faso	2007–2008	Chronic TB cases	13/63 20.6%	13/51 25.5%	NA	NA	*M. avium* 38.5 (5) *M. intracellulare* 30.8 (4) *M. simiae* 23.1(3) *M. sherrisii* 30.8 (4)[a]		4 NTM pts. HIV+	[149] [155][a]
Zambia	2002–2003	96% suspected pulm TB 4% skin inf'n/LAD	Pts:31/180 (17.2%), 72 isolates Controls: 99/385 (25.7%), 99 isolates	P:72/273 26.4% C:99/104	7/29 24.1% 2 pos cultures	Median Pts:35 years (16–80) 55%F Controls: 30 years (15–78) 69%F	*M. intracellulare* 7 (12) *M. avium* 4.7 (8) *M. gordonae* 2.3 (4) *M. peregrinum* 3.5 (6) *M. goodie* 2.9 (5) *M. porcinum* 2.3 (4) Unident 66, unknown 55	16s rRNA seq	71% Pts HIV +; 23% control HIV +	[152]
Zambia	2001	Hospitalized patients	93/167 56%	93/131 70.1%	66/101 65%	37 years 59% F	*M. lentiflavum* 45.1 (42) *M. intracellulare* 40.9 (38) *M. chelonae* 8.6 (8)	16s rRNA seq	79% NTM HIV+	[172]
Zambia	NA	TB suspects (national prevalence survey)	923/6123 (15.1%) 936/6123 (15.3%) (13 coinf with TB)	936/1539 60.8%	1477/100,000 (95% CI 1010–1943) aged >15	51.7% F[a]	NA	MPT 64 Negative	5.85% NTM cases HIV+;regional variation of prevalence	[171]

(continued)

Table 12 (continued)

Country	Time	Population	Proportional data			Mean/median age, %F	Predominant species of total NTM % total (absolute number)	Identification method		Reference
			Prop NTM/ total pts. screened	Prop NTM/Pos cultures	NTM-PD/total NTM					
Zambia	2011–2012	TB suspects	9/100 9%	9/55 16.4%	NA	NA	*M. intracellulare* 4 *M. abscessus* 2 *M. chimaera* 2 *M. bolletii* 1	16s rRNA seq		[179]
Guinea-Bissau	1989–1996	TB suspects	17MAC/814 (2.1%)	17/206 (8%)	6/17[b] (35.3%)	Age NA 64.7% F	MAC only identified	Biochemical tests AccuProbe for MAC and TB	2/17 HIV+	[161]
Mali	2004–2009	AFB Smear + TB (suspects incl chronic TB)	NA	17/142 (12%)	11/142 (8%) “chronic TB”	NA	*M. avium* 64.7 (11) *M. simiae* 11.8 (2)[c]	Nucleic acid probes, 16s rRNA gene seq		[154]
Mozambique	2002–2004	HIV pos TB suspects	3/447 0.7%	3/277 1.1%	NA	33 years 62.25%F	*M. avium* 2 *M. simiae* 1	PRA of hsp65 gene		[158]
Western Kenya	2007–2009	TB suspects	1.7%	15/361 4.2%	NA	Mean 35 Median 32 45.1%F	10 unidentified *M. intracellulare* 3(20%) *M. fortuitum* 1 *M. peregrinum* 1	Genotype CM/AS	41.8% HIV+	[153]
South Africa	2010	Lab Suspected NTM	133/200 66.5%	133/139 95.7%	NA	NA	*M. intra* 45.9% *M. avium ssp.* 11.3% *M. gordonae* 6% *M. kansasii* 4.5%	Genotype CM/AS	Lab based study	[145]
South Africa	1993–1996	NTM pos HIV gold miners	All 243	NA	>80% 66/100,000M. kansasii 12/100,000M. scrof	44 years All M	*M. kansasii* 68% *M. scrofulaceum* 14% MAI 6%	Biochemical tests		[169]
South Africa	2006–2007	NTM pos miners	299/2328 12.8%	299/720 41.5%	NA	44 years 2%F	*M. gordonae* 60/232 *M. kansasii* 50 MAC 38	16s rRNA seq	77% HIV+	[159]

South Africa	1996–1998	Miners with suspected mycobacterial disease	All 118	NA	27% 47.6 per 100,000 employee years	All M	*M. kansasii* 23 *M. scrofulaceum* 7 *M. avium* 1 *M. abscessus* 1	Biochemical tests	34% NTM HIV+	[158]
South Africa	2001–2005	Children TB Suspects (part of TB vaccine trial)	109/1732 6%	109/296 36.8%	95%	NTM 18mths (13–23) TB 10mths (5–16) Gender NA	*M. intracellulare* 47 *M. gastri* 7	PRA hsp 65	All NTM childrenHIV-	[160]
Ethiopia	2011	Children TB suspects	10/101	10/25	1–5 years (43% 6–10 years (46% 11–15 years (23% 43% F	NA	Gastric lavage *M. fortuitum/M. porcinum* *M. triviale* *M. parascrofulaceum/M. seoulense/M. gastri/M. kansasii/M. nebraskense* *M. fortuitum* *M. terrae* Sputum *M. engbackii* *M. confluentis*	Multiplex PCR 16s rDNA seq		[180]

LAD – lymphadenopathy

[a]Mean age not stated but proportions in different age groups categories tabulated

[b]Estimated from information in text

[c]1 each of *M. moriokaense, M. kubicae, M. kumamotonense, M. fortuitum, M. gordonae*

Organisms previously identified as *M. scrofulaceum* in this study were almost all subsequently identified as *M. gordonae* by 16S sequencing.

Another South African study from the Kwa-Zulu province of Natal also included molecular techniques to further identify mycobacterial isolates at a species level. A laboratory-based prospective study of 200 suspected NTM isolates conducted during 2010 (representing 1/6 of the total suspected NTM received by that laboratory) revealed 66.5% confirmed to be NTM, with *Mycobacterium intracellulare* (45.9%), *M. avium* subspecies (11.3%), *M. gordonae* (6.0%) and *M. kansasii* (4.5%) the most common identified [145].

Lower rates of NTM were reported in a cohort of 503 HIV-infected patients with suspected TB in Mozambique [158]. Of 277 isolates tested, only 3 were NTM – 2 *M. avium* and 1 *M. simiae*. These patients had clinical profiles indistinguishable from those of other patients. Similarly, low rates were observed in a cross-sectional study carried out at 10 hospitals in western Kenya 2007–2009; sputa from 872 TB suspects revealed TB (39.7%) and NTM (1.7%) [153].

A cross-sectional study was conducted between November 2012 and January 2013 in self-presenting TB suspects in two hospitals in Tanga, Northeastern Tanzania [148]. The overall frequency of NTM patients was 9.7%. Of the 36 patients with NTM, 80.6% were positive for NTM in three samples, and 19.4% positive in 2/3 samples. HIV co-infection was present in 30.6%. NTM species isolated from the 36 patients using Genotype CM/AS included *M. gordonae* (6), *M. interjectum* (6), *M. intracellulare* (4), *M. scrofulaceum* (3), *M. avium* ssp. (2), *M. fortuitum* (2), *M. kansasii* (1), *M. lentiflavum* (1), *M. simiae* (1), *M. celatum* (1) and *M. marinum* (1). Eight isolates required 16s rRNA gene sequencing and were identified as *M. kumamotonense* ($n = 2$), *M. intracellulare/kansasii, M. intermedium/triplex, M. acapulcensis/favenscens, M. stomatepiae, M. colombiense* and *M. terrae* complex ($n = 1$).

Geographical differences in NTM prevalence are supported by environmental studies. NTM are common in a soil and water in the Karonga District of Malawi [170]. Regional differences in leprosy rates, skin test positivity and BCG protection rates prompted an investigation into the impact of environmental NTM exposure. Higher recovery of NTM was seen from soil in the north than south in the dry season but higher recovery in the south in the wet season. Possible explanations for this include a survival advantage in hot, dry soil imposed by the thick lipid-rich wall of mycobacteria, greater bacterial competition in cooler/wetter conditions, variable detection of NTM in the wet season due to dilution in soil by heavy rain or increased inhibitory compounds in wet soil.

The prevalence of NTM among TB suspects in rural Zambia also varied geographically. In three areas with high rates of HIV co-infection, but very similar TB rates (19–25%), NTM proportions ranged from 78% in Katete and 65% in Sesheke to 21% in Chilonga. Furthermore, the distribution of NTM species was different at the three sites [152].

The potential for cross-species infection or transmission between humans, wildlife and livestock was investigated in Tanzania, revealing a diversity of NTM species in the Serengeti ecosystem, some of which are known disease-causing pathogens. Isolates obtained from sputum of 472 TB suspects, 606 tissues from wildlife and indigenous cattle were identified to species level by 16s rRNA sequencing. A total of fifty-five (55) NTM isolates representing 16 mycobacterial species and 5 isolates belonging to the MTBC were detected. Overall, *Mycobacterium intracellulare* (isolated from human, cattle and wildlife) was the most frequently isolated species (20 isolates, 36.4%) followed by *M. lentiflavum* (11 isolates, 20%), *M. fortuitum* (4 isolates, 7.3%) and *M. chelonae-abscessus* group (3 isolates, 5.5%). In terms of hosts, 36 isolates were from cattle and 12 from humans, the balance being found in various wildlife species [143].

More detailed clinical reports of patients with NTM disease have emerged from Zambia [171, 172] and in children from rural Uganda [151]. Of the 6123 Zambian individuals with presumptive TB, 923 (15.1%) were found to have NTM, 13 (0.2%) were MTB/NTM co-infected, and 338 (5.5%) were contaminated (indeterminate). The prevalence of symptomatic NTM was found to be 1477/100,000 [95% CI 1010–1943], significantly higher than rates reported in many western countries. Smear positivity, history of cough or chest pain and HIV positivity were risk factors for NTM. However, there was a significant disease burden in the HIV-negative population. The prevalence of symptomatic NTM increased with participant's age (both male and female) from 452/100,000 among those aged 15–24 years to 5160/100,000 among those aged 65 years and above. Radiological abnormalities were similar to those reported elsewhere and included evidence of bronchiectasis, nodules, fibrosis or cavitation [171, 172].

The most prevalent species of NTM reported from four regions of Zambia 2009–2012 were *M. intracellulare* followed by *M. lentiflavum* and *M. avium* [150]. Other species included *M. fortuitum M. gordonae, M. kumamotonense, M. indicus pranii, M. peregrinum, M. elephantis, M. flavescens, M. asiaticum, M. bouchedurhonense, M. chimaera, M. europaeum, M. neoaurum and M. nonchromogenicum.*

This was in partial agreement with the findings of the study conducted by Buijtels [152] in the Eastern region of Zambia. Of 180 patients admitted to 3 hospitals in Zambia in 2001 with infective respiratory symptoms and signs, 60 (33%) had only *M. tuberculosis* isolates in their sputum samples, 12 (7%) had *M. tuberculosis* and NTM, and 19 (11%) had only NTM. *Mycobacterium avium* complex dominated samples (15 patients); a further 15 were identified as *M. intracellulare* and 3 as *M. avium*. Other species included *M. gordonae, M. peregrinum, M. goodie, M. porcinum, M. lentiflavum, M. fortuitum, M. neoaurum, M. simiae, M. asiaticum, M. aurum and M. conspicuum.* Although numbers of cases were small, the estimated rate of colonization in this study in the patient population was 9% (14/154), and the rate of disease was ≈2% (3/154) [152].

West Africa

In 1996 the presence of MAC organisms was reported in the sputum of patients from Guinea-Bissau. Of 1000 isolates, 28 were identified initially as MAC using the AccuProbe test and then later using 16s rRNA sequencing as *M. avium* (2), *M. saskatchewanense* and (4), the rest *M. intracellulare* (4 different strain types) [161].

A retrospective study from the Oyo and Osun states of Nigeria looked at 102 patients diagnosed with pulmonary TB between 2007 and 2009. Of those with positive cultures, 9.1% had NTM identified; species included *M. fortuitum*, *M. intracellulare* and *M. chelonae*. Fifty-seven percent of the NTM patients had previously been treated with TB drugs and were classified as treatment failures [173].

A report from the Cross River State in Nigeria (2008–2009) found that 12.4% of AFB-positive patients treated for TB had infections with non-mycobacterial organisms, and 4.1% had NTM and included *M. fortuitum I, M. fortuitum II/M. mageritense, M. abscessus* and *M. avium* ssp. [147].

Among 1603 consecutive suspected new TB cases over 12 months 2010–2011 (age 18 and over) also from Nigeria [146], 444 (28%) were culture positive – 85% MTB complex, 15% NTM. Species isolated included *M. intracellulare* 21 (30.4%), *M. abscessus* 8 (11.6%), *M. fortuitum* 4 (5.8%), *M. scrofulaceum* 3 (4.3%), *M. gordonae* 4 (5.8%), *M. malmoense* 2 (2.9%), *M. kansasii* 1 (1.4%), *M. interjectum* 1 (1.4%), *M. peregrinum* 1 (1.4%) and *M. xenopi* 1 (1.4%) species. Mixed infection of NTM and MTB was found in one case (0.2%) [146]. The high prevalence (15%) of clinical pulmonary NTM was linked to Harmattan dust exposure (Jan–Feb) and to HIV co-infection (38%).

As in other parts of the world, there may be differences in the geographical distribution of NTM in parts of Africa. This is suggested by reports from Nigeria on varying prevalence of NTM among TB suspects, between 4.1% in the South [147] and 23.1–26.6% in the north [174, 175]. Of note however, the time periods of these studies were not comparable, nor culture and identification methods.

In Bamako, Mali, sputum specimens from both newly diagnosed (naïve) and previously treated (chronic) TB cases between 2004 and 2009 were re-evaluated for NTM. Of 61 chronic cases, 17 (12%) were infected with NTM. Eleven of these patients had NTM disease alone (8 *M. avium*, 2 *M. simiae* and 1 *M. palustre*); the other six patients were from the treatment-naive group and had co-infection with MTB. Species identified included *M. avium*, *M. moriokaense, M. fortuitum, M. kumamotonense* and *M. kubicae*. As these six patients were only treated for MTB, it is difficult to know if the NTM isolation was clinically significant, whereas all of the 11 chronic disease patients with NTM met ATS criteria for disease [154].

Similarly, in Burkina Faso, pulmonary TB cases failing standard therapy with positive sputum smear microscopy after 5 months are defined as “chronic” TB patients. A prospective, countrywide investigation conducted in 2007–2008 [149] registered 83 chronic TB cases, of which 63 provided sputum samples for culture, and 13 (20.6%) grew NTM. Four patients were co-infected with HIV, and infecting species included *M. avium* (5), *M. intracellulare* (4), *M. simiae* (3) and *M. chelonae* (1).

As a consequence of these findings, the National Tuberculosis Programme (Ouagadougou, Burkina Faso) started a process to develop national guidelines for NTM management. A further 314 samples from the same centre were examined in 2012 [155]. NTM grew in culture in 36 (11%). Most NTM were identified as MAC (20 isolates). In culture of samples from four of the remaining patients, *M. sherrisii* grew. This is a relatively recently described species, closely related to *M. simiae*, and may have been underestimated in many settings due to misidentification as *M. simiae* using the commercially available line probe assays. The clinical characteristics of the patients infected with *M. sherrisii* were comparable to those infected with other common pathogenic species, such as MAC. Other reports suggest that this species may be relatively more common in Africa countries [163, 176–178].

Children

The prevalence of NTM in Ugandan children has been investigated using a different approach [151]. Two cohorts of BCG vaccinated infants (2500) and adolescents (7500) aged 12–18 years were recruited and followed up for 1–2 years to determine the incidence of TB. The prevalence of NTM among infant TB suspects was 3.7% and adolescents 4.6%. Of 127 isolates obtained, 103 were confirmed mycobacteria, 95 were NTM and 8 MTB. The Genotype CM/AS assay identified 63 of the 95 NTM isolates, whilst 32 remained unidentified. The identified NTM species were *M. fortuitum* (40 isolates, 63.5%), *M. szulgai* (9 isolates, 14.3%), *M. gordonae* (6 isolates, 9.5%), *M. intracellulare* (3 isolates, 4.7%), *M. scrofulaceum* (2 isolates, 3.2%), *M. lentiflavum* (2 isolates, 3.2%) and *M. peregrinum* (1 isolate, 1.6%). Strain typing revealed minimal clustering within species, consistent with environmental sources within the community. In rural agro-pastoral Uganda, a recent study of 310 samples from soil, water and faecal matter from cattle and pigs isolated 48 NTM [142]. The major species identified in that study were 15 (31.2%) *M. avium* complex, 12 (25%) *M. fortuitum-peregrinum* complex, 5 (10.4%) *M. gordonae* and 5 (10.4%) *M. nonchromogenicum.*

A similar study in children <5 years of age was conducted in South Africa (2001–2005) as part of the TB vaccine surveillance programme [160]. NTM were isolated from sputum or gastric lavage in 6% of children investigated for pulmonary TB, with 114 isolates coming from 109 children. Only 2% of these children were co-infected with HIV, and 95% were symptomatic. *M. intracellulare* was the most common pathogen (51) followed by *M. gastri* (7), *M. avium* (5), *M. gordonae* (5), *M. flavescens* (3), *M. scrofulaceum* (3), *M. szulgai* (1) and *M. kansasii* (1). Compared to the children with MTB, the NTM children were older (mean age 18 months; range 13–23) and were more likely to have constitutional symptoms including fever and weight loss and less likely to have a positive PPD and CXR findings consistent with TB [160].

Summary

Information regarding the epidemiology of NTM in Africa lags behind that of developed countries. However, the importance of NTM in confounding TB diagnosis and treatment is appreciated, and with laboratory collaborations, more studies are revealing the magnitude of this effect. Rates of NTM in TB suspects in Africa are commonly between 5% and 15%. The wide variability in reported studies (0.2%–78%) is likely due to varying rates of HIV co-infection, geographical distribution and frequency of NTM in the environment and laboratory methods of culture and identification.

Species distribution is similar to other parts of the world with MAC organisms the most common. However, some species appear unique to Africa (such as *M. sherrisii*) and others more common in some areas (e.g. *M. kansasii* in miners, *M. lentiflavum* in Zambia).

Importance of correct identification of NTM, to avoid confounding with TB, and unnecessary treatment for MDR/XDR TB cannot be overemphasized. The morbidity and mortality of NTM disease in Africa, particularly in association with HIV infection, should also be a public/environmental health priority.

Central and Southern America

Dirk Wagner

Introduction

Data from Central and South Americas are general laboratory-based surveys, mostly from single centre institutions. Since population-based surveys are lacking from this region and in most studies reliable clinical data are missing, estimates of the prevalence and incidence of NTM-PD in these regions are very scarce (182). Data are affected by referral bias and from selection bias of samples that were submitted for culture, e.g. from patients newly diagnosed with TB who remain positive for AFB during treatment, patients that had a history of prior treatment for TB and are newly AFB positive, patients who are contacts of persons with drug-resistant TB, or patients who are part of specific population groups, including health professionals, the homeless, prisoners, indigenous populations, and HIV-positive persons [182]. This leads to an overproportion of or selection of patients with smear-positive NTM-PD disease (Table 13), and – although smear-positive patients have a high probability to indeed have NTM-PD according to ATS-guidelines – to an underestimation of the true incidence and annual prevalence of NTM-PD. Furthermore, data are difficult to compare since the number of repetitive isolations of an NTM from the same patient as well as the exact specimen that was cultured has not always been stated [181, 183]. Figure 16 indicates the countries that have produced articles included in this review.

Table 13 Studies from South America that have published data on isolation of and disease due to NTM

			Various proportional data					Prevalent species				
Study location	Study period	Target population	NTM isolate/culture-positive specimen (%)	Patient with NTM isolate/culture-positive patient (%)	NTM-PD patient/patient with NTM isolate (%)	NTM-PD patient/NTM culture-positive specimens (%)	*Median* or mean age, female %	Proportion of each prevalent NTM species among NTM-PD patients	Additional information [proportion of prevalent species of isolates]	Identification method	Annotation	Reference
Argentina (North of Buenos Aires)	2004–2010	TB suspects (?)	178/2736 (6%)	108/2226 (5%)	54/108 (50%) #	54/178 (30%) #	42 years (35%)	MAC 48%, *M. kansasii* 13%	[MAC 59%, *M. gordonae* 12%, *M. kansasii* 8%]	Biochemical tests and RFLP of hsp65 PCR-product	28% HIV coinfection, 25% with previous TB treatment, 78% respiratory specimen	[196]
Argentina (Cordoba)	1991–2000	TB suspects (?)	32/716 (4%)	NA	NA	NA	NA	NA	[MAC 75%, *M. fortuitum* 16%, *M. chelonae* 6%, *M. kansasii* 3%]	Biochemical typing		[195]
Brazil (Rondônia)	2008–2010	TB suspects	75/444 (17%)	45/444 (10%)	19/24 (79%) #	19/45 (42%) #	50 years (36%)	NA	[MABC 32%, MAC 17,3%, *M. fortuitum* 12%]	RFLP of hsp65 PCR-product	The species of 21% of isolates were not identified	[188]

(continued)

Table 13 (continued)

Study location	Study period	Target population	Various proportional data				*Median* or mean age, female %	Prevalent species	Additional information [proportion of prevalent species of isolates]	Identification method	Annotation	Reference
			NTM isolate/culture-positive specimen (%)	Patient with NTM isolate/culture-positive patient (%)	NTM-PD patient/patient with NTM isolate (%)	NTM-PD patient/NTM culture-positive specimens (%)		Proportion of each prevalent NTM species among NTM-PD patients				
Brazil (Para)	2010–2011	TB suspects	69/281 (25%)	38/281 (14%)	29/38 (76%)[a]	NA	52 years (72%)	*M. massiliense* 45%, MAC 21%, *M. simiae* complex 10%; *M. abscessus* 7%	NA	16S rRNA and hsp65 gene sequencing	All NTM-PD patients were smear-positive and had been treated for presumptive TB	[185]
Brazil (Para)	1999–2010	TB suspects	249/1580 (16%)	128/1580 (8%)	73/128 (57%)[b]	73/249 (29%)[b]	NA (64%)	MAC 29%, *M. massiliense* 27%, *M. simiae* complex 22%, *M. bolletii* 5%, *M. abscessus* 4%	[*M. avium* complex 36%, *M. simiae* complex 22%, *M. massiliense* 19%, *M. fortuitum* (10%), *M. bolletii* 6%	16S rRNA and hsp65 gene sequencing	72/73 NTM-PD patients had been treated for presumptive TB	Fusco da Costa et al. 2012 In: Amal A, editor. Pulm Infect Chapter 3: 37–54, InTech (ISBN: 978–953–51-0286-1)

Brazil (Rio de Janeiro)	1993–2011	TB suspects	NA	174/NA	127/174 (73%)[b]	NA	55 years (38%)	MAC 35%, *M. kansasii* 33%, *M. abscessus* 19%, *M. fortuitum* 9%	Bronchiectasis 22%, COPD 21% [*M. kansasii* 34%, MAC 30%, *M. abscessus* 13%, *M. fortuitum* 8%.]	Until 2004: biochemical tests 2004–2011: RFLP of hsp65 PCR product	58% of 174 patients had been treated for presumptive TB, in patients with no prior TB-treatment: 59% cavitations	[182]
Brazil (Piauí)	2007	TB suspects	9/103 (9%)	NA	NA	NA	64 years (44%)	NA	[*M. gordonae* 33%, *M. kansasii* 22%, *M. abscessus* 22%, *M. smegmatis* 11%, *M. flavescens* 11%]	RFLP of hsp65 PCR-product		[185]
Brazil (Bahia)	1998–2003	TB suspects	NA	19/231 (8%)	NA	14/19 (74%)[b]	49 years (32%)	NA	[*M. chelonae*/*abscessus* 58%; MAC 16%; *M. kansasii* 16%; *M. fortuitum* 11%]	Biochemical typing		[186]
Brazil (Sao Paulo)	2009–2010	TB suspects (?)	109/476 (23%)	NA	NA	NA	NA	NA	[*M. avium* complex 45%, *M. fortuitum* (10%), *M. abscessus* (10%), *M. kansasii* (8%), *M. gordonae* (6%), *M. flavescens* (4%)]	hsp65/IS6110 multiplex PCR RFLP of hsp65 PCR-product		[192]

(continued)

Table 13 (continued)

Study location	Study period	Target population	Various proportional data: NTM isolate/culture-positive specimen (%)	Patient with NTM isolate/culture-positive patient (%)	NTM-PD patient/patient with NTM isolate (%)	NTM-PD patient/NTM culture-positive specimens (%)	*Median* or mean age, female %	Prevalent species: Proportion of each prevalent NTM species among NTM-PD patients	Additional information [proportion of prevalent species of isolates]	Identification method	Annotation	Reference
Brazil (Sao Paulo)	1996–2005	TB suspects (?)	317/1300 (24%) (all specimen) 271/1254 (22%) (respiratory)	216/NA	184/	NA	NA	MAC 63%, *M. fortuitum* 11%	[MAC 56%, *M. gordonae* 12%, *M. fortuitum* 8%, *M. terrae* 3%, *M. chelonae* 3%] (respiratory)	Biochemical tests and RFLP of hsp65 PCR-product		[191]
Brazil (Sao Paulo)	2000–2005	TB -suspects	194/NA	125/NA	24/125 (19%)[b]	NA	(25%)	*M. kansasii* 21%, MAC 20%, *M. fortuitum* 16%	[*M. kansasii* 34%, MAC 17%, *M. fortuitum* 13%, *M. peregrinum* 2%, *M. gordonae* 2%]	Biochemical tests and RFLP of hsp65 PCR-product	42% HIV-positive patients, 52/194 (26%) inconclusive	[190]
Brazil (Sao Paulo)	1995–1998	TB suspects (?)	1604/9381 (17%)	NA	NA	NA	NA	NA	[MAC 58%, *M. kansasii* 18%, *M. gordonae* 7%, *M. chelonae* 5%, *M. fortuitum* 4%]	Biochemical tests and RFLP of hsp65 PCR-product		[189]

MAC Mycobacterium avium complex, *MABC Mycobacterium abscessus* complex, *NB* nodular bronchiectatic type, *FC* fibrocavitary type, *NA* Not analysed/reported

[a]Fulfilment of all criteria of the NTM-PD definition (2007 ATS/IDSA)

[b]Fulfilment of the microbiological criteria of the NTM-PD definition (2007 ATS/IDSA)

South America

NTM isolation Most data for patients infected with NTM in Southern America have been collected in *Brazil*, usually in patients with suspected tuberculosis. Predominantly 16S rRNA- and hsp65 gene fragment sequencing or RFLP of hsp65 PCR product have been used as the detection method (Fig. 17, Table 13). In the states Para, Piaui, Bahia and Rhodônia, between 9% and 25% of the culture-positive specimens grew NTM. The percentage of individual patients with NTM-positive specimens was lower (8–14%), indicating repetitive culture positivity in some NTM patients [182, 184–188]. In Sao Paulo, where several studies reported data from tuberculosis suspects from 1995 to 2010, consistently a fifth of respiratory specimens grew NTM (17–24%) [189–192]. The isolated species differed substantially between studies and regions with a high percentage of rapid-growing mycobacteria being reported in Rhodônia [188] and Bahia [186], whereas the slow-growing mycobacteria MAC and *M. kansasii* were the predominant species in Para [184], Rio de Janeiro [182] or Sao Paulo [189–192]. In adult patients with cystic fibrosis, NTM isolation frequency from sputum was 11% in Campinas (State of Sao Paulo, no species data were reported) [193], and – more recently – 8% in children in Rio de Janeiro [194]; 8 of the 12 patients were infected with *M. abscessus* group, 3 with MAC [194]. In the two studies from *Argentina*, Cordoba (1991–2000) [195] and Northern Buenos Aires (2004–2010) [196], isolation frequency of NTM from mycobacterial cultures or in culture-positive patients was significantly lower (4–6%). In both studies a high percentage of MAC as the predominant species was reported, although the former study only used biochemical methods for species identification.

NTM-PD In general, studies from South and Central America fail to report the proportion of NTM-positive patients meeting the definition of disease according to the 2007 ATS/IDSA guidelines [197]. One exception is a study from Para [187] in which 29 of the 38 (76%) symptomatic patients with radiological abnormalities and with an NTM isolated from respiratory secretions had NTM-PD due to *M. massiliense* (45%), MAC (21%), *M. simiae* complex (10%) and *M. abscessus* (7%) [187]. Several studies [182, 184, 186, 188, 190, 196] have used the microbiological criteria as a surrogate, sometimes with the additional note that all patients were symptomatic. This approach has been validated in Oregon, USA, finding that 86% of patients that met the ATS/IDSA microbiologic criteria for disease also met the full ATS/IDSA criteria [198]. A second older study from Para, which only used the microbiological criteria [184], showed a similar species distribution in these patients (MAC 29%, *M. massiliense* 27%, *M. simiae* complex 22%, *M. bolletii* 5%, *M. abscessus* 4%). In both studies almost all patients were smear-positive and had been treated for presumptive TB. In comparison to other Brazilian states and the reports from Argentina, the higher percentage of *M. massiliense*, a rapid-growing NTM, in patients with NTM-PD is noteworthy. The absence of *M. kansasii* may be explained by pretreatment with a standard TB regimen of rifampin, ethambutol and isoniazid, to which *M. kansasii* is susceptible. Different percentages of pretreated patients

(25% [196], 58% [182], 100% [187]) have likely influenced the reported NTM species distribution. Still geographic factors as well as environmental factors (such as comparably less access to a piped water supply in the state of Para relative to other Brazilian states) have been discussed as influencing exposure risk and disease manifestation [187]. The fact that in the two studies from Para the patients with NTM isolations were predominantly female (64–72%) in contrast to the other studies from South America (25–44%, Table 13) points to different risk factors in the state of Para. People of Pardo heritage (mixed white and indigenous) were over-represented (90% vs background 69.5%). This may be due to increased susceptibility of Pardo individuals to NTM disease or confounded by greater exposure to NTM in agricultural soils or drinking water [187].

Despite the significant limitations (and likely high degree of underestimation) of epidemiologic estimates [181], Prevots and Marras [181] analysed two studies [190, 191]. In the Baixada Santista (coastal) region of Sao Paolo (2000–2005), the annual prevalence for NTM isolation was calculated as 1.3/100.000, and for NTM disease as 0.25/100,000 [181, 190]. A significantly higher annual prevalence for NTM isolation (5.3/100,000) and NTM disease (1/100,000) was calculated for São José do Rio Preto (1996–2005), a more inland region in the state of Sao Paolo [181, 191]. The paucity of data allows no firm conclusions to be drawn about population frequency of NTM-PD in Southern America.

Central America

Very little information has been published from Central American countries (Fig. 16, Table 14). In a study from Mexico, 537/1646 (32%) of clinical isolates from acid fast-positive respiratory secretions from 2008 to 2011 were identified as NTM by hsp65-PCR and restriction enzyme analysis; no data on the species distribution were reported [199]. In a study from Honduras (1997) analysing 235 patients with suspected tuberculosis, 10% of the 100 specimens that grew mycobacteria were NTM; MAC was isolated in three patients, *M. fortuitum* in two. Five of these ten patients were smear-positive, seven had been treated for presumptive tuberculosis, one for more than 20 years [200].

Summary

Information regarding the epidemiology of NTM in South America and more so in Central America are comparable to the limited information found in African countries. Similarly, NTM is confounding TB diagnosis and treatment, but the magnitude of the effect has not been determined sufficiently. A significant difference in species distribution has been reported in Brazil, but no population-based data have

Fig. 16 Countries reporting data summarized in this review. Grey indicates countries without reports, white countries outside of South and Central America

been reported to correctly estimate the NTM-PD disease burden in this region. Efforts to diagnose NTM-PD not only in the tuberculosis suspect will likely increase these numbers. Similar to other regions with ageing populations, the decreasing TB prevalence, increasing prevalence of risk groups and availability of rapid identification methods for NTM-PD will be of increasing importance in the future.

Summary

This review summarized the heterogeneous epidemiology of NTM isolations and NTM disease worldwide. Not only the epidemiology of NTM isolation is heterogeneous, as has been shown by the NTM world map study [13], but also the available data differ from continent to continent, from country to country and within countries. Still current available population-based data imply that NTM-PD is increasing

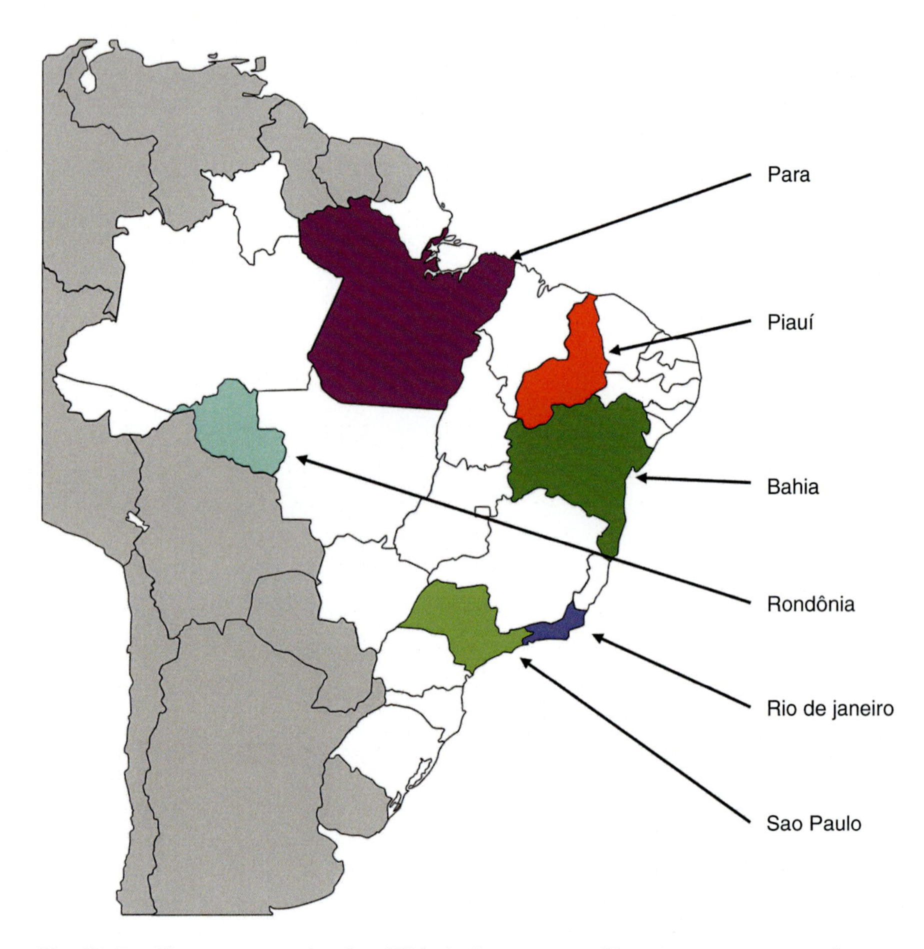

Fig. 17 Brazilian states reporting data. White indicates states without reports; grey countries outside of Brazil

not only in North America but also in Europe, Asia and Australia. Often, NTM is confounding TB diagnosis and treatment, but the magnitude of the effect has not been determined sufficiently. Due to the lack of availability of detection methods but also due to the lack of awareness, NTM-PD is probably still underdiagnosed in many regions worldwide. Our understanding of the epidemiology of NTM worldwide would be considerably improved by undertaking large population-based studies, collecting relevant clinical data and providing information on subsequent outcome. A respective registry has been implemented in the USA but is urgently needed in other continents. New molecular methodology, e.g. whole-genome sequencing, will considerably enhance our appreciation of NTM epidemiology, in terms of both subspeciation of potentially disease-causing organisms, and whether repeated NTM isolates in an individual represent the same or different strains of the organism occurring over time.

Table 14 Studies from Central America that have published data on isolation of and disease due to NTM

			Various proportional data					Prevalent species				
Study location	Study period	Target population	NTM isolate/culture-positive specimen (%)	Patient with NTM isolate/culture-positive patient (%)	NTM-PD patient/patient with NTM isolate (%)	NTM-PD patient/NTM culture-positive specimens (%)	*Median* or mean age, female %	Proportion of each prevalent NTM species among NTM-PD patients	Additional information [proportion of prevalent species of isolates]	Identification method	Annotation	Reference
Mexico	2008–2011	TB suspects	537/1646 (33%)	NA	NA	NA	NA	NA	NA	Biochemical tests, RFLP of hsp65 PCR product and hsp gene sequencing	All isolates were smear-positive	[199]
Honduras	1994–1995	TB suspects	10/100 (10%)	NA	NA	NA	53 years (50%)	NA	[MAC 30%, *M. fortuitum* 20%}]	Biochemical tests, AccuProbe, 16S rRNA gene sequencing		[200]

MAC Mycobacterium avium complex, *MABC Mycobacterium abscessus* complex, *NB* nodular bronchiectatic type, *FC* fibrocavitary type
[a]Fulfilment of all criteria of the NTM-PD definition (2007 ATS/IDSA)
[b]Fulfilment of the microbiological criteria of the NTM-PD definition (2007 ATS/IDSA)

References

England, Wales, Northern Ireland, Ireland, Scotland and Spain

1. Moore JE, Kruijshaar ME, Ormerod LP, Drobniewski F, Abubakar I. Increasing reports of non-tuberculous mycobacteria in England, Wales and Northern Ireland, 1995-2006. BMC Public Health. 2010;10:612.
2. Shah NM, Davidson JA, Anderson LF, Lalor MK, Kim J, Thomas HL, et al. Pulmonary Mycobacterium avium-intracellulare is the main driver of the rise in non-tuberculous mycobacteria incidence in England, Wales and Northern Ireland, 2007-2012. BMC Infect Dis. 2016;16:195.
3. Fowler SJ, French J, Screaton NJ, Foweraker J, Condliffe A, Haworth CS, et al. Nontuberculous mycobacteria in bronchiectasis: Prevalence and patient characteristics. Eur Respir J. 2006;28(6):1204–10.
4. Seddon P, Fidler K, Raman S, Wyatt H, Ruiz G, Elston C, et al. Prevalence of nontuberculous mycobacteria in cystic fibrosis clinics, United Kingdom, 2009. Emerg Infect Dis. 2013;19(7):1128–30.
5. Viviani L, Harrison MJ, Zolin A, Haworth CS, Floto RA. Epidemiology of nontuberculous mycobacteria (NTM) amongst individuals with cystic fibrosis (CF). J Cystic Fibrosis Off J Eur Cystic Fibrosis Soc. 2016;15(5):619–23.
6. Yates MD, Pozniak A, Grange JM. Isolation of mycobacteria from patients seropositive for the human immunodeficiency virus (HIV) in south east England: 1984-92. Thorax. 1993;48(10):990–5.
7. Davies BS, Roberts CH, Kaul S, Klein JL, Milburn HJ. Non-tuberculous slow-growing mycobacterial pulmonary infections in non-HIV-infected patients in south London. Scand J Infect Dis. 2012;44(11):815–9.
8. Bollert FG, Watt B, Greening AP, Crompton GK. Non-tuberculous pulmonary infections in Scotland: a cluster in Lothian? Thorax. 1995;50(2):188–90.
9. Russell CD, Claxton P, Doig C, Seagar AL, Rayner A, Laurenson IF. Non-tuberculous mycobacteria: a retrospective review of Scottish isolates from 2000 to 2010. Thorax. 2014;69(6):593–5.
10. Martin-Casabona N, Bahrmand AR, Bennedsen J, Thomsen VO, Curcio M, Fauville-Dufaux M, et al. Non-tuberculous mycobacteria: patterns of isolation. A multi-country retrospective survey. Int J Tuberc Lung Dis Off J Int Union Against Tuberc Lung Dis. 2004;8(10):1186–93.
11. Kennedy MP, O'Connor TM, Ryan C, Sheehan S, Cryan B, Bredin C. Nontuberculous mycobacteria: incidence in Southwest Ireland from 1987 to 2000. Respir Med. 2003;97(3):257–63.
12. Chong SG, Kent BD, Fitzgerald S, McDonnell TJ. Pulmonary non-tuberculous mycobacteria in a general respiratory population. Ir Med J. 2014;107(7):207–9.
13. Hoefsloot W, van Ingen J, Andrejak C, Angeby K, Bauriaud R, Bemer P, et al. The geographic diversity of nontuberculous mycobacteria isolated from pulmonary samples: an NTM-NET collaborative study. Eur Respir J. 2013;42(6):1604–13.
14. Maiz L, Giron R, Olveira C, Vendrell M, Nieto R, Martinez-Garcia MA. Prevalence and factors associated with nontuberculous mycobacteria in non-cystic fibrosis bronchiectasis: a multicenter observational study. BMC Infect Dis. 2016;16(1):437.
15. Alvaro-Meca A, Rodriguez-Gijon L, Diaz A, Gil A, Resino S. Trends in nontuberculous mycobacterial disease in hospitalized subjects in Spain (1997-2010) according to HIV infection. HIV Med. 2015;16(8):485–93.

Europe Remain (Europe Except England, Wales, Northern Ireland, Ireland, Scotland and Spain)

16. Prevots DR, Marras TK. Epidemiology of human pulmonary infection with nontuberculous mycobacteria: a review. Clin Chest Med. 2015;36(1):13–34.
17. Dabo H, Santos V, Marinho A, et al. Nontuberculous mycobacteria in respiratory specimens: clinical significance at a tertiary care hospital in the north of Portugal. J Bras Pneumol. 2015;41(3):292–4.
18. Jankovic M, Samarzija M, Sabol I, et al. Geographical distribution and clinical relevance of non-tuberculous mycobacteria in Croatia. Int J Tuberc Lung Dis. 2013;17(6):836–41.
19. Maugein J, Dailloux M, Carbonnelle B, Vincent V, Grosset J. Sentinel-site surveillance of Mycobacterium avium complex pulmonary disease. Eur Respir J. 2005;26(6):1092–6.
20. Dailloux M, Abalain ML, Laurain C, et al. Respiratory infections associated with nontuberculous mycobacteria in non-HIV patients. Eur Respir J. 2006;28(6):1211–5.
21. Andrejak C, Thomsen VO, Johansen IS, et al. Nontuberculous pulmonary mycobacteriosis in Denmark: incidence and prognostic factors. Am J Respir Crit Care Med. 2010;181(5):514–21.
22. Ringshausen FC, Apel RM, Bange FC, et al. Burden and trends of hospitalisations associated with pulmonary non-tuberculous mycobacterial infections in Germany, 2005-2011. BMC Infect Dis. 2013;13:231.
23. Ringshausen FC, Wagner D, de Roux A, et al. Prevalence of Nontuberculous Mycobacterial Pulmonary Disease, Germany, 2009-2014. Emerg Infect Dis. 2016;22(6):1102–5.
24. Diel R, Jacob J, Lampenius N, Loebinger M, Nienhaus A, Rabe KF, Ringshausen FC. Burden of non-tuberculous mycobacterial pulmonary disease in Germany. Eur Respir J. 2017:26;49(4).
25. Del Giudice G, Iadevaia C, Santoro G, Moscariello E, Smeraglia R, Marzo C. Nontuberculous mycobacterial lung disease in patients without HIV infection: a retrospective analysis over 3 years. Clin Respir J. 2011;5(4):203–10.
26. Panagiotou M, Papaioannou AI, Kostikas K, et al. The epidemiology of pulmonary nontuberculous mycobacteria: data from a general hospital in Athens, Greece, 2007-2013. Pulm Med. 2014;2014:894976.
27. Molicotti P, Bua A, Cannas S, et al. Identification of non-tuberculous mycobacteria from clinical samples. New Microbiol. 2013;36(4):409–11.
28. Manika K, Tsikrika S, Tsaroucha E, et al. Distribution of nontuberculous mycobacteria in treated patients with pulmonary disease in Greece – relation to microbiological data. Future Microbiol. 2015;10(8):1301–6.
29. Zmak L, Obrovac M, Jankovic Makek M, Sabol I, Katalinic-Jankovic V. Isolation of nontuberculous mycobacteria among tuberculosis patients during a 5-year period in Croatia. Infect Dis (Lond). 2015;47(4):275–6.
30. Gerogianni I, Papala M, Kostikas K, Petinaki E, Gourgoulianis KI. Epidemiology and clinical significance of mycobacterial respiratory infections in Central Greece. Int J Tuberc Lung Dis. 2008;12(7):807–12.
31. Amorim A, Macedo R, Lopes A, Rodrigues I, Pereira E. Non-tuberculous mycobacteria in HIV-negative patients with pulmonary disease in Lisbon, Portugal. Scand J Infect Dis. 2010;42(8):626–8.
32. Hoefsloot W, van Ingen J, de Lange WC, Dekhuijzen PN, Boeree MJ, van Soolingen D. Clinical relevance of Mycobacterium malmoense isolation in The Netherlands. Eur Respir J. 2009;34(4):926–31.
33. Phelippeau M, Dubus JC, Reynaud-Gaubert M, et al. Prevalence of Mycobacterium lentiflavum in cystic fibrosis patients, France. BMC Pulm Med. 2015;15:131.
34. van Ingen J, Boeree MJ, de Lange WC, et al. Mycobacterium xenopi clinical relevance and determinants, the Netherlands. Emerg Infect Dis. 2008;14(3):385–9.

35. van Ingen J, de Zwaan R, Dekhuijzen RP, Boeree MJ, van Soolingen D. Clinical relevance of Mycobacterium chelonae-abscessus group isolation in 95 patients. J Infect. 2009;59(5):324–31.
36. van Ingen J, Hoefsloot W, Dekhuijzen PN, Boeree MJ, van Soolingen D. The changing pattern of clinical Mycobacterium avium isolation in the Netherlands. Int J Tuberc Lung Dis. 2010;14(9):1176–80.
37. Soetaert K, Vluggen C, Andre E, Vanhoof R, Vanfleteren B, Mathys V. Frequency of Mycobacterium chimaera among Belgian patients, 2015. J Med Microbiol. 2016;65(11):1307–10.
38. Zlojtro M, Jankovic M, Samarzija M, et al. Nosocomial pseudo-outbreak of Mycobacterium gordonae associated with a hospital's water supply contamination: a case series of 135 patients. J Water Health. 2015;13(1):125–30.
39. Qvist T, Gilljam M, Jonsson B, et al. Epidemiology of nontuberculous mycobacteria among patients with cystic fibrosis in Scandinavia. J Cyst Fibros. 2015;14(1):46–52.
40. Qvist T, Taylor-Robinson D, Waldmann E, et al. Comparing the harmful effects of nontuberculous mycobacteria and Gram negative bacteria on lung function in patients with cystic fibrosis. J Cyst Fibros. 2016;15(3):380–5.
41. Viviani L, Harrison MJ, Zolin A, Haworth CS, Floto RA. Epidemiology of nontuberculous mycobacteria (NTM) amongst individuals with cystic fibrosis (CF). J Cyst Fibros. 2016;15(5):619–23.
42. Andrejak C, Nielsen R, Thomsen VO, Duhaut P, Sorensen HT, Thomsen RW. Chronic respiratory disease, inhaled corticosteroids and risk of non-tuberculous mycobacteriosis. Thorax. 2013;68(3):256–62.
43. Roux AL, Catherinot E, Ripoll F, et al. Multicenter study of prevalence of nontuberculous mycobacteria in patients with cystic fibrosis in france. J Clin Microbiol. 2009;47(12):4124–8.
44. Martin-Casabona N, Bahrmand AR, Bennedsen J, et al. Non-tuberculous mycobacteria: patterns of isolation. A multi-country retrospective survey. Int J Tuberc Lung Dis. 2004;8(10):1186–93.
45. Hoefsloot W, van Ingen J, Andrejak C, et al. The geographic diversity of nontuberculous mycobacteria isolated from pulmonary samples: an NTM-NET collaborative study. Eur Respir J. 2013;42(6):1604–13.
46. van der Werf MJ, Kodmon C, Katalinic-Jankovic V, et al. Inventory study of non-tuberculous mycobacteria in the European Union. BMC Infect Dis. 2014;14:62.
47. Gitti Z, Mantadakis E, Maraki S, Samonis G. Clinical significance and antibiotic susceptibilities of nontuberculous mycobacteria from patients in Crete, Greece. Future Microbiol. 2011;6(9):1099–109.
48. van Ingen J, Bendien SA, de Lange WC, et al. Clinical relevance of non-tuberculous mycobacteria isolated in the Nijmegen-Arnhem region, The Netherlands. Thorax. 2009;64(6):502–6.
49. Rindi L, Garzelli C. Increase in non-tuberculous mycobacteria isolated from humans in Tuscany, Italy, from 2004 to 2014. BMC Infect Dis. 2016;16:44.
50. Marusic A, Katalinic-Jankovic V, Popovic-Grle S, et al. Mycobacterium xenopi pulmonary disease – epidemiology and clinical features in non-immunocompromised patients. J Infect. 2009;58(2):108–12.
51. Svensson E, Jensen ET, Rasmussen EM, Folkvardsen DB, Norman A, Lillebaek T. Mycobacterium chimaera in Heater-Cooler Units in Denmark Related to Isolates from the United States and United Kingdom. Emerg Infect Dis. 2017;23(3):507–9.
52. Chand M, Lamagni T, Kranzer K, et al. Insidious Risk of Severe Mycobacterium chimaera Infection in Cardiac Surgery Patients. Clin Infect Dis. 2017;64(3):335–42.
53. Trudzinski FC, Schlotthauer U, Kamp A, et al. Clinical implications of Mycobacterium chimaera detection in thermoregulatory devices used for extracorporeal membrane oxygenation (ECMO), Germany, 2015 to 2016. Eurosurveillance. 2016;21(46):30398.
54. Schreiber PW, Kuster SP, Hasse B, et al. Reemergence of Mycobacterium chimaera in Heater-Cooler Units despite Intensified Cleaning and Disinfection Protocol. Emerg Infect Dis. 2016;22(10):1830–3.
55. Kohler P, Kuster SP, Bloemberg G, et al. Healthcare-associated prosthetic heart valve, aortic vascular graft, and disseminated Mycobacterium chimaera infections subsequent to open heart surgery. Eur Heart J. 2015;36(40):2745–53.

56. Sax H, Bloemberg G, Hasse B, et al. Prolonged Outbreak of Mycobacterium chimaera Infection After Open-Chest Heart Surgery. Clin Infect Dis. 2015;61(1):67–75.
57. Haller S, Holler C, Jacobshagen A, et al. Contamination during production of heater-cooler units by Mycobacterium chimaera potential cause for invasive cardiovascular infections: results of an outbreak investigation in Germany, April 2015 to February 2016. Euro Surveill. 2016;21(17):pii=30215.
58. Sommerstein R, Ruegg C, Kohler P, Bloemberg G, Kuster SP, Sax H. Transmission of Mycobacterium chimaera from Heater-Cooler Units during Cardiac Surgery despite an Ultraclean Air Ventilation System. Emerg Infect Dis. 2016;22(6):1008–13.

Asia

59. Simons S, van Ingen J, Hsueh PR, Van Hung N, Dekhuijzen PN, Boeree MJ, van Soolingen D. Nontuberculous mycobacteria in respiratory tract infections, eastern Asia. Emerg Infect Dis. 2011;17:343–9.
60. Okada H, SHigematsu I, et al. Epidemiological study of atypical mycobacteria in Japan. Jpn Med J. 1960;11:14–24.
61. Okada H. Epidemiological study on atypical mycobacteria infection in Japan. Jpn Med J. 1962;10:22–9.
62. Namkoong H, Kurashima A, Morimoto K, Hoshino Y, Hasegawa N, Ato M, Mitarai S. Epidemiology of Pulmonary Nontuberculous Mycobacterial Disease, Japan(1). Emerg Infect Dis. 2016;22:1116–7.
63. Tsukamura M, Shimoide H, Kita N, Kawakami K, Ito T, Nakajima N, Kondo H, Yamamoto Y, Matsuda N, Tamura M, Yoshimoto K, Shirota N, Kuze A. Epidemiological and bacteriological studies on atypical mycobacteriosis in Japan (author's transl). Kekkaku. 1980;55:273–80.
64. Tsukamura M, Shimoide H, Kita N, Segawa J, Ito T. A study on the frequency of 'atypical' Mycobacteria and of 'atypical' mycobacterioses in Japanese National Chest Hospitals. (Report of the year 1971-1972). Kekkaku. 1973;48:203–11.
65. Morimoto K, Iwai K, Uchimura K, Okumura M, Yoshiyama T, Yoshimori K, Ogata H, Kurashima A, Gemma A. Kudoh S. A steady increase in nontuberculous mycobacteriosis mortality and estimated prevalence in Japan. Ann Am Thorac Soc. 2014;11:1–8.
66. Morimoto K, Hasegawa N, Izumi K, Namkoong H, Uchimura K, Yoshiyama T, Hoshino Y, Kurashima A, Sokunaga J, Shibuya S, Shimojima M, Ato M, Mitarai S. A Laboratory-based Analysis of Nontuberculous Mycobacterial Lung Disease in Japan from 2012 to 2013. Ann Am Thorac Soc. 2017;14(1):49–56.
67. Ide S, Nakamura S, Yamamoto Y, Kohno Y, Fukuda Y, Ikeda H, Sasaki E, Yanagihara K, Higashiyama Y, Hashiguchi K, Futsuki Y, Inoue Y, Fukushima K, Suyama N, Kohno S. Epidemiology and clinical features of pulmonary nontuberculous mycobacteriosis in Nagasaki, Japan. PLoS One. 2015;10:e0128304.
68. Ito Y, Hirai T, Maekawa K, Fujita K, Imai S, Tatsumi S, Handa T, Matsumoto H, Muro S, Niimi A, Mishima M. Predictors of 5-year mortality in pulmonary Mycobacterium avium-intracellulare complex disease. Int J Tuberc Lung Dis. 2012;16:408–14.
69. Ito Y, Hirai T, Fujita K, Maekawa K, Niimi A, Ichiyama S, Mishima M. Increasing patients with pulmonary Mycobacterium avium complex disease and associated underlying diseases in Japan. J Infect Chemother Off J Jpn Soc Chemother. 2015;21:352–6.
70. Hayashi M, Takayanagi N, Kanauchi T, Miyahara Y, Yanagisawa T, Sugita Y. Prognostic factors of 634 HIV-negative patients with Mycobacterium avium complex lung disease. Am J Respir Crit Care Med. 2012;185:575–83.
71. The Co-operative Study Group of the Japanese National Sanatoria on Atypical Mycobacteria. A study on the frequency of 'atypical' mycobacteria in Japanese National Sanatoria. Tubercle. 1970;51:270–9.

72. Tsukamura M, Kita N, Shimoide H, Nagasawa S, Arakawa H, Shinoda A, Kuze A, Matsuura K, Yoshimoto K, Wada R, et al. Studies on lung disease due to non-tuberculous mycobacteria in Japan (Report for the year 1984 of the Mycobacteriosis Research Group of the Japanese National Chest Hospitals) – the same trend in the epidemiological state as seen in the preceding year. Kekkaku. 1986;61:277–84.
73. Satoh S. Pulmonary nontuberculous mycobacterial disease: an update. Curr Med. 2008;56:317–24.
74. Saito H, Tomioka H, Sato K, Tasaka H, Tsukamura M, Kuze F, Asano K. Identification and partial characterization of Mycobacterium avium and Mycobacterium intracellulare by using DNA probes. J Clin Microbiol. 1989;27:994–7.
75. Cassidy PM, Hedberg K, Saulson A, McNelly E, Winthrop KL. Nontuberculous mycobacterial disease prevalence and risk factors: a changing epidemiology. Clin Infect Dis Off Publ Infect Dis Soc Am. 2009;49:e124–9.
76. Prevots DR, Shaw PA, Strickland D, Jackson LA, Raebel MA, Blosky MA, Montes de Oca R, Shea YR, Seitz AE, Holland SM, Olivier KN. Nontuberculous mycobacterial lung disease prevalence at four integrated health care delivery systems. Am J Respir Crit Care Med. 2010;182:970–6.
77. Park YS, Lee CH, Lee SM, Yang SC, Yoo CG, Kim YW, Han SK, Shim YS, Yim JJ. Rapid increase of non-tuberculous mycobacterial lung diseases at a tertiary referral hospital in South Korea. Int J Tuberc Lung Dis. 2010;14:1069–71.
78. Koh WJ, Chang B, Jeong BH, Jeon K, Kim SY, Lee NY, Ki CS, Kwon OJ. Increasing Recovery of Nontuberculous Mycobacteria from Respiratory Specimens over a 10-Year Period in a Tertiary Referral Hospital in South Korea. Tuberc Respir Dis. 2013;75:199–204.
79. Jang MA, Koh WJ, Huh HJ, Kim SY, Jeon K, Ki CS, Lee NY. Distribution of nontuberculous mycobacteria by multigene sequence-based typing and clinical significance of isolated strains. J Clin Microbiol. 2014;52:1207–12.
80. Kim HS, Lee Y, Lee S, Kim YA, Sun YK. Recent trends in clinically significant nontuberculous Mycobacteria isolates at a Korean general hospital. Ann Lab Med. 2014;34:56–9.
81. Lee SK, Lee EJ, Kim SK, Chang J, Jeong SH, Kang YA. Changing epidemiology of nontuberculous mycobacterial lung disease in South Korea. Scand J Infect Dis. 2012;44:733–8.
82. Ryoo SW, Shin S, Shim MS, Park YS, Lew WJ, Park SN, Park YK, Kang S. Spread of nontuberculous mycobacteria from 1993 to 2006 in Koreans. J Clin Lab Anal. 2008;22:415–20.
83. Koh WJ, Kwon OJ, Jeon K, Kim TS, Lee KS, Park YK, Bai GH. Clinical significance of nontuberculous mycobacteria isolated from respiratory specimens in Korea. Chest. 2006;129:341–8.
84. Lai CC, Tan CK, Lin SH, Liu WL, Liao CH, Huang YT, Hsueh PR. Clinical significance of nontuberculous mycobacteria isolates in elderly Taiwanese patients. Eur J Clin Microbiol Infect Dis Off Publ Eur Soc Clin Microbiol. 2011;30:779–83.
85. Ding LW, Lai CC, Lee LN, Hsueh PR. Disease caused by non-tuberculous mycobacteria in a university hospital in Taiwan, 1997-2003. Epidemiol Infect. 2006;134:1060–7.
86. Shen MC, Lee SS, Huang TS, Liu YC. Clinical significance of isolation of Mycobacterium avium complex from respiratory specimens. J Formosan Med Assoc Taiwan yi zhi. 2010;109:517–23.
87. Lai CC, Tan CK, Chou CH, Hsu HL, Liao CH, Huang YT, Yang PC, Luh KT, Hsueh PR. Increasing incidence of nontuberculous mycobacteria, Taiwan, 2000-2008. Emerg Infect Dis. 2010;16:294–6.
88. Wang CC, Lin MC, Liu JW, Wang YH. Nontuberculous mycobacterial lung disease in southern Taiwan. Chang Gung Med J. 2009;32:499–508.
89. Shu CC, Lee CH, Hsu CL, Wang JT, Wang JY, Yu CJ, Lee LN. Clinical characteristics and prognosis of nontuberculous mycobacterial lung disease with different radiographic patterns. Lung. 2011;189:467–74.
90. Jing H, Wang H, Wang Y, Deng Y, Li X, Liu Z, Graviss EA, Ma X. Prevalence of nontuberculous mycobacteria infection, China, 2004-2009. Emerg Infect Dis. 2012;18:527–8.
91. Shao Y, Chen C, Song H, Li G, Liu Q, Li Y, Zhu L, Martinez L, Lu W. The epidemiology and geographic distribution of nontuberculous mycobacteria clinical isolates from sputum samples in the eastern region of China. PLoS Negl Trop Dis. 2015;9:e0003623.

92. Wang X, Li H, Jiang G, Zhao L, Ma Y, Javid B, Huang H. Prevalence and drug resistance of nontuberculous mycobacteria, northern China, 2008-2011. Emerg Infect Dis. 2014;20:1252–3.
93. Xu K, Bi S, Ji Z, Hu H, Hu F, Zheng B, Wang B, Ren J, Yang S, Deng M, Chen P, Ruan B, Sheng J, Li L. Distinguishing nontuberculous mycobacteria from multidrug-resistant Mycobacterium tuberculosis, China. Emerg Infect Dis. 2014;20:1060–2.
94. Jing H, Tan W, Deng Y, Gao D, Li L, Lu Z, Graviss EA, Ma X. Diagnostic delay of pulmonary nontuberculous mycobacterial infection in China. Multidisciplinary Respir Med. 2014;9:48.
95. Wu J, Zhang Y, Li J, Lin S, Wang L, Jiang Y, Pan Q, Shen X. Increase in nontuberculous mycobacteria isolated in Shanghai, China: results from a population-based study. PLoS One. 2014;9:e109736.
96. Bi S, Xu KJ, Ji ZK, Zheng BW, Sheng JF. Sentinel site surveillance of nontuberculous mycobacteria pulmonary diseases in Zhejiang, China, 2011-2013. Braz J Infect Dis Off Publ Braz Soc Infect Dis. 2015;19:670–1.
97. Wang HX, Yue J, Han M, Yang JH, Gao RL, Jing LJ, Yang SS, Zhao YL. Nontuberculous mycobacteria: susceptibility pattern and prevalence rate in Shanghai from 2005 to 2008. Chin Med J. 2010;123:184–7.
98. Yu XL, Lu L, Chen GZ, Liu ZG, Lei H, Song YZ, Zhang SL. Identification and characterization of non-tuberculous mycobacteria isolated from tuberculosis suspects in Southern-central China. PLoS One. 2014;9:e114353.
99. Yu X, Liu P, Liu G, Zhao L, Hu Y, Wei G, Luo J, Huang H. The prevalence of non-tuberculous mycobacterial infections in mainland China: Systematic review and meta-analysis. J Infect. 2016;73:558–67.
100. Jain S, Sankar MM, Sharma N, Singh S, Chugh TD. High prevalence of non-tuberculous mycobacterial disease among non-HIV infected individuals in a TB endemic country – experience from a tertiary center in Delhi, India. Pathog Global Health. 2014;108:118–22.
101. Jesudason MV, Gladstone P. Non tuberculous mycobacteria isolated from clinical specimens at a tertiary care hospital in South India. Indian J Med Microbiol. 2005;23:172–5.
102. Shenai S, Rodrigues C, Mehta A. Time to identify and define non-tuberculous mycobacteria in a tuberculosis-endemic region. Int J Tuberc Lung Dis. 2010;14:1001–8.
103. Ahmed I, Jabeen K, Hasan R. Identification of non-tuberculous mycobacteria isolated from clinical specimens at a tertiary care hospital: a cross-sectional study. BMC Infect Dis. 2013;13:493.
104. Bicmen C, Coskun M, Gunduz AT, Senol G, Cirak AK, Tibet G. Nontuberculous mycobacteria isolated from pulmonary specimens between 2004 and 2009: causative agent or not? New Microbiol. 2010;33:399–403.
105. Braun E, Sprecher H, Davidson S, Kassis I. Epidemiology and clinical significance of non-tuberculous mycobacteria isolated from pulmonary specimens. Int J Tuberc Lung Dis. 2013;17:96–9.
106. Varghese B, Memish Z, Abuljadayel N, Al-Hakeem R, Alrabiah F, Al-Hajoj SA. Emergence of clinically relevant Non-Tuberculous Mycobacterial infections in Saudi Arabia. PLoS Negl Trop Dis. 2013;7:e2234.
107. Al-Harbi A, Al-Jahdali H, Al-Johani S, Baharoon S, Bin Salih S, Khan M. Frequency and clinical significance of respiratory isolates of non-tuberculous mycobacteria in Riyadh, Saudi Arabia. Clin Respir J. 2016;10:198–203.
108. Al-Mahruqi SH, van-Ingen J, Al-Busaidy S, Boeree MJ, Al-Zadjali S, Patel A, Richard-Dekhuijzen PN, van-Soolingen D. Clinical relevance of nontuberculous Mycobacteria, Oman. Emerg Infect Dis. 2009;15:292–4.
109. Velayati AA, Farnia P, Mozafari M, Malekshahian D, Seif S, Rahideh S, Mirsaeidi M. Molecular epidemiology of nontuberculous mycobacteria isolates from clinical and environmental sources of a metropolitan city. PLoS One. 2014;9:e114428.
110. Velayati AA, Farnia P, Mozafari M, Mirsaeidi M. Nontuberculous Mycobacteria Isolation from Clinical and Environmental Samples in Iran: Twenty Years of Surveillance. Biomed Res Int. 2015;2015:254285.

111. Nasiri MJ, Dabiri H, Darban-Sarokhalil D, Hashemi Shahraki A. Prevalence of Non-Tuberculosis Mycobacterial Infections among Tuberculosis Suspects in Iran: Systematic Review and Meta-Analysis. PLoS One. 2015;10:e0129073.

Oceania

112. Yew W-W, Chiang C-Y, Lumb R, Islam T. Are pulmonary non-tuberculous mycobacteria of concern in the Western Pacific Region? Int J Tuberc Lung Dis. 2015;19(5):499–500.
113. Phelippeau M, Osman D, Musso D, Drancourt M. Epidemiology of nontuberculous mycobacteria in French Polynesia. J Clin Microbiol. 2015;53(12):3798–804.
114. Brown-Elliott B, Wallace R, Tichindelean C, Sarria J, McNulty S, Vasireddy R, et al. Five-Year Outbreak of Community- and Hospital-Acquired *Mycobacterium porcinum* Infections Related to Public Water Supplies. J Clin Microbiol. 2011;49(12):4231–8.
115. Shojaei H, Heidarieh P, Hshemi A, Feizabadi M, DaeiNaser A. Species identification of neglected nontuberculous mycobacteria in a developing country. Jpn I Infect Dis. 2011;64:265–71.
116. Perez-Martinez I, Aguilar-Ayala DA, Fernandez-Rendon E, Carrillo-Sanchez AK, Helguera-Repetto AC, Rivera-Gutierrez S, et al. Occurrence of potentially pathogenic nontuberculous mycobacteria in Mexican household potable water: a pilot study. BMC Res Notes. 2013;6:531.
117. Tsai C, Shiau M, Chang Y, Wang Y, JHuang T, Liaw Y, et al. Trends of mycobacterial clinical isolates in Taiwan. Trans R Soc Trop Med Hyg. 2011;105:148–52.
118. Lai C, Hsueh P. Diseases caused by nontuberculous mycobacteria in Asia. Future Microbiol. 2014;9:93–106.
119. Thomson R, Carter R, Tolson C, Huygens F, Hargreaves M. Factors associated with the isolation of Nontuberculous mycobacteria (NTM) from a large municipal water system in Brisbane, Australia. BMC Microbiol. 2013;13:89.
120. Thomson RM, Tolson CE, Carter R, Huygens F, Hargreaves M. Heterogeneity of clinical and environmental isolates of Mycobacterium fortuitum using repetitive element sequence-based PCR: municipal water an unlikely source of community-acquired infections. Epidemiol Infect. 2014;142(10):2057–64
121. Thomson R, Tolson C, Sidjabat H, Huygens F, Hargreaves M. *Mycobacterium abscessus* isolated from municipal water – a potential source of human infection. BMC Infect Dis. 2013;13(1):241.
122. Ley S, Carter R, Millan K, Phuanukoonnon S, Pandey S, Coulter C, et al. Non-tuberculous mycobacteria: baseline data from three sites in Papua New Guinea, 2010–2012. West Pac Surveill Response J WPSAR. 2015;6(4):24–9.
123. Haverkort F. Australian Mycobacterium Reference Laboratory Network. National atypical mycobacteria survey, 2000. Commun Dis Intell. 2003;27:180–9.
124. Pang SC. *Mycobacterium kansasii* infections in Western Australia. Respir Med. 1991;85:213–8.
125. O'Brien D, Currie B, Krause V. Nontuberculous mycobacterial disease in northern Australia: a case series and review of the literature. Clin Infect Dis. 2000;31(4):958–67.
126. Freeman J, Morris A, Blackmore T, Hammer C, Munroe S, McKnight L. Incidence of nontuberculous mycobacterial disease in New Zealand, 2004. N Z Med J. 2007;120(1256):50–6.
127. Thomson R. Changing epidemiology of pulmonary nontuberculous mycobacteria infections. EID. 2010;16(10):1576–82.
128. Tebruegge M, Pantazidou A, MacGregor D, Gonis G, Leslie D, Sedda L, et al. Nontuberculous Mycobacterial Disease in Children – Epidemiology, Diagnosis & Management at a Tertiary Center. PLoS One. 2016;11(1):e0147513.
129. State of Queensland (Queensland Health). 2016. Nontuberculous mycobacterium 2015. Available from URL https://www.health.qld.gov.au/publications/clinical-practice/guidelines-procedures/diseases-infection/diseases/tuberculosis/report-tb-ntm-2015.pdf.
130. Chou M, Clements A, Thomson R. A spatial epidemiological analysis of nontuberculous mycobacterial infections in Queensland, Australia. BMC Infect Dis. 2014;14:279.

131. Corbett E, Blumberg L, Churchyard G, Moloi N, Mallory K, Clayton T, et al. Nontuberculous mycobacteria. Defining disease in a prospective cohort of South African miners. Am J Resp Crit Care Med. 1999;160:15–21.
132. Kubin M, Švanclová E, Medek B, Chobot S, Olšovský Ž. *Mycobacterium kansasii* infection in an endemic area of Czechoslovakia. Tubercle. 1980;61(4):207–12.
133. Kwenda G, Churchyard GJ, Thorrold C, Heron I, Stevenson K, Duse AG, et al. Molecular characterisation of clinical and environmental isolates of *Mycobacterium kansasii* isolates from South African gold mines. J Water Health. 2015;13(1):190–202.
134. Iinuma Y, Ichiyama SS, Tsukamura M, Hasegawa Y, Shimokata K, Kawahara S, Matsushima T. Large-restriction-fragment analysis of *Mycobacterium kansasii* genomic DNA and its application in molecular typing. J Clin Microbiol. 1997;35:596–9.
135. Kazda J. The ecology of mycobacteria. Dordrecht: Springer; 2009.
136. Smith D, SIms B, O'Neill D. Testing and evaluation of agricultural machinery and equipment. Rome: Food and Agriculture Organisation of the United Nations; 1994.
137. Jacobs JMR, Sturgis B, Wood R. Influence of Environmental Gradients on the Abundance and Distribution of *Mycobacterium* spp. in a Coastal Lagoon Estuary. Appl Environ Microbiol. 2009;75:7378–84.
138. By Historicair – Own work. Based on Australian Government Department of Resources, Energy and Tourism 2008 report + File:Oil drop.svg by User:Slashme, CC BY-SA 3.0, https://commons.wikimedia.org/w/index.php?curid=15478547.

Africa

139. Botha L, Gey van Pittius NC, van Helden PD. Mycobacteria and Disease in Southern Africa. Transbound Emerg Dis. 2013;60:147–56.
140. Gcebe N, Rutten V, Gey van Pittius NC, Michel A. Prevalence and Distribution of Non-Tuberculous Mycobacteria (NTM) in Cattle, African Buffaloes (Syncerus caffer) and their Environments in South Africa. Transbound Emerg Dis. 2013;60:74–84.
141. September S, Brozel V, Venter S. Diversity of nontuberculoid Mycobacterium species in biofilms of urban and semiurban drinking water distribution systems. Appl Environ Microbiol. 2004;70(12):7571–3.
142. Kankya C, Muwonge A, Djønne B, Munyeme M, Opuda-Asibo J, Skjerve E, et al. Isolation of non-tuberculous mycobacteria from pastoral ecosystems of Uganda: Public Health significance. BMC Public Health. 2011;11(1):1–9.
143. Katale BZ, Mbugi EV, Botha L, Keyyu JD, Kendall S, Dockrell HM, et al. Species diversity of non-tuberculous mycobacteria isolated from humans, livestock and wildlife in the Serengeti ecosystem, Tanzania. BMC Infect Dis. 2014;14:616.
144. Kwenda G, Churchyard GJ, Thorrold C, Heron I, Stevenson K, Duse AG, et al. Molecular characterisation of clinical and environmental isolates of *Mycobacterium kansasii* isolates from South African gold mines. J Water Health. 2015;13(1):190–202.
145. Sookan L, Coovadia Y. A laboratory-based study to identify and speciate non-tuberculous mycobacteria isolated from specimens submitted at a central tuberculosis laboratory from throughout KwaZulu-Natal Province, South Africa. S Afr Med J. 2014;104(11):766–88.
146. Aliyu G, El-Kamary SS, Al A, Brown C, Tracy K, Hungerford L, et al. Prevalence of Non-Tuberculous Mycobacterial Infections among Tuberculosis Suspects in Nigeria. PLoS One. 2013;8(5):e63170.
147. Pokam BT, Asuquo AE. Acid-Fast Bacilli Other than Mycobacteria in Tuberculosis Patients Receiving Directly Observed Therapy Short Course in Cross River State, Nigeria. Tuberc Res Treat. 2012;2012:301056.
148. Hoza AS, Mfinanga SGM, Rodloff AC, Moser I, König B. Increased isolation of nontuberculous mycobacteria among TB suspects in Northeastern, Tanzania: public health and diagnostic implications for control programmes. BMC Res Notes. 2016;9:109.

149. Badoum G, Saleri N, Dembélé MS, Ouedraogo M, Pinsi G, Boncoungou K, et al. Failing a re-treatment regimen does not predict MDR/XDR tuberculosis: is "blind" treatment dangerous? Eur Respir J. 2011;37(5):1283.
150. Mwikuma G, Kwenda G, Hang'ombe BM, Simulundu E, Kaile T, Nzala S, et al. Molecular identification of non-tuberculous mycobacteria isolated from clinical specimens in Zambia. Ann Clin Microbiol Antimicrob. 2015;14:1.
151. Asiimwe BB, Bagyenzi GB, Ssengooba W, Mumbowa F, Mboowa G, Wajja A, et al. Species and genotypic diversity of non-tuberculous mycobacteria isolated from children investigated for pulmonary tuberculosis in rural Uganda. BMC Infect Dis. 2013;13:88.
152. Buijtels P, van der Sande M, de Graaff C, Parkinson S, Verbrugh H, Petit P, et al. Nontuberculous mycobacteria, Zambia. EID. 2009;15(2):242–9.
153. Nyamogoba HD, Mbuthia G, Mining S, Kikuvi G, Biegon R, Mpoke S, et al. HIV co-infection with tuberculous and non-tuberculous mycobacteria in western Kenya: challenges in the diagnosis and management. Afr Health Sci. 2012;12(3):305–11.
154. Maiga M, Siddiqui S, Diallo S, Diarra B, Traoré B, Shea YR, et al. Failure to Recognize Nontuberculous Mycobacteria Leads to Misdiagnosis of Chronic Pulmonary Tuberculosis. PLoS One. 2012;7(5):e36902.
155. Borroni E, Badoum G, Cirillo D, Matteelli A, Moyenga I, Ouedraogo M, et al. *Mycobacterium sherrisii* Pulmonary Disease, Burkina Faso. Emerg Infect Dis. 2015;21(11):2093.
156. Corbett E, Blumberg L, Churchyard G, Moloi N, Mallory K, Clayton T, et al. Nontuberculous mycobacteria. Defining disease in a prospective cohort of South African miners. Am J Resp Crit Care Med. 1999;160:15–21.
157. Ballard J, Turenne CY, Wolfe JN, Reller LB, Kabani A. Molecular characterization of non-tuberculous mycobacteria isolated from human cases of disseminated disease in the USA, Thailand, Malawi, and Tanzania. J Gen Appl Microbiol. 2007;53(2):153–7.
158. Nunes EA, De Capitani EM, Coelho E, Panunto AC, Joaquim OA, MdC R. Mycobacterium tuberculosis and nontuberculous mycobacterial isolates among patients with recent HIV infection in Mozambique. J Bras Pneumol. 2008;34:822–8.
159. van Halsema CL, Chihota VN, Gey van Pittius NC, Fielding KL, Lewis JJ, van Helden PD, et al. Clinical Relevance of Nontuberculous Mycobacteria Isolated from Sputum in a Gold Mining Workforce in South Africa: An Observational, Clinical Study. Biomed Res Int. 2015;2015:959107.
160. Hatherill M, Hawkridge T, Whitelaw A, Tameris M, Mahomed H, Moyo S, et al. Isolation of Non-Tuberculous Mycobacteria in Children Investigated for Pulmonary Tuberculosis. PLoS One. 2006;1(1):e21.
161. Koivula T, Hoffner S, Winqvist N, Naucl A, Dias F, et al. *Mycobacterium avium* Complex Sputum Isolates from Patients with Respiratory Symptoms in Guinea-Bissau. J Infect Dis. 1996;173(1):263–5.
162. Muwonge A, Kankya C, Godfroid JD, Ayanaw T, Munyeme M, Skjerve E. Prevalence and associated risk factors of mycobacterial infections in slaughter pigs from Mubende district in Uganda. Trop Anim Health Prod. 2010;42(5):905–13.
163. Loulergue P, Lamontagne F, Vincent V, Rossier A, Pialoux G. *Mycobacterium sherrisii*: a new opportunistic agent in HIV infection? AIDS. 2007;21:893–4.
164. Zykov M, Boulet H, Gaya N. Non-tuberculosis mycobacteria in Africa. I. Isolation and identification. Bull World Health Organisation. 1967;37:927–38.
165. Fourie PB, Gatner EMS, Glatthaar E, Kleeberg HH. Follow-up tuberculosis prevalence survey of Transkei. Tubercle. 1980;61(2):71–9.
166. Arabin G, Gartig D, Kellberg H. First tuberculosis prevalence survey in KwaZulu. S Afr Med J. 1979;56:434–8.
167. Corbett E, Churchyard G, Clayton TC, Williams B, Mulder D, Hayes R, et al. HIV infection and silicosis: the impact of two potent risk factors on the incidence of mycobacterial disease in South African miners. AIDS. 2000;14(17):2759–68.
168. Corbett E, Churchyard G, Hay M, Herselman P, Clayton T, Williams B, et al. The Impact of HIV Infection on *Mycobacterium kansasii* Disease in South African Gold Miners. AJRCCM. 1999;160:10–4.

169. Corbett EL, Hay M, Churchyard GJ, Herselman P, Clayton T, Williams BG, et al. *Mycobacterium kansasii* and *M. scrofulaceum* isolates from HIV-negative South African gold miners: incidence, clinical significance and radiology. Int J Tubercle Lung Dis. 1999;3(6):501–7.
170. Chilima BZ, Clark IM, Floyd S, Fine PEM, Hirsch PR. Distribution of Environmental Mycobacteria in Karonga District, Northern Malawi. Appl Environ Microbiol. 2006;72(4):2343–50.
171. Chanda-Kapata P, Kapata N, Klinkenberg E, Mulenga L, Tembo M, Katemangwe P, et al. Non-tuberculous mycobacteria (NTM) in Zambia: prevalence, clinical, radiological and microbiological characteristics. BMC Infect Dis. 2015;15:500.
172. Buijtels PCAM, Van Der Sande MAB, Parkinson S, Verbrugh HA, Petit PLC, Van Soolingen D. Isolation of non-tuberculous mycobacteria at three rural settings in Zambia; a pilot study. Clin Microbiol Infect. 2010;16(8):1142–8.
173. Daniel O, Osman E, Adebiyi P, Mourad G, Declarcq E, Bakare R. Non tuberculosis mycobacteria isolates among new and previously treated pulmonary tuberculosis patients in Nigeria. Asian Pac J Trop Dis. 2011;1(2):113–5.
174. Allanana J, Ikeh E, Bello C. *Mycobacterium* species in clinical specimens in Jos Nigeria. Niger J Med. 1991;2:111–2.
175. Mawak J, Gomwalk N, Bello C, Kandakai-Olukemi Y. Human pulmonary infections with bovine and environmental (atypical mycobacteria) in Jos. Nigeria Ghana Med J. 2006;40:132–6.
176. Crump J, van Ingen J, Morrissey A, Boeree M, Mavura D, Swal B, et al. Invasive disease caused by Nontuberculous mycobacteria, Tanzania. EID. 2009;15(1):53–5.
177. Selvarangan R, Wu W-K, Nguyen TT, Carlson LDC, Wallis CK, Stiglich SK, et al. Characterization of a Novel Group of Mycobacteria and Proposal of *Mycobacterium sherrisii* sp. nov. J Clin Microbiol. 2004;42(1):52–9.
178. Tortoli E, Galli L, Andebirhan T, Baruzzo S, Chiappini E, de Martino M, et al. The first case of *Mycobacterium sherrisii* disseminated infection in a child with AIDS. AIDS. 2007;21:1496–8.
179. Malama S, Munyeme M, Mwanza S, Muma JB. Isolation and characterization of non tuberculous mycobacteria from humans and animals in Namwala District of Zambia. BMC Res Notes. 2014;7:622.
180. Workalemahu B, Berg S, Tsegaye W, Abdissa A, Girma T, Abebe M, et al. Genotype diversity of Mycobacterium isolates from children in Jimma, Ethiopia. BMC Res Notes. 2013;6(1):352.

Central and Southern America

181. Prevots DR, Marras TK. Epidemiology of human pulmonary infection with nontuberculous mycobacteria: a review. Clin Chest Med. 2015;36(1):13–34.
182. de Mello KGC, Mello FCQ, Borga L, Rolla V, Duarte RS, Sampaio EP, et al. Clinical and therapeutic features of pulmonary nontuberculous mycobacterial disease, Brazil, 1993-2011. Emerg Infect Dis. 2013;19(3):393–9.
183. Nunes-Costa D, Alarico S, Dalcolmo MP, Correia-Neves M, Empadinhas N. The looming tide of nontuberculous mycobacterial infections in Portugal and Brazil. Tuberculosis. 2016;96:107–19.
184. da Costa ARF, Lima KVB, Sales LHM, de Sousa MS, Lopes ML, Suffys PN. Pulmonary nontuberculous mycobacterial infections in the state of para, an endemic region for tuberculosis in North of Brazil [Internet]. INTECH Open Access Publisher; 2012 [cited 2016 Nov 19]. Available from: http://cdn.intechopen.com/pdfs/32001.pdf.
185. das Graças Motta e Bona M, Leal MJS, Martins LMS, da Silva RN, de Castro JAF, do Monte SJH. Restriction enzyme analysis of the hsp65 gene in clinical isolates from patients sus-

pected of having pulmonary tuberculosis in Teresina, Brazil. J Bras Pneumol Publicacao Soc Bras Pneumol E Tisilogia 2011;37(5):628–635.
186. Matos ED, Santana MA, de Santana MC, Mamede P, de Lira Bezerra B, Panão ED, et al. Nontuberculosis mycobacteria at a multiresistant tuberculosis reference center in Bahia: clinical epidemiological aspects. Braz J Infect Dis Off Publ Braz Soc Infect Dis. 2004;8(4):296–304.
187. da Costa ARF, Falkinham JO, Lopes ML, Barretto AR, Felicio JS, Sales LHM, et al. Occurrence of Nontuberculous Mycobacterial Pulmonary Infection in an Endemic Area of Tuberculosis. PLoS Negl Trop Dis [Internet]. 2013 [cited 2016 Nov 19];7(7). Available from: http://www.ncbi.nlm.nih.gov/pmc/articles/PMC3715520/.
188. Mendes de Lima CA, Gomes HM, Oelemann MAC, Ramos JP, Caldas PC, Campos CED, et al. Nontuberculous mycobacteria in respiratory samples from patients with pulmonary tuberculosis in the state of Rondônia, Brazil. Mem Inst Oswaldo Cruz. 2013;108(4):457–62.
189. Chimara E, Giampaglia CMS, Martins MC, da Telles MA, Ueki SYM, Ferrazoli L. Molecular characterization of *Mycobacterium kansasii* isolates in the State of São Paulo between 1995-1998. Mem Inst Oswaldo Cruz. 2004;99(7):739–43.
190. Zamarioli LA, Coelho AGV, Pereira CM, Nascimento ACC, Ueki SYM, Chimara E. Descriptive study of the frequency of nontuberculous mycobacteria in the Baixada Santista region of the state of São Paulo, Brazil. J Bras Pneumol Publicacao Of Soc Bras Pneumol E Tisilogia. 2008;34(8):590–4.
191. da Silveira Paro Pedro H, MIF P, Maria do Rosário Assad G, SYM U, Chimara E. Nontuberculous mycobacteria isolated in São José do Rio Preto, Brazil between 1996 and 2005. J Bras Pneumol Publicacao Of Soc Bras Pneumol E Tisilogia. 2008;34(11):950–5.
192. Bensi EPA, Panunto PC, Ramos M de C. Incidence of tuberculous and non-tuberculous mycobacteria, differentiated by multiplex PCR, in clinical specimens of a large general hospital. Clin Sao Paulo Braz. 2013;68(2):179–84.
193. Paschoal IA, de Oliveira Villalba W, Bertuzzo CS, Cerqueira EMFP, Pereira MC. Cystic fibrosis in adults. Lung. 2007;185(2):81–7.
194. Cândido PHC, Nunes L de S, Marques EA, Folescu TW, Coelho FS, de Moura VCN, et al. Multidrug-resistant nontuberculous mycobacteria isolated from cystic fibrosis patients. J Clin Microbiol. 2014;52(8):2990–7.
195. Barnes AI, Rojo S, Moretto H. Prevalence of mycobacteriosis and tuberculosis in a reference hospital, Cordoba province. Rev Argent Microbiol. 2004;36(4):170–3.
196. Imperiale B, Zumárraga M, Gioffré A, Di Giulio B, Cataldi A, Morcillo N. Disease caused by non-tuberculous mycobacteria: diagnostic procedures and treatment evaluation in the North of Buenos Aires Province. Rev Argent Microbiol. 2012;44(1):3–9.
197. Griffith DE, Aksamit T, Brown-Elliott BA, Catanzaro A, Daley C, Gordin F, et al. An Official ATS/IDSA Statement: Diagnosis, Treatment, and Prevention of Nontuberculous Mycobacterial Diseases. Am J Respir Crit Care Med. 2007;175(4):367–416.
198. Winthrop KL, McNelley E, Kendall B, Marshall-Olson A, Morris C, Cassidy M, et al. Pulmonary Nontuberculous Mycobacterial Disease Prevalence and Clinical Features. Am J Respir Crit Care Med. 2010;182(7):977–82.
199. Escobar-Escamilla N, Ramírez-González JE, González-Villa M, Torres-Mazadiego P, Mandujano-Martínez A, Barrón-Rivera C, et al. Hsp65 phylogenetic assay for molecular diagnosis of nontuberculous mycobacteria isolated in Mexico. Arch Med Res. 2014;45(1):90–7.
200. Pineda-Garcia L, Ferrera A, Galvez CA, Hoffner SE. Drug-Resistant Mycobacterium tuberculosis and Atypical Mycobacteria Isolated From Patients With Suspected Pulmonary Tuberculosis in Honduras. Chest. 1997;111(1):148–53.

Diagnosis of NTM Disease: Pulmonary and Extrapulmonary

Jeremy M. Clain and Timothy R. Aksamit

Introduction

For the past decade, the diagnosis of NTM disease has been guided by criteria laid out in an official statement of the American Thoracic Society (ATS) and the Infectious Diseases Society of America (IDSA) [1]. Pulmonary and extrapulmonary NTM disease are considered separately, but in all cases the clinical and radiographic context of laboratory findings is paramount. This chapter serves as a summary of current diagnostic principles and common diagnostic dilemmas in NTM disease.

Diagnosis of Pulmonary NTM Disease

General Guiding Principles

The diagnosis of NTM pulmonary disease requires consideration of clinical, radiographic, and microbiologic factors, each of which is subject to ambiguity. The complexity of the diagnostic criteria for disease derives from two main sources of uncertainty. First, NTM pulmonary disease is quite uncommon, relative to the universality of NTM exposure. Since environmental reservoirs of NTM are numerous, ubiquitous, and unavoidable, all individuals come into frequent contact with NTM. Therefore, NTM organisms cultured from respiratory sites often reflect transient exposure to NTM and not pathologic disease. The second main source of uncertainty regarding the diagnosis of NTM pulmonary disease is the fact that respiratory samples are nonsterile. With environmental reservoirs of NTM that prominently

J. M. Clain (✉) · T. R. Aksamit
Mayo Clinic, Pulmonary Disease and Critical Care Medicine, Rochester, MN, USA
e-mail: clain.jeremy@mayo.edu; aksamit.timothy@mayo.edu

D. E. Griffith (ed.), *Nontuberculous Mycobacterial Disease*, Respiratory Medicine,
https://doi.org/10.1007/978-3-319-93473-0_9

Clinical symptoms that are attributable to NTM lung disease, and not to comorbid conditions or alternative diagnoses, which may include:

- Pulmonary symptoms: cough, dyspnea, sputum production, hemoptysis, chest pain
- Constitutional symptoms: fatigue, fevers, night sweats, and weight loss

AND

Radiographic findings that are consistent with NTM lung disease, which may include:

- Fibrocavitary lesions
- Regions of bronchiectasis, nodular infiltrates, consolidation, and tree-in-bud opacities

AND

Microbiologic evidence that is suggestive of NTM infection, which should include:

- Positive culture results from at least two separate sputum samples

  OR

- Positive culture result from at least one bronchial wash or bronchoalveolar lavage

  OR

- Lung biopsy with mycobacterial histopathologic features and positive culture for NTM either from the biopsy specimen or from a single respiratory sample

Fig. 1 Summary of criteria required for the diagnosis of nontuberculous mycobacterial pulmonary disease

include tap water, this lack of sterility means that there is a considerable possibility that a positive NTM culture result reflects contamination, and not disease.

The general diagnostic criteria for NTM pulmonary disease, based on the ATS/IDSA statement, are summarized in Fig. 1 [1]. Simply stated, these criteria demand confirmatory microbiologic evidence of NTM infection in the context of clinical and radiographic findings suggestive of NTM pulmonary disease. Though these criteria serve as guides to the diagnosis of all NTM pulmonary disease, it is imperative to carefully consider the particular NTM organism encountered when contemplating a diagnosis of NTM pulmonary disease. Since these criteria are largely based on the experience with *M. avium complex* (MAC), they are most readily applied to the diagnosis of MAC NTM disease but are also extrapolated to other NTM isolates. When considering the possibility of a diagnosis of NTM pulmonary disease due to non-MAC organisms, the epidemiology and the virulence of the NTM in question need to be taken into account, and expert consultation is often required [2–5]. Most significantly, isolation of certain low-virulence organisms (such as *M. gordonae*) is rarely associated with clinical disease, even when other clinical and radiographic criteria are met [2].

Clinically, the signs and symptoms of NTM pulmonary disease are varied and nonspecific. Most affected patients will have chronic or relapsing cough. Some will

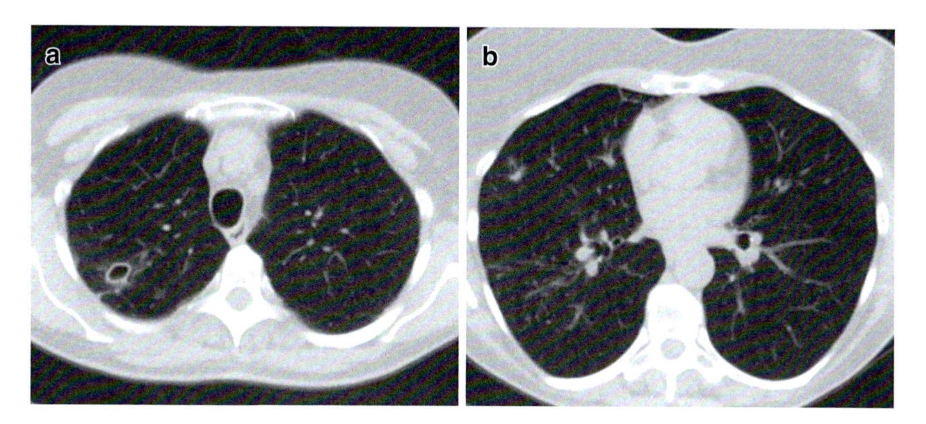

Fig. 2 Representative CT chest findings from a patient with nontuberculous mycobacterial pulmonary disease. In panel (**a**), a fibrocavitary lesion is seen at the right upper lobe. In panel (**b**), mild bronchiectasis and nodular and tree-in-bud infiltrates are demonstrated in the right lower lobe, along with consolidative infiltrates at the right middle lobe and lingua

have additional pulmonary symptoms, which can include progressive dyspnea, sputum production, hemoptysis, and chest pain. Constitutional symptoms also occur, which can include fatigue, fevers, night sweats, and weight loss. A major challenge is differentiating symptoms due to potential NTM pulmonary disease from those due to underlying structural lung disease, particularly bronchiectasis or COPD. However, it has been suggested that NTM pulmonary disease should be considered in any patient with structural lung disease whose symptoms do not respond to treatment of their underlying condition [6].

Radiographically, the cardinal features of NTM pulmonary disease are fibrocavitary lesions or regions of bronchiectasis, nodular infiltrates, consolidation, and tree-in-bud opacities (Fig. 2) [1]. No radiographic pattern is specific for NTM pulmonary disease. The finding of tree-in-bud opacities on chest CT often raises concern for NTM, but one case series showed that among patients in whom the cause of tree-in-bud opacities could be established, less than half were due to NTM pulmonary disease [7]. Certain radiographic patterns are unusual in NTM pulmonary disease, including pleural disease, ground-glass opacities, prominent thoracic adenopathy, and air-fluid levels in cavitary lesions.

Mycobacterial culture data is an absolute requirement for the diagnosis of NTM pulmonary disease, even in the patient with particularly suggestive clinical and radiographic features. In the patient who is able to produce sputum, either spontaneously or via induction with nebulized saline, at least three specimens should be obtained for AFB smear and mycobacterial culture. The diagnosis of NTM pulmonary disease is best supported by positive culture results on at least two of the three specimens. The basis of this two out of three sputum requirement dates back to retrospective observations made from a single center with respect to patients in whom *M. avium complex* was cultured in sputum through the 1980s [8]. Among these patients, only 2% of those with a single positive sputum culture demonstrated significant radiographic changes over the course of at least 12 months of follow-up. In contrast, 98% of those with two or more positive sputum cultures experienced radiographic progression, with new cavitary or infiltrative lesions noted. Furthermore, 97% of this latter group had positive cultures observed on at least two of their first three sputum samples.

Among patients unable to produce sputum in whom there is a high index of suspicion for NTM pulmonary disease based on clinical and radiographic findings, culture results from a single bronchoscopic wash or bronchoalveolar lavage (BAL) can be considered sufficient microbiologic evidence of NTM pulmonary disease. However, care should be taken in interpreting the results of bronchoscopic studies. The use of a single bronchial washing or lavage sample to establish the diagnosis of NTM pulmonary disease is most appropriate in the patient who is unable to produce sputum. Bronchoscopy should be followed up by sputum collection in a patient in whom a positive NTM culture was unexpected. Additionally, great care should be taken when ascribing NTM pulmonary disease to an unusual NTM organism on the basis of a single bronchoscopic culture, and expert consultation in this situation is generally required.

When judging the significance of microbiologic results, consideration should also be given to quantitative measures. Low numbers of NTM organisms isolated in respiratory cultures, whether from sputum or bronchoscopic samples, should cast doubt on a potential diagnosis of NTM pulmonary disease. In contrast, positive smear results for acid-fast bacilli and/or large numbers of NTM organisms on culture should increase suspicion for NTM pulmonary disease.

Ultimately, the diagnosis of NTM pulmonary disease is most straightforward when a patient exhibits clear signs and symptoms of disease, has chest imaging demonstrating classic patterns of NTM infection, and submits multiple respiratory samples that grow common, disease-generating NTM on mycobacterial culture. Situations in which a diagnosis is not so clear are considered below.

The Significance of a Single Sputum Culture Positive for NTM

In general, a single respiratory culture that is positive for NTM is an indeterminate finding, which could be due to contamination, transient presence of an NTM organism in the airways following environmental exposure, or true NTM pulmonary disease [9]. Only a minority of patients with a single sputum culture positive for NTM are ultimately diagnosed with NTM pulmonary disease, and so care should be used not to over interpret the meaning of a single positive culture [10–12]. For the most part, a single sputum culture positive for NTM is an indication to follow a patient closely and to collect multiple additional respiratory samples. However, as with most aspects of diagnosis of NTM pulmonary disease, the approach needs to be individualized, and consideration should be given to the specific organism cultured as well as the patient's ability to produce multiple, high-quality respiratory samples. It is not uncommon for experienced clinicians to pursue treatment for NTM pulmonary disease in the absence of multiple positive cultures [13]. Making the diagnosis of NTM pulmonary disease with only a single sputum culture positive for NTM is most appropriate in the setting of a particularly virulent organism, such as *M. kansasii, M. szulgai*, and *M. malmoense*. However, in most cases, the indolent course of NTM pulmonary disease allows for sufficient time to carefully and longitudinally establish a microbiologic diagnosis with multiple samples.

The Significance of an Unexpected NTM Culture Result While Treating a Different Mycobacterial Disease

Patients being treated for pulmonary mycobacterial disease typically undergo multiple sputum cultures during the course of treatment. It is now recognized that isolation of unexpected NTM species among these routine surveillance cultures is not uncommon. Rates of unexpected NTM isolation among patients undergoing treatment of pulmonary TB have been reported at 7–14% among populations in the USA, Canada, Korea, and Taiwan [14–17]. In all of these populations, the incidence of NTM pulmonary disease was felt to be uncommon, but close follow-up of such patients beyond completion of TB therapy is thought to be warranted. An analogous situation has been described during the treatment of MAC, whereby surveillance cultures unexpectedly grow out *M. abscessus* subspecies abscessus [18]. Here again, the clinical significance of the unexpected NTM culture result was uncertain upon the initial finding; only a minority of patients with *M. abscessus* isolated on sputum culture were ultimately felt to have *M. abscessus* pulmonary disease in addition to the MAC pulmonary disease for which they were being treated. Current experience suggests that the isolation of an unexpected NTM species while treating a different mycobacterial disease should prompt close follow-up, with serious consideration as to the possibility of the development of a second NTM pulmonary disease.

When to Pursue Bronchoscopy

Bronchoscopy can be a useful tool in the diagnosis of NTM pulmonary disease. It is usually unnecessary in the setting of fibrocavitary disease, where high bacterial burdens generally facilitate making the microbiologic diagnosis via serial sputum samples. It is not uncommon that patients with nodular-bronchiectatic disease cannot produce adequate sputum samples, and bronchoscopy should generally be reserved for such patients. Prior to pursuing bronchoscopy, the clinician should consider how results will affect management. If a decision has already been made not to treat a patient for NTM pulmonary disease at a particular point in time even with a positive bronchoscopic NTM culture result, bronchoscopy should be deferred. However, if alternative diagnoses are being entertained, including bacterial or fungal infections, bronchoscopy may be justified even if NTM pulmonary disease would not immediately be treated. While bronchoalveolar lavage is likely more sensitive than sputum culture in most patients with nodular-bronchiectatic disease, basing the microbiologic diagnosis of NTM pulmonary disease on a single bronchoscopic sample is only appropriate in the patient who is unable to produce sputum. If a bronchoscopic sample unexpectedly yielded NTM on culture, efforts should subsequently be made to obtain three sputum samples for further evaluation.

When to Pursue Pulmonary Biopsy

Lung biopsy is not usually necessary to establish a diagnosis of NTM pulmonary disease, and the risks of biopsy to establish a diagnosis are frequently not justified. However, there are two scenarios in which lung biopsy may be indicated: (1) when an alternative disease (often malignancy) cannot otherwise be excluded and (2) when there is a very high index of suspicion for NTM pulmonary disease, but sputum and/or bronchial wash results have been nondiagnostic. When a lung biopsy is obtained, findings of histopathologic features of mycobacterial infection (granulomatous inflammation and/or AFB-positive organisms) with NTM isolated on culture from the sample or from a single separate sputum culture or bronchial washing are sufficient for diagnosis.

The Decision to Treat Pulmonary NTM Disease

In general, establishing a diagnosis of NTM pulmonary disease should not automatically trigger initiation of directed anti-mycobacterial therapy. The decision as to whether a patient requires therapy depends on a number of factors, including the virulence of the specific organism cultured, the patient's immune status, the severity of symptoms attributable to NTM infection, the presence or absence of cavitary lesions, the rate of clinical and radiographic progression, and the patient's ability to tolerate therapy. Antibiotic therapy for NTM pulmonary disease invariably requires prolonged exposure to multiple agents, with attendant risks of significant adverse effects, and so the decision to pursue treatment should not be taken lightly. Prior to initiating therapy for NTM pulmonary disease, careful consideration should be given to the balance of risks and benefits involved in pursuing an extended treatment regimen. Ideally, this calculus includes clear definitions of a patient's goals, preferences, and expectations.

Diagnosis of Extrapulmonary NTM Disease

The diagnosis of NTM disease outside of the lung depends on direct sampling of affected spaces. A high index of suspicion, informed by consideration of host risk factors, is required to pursue appropriate diagnostic sampling.

Lymphadenitis

Localized cervical lymphadenitis is the most common form of NTM disease in immunocompetent children but is rarely seen in immunocompetent adults. The vast majority of cases occur in children younger than 5 years, with typical presentation involving painless, unilateral lymph node swelling that persists for weeks to months, without any associated systemic symptoms. Though spontaneous resolution of the lymph node swelling can occur, progression, liquefaction, and prolonged drainage are common.

The main differential diagnostic consideration is *M. tuberculosis*-associated lymphadenitis, which can present in an identical fashion. Other infectious organisms should also be considered, as should lymphoma. Diagnosis relies on lymph node sampling, either by fine needle aspirate or excisional biopsy, with excisional biopsy carrying the advantage of being potentially curative. Definitive diagnosis is based on recovery of NTM organisms from lymph node culture or by identification of NTM via molecular probe. It should be noted that sensitivity of lymph node culture is suboptimal, even in the setting of histopathology showing caseating granulomas with AFB, and so molecular probes can be very helpful in differentiating between lymphadenitis due to NTM (particularly MAC) and *M. tuberculosis*.

Lymphadenitis caused by NTM is also discussed in the chapter "Nontuberculous Mycobacterial Disease in Pediatric Populations".

Skin and Soft Tissue Disease

Though cutaneous exposures to NTM are commonplace, NTM infections of the skin and soft tissues due are relatively rare. A population-based study in Olmsted County, Minnesota, described an incidence of skin and soft tissue NTM disease of 2.0 per 100,000 person-years, though this would be expected to vary geographically due to environmental considerations [19]. As with other forms of NTM disease, a diagnosis of NTM infection of the skin and soft tissue requires consideration of the clinical context and supporting data, in addition to culture results. Isolation of NTM from skin or wound specimens does not by itself establish a diagnosis of NTM infection, and it is necessary to consider whether a patient has risk factors for NTM skin and soft tissue disease and whether the histopathology of collected specimens is compatible with the diagnosis.

Cutaneous NTM infections manifest with lesions that evolve slowly over the course of weeks to months, taking many forms, including abscesses, ulcers, nodules, papules, and rashes [19–21]. The hands and forearms are the sites most commonly affected, and pain is the most common symptom, though some will complain of pruritus, swelling, or drainage [20]. Causative organisms are most often *M. marinum* or one of the rapidly growing mycobacterium (e.g., *M. abscessus*, *M. chelonae*, *M. fortuitum*) [19, 20].

The major risk factors for NTM skin and soft tissue disease are immunosuppression, trauma, surgery, and cosmetic procedures. A history of solid organ transplantation and treatment with tumor necrosis factor-alpha inhibitors are key immunosuppressive states associated with NTM skin and soft tissue disease [22–24]. Trauma, surgery, and relevant cosmetic procedures (including tattoos, piercings, and pedicures) are all risk factors for NTM infection due to disruption of the skin's barrier protection, and initial infection general occurs at the site of this disruption [24]. Exposure to tap water has been closely associated with many occurrences of NTM skin and soft tissue infections, including nosocomial NTM infections [1].

Osteomyelitis

Infection of bone with NTM is an uncommon cause of osteomyelitis, and diagnosis is often delayed due to clinicians' low index of suspicion for mycobacterial infection. In immunocompetent hosts, most NTM bone disease develops as a consequence of direct inoculation due to trauma or surgery, with symptoms of pain and local inflammation developing slowly over weeks to months [25, 26]. In immunocompromised patients, hematogenous spread to bone is more likely, and osteomyelitis is often a manifestation of disseminated disease [27]. Radiographic findings of NTM osteomyelitis are nonspecific. MRI is often sought in the workup to help establish the likelihood of infection, but radiographic features alone cannot generally distinguish among infections due to bacteria, *M. tuberculosis*, and NTM [28]. The diagnosis of osteomyelitis due to NTM depends on bone biopsy with associated mycobacterial culture and histopathologic examination.

Disseminated Disease

Disseminated disease due to NTM is almost exclusively seen among patients with advanced HIV or other profound immunosuppression and is almost always due to MAC. In the appropriate clinical context, disseminated NTM should be considered in HIV-infected individuals with CD4+ T-cell counts less than 50 cells/μl, though most cases occur in the setting of CD4+ T-cell counts less than 25 cells/μl. Symptoms of disseminated NTM are nonspecific (particularly in the context of advanced HIV) and include fever, night sweats, weight loss, abdominal pain, and diarrhea. Findings on routine laboratory studies are also nonspecific but may include anemia and elevated lactate dehydrogenase and alkaline phosphatase. The diagnosis of disseminated NTM is usually made via positive blood culture, though isolation of NTM organisms from lymph node, bone marrow, liver, or another normally sterile compartment would also confirm the diagnosis. While mycobacterial loads are typically higher in bone marrow than in blood samples, there is little difference in culture yield from the two sites, and bone marrow sampling rarely provides more timely results [29, 30]. Therefore, blood culture for mycobacterium should be the initial diagnostic step in most HIV-infected patients in whom disseminated NTM is being entertained.

Disseminated disease due to NTM is exceedingly rare in immunosuppressed states other than HIV infection, though it has been reported in patients following solid organ transplantation; in patients with hematologic malignancies, anti-cytokine antibodies, and rare genetic disorders; and in patients treated with corticosteroids. In such patients, the diagnosis of disseminated disease is based on isolation of NTM from culture of blood or other normally sterile compartment, just as in HIV-infected patients. However, as compared with patients with advanced HIV, those patients not infected with HIV are more likely to have disseminated disease due to an NTM other than MAC.

Bibliography

1. Griffith DE, Aksamit T, Brown-Elliott BA, et al. An official ATS/IDSA statement: diagnosis, treatment, and prevention of nontuberculous mycobacterial diseases. Am J Respir Crit Care Med. 2007;175(4):367–416.
2. van Ingen J, Bendien SA, de Lange WC, et al. Clinical relevance of non-tuberculous mycobacteria isolated in the Nijmegen-Arnhem region, The Netherlands. Thorax. 2009;64(6):502–6.
3. Griffith DE. Nontuberculous mycobacterial lung disease. Curr Opin Infect Dis. 2010;23(2):185–90.
4. Hoefsloot W, van Ingen J, Andrejak C, et al. The geographic diversity of nontuberculous mycobacteria isolated from pulmonary samples: an NTM-NET collaborative study. Eur Respir J. 2013;42(6):1604–13.
5. Philley JV, DeGroote MA, Honda JR, et al. Treatment of non-Tuberculous mycobacterial lung disease. Curr Treat Options Infect Dis. 2016;8(4):275–96.
6. Field SK, Cowie RL. Lung disease due to the more common nontuberculous mycobacteria. Chest. 2006;129(6):1653–72.
7. Miller WT Jr, Panosian JS. Causes and imaging patterns of tree-in-bud opacities. Chest. 2013;144(6):1883–92.
8. Tsukamura M. Diagnosis of disease caused by Mycobacterium avium complex. Chest. 1991;99(3):667–9.
9. Griffith DE, Aksamit TR. Understanding nontuberculous mycobacterial lung disease: it's been a long time coming. F1000Res. 2016;5:2797.
10. Koh WJ, Kwon OJ, Jeon K, et al. Clinical significance of nontuberculous mycobacteria isolated from respiratory specimens in Korea. Chest. 2006;129(2):341–8.
11. Martiniano SL, Sontag MK, Daley CL, Nick JA, Sagel SD. Clinical significance of a first positive nontuberculous mycobacteria culture in cystic fibrosis. Ann Am Thorac Soc. 2014;11(1):36–44.
12. McShane PJ, Glassroth J. Pulmonary disease due to nontuberculous mycobacteria: current state and new insights. Chest. 2015;148(6):1517–27.
13. Plotinsky RN, Talbot EA, von Reyn CF. Proposed definitions for epidemiologic and clinical studies of Mycobacterium avium complex pulmonary disease. PLoS One. 2013;8(11):e77385.
14. Kendall BA, Varley CD, Hedberg K, Cassidy PM, Winthrop KL. Isolation of non-tuberculous mycobacteria from the sputum of patients with active tuberculosis. Int J Tuberc Lung Dis. 2010;14(5):654–6.

15. Damaraju D, Jamieson F, Chedore P, Marras TK. Isolation of non-tuberculous mycobacteria among patients with pulmonary tuberculosis in Ontario, Canada. Int J Tuberc Lung Dis. 2013;17(5):676–81.
16. Jun HJ, Jeon K, Um SW, Kwon OJ, Lee NY, Koh WJ. Nontuberculous mycobacteria isolated during the treatment of pulmonary tuberculosis. Respir Med. 2009;103(12):1936–40.
17. Huang CT, Tsai YJ, Shu CC, et al. Clinical significance of isolation of nontuberculous mycobacteria in pulmonary tuberculosis patients. Respir Med. 2009;103(10):1484–91.
18. Griffith DE, Philley JV, Brown-Elliott BA, et al. The significance of Mycobacterium abscessus subspecies abscessus isolation during Mycobacterium avium complex lung disease therapy. Chest. 2015;147(5):1369–75.
19. Wentworth AB, Drage LA, Wengenack NL, Wilson JW, Lohse CM. Increased incidence of cutaneous nontuberculous mycobacterial infection, 1980 to 2009: a population-based study. Mayo Clin Proc. 2013;88(1):38–45.
20. Dodiuk-Gad R, Dyachenko P, Ziv M, et al. Nontuberculous mycobacterial infections of the skin: a retrospective study of 25 cases. J Am Acad Dermatol. 2007;57(3):413–20.
21. Uslan DZ, Kowalski TJ, Wengenack NL, Virk A, Wilson JW. Skin and soft tissue infections due to rapidly growing mycobacteria: comparison of clinical features, treatment, and susceptibility. Arch Dermatol. 2006;142(10):1287–92.
22. Doucette K, Fishman JA. Nontuberculous mycobacterial infection in hematopoietic stem cell and solid organ transplant recipients. Clin Infect Dis. 2004;38(10):1428–39.
23. Winthrop KL, Baxter R, Liu L, et al. Mycobacterial diseases and antitumour necrosis factor therapy in USA. Ann Rheum Dis. 2013;72(1):37–42.
24. Atkins BL, Gottlieb T. Skin and soft tissue infections caused by nontuberculous mycobacteria. Curr Opin Infect Dis. 2014;27(2):137–45.
25. Elsayed S, Read R. Mycobacterium haemophilum osteomyelitis: case report and review of the literature. BMC Infect Dis. 2006;6:70.
26. Bi S, Hu FS, Yu HY, et al. Nontuberculous mycobacterial osteomyelitis. Infect Dis (Lond). 2015;47(10):673–85.
27. Hirsch R, Miller SM, Kazi S, Cate TR, Reveille JD. Human immunodeficiency virus-associated atypical mycobacterial skeletal infections. Semin Arthritis Rheum. 1996;25(5):347–56.
28. Theodorou DJ, Theodorou SJ, Kakitsubata Y, Sartoris DJ, Resnick D. Imaging characteristics and epidemiologic features of atypical mycobacterial infections involving the musculoskeletal system. AJR Am J Roentgenol. 2001;176(2):341–9.
29. Kilby JM, Marques MB, Jaye DL, Tabereaux PB, Reddy VB, Waites KB. The yield of bone marrow biopsy and culture compared with blood culture in the evaluation of HIV-infected patients for mycobacterial and fungal infections. Am J Med. 1998;104(2):123–8.
30. Hafner R, Inderlied CB, Peterson DM, et al. Correlation of quantitative bone marrow and blood cultures in AIDS patients with disseminated Mycobacterium avium complex infection. J Infect Dis. 1999;180(2):438–47.

Nontuberculous Mycobacterial Disease Management Principles

Timothy R. Aksamit and David E. Griffith

Introduction

The reader will undoubtedly notice that throughout this volume, there are repeated caveats about the complexities, incongruities, paradoxes, and frustrations involved with nontuberculous mycobacterial (NTM) disease management, especially pulmonary disease (PD). This chapter is an attempt to address as many of the difficult management problems as possible in one section. Obviously, many areas in this discussion lack a firm evidence base and are still controversial and hotly debated. Where possible we offer corroborating evidence for recommendations but also frequently call upon almost three decades of experience managing NTM patients. It should also be noted that perspectives and recommendations made in this chapter are primarily made from a North American perspective and may or may not apply equally to other areas in the world, especially in developing countries. We fully recognize that the recommendations in this chapter will not be universally endorsed, but they will hopefully serve as a starting point for readers to explore difficult management decisions in more depth. While the initial admonition for this volume is not to use it as a quick "how-to" guide, this chapter is the exception. We hope the reader will find the recommendations in this chapter helpful in the practical day-to-day management of NTM disease patients.

T. R. Aksamit (✉)
Mayo Clinic, Pulmonary Disease and Critical Care Medicine, Rochester, MN, USA
e-mail: aksamit.timothy@mayo.edu

D. E. Griffith
University of Texas Health Science Center, Tyler, TX, USA
e-mail: david.griffith@uthct.edu

D. E. Griffith (ed.), *Nontuberculous Mycobacterial Disease*, Respiratory Medicine,
https://doi.org/10.1007/978-3-319-93473-0_10

Diagnosis

The first essential element in the management of NTM patients is adequate mycobacteriology laboratory support. All patient management decisions discussed below require accurate NTM identification and in vitro susceptibility testing results [1].

The management of patients with NTM PD begins with confidently establishing the diagnosis using published diagnostic criteria [1]. This topic is discussed in detail in chapter "Diagnosis of NTM Disease: Pulmonary and Extrapulmonary". With very rare exception, we do not recommend or endorse empiric treatment for NTM infection, especially NTM PD. Meeting diagnostic criteria for NTM PD is the essential first step for confidently approaching subsequent management decisions such as whether to proceed with an often complicated and prolonged NTM PD treatment course. We also understand the limitations of the diagnostic criteria which are clearly not applicable to all NTM species isolated from respiratory specimens. It is absolutely essential for clinicians to be familiar with the virulence and disease-causing potential of NTM species to intelligently apply the NTM PD diagnostic criteria [2]. And while newer serologic tests are being developed to assist in the diagnosis of NTM PD (especially MAC), there is a large unmet need for additional meaningful biomarkers and diagnostic tools, representing an area under active investigation [3–8].

Diagnosis

- Diagnosis of NTM PD requires clinical, microbiologic, and radiographic confirmation
- Empiric treatment of NTM PD without confirmed diagnosis is not recommended

NTM Pulmonary Disease and Tuberculosis

Pulmonary tuberculosis (TB) is invariably mentioned as the first differential diagnostic consideration for NTM PD, so it seems reasonable to discuss the four important intersections between the two disease processes here.

First, patients with cavitary NTM PD present with chest radiographs consistent with reactivation pulmonary TB disease and with sputum that is often acid-fast bacilli (AFB) smear positive. Problems expeditiously differentiating TB and NTM PD in this type of patient are greatly alleviated by the widespread availability of sputum nucleic acid amplification testing (NAAT) for TB. In the setting of sputum AFB smear positivity, a negative NAAT carries a high negative predictive value and can in most, but not all, instances exclude a diagnosis of TB [9]. However, there may be instances where patients are at high risk for having TB based on the clinical set-

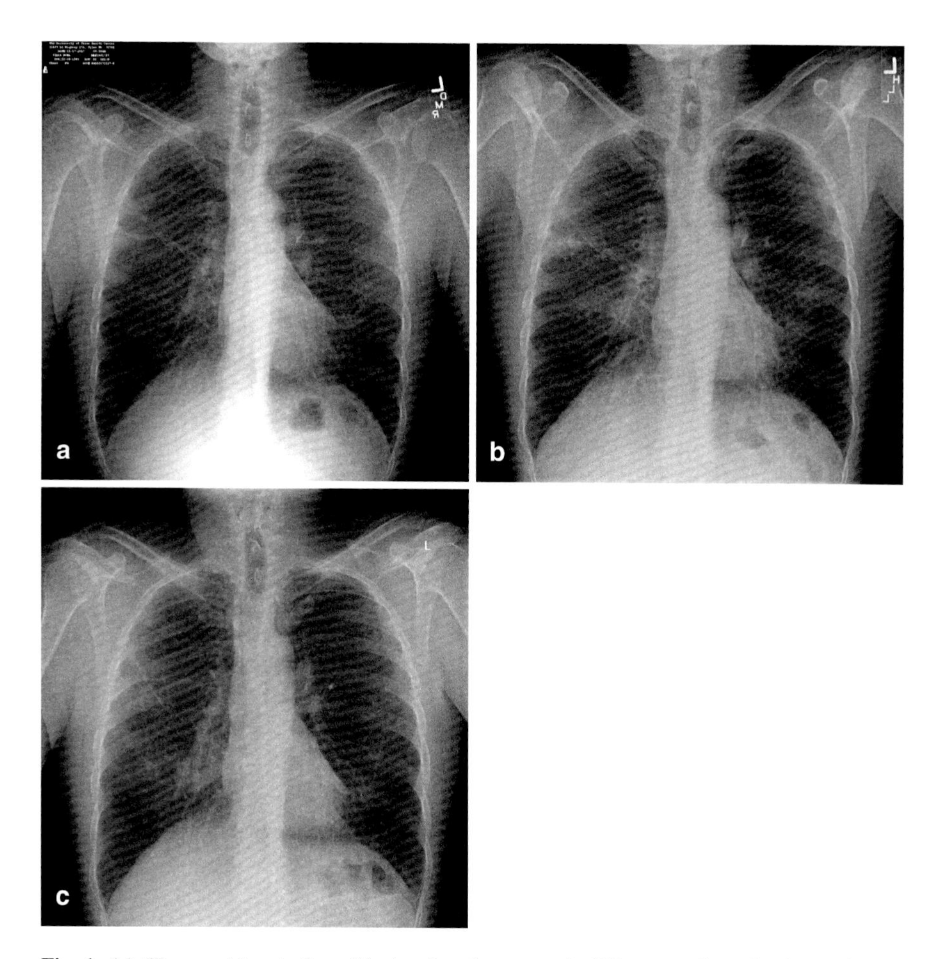

Fig. 1 (**a**) 57-year-old male from Mexico found as part of a TB contact investigation to have a positive QuantiFERON test. Patient with chronic cough and abnormal chest radiograph. Initial sputum AFB smear positive but nucleic acid amplification test (NAAT) negative. Patient started on multidrug anti-tuberculosis therapy. (**b**): After 3 months of multidrug anti-tuberculosis therapy, no symptomatic improvement and progression of radiographic abnormalities. Multiple sputum specimens culture positive for MAC. (**c**) After 3 months of guideline-based MAC therapy including macrolide, patient symptomatically better with radiographic improvement and conversion of sputum to culture negative

ting where NAAT is either not available or negative, and the initial organism identification by some other means such as high-performance liquid chromatography (HPLC) suggests an NTM pathogen. In that circumstance we believe it is prudent to treat for TB while waiting for microbiologic confirmation of the diagnosis (Fig. 1).

A second more unusual occurrence is the isolation of MTB during the course of NTM PD therapy in a patient with established NTM disease. This circumstance represents a more difficult diagnostic challenge. Experience in the USA suggests that most *M. tuberculosis* (MTB) isolated during NTM PD therapy represent specimen laboratory contamination. We have, however, also rarely seen newly acquired

TB in NTM PD patients who were close or household contacts to TB cases. In either case, genotyping of the MTB isolate is essential for determining its significance. The processing lab must be notified that specimen contamination is suspected which should trigger the appropriate investigation including genotyping of *M tuberculosis* isolates processed in proximity to the specimen in question. Robust communication with laboratory colleagues will facilitate efficient and quality decision-making on behalf of the patient. Unfortunately, redirecting therapy against TB, at least temporarily, may be unavoidable. This analysis is significantly more complicated in regions outside the USA with high TB incidence and high risk for TB transmission.

A third and more common occurrence is the isolation of NTM respiratory pathogens during the course of TB therapy [10–12]. There are limited data suggesting that most NTM isolated in this circumstance are not clinically significant; however, that determination must be made on an individual basis. We generally recommend that NTM isolated during TB therapy do not require a change in therapy and that TB therapy should be completed before addressing the clinical significance of the NTM isolate. Obviously, patients must have ongoing evaluation to determine if the isolated NTM is adversely affecting the patient's course, thereby necessitating redirection of therapy to include coverage of the NTM pathogen. Most patients undergoing TB therapy in the USA are managed by public health entities, and evaluating the significance of NTM isolates is beyond their scope and resources which means that these patients will need evaluation in the community to determine the significance of NTM isolates. In that context, patients should be referred to appropriate specialists to address and possibly treat the NTM PD after TB treatment has been completed.

A fourth situation can occur when biopsies of an extrapulmonary organ unexpectedly yield an AFB smear-positive specimen. Too often cultures are not obtained during these procedures so that culture confirmation of the organism identity is not available. In this situation, patients are often assumed to have TB, especially if there is also a positive tuberculin skin test or interferon gamma release assay. Recently, molecular testing of the AFB smear-positive tissue has been employed through a service offered by the Centers for Disease Control and Prevention (CDC) that can identify MTB as well as NTM organisms from preserved tissue specimens [13]. The invaluable information obtained is not necessarily diagnostic of active mycobacterial disease and must be interpreted in the context of the patient's overall clinical status.

NTM PD and Tuberculosis

- Differentiation of TB disease from NTM PD may be challenging and requires careful clinical assessment, collaboration with laboratory colleagues, and use of available molecular diagnostic tools.
- NTM PD during treatment of TB pulmonary disease and, to lesser extent, TB pulmonary disease during NTM PD treatment can occur.

- Treatment of TB is the highest priority with need for referral (from public health) of coinfected NTM PD patients after completing TB treatment to appropriate pulmonary or infectious disease specialists.

Respiratory Comorbidities

The most important initial NTM lung disease management effort is typically how best to address underlying and often complex comorbidities. The most common of these comorbidities are bronchiectasis, chronic obstructive lung disease (COPD), sinus disease, and gastroesophageal reflux disease (GERD). These diagnoses are covered in more detail in chapter "Management of Lung Diseases Associated with NTM Infection", but some key points are worth emphasizing.

It was observed more than four decades ago that some patients with non-cavitary *Mycobacterium avium* complex (MAC) disease experienced conversion of sputum AFB cultures to negative with "pulmonary toilet" measures, usually chest percussion and postural drainage [14]. At the time of that observation, patients with bronchiectasis and nodular bronchiectatic (NB) MAC lung disease were not recognized as such. More recent studies have confirmed this phenomenon [15–17]. It is our experience that treatment of bronchiectasis can be symptomatically transformative for some patients. The management of bronchiectasis symptoms, which frequently mimic and are indistinguishable from the symptoms of NTM PD, can also help clinicians decide whether or not to start NTM therapy. Patients with mild and/or indolent NTM PD may experience sufficient symptomatic improvement with airway clearance measures and treatment of other comorbidities that NTM therapy may not be necessary even though NTM PD is still present.

Controversy exists about the utility of efforts to determine the etiology of bronchiectasis in patients without an obvious explanation. A careful and detailed history is unquestionably the most important aspect of any effort to determine the etiology of bronchiectasis. Recent guidelines have recommended core laboratory evaluation for all patients with additional evaluations based on clinical suspicion of specific contributing etiologies [18, 19]. The one consensus indication for a more extensive etiologic search is young age at onset of bronchiectasis. In our opinion this indication includes patients diagnosed with bronchiectasis at a young age but also older bronchiectasis patient with symptoms that began at a young age (Fig. 2). Severe and/or extensive bronchiectasis would be other indications for considering an etiologic evaluation (Figs. 3 and 4). The extent of the evaluation is influenced by cost for the testing and the potential impact of a positive test including the availability of interventions for a specific diagnosis and genetic counselling for family members.

The management of COPD may also provide significant symptomatic benefit to the patient; however, it is less likely to allow NTM disease treatment avoidance because COPD patients are more likely to have cavitary NTM PD which requires aggressive treatment [17, 20–22]. Regardless, the potential symptomatic benefit from COPD treatment is a worthwhile goal of its own. It should be noted that

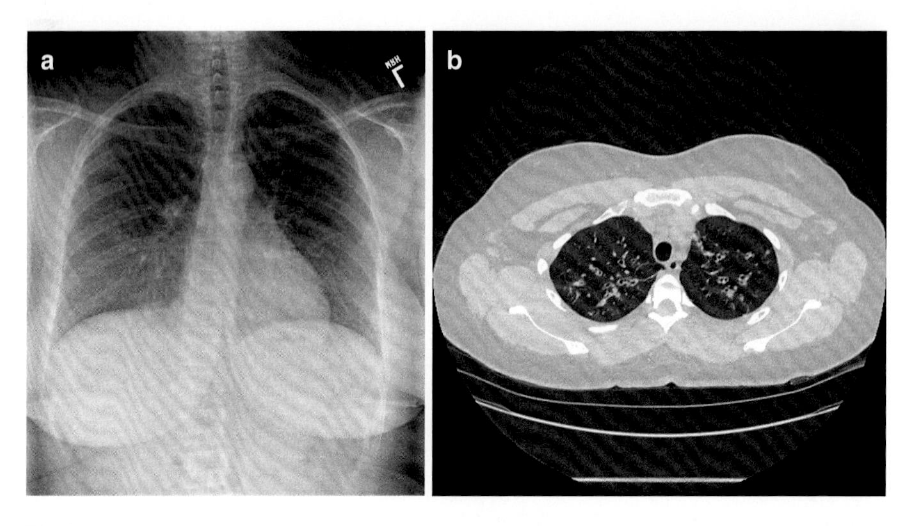

Fig. 2 (**a**) PA chest radiograph from 32-year-old female with history of recurrent pneumonia since childhood. Undergoing infertility evaluation. Sputum culture positive for MAC, *M. abscessus*, three strains of *Pseudomonas*. Chest radiograph with abnormalities consistent with diffuse bronchiectasis. (**b**) Chest CT slice from the same patient showing extensive upper lobe bronchiectasis consistent with cystic fibrosis. Sweat chloride level > 90 mmol/L

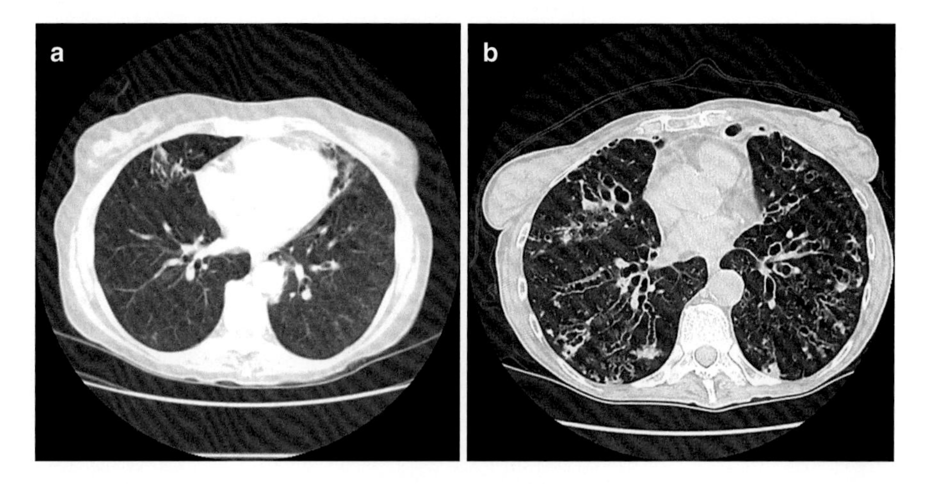

Fig. 3 (**a**) Chest CT slice from 69-year-old patient, never smoker, with bronchiectasis in 2006 showing bilateral bronchiectasis. Patient with history of recurrent respiratory infections including pneumonia since early adulthood. (**b**) Chest CT slice at comparable level from the same patient in 2012 showing progression of bronchiectasis. Patient's sputum culture positive for MAC, *M. abscessus*, *Nocardia*, *Pseudomonas*, *Stenotrophomonas*, *Burkholderia*, and *Aspergillus*. Alpha-one-antitrypsin level < 30 mg/dl, Zz phenotype

appropriate treatment of advanced COPD may involve use of inhaled corticosteroids (ICS), but that ICS may increase the risk of NTM lung disease [23, 24] Our approach is to avoid ICS in the management of NTM PD in the absence of a COPD phenotype that specifically benefits from ICS administration.

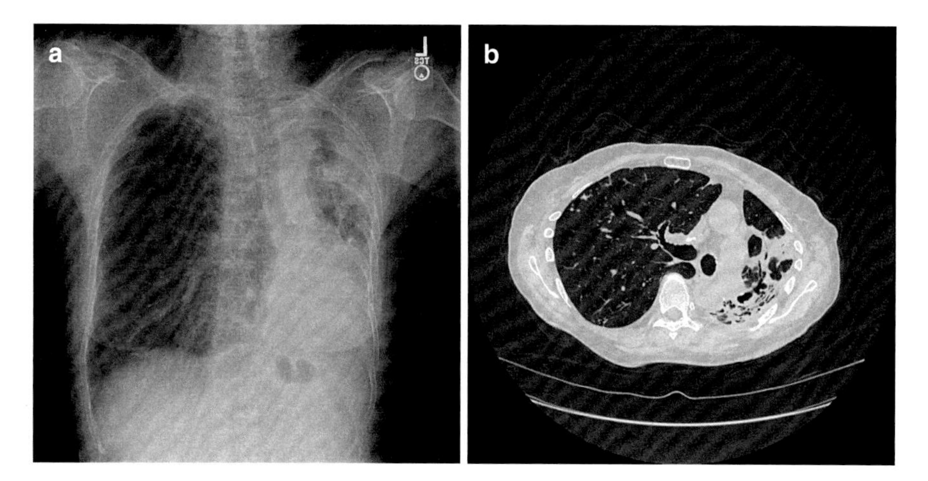

Fig. 4 (**a**) Chest radiograph from 80-year-old patient with history of recurrent respiratory infections including pneumonia since childhood. Mother and maternal grandmother with chronic cough without diagnosis. Patient diagnosed with MAC disease 6 years prior with multiple subsequent treatment efforts including left lower lobe lobectomy. (**b**) Chest CT slice from the same patient showing extensive cavitary consolidation in the remaining left lung. Sputum cultures positive for MAC, *M. abscessus*, *Pseudomonas*, and *Aspergillus*. IgG level < 400 mg/dl

A second important consideration for cavitary NTM PD patients is the potential for surgical intervention, which can dramatically improve NTM disease outcomes. Surgery for NTM disease is discussed in detail in chapter "Surgical Management of NTM Diseases". Medical therapy alone for NTM PD has a low likelihood of sterilizing large cavities and converting associated positive sputum cultures when present. Best outcomes, defined as culture conversion, frequently require a combination of medical therapy and surgical intervention. It should be further emphasized that medical therapy for NTM PD should be given pre- and postoperatively when adjunctive lung resection for cavitary NTM PD is considered. Optimizing pulmonary function is critical for a successful surgical outcome and requires comprehensive and coordinated care by a multidisciplinary perioperative program including an experienced mycobacterial disease thoracic surgeon [25]. In our opinion, surgery is sufficiently effective that it should be considered, if only briefly, for all NTM PD patients and especially those with severe or treatment refractory disease.

Respiratory Comorbidities

- Symptoms of NTM PD are not specific and similar to other comorbidities such as bronchiectasis or COPD.
- Treatment of comorbidities is of paramount importance to facilitate etiologic discernment of nonspecific symptoms.
- Consider etiologic evaluation for young bronchiectasis patients and those with long-standing symptoms or severe disease.

- Bronchial hygiene is a core component of treatment programs for NTM PD patients understanding that most patients with NTM PD also have bronchiectasis.
- Minimize exposure to ICS unless justified by comorbidities.
- Select NTM PD patients may benefit from adjunctive surgical therapy.

Starting NTM PD Therapy

After optimizing management of associated comorbidities, the next decision for the NTM PD patient is whether to begin therapy. For cavitary NTM PD patients, that decision is not difficult due to the inarguable risk for disease progression with attendant morbidity and mortality [20, 21]. Thus, the risk/benefit assessment overwhelmingly favors initiating therapy at the time of NTM cavitary PD diagnosis.

For patients with NTM NB PD, especially NB MAC PD, that decision is frequently complicated and requires a more deliberate approach. It is evident that some, perhaps many patients with NB MAC PD have indolent disease so that MAC isolation from a respiratory specimen, whether or not diagnostic criteria are met, does not reflexively or automatically require initiation of therapy. While many questions remain about the natural history of MAC PD, it is clear that not all patients with MAC isolated from respiratory specimens subsequently or inevitably have progressive MAC lung disease that requires treatment [26]. There is even less data regarding the natural history for other NTM PD. Nonetheless, the general concept of a variable natural history likely holds true for NTM PD other than MAC as well as MAC PD.

The critical element in the decision to start or withhold therapy is a careful risk/benefit determination for an individual patient. For mild NB MAC PD, the risks of treatment with uncertain benefit that potentially exposes patients to medication toxicity and side effects must be balanced against possible undertreatment of progressive disease, which exposes patients to disease morbidity. A common scenario is the patient with persistently positive sputum AFB cultures for MAC who has minimal symptoms and stable NB radiographic abnormalities. There is a consensus that the benefit of therapy for this type of patient would not likely outweigh the risks of MAC therapy. Fortunately, NB MAC lung disease is sufficiently indolent that careful longitudinal appraisal without therapy is safe and presents little risk for rapid progression of MAC PD or later hindrance to favorable therapeutic response. Should MAC therapy be held, it is imperative that macrolides are not used for exacerbations of bronchiectasis or other indications so as to mitigate the risk of developing macrolide-resistant MAC.

The next critical element is persistence in longitudinal follow-up. Those patients not started on therapy must be followed indefinitely as there is increasing evidence for significant risk of irreversible radiographic progression and pulmonary function decline in some patients not on therapy even if guideline-based treatment is started at a later date [27]. Our approach to initiating MAC lung disease therapy has rested on three essential factors, patient symptoms, microbiologic results, and radiographic findings. The most important factor is the radiographic appearance, especially the development of cavitation, which would strongly favor initiation of therapy regardless of symptomatic or microbiologic stability.

There are few objective markers of NTM PD disease progression. A recent study compared the clinical characteristics of MAC PD patients who had a progressive course resulting in treatment initiation within 3 years of diagnosis with patients who exhibited a stable course for at least 3 years [26]. Compared to stable MAC PD, patients with progressive MAC PD had lower body mass index (BMI) and more systemic symptoms, positive sputum AFB smears, and fibrocavitary radiographic findings. Hopefully, other biomarkers of disease progression will emerge to supplement or even supplant the current dependence on these clinical, microbiologic, and radiographic criteria.

The intensity and frequency of clinical scrutiny is also function of the specific NTM causing the patient's lung disease. It has been recognized for many years that there is a spectrum of virulence among nontuberculous mycobacterial lung pathogens (chapters "*Mycobacterium avium* Complex Disease", "NTM Disease Caused by *M. kansasii*, M. *xenopi*, *M. malmoense* and Other Slowly Growing NTM", and "Disease Caused by *Mycobacterium abscessus* and Other Rapidly Growing Mycobacteria (RGM)") [2]. It is not as clear if other NTM PD pathogens in the setting of NB disease such as *M. abscessus*, *M. kansasii*, or *M. xenopi* behave as benignly as MAC. It is incumbent upon the treating physician to be familiar with the relative virulence of common NTM pathogens in general as well as locally isolated NTM species [28, 29]. For instance, a clinician in Central Texas would need to be familiar with the virulence and natural history of *M. simiae* disease, whereas physicians in the Northern United States must be familiar with the disease-causing potential and natural history of *M. xenopi*.

No single or simple algorithm is adequate for determining the intensity of follow-up for all patients. The physician must be familiar with the virulence of the NTM pathogen in question and the pattern of disease stability and/or progression for each patient rather than depending on arbitrary recommendations for the frequency of clinical, microbiologic, and radiographic follow-up. For patients who meet NTM PD diagnostic criteria but who do not start therapy, we recommend 3–6-month (or sooner if symptoms worsen) follow-up pulmonary visits with sputum collection and radiographic assessment over at least a 24-month period. A 24-month period is generally adequate for determining which ostensibly stable patients will need therapeutic intervention. If therapy is not started in that time, we recommend at least yearly follow-up thereafter. We again stress that individualized patient assessment schedules are required. It is quite possible that management of other comorbid conditions such as bronchiectasis will dictate more frequent physician visits. We strongly urge indefinite follow-up for these patients as there is no recognized or consensus statute of limitations for when progressive NTM PD might develop after isolation of NTM from a respiratory specimen.

Patient participation is mandatory and an essential aspect of these considerations. A frank discussion with the patient should not be limited to a list of medication toxicities but also weighing the possibility of progressive lung disease if therapy is held.

Too often patients are simply told that "the treatment is worse than the disease" to prejudice the patient against treatment. Our experience is that most patients are quite willing to tolerate some diagnostic uncertainties while knowing that they are part of a careful and deliberate long-term evaluation. Patients must trust that the process will not push them into unnecessary therapy nor abandon them to untreated disease progression. In our experience a major advantage of this deliberate approach is that by the time it is clear to the physician that treatment initiation is necessary, it is also usually clear to the patient as well. Attaining confidence in the need for therapy is absolutely essential for patient adherence with extended anti-mycobacterial treatment regimens. In our experience the sequential and incremental introduction of (oral) NTM medications has generally enhanced tolerance at the beginning of treatment in contrast to starting all NTM medications at full doses all at once. Buildup to full dosing and starts of new medication are frequently recommended at 2–3-day intervals so as to reach full dosing of all medications in an average of 2–3 weeks. Use of probiotics has been associated with a reduction of antibiotic-associated diarrhea and, in our experience, may improve gastrointestinal tolerance of medications [30].

Starting NTM PD Therapy

- The decision to start treatment or observe for NTM PD is complex and unique to each patient, comorbidities, and overall risk-benefit assessment.
- The sequential and incremental introduction of NTM medications may improve tolerance.
- If treatment for NTM PD is not started, regular and longitudinal follow-up is essential.

Choosing Anti-mycobacterial Treatment Regimens

Two of the greatest challenges for choosing NTM treatment regimens are understanding NTM drug resistance mechanisms and recognizing the limitations of in vitro susceptibility testing for guiding NTM anti-mycobacterial therapy. The latter consideration is neither intuitive nor facile and is perhaps the greatest source of frustration among clinicians treating patients with NTM disease. This topic is so important that it is covered in detail in two chapters in this volume (chapters "Laboratory Diagnosis and Antimicrobial Susceptibility Testing of Nontuberculous Mycobacteria" and "Drug Susceptibility Testing of Nontuberculous Mycobacteria"). The reader is strongly urged to read both of these discussions as they have somewhat different perspectives and emphasis. Even so it is worth reiterating that in vitro susceptibility testing for many NTM pathogens is frequently not a reliable guide to effective anti-mycobacterial drug choices and clinical responses (Table 1). The most important and common example is MAC where only macrolide and amikacin in vitro susceptibilities predict in vivo treatment response. This observation is so

Table 1 Association of NTM in vitro susceptibilities and in vivo clinical response

Correlation between treatment response and in vitro susceptibilities
MAC (macrolide, amikacin)
M. kansasii (rifampin, macrolide, isoniazid, ethambutol, fluoroquinolone, streptomycin, sulfamethoxazole)
M. marinum (rifampin)
M. szulgai (macrolide)
M. fortuitum (no macrolide, multiple antibiotics)
M. chelonae (macrolide, multiple antibiotics)
M. abscessus subsp. *abscessus* (macrolide if erm gene not active)
M. abscessus subsp. *massiliense* (macrolide if erm gene not active)
Limited or no correlation between treatment response in vitro susceptibility
M. xenopi
M. malmoense
M. simiae
M. abscessus (active erm gene)

Table 2 Nontuberculous mycobacteria subject to emergence of acquired mutational resistance while on therapy

M. avium complex:
(a) 23S rRNA gene (macrolides)
(b) 16S rRNA gene (amikacin)
M. kansasii: rpo β gene (rifamycins)
M. abscessus: 23S rRNA gene (macrolides)

important and so frequently ignored or misunderstood that it bears reinforcing. Awareness of the potential for acquired macrolide resistance for MAC means inclusion of adequate companion medications (usually ethambutol) to prevent acquired macrolide resistance which is associated with significantly worse clinical outcomes compared with macrolide-susceptible MAC isolates (Table 2; Fig. 5) [31].

Formulating an adequate treatment regimen for a specific NTM pathogen and achieving therapeutic success require familiarity with both innate and acquired resistance mechanisms for that pathogen [32]. Conversely, a lack of familiarity with these mechanisms is not likely to be associated with treatment success and may in fact exacerbate or worsen the patient's status. While specific NTM PD regimens including oral, inhaled, and/or parenteral antimicrobial agents vary considerably across different NTM species, the concept of using multidrug regimens for avoidance of acquired drug resistance is universally applicable.

Choosing Anti-mycobacterial Treatment Regimens

- Discordance between in vitro susceptibilities and in vivo treatment responses for NTM is common.
- Understanding the mechanisms of drug resistance in NTM is important for successful treatment of NTM.

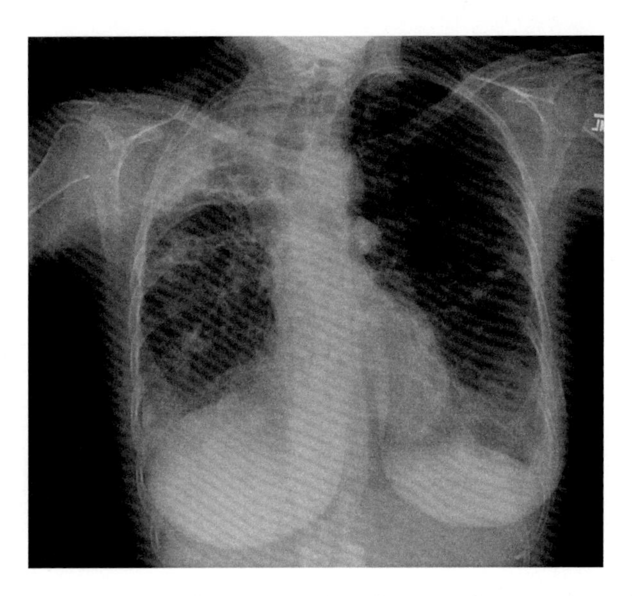

Fig. 5 68-year-old patient diagnosed with MAC lung disease. Patient started on guideline-based therapy including macrolide, ethambutol, and rifampin. Ethambutol discontinued after "resistant" MIC for ethambutol reported by reference laboratory. Moxifloxacin substituted for ethambutol because of "susceptible" MIC reported by the reference lab. Rifampin stopped due to patient intolerance. After 12 months of therapy with macrolide and moxifloxacin, the patient had progressive cavitary destruction of the right upper lobe and a macrolide-resistant MAC isolate

Patient Evaluation During NTM PD Therapy Including Response to Therapy

Once the patient has been started on an appropriate anti-mycobacterial treatment regimen, there are multiple potential impediments to the completion of adequate therapy. Most of these impediments are related to the long duration of treatment and the need for multiple potentially toxic anti-mycobacterial medications. Other impediments are generally related to the poor overall physical status of NTM patients due to other comorbidities, usually bronchiectasis or chronic obstructive lung disease. These comorbidities can obfuscate or confuse symptoms of mycobacterial disease. In one recent study, it was found that even for patients who responded well to anti-mycobacterial therapy, there was invariably at least one bronchiectasis exacerbation while on anti-mycobacterial therapy [33]. Some of the potential obstacles to successful anti-mycobacterial therapy are listed in Table 3.

It is noteworthy that a major impediment to successful therapy is a lack of adherence by treating physicians to recommended NTM PD guideline-based treatment [34, 35]. We readily concede that current treatment guidelines are suboptimal and frequently do not result in treatment success. We just as readily suggest that nonadherence to the treatment guidelines does not improve the chances for treatment success and may adversely affect a patient's disease course and prognosis [34, 35].

Table 3 Impediments to effective NTM pulmonary disease therapy

Long duration of therapy required
Multiple drugs necessary
Ubiquitous presence of NTM in the environment resulting in ongoing exposure
Innate antibiotic resistance mechanisms
Poor correlation between in vitro susceptibility and clinical (in vivo) response
Acquired antibiotic resistance
Marginally effective drugs
Poor correlation between pharmacokinetic and pharmacodynamic indices and clinical response
Comorbidities (bronchiectasis/COPD)
Clinical symptoms nonspecific and exacerbations frequent
Predisposition for NTM reinfection with ongoing exposure
As yet unidentified factors inhibiting therapeutic response
Clinician challenges
Lack of familiarity or adherence with published guidelines

Given the combination of coexisting pulmonary comorbidities and multiple potentially toxic medications, it is not surprising that NTM PD patients frequently experience problems with anti-mycobacterial therapy requiring therapeutic adjustments. In our experience, the majority of adjustments are needed at the front end of the NTM PD regimen. One example of an effective adjustment is the improved tolerance of intermittent (three times weekly) medication compared with daily medication administration for NB MAC PD [36, 37]. Changes in mycobacterial treatment required due to gastrointestinal drug intolerances generally become less likely once patients are well into the treatment course.

Exceptions for late intolerances include hearing loss with the longer-term use of amikacin, and to lesser extent macrolide, and optic nerve toxicity with ethambutol. It is essential for the clinician to manage these patients in such a way as to maintain as many "first-line" drugs as possible in the patient's treatment regimen as there are few effective alternatives. We recommend audiograms for patients on parenteral amikacin and visual and color vision testing for patients on ethambutol in accordance with monitoring recommendations from recent TB guidelines [38].

Macrolides are the most important anti-mycobacterial component in MAC and macrolide-susceptible *M. abscessus* treatment regimens. Simply stated, there is not a comparably effective replacement in either circumstance. It is clear that maintaining a macrolide in the treatment regimen in these situations has the highest priority. Patients who are intolerant of one macrolide (clarithromycin or azithromycin) because of drug toxicity can frequently tolerate the other. Even for mild hypersensitivity responses, there is not complete cross-reaction between the two drugs so that patients with a rash on one macrolide should be challenged (under appropriate observation and monitoring) with the other macrolide. Although not as critical for successful therapy, patients with intolerance to one rifamycin (rifampin or rifabutin), including mild hypersensitivity reactions, may tolerate the other rifamycin. Rifampin-related hypersensitivity reactions can also be addressed through established rifampin desensitization protocols [39]. This step would be especially important for maintaining rifampin in *M. kansasii* (rifampin susceptible) treatment regimens.

Patient Evaluation During NTM PD Therapy and Assessment of Response to Therapy – 1

- Established guidelines should be followed to optimize chances of successful NTM PD treatment outcomes.
- Monitoring while on NTM PD treatment with blood work, visual assessments, and hearing/vestibular testing is required without exception and should be tailored individually to patients and associated comorbidities as well as specific anti-mycobacterial regimen (see text).

In MAC disease, the most important of the macrolide companion drugs is ethambutol. While ethambutol is not a potent anti-MAC drug per se, it has been shown to protect against the emergence of acquired mutational macrolide resistance [40]. If ethambutol is lost in the treatment regimen, it cannot be easily replaced. Intermittent ethambutol administration appears to be associated with less ocular toxicity than daily ethambutol. It was also shown that older patients taking ethambutol who develop ocular symptoms frequently have explanations for those symptoms other than ethambutol. Even though many, perhaps most, ophthalmologists currently in practice are unfamiliar with ethambutol ocular toxicity, it is our practice to heed the advice of an ophthalmology consultant about discontinuation of ethambutol in a patient with new or worsening ocular symptoms. Patients with ethambutol hypersensitivity reactions can be successfully desensitized via published protocols so that ethambutol can remain in the treatment regimen [39].

Two particularly poor ethambutol replacement strategies for MAC disease are the substitution of fluoroquinolone for ethambutol (macrolide + fluoroquinolone ± rifamycin) or the use of macrolide with only a rifamycin (Fig. 5). The rifamycins decrease macrolide serum levels, and the fluoroquinolones do not protect against the emergence of acquired mutational macrolide resistance. Possible substitutions include parenteral amikacin, inhaled amikacin, and clofazimine although there is little data to support this recommendation.

Patient Evaluation During NTM PD Therapy and Assessment of Response to Therapy – 2

- The use of fluoroquinolones for the treatment of MAC, either alone or in combination with macrolide, is not recommended.
- Ethambutol as a companion drug in MAC treatment regimens protects against development of macrolide resistance.

- Rifamycins should not be a single companion drug to macrolide in MAC PD treatment regimens.
- Alternative MAC companion medications to protect macrolide include amikacin and clofazimine.

One trend in MAC PD therapy deserves special mention, that is, the use of inhaled generic amikacin [41–44]. Our impression is that inhalation of a parenteral amikacin preparation is widely prescribed in the USA, although there is little published clinical experience with inhaled amikacin. There is not an FDA-approved inhalation form of amikacin, so there is no standardization in dosing, delivery, or administration. We believe that amikacin is effective against MAC when given parenterally, but its effectiveness by inhalation is less certain. Variable and heterogenous lung deposition and concentrations conceivably could promote acquired amikacin resistance or even acquired macrolide resistance if it is the only companion drug for macrolide-susceptible isolates. It is understandable that patients and clinicians prefer amikacin inhalation to parenteral amikacin administration, but both should be aware that there is no proof that they are comparably effective.

In contrast to inhaled generic amikacin, results from a phase II study using liposomal amikacin for inhalation for refractory MAC PD were promising [45]. A second Phase III study has been published and confirms the effectiveness of inhaled liposomal amikacin for producing sputum conversion in refractory MAC lung disease [46]. These studies with provide sufficient safety and efficacy information for establishing appropriate placement of inhaled liposomal amikacin in treatment regimens for MAC.

Treatment of other NTM PD pathogens such as *M. abscessus*, *M. kansasii,* and *M. xenopi* is discussed in detail in chapters "NTM Disease Caused by *M. kansasii*, M. *xenopi*, *M. malmoense* and Other Slowly Growing NTM" and "Disease Caused by *Mycobacterium abscessus* and Other Rapidly Growing Mycobacteria (RGM)". Some of the concepts discussed for MAC PD are pertinent to these pathogens although each NTM pathogen presents its own challenges and obstacles to successful therapy, such as the need for parenteral therapy for *M. abscessus*. Our impression is that for most non-MAC NTM pathogens, expert consultation is sought more frequently than for MAC PD so that familiarity with each one may not be as critically important as with MAC for the general pulmonary or infectious disease specialist.

When the patient and the clinician embark on a treatment course for NTM PD, there must be a clear understanding that treatment difficulties and medication intolerances will inevitably arise but can generally be managed with modifications of drug doses or dosing intervals. In our experience, modification of treatment regimens during the course of NTM PD therapy is the rule rather than the exception. Given the paucity of effective drugs for treating NTM PD pathogens, premature abandonment of "first-line" therapy or specific components of that therapy will usually not result in a successful treatment outcome and adversely impact long-term prognosis. Some frequently encountered medication-related problems are outlined in Table 5.

Just as there are three components (clinical, radiographic, and microbiologic) for establishing MAC PD diagnosis and for deciding when to begin NTM PD therapy, there are also three major components for evaluating treatment response. Clinically it is expected that patients would have symptomatic improvement, with symptoms such as cough, sputum production, fatigue, and weight loss. The use and role of a recently developed quality of life (QOL) instrument developed for NTM in clinical practice remains to be fully clarified [47]. In our experience, improvements in microbiologic status (i.e., culture conversion) parallel improvements in clinical symptoms and stabilization of radiographic abnormalities [33]. The universal coexistence of pulmonary comorbidities such as bronchiectasis and chronic airflow obstruction frequently make this assessment difficult and unsatisfying for the patient. As discussed above the overlapping symptoms between comorbid pulmonary conditions and NTM PD make optimal management of those comorbid conditions absolutely critical for adequate NTM PD assessment. Clinicians should be particularly mindful to recognize undertreated comorbidities if NTM microbiologic improvements occur in the setting of progressive clinical symptoms.

We recommend clinic visits at least every 2–3 months while on therapy to evaluate patient medication tolerance, treatment response, and toxicity monitoring understanding that the frequency may need to be more often for some patients on parenteral-based regimens and/or with substantial intolerances. Patients should be regularly and systematically questioned at each visit as to any symptoms including common drug toxicity symptoms (Table 4). As noted above, we also recommend visual acuity and color vision testing for all patients who are on ethambutol on a regular 2- to 3-month interval basis or sooner if symptoms develop. For patients on a rifamycin and/or macrolide, we also recommend a complete blood count and chemistry panel including liver enzymes at each visit although the utility of this approach has not been rigorously evaluated. Patients who are receiving an intravenous aminoglycoside should have baseline audiometry and vestibular function with follow-up monitoring studies guided by patient's symptoms and published guidelines. Initial clinical assessment and follow-up of hearing and vestibular function are also required for those patients receiving inhaled aminoglycoside [48]. Baseline EKG is generally warranted in all patients starting macrolides to assess for significant baseline EKG abnormalities. A role for the ongoing monitoring of patients on macrolide with EKG's is not established nor, in our opinion, justified at this point unless there are abnormalities at baseline or if cardiac rhythm risk factors are present. The presence of comorbid pulmonary conditions especially bronchiectasis can also complicate the interpretation of radiographic response to therapy. Patients with bronchiectasis with or without NTM lung disease frequently have waxing and waning densities associated with secretion retention in the airways as well as acute inflammatory responses that can be associated with bronchiectasis exacerbations. It is our experience that acute symptomatic and radiographic changes with the development of purulent sputum production most often reflect a bronchiectasis exacerbation rather than failure of a NTM PD regimen. These varying radiographic features of bronchiectasis often cloud the radiographic assessment of response to NTM PD therapy. Once again the aggressive management of the underlying pulmonary comorbidity can greatly facilitate interpretation of the patient's response to anti-mycobacterial therapy.

Table 4 Common medication side effects

Medication	Common side effects
Rifamycins Rifampin (Rifadin™, Rimactane™) Rifabutin (Mycobutin™)	Red, brown, or orange discoloration of urine, feces, saliva, sweat, or tears Diarrhea Upset stomach Rash
Ethambutol (Myambutol™)	Vision changes Numbness, tingling in hands and feet Rash
Macrolides Clarithromycin (Biaxin™) Azithromycin (Zithromax™)	Upset stomach Unusual taste in mouth Hearing changes Diarrhea Rash
Aminoglycosides Amikacin (Amikin™)	Hearing changes Nausea Muscle weakness Rash Kidney problems
Quinolones Ciprofloxacin (Cipro™) Levofloxacin (Levaquin™) Moxifloxacin (Avelox™)	Upset stomach Rash Diarrhea Headache Dizziness
Tetracyclines Minocycline (Minocin™) Doxycycline (Vibramycin™) Tigecycline injection (Tygacil™)	Sun sensitivity Nausea Diarrhea Dizziness Rash
Sulfamethoxazole/trimethoprim Bactrim™	Itching Loss of appetite Rash

Radiographic abnormalities are not expected to resolve completely even after successful completion of a full course of NTM PD therapy. In most instances improved, albeit not resolved, radiographic abnormalities parallel improvements in clinical symptoms and microbiologic responses while on NTM PD therapy.

We recognize that there is not a universally endorsed approach to the radiographic follow-up for NTM patients on therapy. Patient's generally have both plain chest radiography and chest CT scans at the initiation of therapy. For those patients with obvious abnormalities on plain chest radiograph, it may be adequate to obtain serial follow-up radiographs for detecting significant radiographic changes. For other patients the abnormalities are more subtle and may require serial (low dose) chest CT scans to detect changes. Patients are very cognizant about radiation exposure with chest CT scans, and our general approach is to limit the number of chest CT scans as much as possible. Ideally after starting therapy, patients would only require one follow-up chest CT scan at the termination of therapy.

The microbiologic analysis is ostensibly the most straightforward, but even that analysis has important caveats and pitfalls. Arguably the most important single metric for evaluating mycobacterial treatment response is conversion of the patient's sputum to AFB culture negative. There is ongoing debate about the relative importance of patient symptomatic and radiographic responses and for determining the efficacy of therapy. However, it would be difficult to argue that an anti-mycobacterial intervention was effective without AFB sputum culture conversion to negative. We recommend collection of sputum for AFB analysis on a monthly basis so as to establish sustained sputum culture conversion as early as possible while patients are undergoing treatment but no less than every 2–3 months. We recommend expectorated sputum collection if possible or sputum induction during clinic visits if expectorated sputum is not available. Monitoring of patients with additional sputum culture collections is also recommended during NTM therapy (e.g., MAC) understanding that a second NTM species (e.g., *M. abscessus*) may appear [49]. We do not recommend routine bronchoscopy for specimen collection while patients are undergoing therapy or at the end of therapy even if the original diagnosis was made based on a bronchoscopic specimen. The legitimacy of serial bronchoscopic cultures for evaluating treatment response including the impact on the treatment success criterion of 12 months sputum culture negativity has not been rigorously tested.

Patient Evaluation During NTM PD Therapy and Assessment of Response to Therapy – 3

- The frequency of regular clinic visits and sputum collections during NTM PD therapy are dependent on patient factors and specific medications of the NTM PD regimen.
- A second NTM species, sometimes needing treatment, may occur during NTM PD therapy.
- Radiographic abnormalities are not expected to resolve completely even when successful NTM PD therapy is completed.

Treatment Endpoints

The current primary NTM PD treatment endpoint most often is microbiologic. Patients should achieve 12 months of negative sputum AFB cultures while on therapy. This treatment duration was chosen because it was observed in the past to be associated with sustained sputum culture negativity after discontinuation of MAC therapy [1]. A major confounding factor for this treatment success criterion is the observation that some patients with NB MAC lung disease will have microbiologic recurrence either while still receiving therapy (after sputum conversion) or after completing adequate therapy due to new or unique MAC genotypes that are different from the initial MAC genotype at the time of diagnosis. The source of these

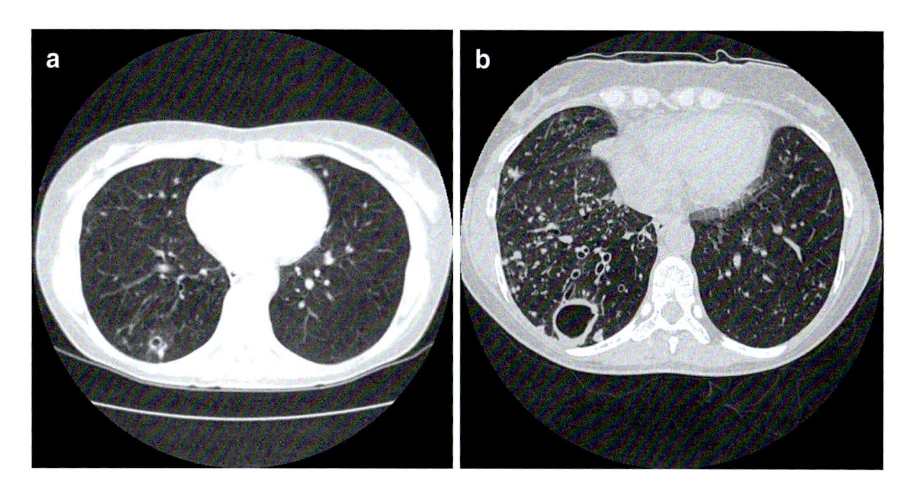

Fig. 6 (**a**) 58-year-old patient with bronchiectasis and MAC lung disease on guideline-based therapy including macrolide with initial symptomatic, radiographic, and microbiologic improvement. (**b**) 58-year-old MAC lung disease patient improving on guideline-based therapy who subsequently had symptomatic regression associated with enlarging cavitary lesion on chest imaging and repeated isolation of *M. abscessus* from sputum

new MAC genotypes is not clear. They may represent reinfection or it is possible that the original MAC infection was polyclonal and the new genotypes were preferentially selected to grow during therapy. It is noteworthy however that when these new genotypes occur during therapy, they are invariably also macrolide susceptible in vitro although the patient was invariably receiving as part of the original guideline-based MAC therapy [36].

Likewise, the goal of 12 months of treatment after sputum conversion is often not practically possible when using NTM PD regimens that incorporate one or more parenteral agents when alternative transition oral or inhaled medications are not available. This practical limitation adds to the risk-benefit assessment complexities not only about starting treatment but also about determining the duration of therapy. In some instances treatment holidays from parenterally based regimens should be considered, even if sputum culture conversion has not occurred. It has been our experience that patients and treating physicians are best served by articulating expectations and endpoints a priori before the start of a course of NTM therapy.

Patients also, as noted above, sometimes have co-isolation of more than one NTM species. A common scenario is the isolation of *M. abscessus* during the course of treatment for MAC. Most of these patients have a limited number of *M. abscessus* isolates without an indication of progressive *M. abscessus* lung disease. No alteration in MAC therapy is necessary for these patients. Some patients however will have repeated *M. abscessus* isolation associated with progressive radiographic abnormalities including cavitation and worsening symptoms (Fig. 6). For these patients therapy may need to be altered to treat *M. abscessus* as well as MAC. This therapeutic shift is difficult because there are few agents with activity against both MAC and *M. abscessus*. Determining the significance of *M. abscessus* isolates during the treatment of MAC requires very close clinical, microbiologic, and radiographic assessment to dissect confidently the role of *M. abscessus* as the cause of clinical deterioration or progressive disease. Unfortunately there are no surrogate markers to aid in that determination so that the clinician must depend on clinical indicators in conjunction with the radiographic as well as microbiologic findings.

A second and equally difficult scenario is the isolation of other known respiratory pathogens such as fungi and *Nocardia* species during the course of NTM PD therapy. The isolation of fungi, especially *Aspergillus* species, is common in patients with cavitary disease presumably due to colonization of cavities with the fungus as well as the use of chronic antimicrobials. In most patients it has been our experience that the isolation of *Aspergillus* was not clinically significant. It is also evident, however, that there are patients with extensive cavitary disease who do not respond well to anti-mycobacterial therapy and are felt to have progressive cavitary fungal disease (Fig. 7) [50, 51]. Unfortunately, we are not aware of a simple diagnostic approach for determining the significance of *Aspergillus* in this circumstance, although measuring serum Aspergillus precipitins or other specific biomarkers has been touted in this role [52]. Because of the usually severe nature of the cavitary disease in these patients, this diagnostic uncertainty means that by default, patients will receive antifungal therapy. The assessment of positive fungal cultures is all the more important because antifungal and anti-mycobacterial therapies are often incompatible due to drug-drug interactions especially with combinations of rifamycins and macrolides. This is clearly an area that urgently needs more research. For now, clinicians must rely on a multidisciplinary approach to the patients and careful assessment of the effects of all interventions.

Similarly, the isolation of *Norcadia* species is not rare in NB NTM PD patients [53]. Prior to the recent expansion in bronchiectasis interest, it was assumed that the isolation of *Nocardia* from respiratory specimens was an ominous and always clinically significant finding and indicative of immune deficiency. More recently, it has become clear that bronchiectasis patients can sometimes have *Nocardia* isolated from sputum without evidence of immune deficiency or progressive *Nocardia* lung disease. The current situation of *Nocardia* isolated in sputum or other respiratory specimens is reminiscent of the situation 30 years ago with bronchiectasis and NTM. As with *Aspergillus*, this is also an area that urgently needs more research.

In contrast to the utility of the microbiologic endpoint, the role of clinical and radiographic changes as endpoints for treatment of NTM PD is even less clear. Certainly both symptomatic and radiographic endpoints are important, but they are simply more difficult to evaluate given their nonspecific character in relation to comorbid medical conditions and lack of validated instruments and endpoints. The culprit again is the common coexistence of underlying lung disease especially bronchiectasis. Bronchiectasis is associated with permanent radiographic abnormalities and shifting patterns of secretion retention with waxing and waning infil-

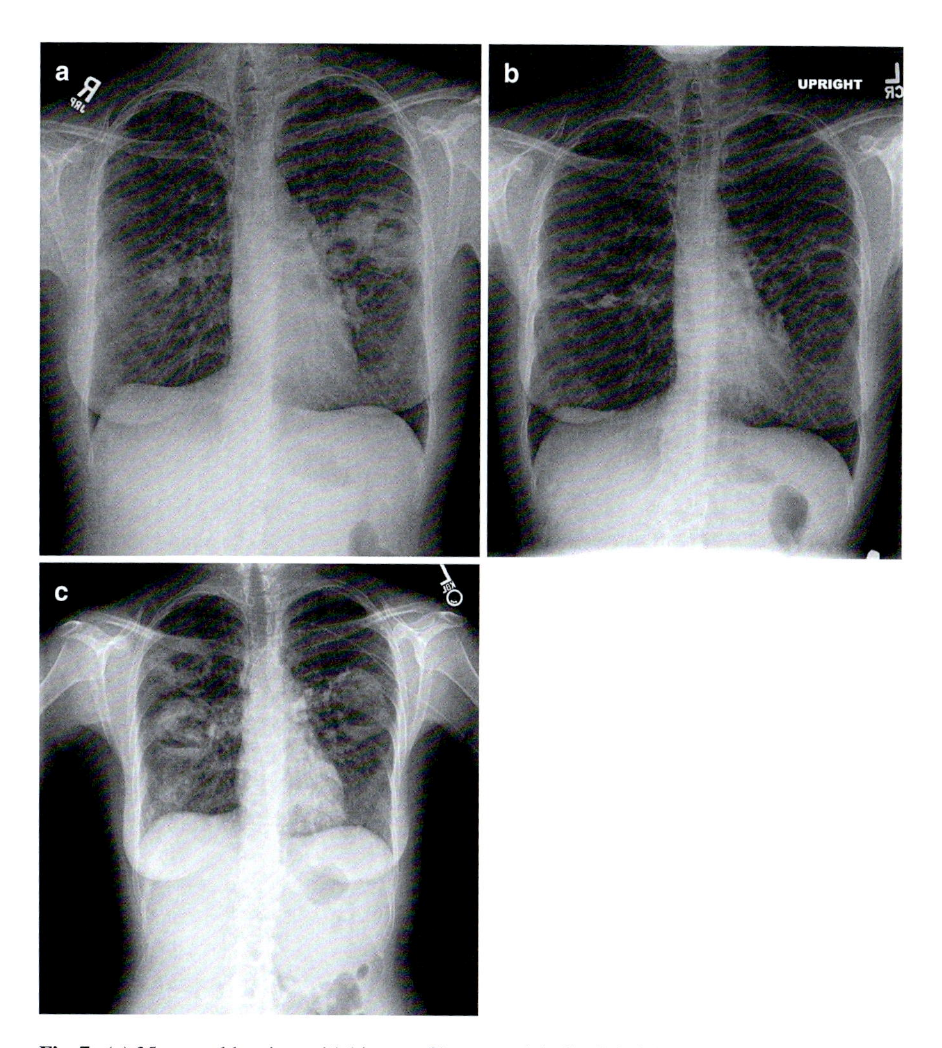

Fig. 7 (**a**) 35-year-old patient with history of immunoglobulin G deficiency and steroid-dependent asthma diagnosed with cavitary MAC lung disease. (**b**) Patient treated with guideline-based therapy including macrolide and parenteral amikacin with symptomatic, radiographic, and microbiologic improvement. (**c**) After initial radiographic improvement on guideline-based MAC therapy including macrolide and amikacin, patient developed rapidly enlarging bilateral cavities associated with repeated isolation of *Aspergillus* from sputum and bronchoscopy

trates. Chest radiographs are certain to remain abnormal throughout therapy and at the completion of MAC lung disease treatment. This is highlighted further with the understanding that all tree-in-bud infiltrates are not NTM PD related [54]. Without a clear-cut or unequivocal indication of mycobacterial disease progression, it would be difficult to justify treatment extension after the patient meets the microbiologic endpoint (culture conversion) for treatment success. Similarly while it is hoped that patients have overall symptomatic improvement, it would also be difficult to justify treatment extension without clear-cut or unequivocal indicators of symptomatic progression due to MAC disease. These areas exemplify the importance of close familiarity by the clinician with the MAC lung disease patient. As importantly in our experience, robust laboratory support and interactions between clinician and laboratorians will further optimize successful treatment outcomes. Assessing treatment response for MAC lung disease is clearly not as simple as the assessment of tuberculosis lung disease treatment response. Pulmonary function testing including forced expiratory volume at 1 s (FEV1) and forced vital capacity (FVC) is not considered to be sensitive or specific enough to use routinely as an endpoint in the assessment of response to NTM treatment. Often, abnormalities of pulmonary function are more reflective of the status of comorbidities rather than NTM PD.

It is worth emphasizing that treatment endpoints of NTM PD are not static universal endpoints and may vary as well through the initial evaluation period and the treatment course or in the follow-up period. For example, it may be decided based on status of disease severity, drug tolerance (or lack thereof), or other contributing factors that the goal of therapy is to control symptoms but not to achieve sputum culture conversion and or radiographic improvement. Goals of treatment and treatment endpoints are particularly important to discuss with patients prior to the start of treatment as well as during treatment. In this instance, modifications in the treatment regimen in intensity and duration may be appropriate. Alternatively, escalation of treatment may also be warranted if anticipated endpoints at the beginning of treatment are not being met with initial treatment regimens. One such escalation is adjunctive surgical resection of involved lung. This strategy is clearly effective for selected patients with localized disease and adequate cardiopulmonary reserve in the hands of experienced surgeons. Surgical intervention should be considered for all NTM PD patients who do not respond adequately to first-line anti-mycobacterial therapy and have surgically favorable lung disease involvement with the understanding that many such patients will still not be appropriate surgical candidates. Specific modifications of de-escalation or escalation in NTM PD therapy regimens are covered in other chapters addressing specific pathogens or interventions.

Treatment Endpoints

- Goals of treatment and treatment endpoints should be discussed with patients prior to the start of therapy as well as during therapy.
- Findings of a second NTM species or other pathogens (e.g.. *Aspergillus* or *Nocardia*) during NTM PD are not uncommon.
- The significance of co-pathogens and need to treat requires complex assessments of clinical, microbiologic, and radiographic factors.

Post-therapy Evaluation

After successful treatment of MAC lung disease is completed, we recommend continued surveillance of sputum with AFB cultures obtained once in 2–3 months and then periodically (up to every 2–3 months) for at least 2–3 years along with longitudinal assessment of clinical symptoms and radiographic abnormalities. Even if the patient is successfully treated for MAC microbiologically, there is an approximate 50% chance of microbiologic recurrence after completion of therapy [36]. Most of these microbiologic recurrences are due to new or unique genotypes compared with the original genotype isolated from the patient [36]. When microbiologic recurrences occur that are due to new or unique genotypes, our approach is the same as for any patient with a new isolation of MAC. As previously mentioned above, macrolide susceptibility is generally expected to be preserved for recurrent MAC lung disease although recheck of macrolide susceptibility on the recurrent isolates is warranted. If the microbiologic recurrence is due to a MAC genotype identical to the originally isolated genotype, we consider those patients as having true disease relapse and generally reinstitute therapy at that point. Subsequent rates of microbiologic responses to repeat treatment courses have been observed to be less relative to initial treatment responses even with preserved macrolide susceptibility [1]. These considerations underscore the critical importance of having adequate laboratory support to make the appropriate determinations about the significance of microbiologic recurrence isolates. If, however, genotyping of microbiologic recurrence isolates is not available, we recommend the same systematic assessment (symptomatic, microbiologic, and radiographic) as would be applicable after an initial isolation of MAC in a respiratory specimen.

The frequency of radiographic follow-up is less certain. Our approach is to obtain plain chest radiographs as often as possible reserving (low dose) chest CT scans for specific questions that might arise. Patients are frequently and understandably quite interested in limiting radiation exposure but also usually quite willing to cooperate with appropriately justified requests for CT scans. Clearly the frequency and type of radiographic follow-up will depend on the patient's clinical status.

We view these patients as lifelong patients not only because of the need to manage their bronchiectasis and comorbidities but because there is no endpoint for exposure to ubiquitous NTM pathogens and presumably to acquisition and reacquisition of these pathogens in the lungs of patients. Moreover, the impact of residential environmental NTM mitigation including but not limited to increasing water heater temperature, the design of showerheads, and showering habits is unknown as to risk of recurrent NTM lung disease infection rates (see chapter "Environmental Niches for NTM and Their Impact on NTM Disease") [55].

Certainly there are patients that were not started on anti-mycobacterial therapy but have persistently positive sputum AFB cultures and stable clinical and radiographic status who require long-term monitoring for evidence of progressive

Table 5 Algorithm for managing NTM lung disease

Appropriate treatment of underlying pulmonary co-morbidities

↓

Identification of NTM pathogen

↓

Does the patient meet NTM lung disease diagnostic criteria?

↙ ↘

Yes — No: continue F/U indefinitely

Is the patient a candidate for treatment?

- How symptomatic is the patient?
- Is the NTM lung disease associated with cavities?
- What are the patient's pulmonary co-morbidities and are they compensated?
- What is the patient's short-and long-term prognosis?
- What does the patient want to do?

↙ ↘

Yes — No: continue F/U indefinitely

↓

Begin therapy according to published guidelines

mycobacterial disease. We feel strongly that close familiarity with the course of an individual's bronchiectasis and mycobacterial disease facilitates the evaluation of mycobacterial disease activity during the waxing and waning course of underlying lung disease, especially bronchiectasis. An overview of NTM lung disease management recommendations is provided in Table 5.

Post-therapy Evaluation

- Microbiologic recurrence of NTM PD after successful therapy is not uncommon.
- Patients with NTM PD need indefinite follow-up and should be considered lifelong patients.

Extrapulmonary NTM Disease

Early suspicion of NTM infection is the critical element for timely diagnosis of extrapulmonary NTM disease. It is typical for patients with NTM wound infections to have several courses of unsuccessful antibacterial antibiotic therapy before undergoing diagnostic evaluation for NTM pathogens. In some circumstances the typical features of NTM infection are present, such as the purplish nodules of cutaneous and disseminated NTM disease (Fig. 8). In that circumstance there should be an immediate diagnostic effort to culture for NTM. In other less clear or nonspecific circumstances, it is understandable that bacterial pathogens would be therapeutically addressed first, but hopefully, AFB cultures would also be sent to expedite the diagnosis and facilitate initiation of anti-mycobacterial therapy. Findings of granulomatous changes on histopathology, with or without the presence of AFB organisms on AFB stain, should promptly raise suspicion for NTM (or TB) infection.

The general principles of NTM therapy as outlined for NTM PD also generally apply for extrapulmonary NTM disease as well albeit frequently with shorter duration of therapy than as recommended for NTM PD [1].

Following the patient with extrapulmonary NTM disease is also challenging. Patients with cutaneous/disseminated disease often have the appearance of new nodular lesions even while undergoing appropriate therapy for the mycobacterial pathogen. For these patients it is critical to aspirate new lesions to determine if there is active infection in the lesions, possibly due to the emergence of resistance to the antibiotic regimen, or if they are sterile and immunologically mediated as part of the paradoxical inflammatory response to therapy. Similarly patients with mycobacterial lymphadenopathy due to either TB or NTM may show lymph node enlargement or the appearance of new lymph nodes even while on appropriate treatment. Again it is critical that new or enlarging lesions are aspirated for AFB culture to determine

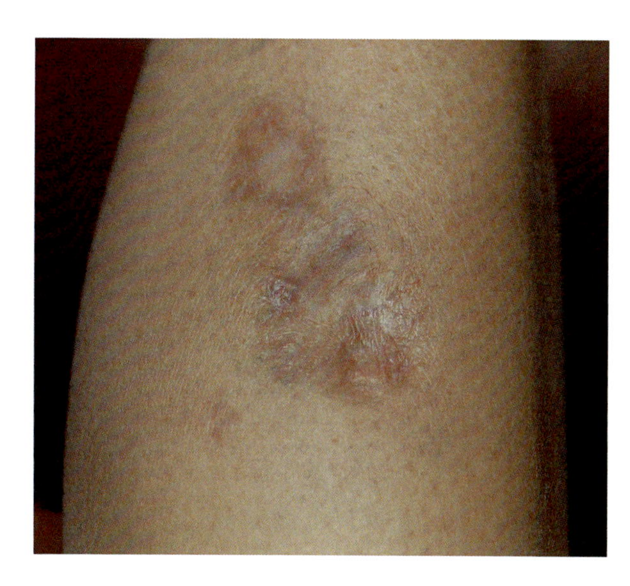

Fig. 8 Discolored (purple) appearing cutaneous nodule that appeared after penetrating skin trauma. Aspiration of the nodule yielded positive cultures for *M. abscessus*

if there is ongoing infection versus an immunologically mediated paradoxical response (e.g., IRIS – immune reconstitution inflammatory syndrome).

For patients who have disseminated NTM infection, it is critically important to correct any modifiable underlying immune suppression. The antibiotics used for treating NTM disease are not as potent as the antibiotics used for treating TB so that without reversal or correction of severe underlying immunosuppression, the chances for long-term treatment success are low. The finding of disseminated NTM should prompt the clinician to assess the patient and exclude an immunodeficiency. Patients with NTM PD at large are not expected to have other immunocompromised infections or substantial immune defects although this is an area under active investigation [56, 57].

For patients who have NTM infection associated with a foreign body, it is also critically important to remove the foreign body for adequate treatment of the mycobacterial infection. Sometimes removal of these foreign bodies is relatively simple such as removal of an indwelling venous catheter, a breast augmentation device, or a peritoneal dialysis catheter. For other foreign bodies such as joint prostheses, removal of the foreign body is more problematic but no less necessary. While there may be instances of successful treatment of joint space mycobacterial infections without removal of a prosthetic joint, there are also instances where failure to remove the joint initially significantly prolongs the duration of antibiotic therapy and does not prevent the eventual removal of the joint prosthesis. We feel the evidence strongly supports prosthetic joint removal at the initiation of the anti-mycobacterial therapy. Iatrogenic or nosocomial extrapulmonary NTM infections often can be traced back to contamination from (tap) water supplies.

Following patients with extrapulmonary NTM disease is often more difficult than following patients with NTM lung disease. Aside from the paradoxical therapeutic responses that may occur with cutaneous and disseminated disease noted above, there are also limited methods of treatment evaluation for other sites of infection. One particularly troubling area is joint space infection. These areas are especially difficult to visualize radiographically and are problematic for obtaining serial specimens for culture. We find objective assessment of treatment response with these infections challenging and all too often frustrating. These types of infections are perhaps even more desperately in need of surrogate disease activity biomarkers than NTM PD.

Summary

In summary, management of pulmonary and extrapulmonary NTM disease requires the clinician to be experienced and familiar with the protean and complex clinical, microbiologic, and radiographic manifestations of NTM disease. Input from patients as stakeholders a priori in the planning of and articulation of the treatment goals of prolonged multidrug anti-mycobacterial treatment regimens is of paramount importance. Close relationships between the treating clinicians and their respective colleagues in the microbiology laboratory and surgery consultation services are equally critical for optimizing treatment success. We are reminded daily of the shortcomings in our treatment options for NTM infections [58]. Currently, success rates for treating macrolide-susceptible MAC are not as good as our success with multidrug-resistant TB, while the success rate for treating macrolide-resistant MAC is closer to the success rate of extensively drug-resistant TB (XDR-TB). We remain optimistic that this situation will improve in the near future given the current international interest and research momentum in NTM diseases.

Bibliography

1. Griffith DE, Aksamit T, Brown-Elliott BA, et al. An official ATS/IDSA statement: diagnosis, treatment, and prevention of nontuberculous mycobacterial diseases. Am J Respir Crit Care Med. 2007;175(4):367–416.
2. van Ingen J, Griffith DE, Aksamit TR, Wagner D. Chapter 3. Pulmonary diseases caused by non-tuberculous mycobacteria. Eur Respir Monogr. 2012;58:25–37.
3. Jeong BH, Kim SY, Jeon K, Lee SY, Shin SJ, Koh WJ. Serodiagnosis of Mycobacterium avium complex and Mycobacterium abscessus complex pulmonary disease by use of IgA antibodies to glycopeptidolipid core antigen. J Clin Microbiol. 2013;51(8):2747–9.
4. Shu CC, Ato M, Wang JT, et al. Sero-diagnosis of Mycobacterium avium complex lung disease using serum immunoglobulin a antibody against glycopeptidolipid antigen in Taiwan. PLoS One. 2013;8(11):e80473.
5. Kitada S, Kobayashi K, Ichiyama S, et al. Serodiagnosis of Mycobacterium avium-complex pulmonary disease using an enzyme immunoassay kit. Am J Respir Crit Care Med. 2008;177(7):793–7.
6. Kitada S, Maekura R, Toyoshima N, et al. Use of glycopeptidolipid core antigen for serodiagnosis of mycobacterium avium complex pulmonary disease in immunocompetent patients. Clin Diagn Lab Immunol. 2005;12(1):44–51.
7. Qvist T, Pressler T, Taylor-Robinson D, Katzenstein TL, Hoiby N. Serodiagnosis of Mycobacterium abscessus complex infection in cystic fibrosis. Eur Respir J. 2015;46(3):707–16.
8. Kitada S, Levin A, Hiserote M, et al. Serodiagnosis of Mycobacterium avium complex pulmonary disease in the USA. Eur Respir J. 2013;42(2):454–60.
9. Peralta G, Barry P, Pascopella L. Use of Nucleic Acid Amplification Tests in Tuberculosis Patients in California, 2010-2013. Open Forum Infect Dis. 2016;3(4):ofw230.
10. Kendall BA, Varley CD, Hedberg K, Cassidy PM, Winthrop KL. Isolation of non-tuberculous mycobacteria from the sputum of patients with active tuberculosis. Int J Tuberc Lung Dis. 2010;14(5):654–6.
11. Huang CT, Tsai YJ, Shu CC, et al. Clinical significance of isolation of nontuberculous mycobacteria in pulmonary tuberculosis patients. Respir Med. 2009;103(10):1484–91.
12. Chung MJ, Lee KS, Koh WJ, et al. Drug-sensitive tuberculosis, multidrug-resistant tuberculosis, and nontuberculous mycobacterial pulmonary disease in nonAIDS adults: comparisons of thin-section CT findings. Eur Radiol. 2006;16(9):1934–41.
13. Yakrus MA, Metchock B, Starks AM. Evaluation of a u.S. public health laboratory service for the molecular detection of drug resistant tuberculosis. Tuberc Res Treat. 2015;2015:701786.
14. Ahn CH, Lowell JR, Onstad GD, Ahn SS, Hurst GA. Elimination of Mycobacterium intracellulare from sputum after bronchial hygiene. Chest. 1979;76(4):480–2.
15. Prince D, Peterson D, Steiner R, et al. Infection with Mycobacterium avium complex in patients without predisposing conditions. N Engl J Med. 1989;321:863–8.
16. Koh WJ, Kwon OJ, Jeon K, et al. Clinical significance of nontuberculous mycobacteria isolated from respiratory specimens in Korea. Chest. 2006;129(2):341–8.

17. Lee G, Lee KS, Moon JW, et al. Nodular bronchiectatic Mycobacterium avium complex pulmonary disease. Natural course on serial computed tomographic scans. Ann Am Thorac Soc. 2013;10(4):299–306.
18. Polverino E, Goeminne PC, McDonnell MJ, et al. European Respiratory Society guidelines for the management of adult bronchiectasis. Eur Respir J. 2017;50(3):1700629.
19. Martinez-Garcia MA, Maiz L, Olveira C, et al. Spanish guidelines on the evaluation and diagnosis of bronchiectasis in adults. Arch Bronconeumol. 2018;54:79.
20. Hayashi M, Takayanagi N, Kanauchi T, Miyahara Y, Yanagisawa T, Sugita Y. Prognostic factors of 634 HIV-negative patients with Mycobacterium avium complex lung disease. Am J Respir Crit Care Med. 2012;185(5):575–83.
21. Lee MR, Yang CY, Chang KP, et al. Factors associated with lung function decline in patients with non-tuberculous mycobacterial pulmonary disease. PLoS One. 2013;8(3):e58214.
22. Aksamit TR. Mycobacterium avium complex pulmonary disease in patients with pre-existing lung disease. Clin Chest Med. 2002;23(3):643–53.
23. Andrejak C, Nielsen R, Thomsen VO, Duhaut P, Sorensen HT, Thomsen RW. Chronic respiratory disease, inhaled corticosteroids and risk of non-tuberculous mycobacteriosis. Thorax. 2013;68(3):256–62.
24. Brode SK, Campitelli MA, Kwong JC, et al. The risk of mycobacterial infections associated with inhaled corticosteroid use. Eur Respir J. 2017;50(3):1700037.
25. Mitchell JD, Bishop A, Cafaro A, Weyant MJ, Pomerantz M. Anatomic lung resection for nontuberculous mycobacterial disease. Ann Thorac Surg. 2008;85(6):1887–92. discussion 1892-1883
26. Hwang JA, Kim S, Jo KW, Shim TS. Natural history of Mycobacterium avium complex lung disease in untreated patients with stable course. Eur Respir J. 2017;49(3):1600537.
27. Kim SJ, Park J, Lee H, et al. Risk factors for deterioration of nodular bronchiectatic Mycobacterium avium complex lung disease. Int J Tuberc Lung Dis. 2014;18(6):730–6.
28. Ford ES, Horne DJ, Shah JA, Wallis CK, Fang FC, Hawn TR. Species-specific risk factors, treatment decisions, and clinical outcomes for laboratory isolates of less common Nontuberculous mycobacteria in Washington state. Ann Am Thorac Soc. 2017;14(7):1129–38.
29. Hoefsloot W, van Ingen J, Andrejak C, et al. The geographic diversity of nontuberculous mycobacteria isolated from pulmonary samples: an NTM-NET collaborative study. Eur Respir J. 2013;42(6):1604–13.
30. Hempel S, Newberry SJ, Maher AR, et al. Probiotics for the prevention and treatment of antibiotic-associated diarrhea: a systematic review and meta-analysis. JAMA. 2012;307(18):1959–69.
31. Griffith DE, Brown-Elliott BA, Langsjoen B, et al. Clinical and molecular analysis of macrolide resistance in Mycobacterium avium complex lung disease. Am J Respir Crit Care Med. 2006;174(8):928–34.
32. van Ingen J, Boeree MJ, van Soolingen D, Mouton JW. Resistance mechanisms and drug susceptibility testing of nontuberculous mycobacteria. Drug Resist Updat. 2012;15(3):149–61.
33. Griffith DE, Adjemian J, Brown-Elliott BA, et al. Semiquantitative culture analysis during therapy for Mycobacterium avium complex lung disease. Am J Respir Crit Care Med. 2015;192(6):754–60.
34. Adjemian J, Prevots DR, Gallagher J, Heap K, Gupta R, Griffith D. Lack of adherence to evidence-based treatment guidelines for nontuberculous mycobacterial lung disease. Ann Am Thorac Soc. 2014;11(1):9–16.
35. van Ingen J, Wagner D, Gallagher J, et al. Poor adherence to management guidelines in nontuberculous mycobacterial pulmonary diseases. Eur Respir J 2017;49(2).
36. Wallace RJ Jr, Brown-Elliott BA, McNulty S, et al. Macrolide/Azalide therapy for nodular/Bronchiectatic: Mycobacterium avium complex lung disease. Chest. 2014;146:276.
37. Jeong BH, Jeon K, Park HY, et al. Intermittent antibiotic therapy for nodular bronchiectatic Mycobacterium avium complex lung disease. Am J Respir Crit Care Med. 2015;191(1):96–103.
38. Nahid P, Dorman SE, Alipanah N, et al. Official American Thoracic Society/Centers for Disease Control and Prevention/Infectious Diseases Society of America clinical practice guidelines: treatment of drug-susceptible tuberculosis. Clin Infect Dis. 2016;63(7):e147-e195.

39. Matz J, Borish LC, Routes JM, Rosenwasser LJ. Oral desensitization to rifampin and ethambutol in mycobacterial disease. Am J Respir Crit Care Med. 1994;149(3 Pt 1):815–7.
40. Bermudez LE, Nash KA, Petrofsky M, Young LS, Inderlied CB. Effect of ethambutol on emergence of clarithromycin-resistant Mycobacterium avium complex in the beige mouse model. J Infect Dis. 1996;174(6):1218–22.
41. Olivier KN, Shaw PA, Glaser TS, et al. Inhaled amikacin for treatment of refractory pulmonary nontuberculous mycobacterial disease. Ann Am Thorac Soc. 2014;11(1):30–5.
42. Yagi K, Ishii M, Namkoong H, et al. The efficacy, safety, and feasibility of inhaled amikacin for the treatment of difficult-to-treat non-tuberculous mycobacterial lung diseases. BMC Infect Dis. 2017;17(1):558.
43. Davis KK, Kao PN, Jacobs SS, Ruoss SJ. Aerosolized amikacin for treatment of pulmonary Mycobacterium avium infections: an observational case series. BMC Pulm Med. 2007;7:2.
44. Safdar A. Aerosolized amikacin in patients with difficult-to-treat pulmonary nontuberculous mycobacteriosis. Eur J Clin Microbiol Infect Dis. 2012;31(8):1883–7.
45. Olivier KN, Griffith DE, Eagle G, et al. Randomized trial of liposomal Amikacin for inhalation in Nontuberculous mycobacterial lung disease. Am J Respir Crit Care Med. 2017;195(6):814–23.
46. Griffith DE, Eagle G, Thomson R, Aksamit TR, Hasegawa N, Morimoto K, Addrizzo-Harris DJ, O'Donnell AE, Marras TK, Flume PA, Loebinger MR, Morgan L, Codecasa LR, Hill AT, Ruoss SJ, Yim JJ, Ringshausen FC, Field SK, Philley JV, Wallace RJ Jr, van Ingen J, Coulter C, Nezamis J, Winthrop KL; CONVERT Study Group. Amikacin Liposome Inhalation Suspension for Treatment-Refractory Lung Disease Caused by Mycobacterium avium Complex (CONVERT): a Prospective, Open-Label, Randomized Study. Am J Respir Crit Care Med. 2018.
47. Henkle E, Aksamit T, Barker A, et al. Patient-centered research priorities for pulmonary Nontuberculous mycobacteria (NTM) infection. An NTM research consortium workshop report. Ann Am Thorac Soc. 2016;13(9):S379–84.
48. Blumberg HM, Burman WJ, Chaisson RE, et al. American Thoracic Society/Centers for Disease Control and Prevention/Infectious Diseases Society of America: treatment of tuberculosis. Am J Respir Crit Care Med. 2003;167(4):603–62.
49. Griffith DE, Philley JV, Brown-Elliott BA, et al. The Significance of Mycobacterium abscessus Subspecies abscessus Isolation During Mycobacterium avium Complex Lung Disease Therapy. Chest. 2015;147(5):1369–75.
50. Kunst H, Wickremasinghe M, Wells A, Wilson R. Nontuberculous mycobacterial disease and Aspergillus-related lung disease in bronchiectasis. Eur Respir J. 2006;28(2):352–7.
51. Zoumot Z, Boutou AK, Gill SS, et al. Mycobacterium avium complex infection in non-cystic fibrosis bronchiectasis. Respirology. 2014;19(5):714–22.
52. Pinel C, Fricker-Hidalgo H, Lebeau B, et al. Detection of circulating Aspergillus fumigatus galactomannan: value and limits of the Platelia test for diagnosing invasive aspergillosis. J Clin Microbiol. 2003;41(5):2184–6.
53. Woodworth MH, Saullo JL, Lantos PM, Cox GM, Stout JE. Increasing Nocardia incidence associated with bronchiectasis at a tertiary care center. Ann Am Thorac Soc. 2017;14(3):347–54.
54. Miller WT Jr, Panosian JS. Causes and imaging patterns of tree-in-bud opacities. Chest. 2013;144(6):1883–92.
55. Falkinham JO 3rd. Reducing human exposure to Mycobacterium avium. Ann Am Thorac Soc. 2013;10(4):378–82.
56. Kim RD, Greenberg DE, Ehrmantraut ME, et al. Pulmonary nontuberculous mycobacterial disease: prospective study of a distinct preexisting syndrome. *American journal of respiratory and critical care medicine*. 2008;178(10):1066–74.
57. Chan ED, Iseman MD. Underlying host risk factors for nontuberculous mycobacterial lung disease. Semin Respir Crit Care Med. 2013;34(1):110–23.
58. Wolinsky E. Nontuberculous mycobacteria and associated diseases. Am Rev Respir Dis. 1979;119:107–59.

Mycobacterium avium Complex Disease

Michael R. Holt and Charles L. Daley

Mycobacterium avium Complex

Mycobacterium avium complex (MAC) conventionally consisted of two named species, *M. avium* and *M. intracellulare*, which are responsible for most human MAC infections. Since 2004, the number of named MAC species has grown to at least ten by virtue of advances in molecular taxonomy. *M. chimaera*, which was previously indistinguishable from *M. intracellulare*, stands apart from the newly named species as a significant human pathogen [1]. *M. chimaera* has been responsible for a global outbreak of extrapulmonary infection complicating the use of certain heater-cooler units during open cardiac surgery [2]. The other MAC species were named after being isolated from human pulmonary or extrapulmonary clinical isolates: *M. colombiense*, *M. arosiense*, *M. marseillense*, *M. timonense*, *M. bouchedurhonense*, *M.vulneris*, and *M. yongonense* [3–7].

M. avium comprises four subspecies [8]. *M. avium* subsp. *avium* and *M. avium* subsp. *silvaticum* are responsible for tuberculosis-like illness in birds, the latter almost exclusively affecting wood pigeons. *M. avium* subsp. *paratuberculosis* usually causes a chronic, progressive, granulomatous enteritis in ruminants. *M. avium* subsp. *hominissuis*, which causes porcine and human infection, is the predominant cause of human *M. avium* disease [9].

The importance of accurate speciation of MAC is illustrated by species-related variation of propensity to cause lung disease, severity of disease, and prognosis [10, 11].

M. R. Holt · C. L. Daley (✉)
Division of Mycobacterial and Respiratory Infections, National Jewish Health, Denver, CO, USA
e-mail: holtm@njhealth.org; daleyc@njhealth.org

D. E. Griffith (ed.), *Nontuberculous Mycobacterial Disease*, Respiratory Medicine,
https://doi.org/10.1007/978-3-319-93473-0_11

Epidemiology and Ecology of MAC

MAC is the principal cause of nontuberculous mycobacterial (NTM) pulmonary disease in most parts of the world [12]. The considerable geographic diversity of MAC was demonstrated by a global study of pulmonary isolates provided by partners in the Nontuberculous Mycobacteria Network European Trials Group [13]. The MAC species were the most commonly isolated NTM. When considered by continent, MAC constituted the highest proportion of NTM in Australia (71.1%) and the lowest proportion in South America (31.3%). The predominant species was *M. avium* in Europe, Asia, South America, and North America. *M. intracellulare* was strikingly predominant in South Africa and the state of Queensland in Australia. Isolation of other MAC species was greatest in Asia.

The geographic diversity of MAC species suggests distinct environmental reservoirs. Unlike *M. abscessus*, which has been associated with possible human-to-human transmission in patients with cystic fibrosis, MAC infection is acquired from environmental sources [14, 15]. Putative routes of infection include ingestion of contaminated water and inhalation of MAC-containing aerosols. *M. avium*, which demonstrates resistance to disinfectants, proliferates within water distribution systems and is enriched to high levels in showerhead biofilms [16–18]. Various molecular analyses have matched clinical *M. avium* isolates to isolates obtained from potting soil aerosols, bathroom plumbing fixtures, and household water [19–22]. In contrast, *M. intracellulare* is not found in water. Although household water isolates from the US mainland were previously identified as *M. intracellulare*, molecular analysis has revealed these to be *M. chimaera* [23]. The environmental reservoir of *M. intracellulare* is likely to be soil. *M. intracellulare* has been isolated from potting soils, and the risk of *M. intracellulare* infection has been linked to topsoil depth, although these studies did not differentiate *M. intracellulare* from *M. chimaera* [19, 24]. *M. intracellulare* was isolated from soil and distinguished from *M. chimaera* in a Hawaiian study [25]. This study also revealed that *M. chimaera* is the predominant species isolated from patient sputum and random water and soil samples in Hawaii. The pathogenic potential of waterborne *M. chimaera* was made evident by reports of infection complicating open cardiac surgery in 2015 [26, 27]. Subsequent investigation linked infection to aerosols produced by a certain brand of heater-cooler unit [2]. Whole-genome sequencing and phylogenetic analysis demonstrated international relatedness of isolates, suggesting a point source of contamination at the manufacturing facility [28].

Virulence of MAC

MAC virulence demonstrates inter- and intraspecies heterogeneity. *M. avium* invades human intestinal cells more efficiently than *M. intracellulare* and is more likely to be associated with pulmonary disease when isolated from sputum [29, 30]. *M. intracellulare* has been associated with a more severe disease phenotype and worse prognosis [10]. Genetic analyses of *M. avium* have identified strains of high virulence and clustering of strains according to propensity to cause progressive pulmonary disease [31–33].

Membrane glycopeptidolipids (GPLs) are significant determinants of MAC virulence. The historical classification of MAC by serotype was based on the antigenicity of polar GPLs, which varies between species and strains. Apolar GPLs do not exhibit this variation. Polar GPLs specific to the pathogenic serotype 4, now known to be *M. avium*, enhance monocyte phagocytosis but impair phagosome-lysosome fusion, establishing an intracellular niche for growth [34, 35]. Apolar GPLs may contribute to virulence by impairing phagosome acidification and delaying phagosome-lysosome fusion [36]. GPLs have also been linked to pathogenicity in studies of MAC colony morphology. Rough colonies, which are less pathogenic than smooth colonies, lack genes for GPL synthesis [37, 38]. Furthermore, GPL biosynthesis appears to be required for the production of *M. avium* biofilms [39]. Other potential *M. avium* virulence factors include production of hemolysin, enhancement of HIV transcription, and the ability to diminish macrophage responsiveness to interferon gamma (IFN-γ) [40–42].

The ubiquity of MAC is at odds with the rarity of MAC disease, suggesting a role for host susceptibility. Conditions known to confer susceptibility to MAC pulmonary disease include immune dysfunction; pre-existing structural lung disease; and genetic conditions that predispose to pulmonary infection, such as cystic fibrosis and primary ciliary dyskinesia [43]. Other risk factors include gastroesophageal reflux disease, prolonged use of inhaled corticosteroids, and immunosuppressant therapy [44–47]. Disseminated MAC disease usually complicates conditions that cause failure of acquired immunity, such as HIV infection and inherited defects of IFN-γ signaling [48].

MAC Pulmonary Disease

Diagnosis and Spectrum of Disease

The diagnosis of MAC pulmonary disease requires the synthesis of clinical, microbiologic, and radiographic findings defined by the American Thoracic Society (ATS) and Infectious Diseases Society of America (IDSA) (Table 1) [49]. Clinical features of disease include cough, sputum production, hemoptysis, fevers, night sweats, fatigue, and weight loss [49]. These symptoms often evolve in an insidious and non-specific fashion, delaying the diagnosis by months or years.

The microbiologic criteria may possess the greatest discriminatory value. During follow-up of Japanese patients for at least 12 months, the finding of a new cavity or infiltrative lesion on chest X-ray occurred in 2% of patients with a single isolation of MAC and 98% of patients with two or more isolations [50]. These patients were characterized by high morbidity and required hospitalization for at least 6 months. As a consequence, the results suggest a rate of radiographic progression in excess of

Table 1 Diagnostic criteria for NTM lung disease

Clinical criteria	Radiographic criteria	Microbiologic criteria
Symptoms consistent with disease (e.g., cough, dyspnea, hemoptysis, fevers, night sweats, weight loss, and/or fatigue) and appropriate exclusion of other diagnoses	Chest radiograph demonstrating cavities or nodular opacities; or high-resolution CT scan demonstrating multifocal bronchiectasis with multiple small nodules	Culture of NTM from sputum expectorated on at least two separated occasions, a single bronchoscopic wash or lavage, or any single specimen in combination with the lung biopsy finding of granulomatous inflammation

Source: Griffith et al. [49]

that expected for patients with nodular bronchiectatic disease [51]. The poor predictive value of a single positive culture is due to the high likelihood of specimen contamination with environmental MAC. In order to maximize diagnostic confidence, sampling of sputum expectorated on three separate mornings is preferred [49]. Sputum induction with hypertonic saline may be used to noninvasively obtain samples from patients who are unable to expectorate spontaneously. Bronchoscopic bronchoalveolar lavage is more likely than sputum sampling to obtain a smear-positive specimen, but its yield of culture-positive samples is not consistently superior [52, 53]. Testing of isolates for clarithromycin susceptibility is recommended for disease-causing isolates that are either treatment-naïve or refractory to macrolide-based therapy [49].

Two radiographic patterns of MAC pulmonary disease predominate [54]. The fibrocavitary pattern usually manifests as upper lobe cavities in men with smoking-related lung disease (Fig. 1). Bronchiectasis with nodules, including tree-in-bud changes, is the typical pattern in patients without pre-existing structural lung disease (Fig. 2). Nodular bronchiectasis that principally affects the right middle lobe and/or lingula tends to afflict otherwise healthy postmenopausal women [55]. This has become the most prominent pattern of disease in a number of countries and has been associated with tall stature, slender body habitus, and increased frequency of connective tissue disease features, such as scoliosis, pectus excavatum, and mitral valve prolapse [56–59]. Although the basis for disease susceptibility in these patients is yet to be defined, recent evidence supports multigenic etiology [60]. Polymorphisms in the cystic fibrosis transmembrane regulator (CFTR) gene and genes governing connective tissue, ciliary and immune functions have been implicated.

A serodiagnostic biomarker is required to simplify the diagnosis and monitoring of MAC pulmonary disease. Kitada and colleagues evaluated an enzyme immunoassay that measures immunoglobulin A antibodies directed against the lipopeptide core of MAC GPL [61]. The performance of this assay was assessed in Asian individuals with MAC pulmonary disease, MAC contamination, pulmonary tuberculosis, other lung diseases, and no disease [62]. At the optimum cutoff point of 0.7 U/mL, sensitivity and specificity for MAC pulmonary disease were 84.3% and 100%, respectively. The levels of antibody correlated with the radiographic burden of disease and were significantly higher in patients whose pattern of disease was nodular bronchiectatic rather than fibrocavitary. The cutoff point and utility of the assay

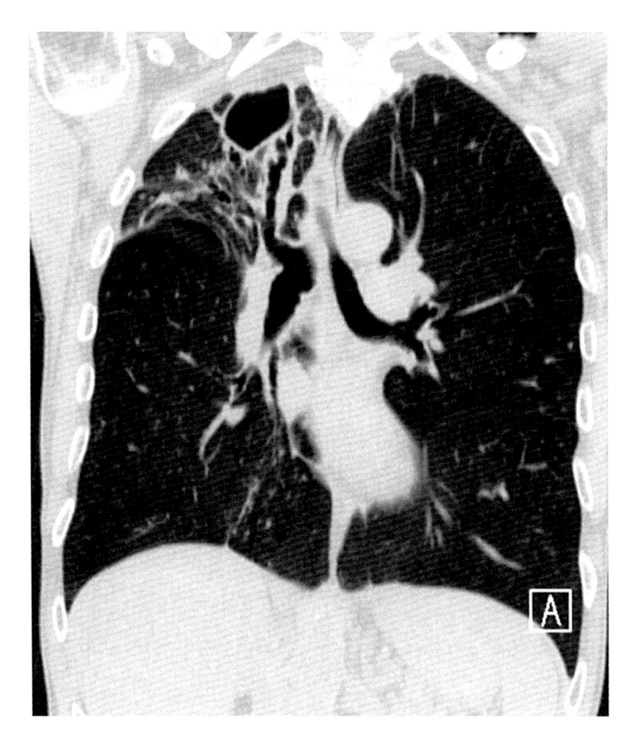

Fig. 1 Chest computed tomography image of a 65-year-old man with *M. intracellulare* and long smoking history. Note the emphysema and right upper lobe fibrocavitary disease with severe volume loss in the right upper lobe

differed when evaluated in a predominantly Caucasian US cohort, which included patients of greater clinical stability. At a cutoff point of 0.3 U/mL, sensitivity and specificity were 70.1% and 93.9%, respectively [63]. A positive sputum culture within 6 months before testing was associated with improved sensitivity. Although assays such as this hold promise, they are not yet generally available and require further development before being implemented in clinical practice.

MAC pulmonary disease is distinguished from "hot tub lung," a condition similar to hypersensitivity pneumonitis. Cases of "hot tub lung" have been linked to inhalation of MAC-containing aerosols generated by hot tubs, showers, and water-based metalworking [64–66]. Common clinical features are dyspnea, cough, and fever. Computed tomography (CT) scan findings include ground-glass pulmonary infiltrates, centrilobular nodules, and gas trapping (Fig. 3) [67]. Pulmonary specimens are culture-positive for MAC and exhibit granulomatous inflammation on histological analysis. The principal management of "hot tub lung" is avoidance of causative exposures. Optimal medical therapy has not been established. Favorable outcomes have been reported without medical therapy and with separate or combined administration of corticosteroid and anti-mycobacterial therapies [65, 68]. Prednisone, weaned from 1 to 2 mg/kg daily over 4–8 weeks, is recommended for

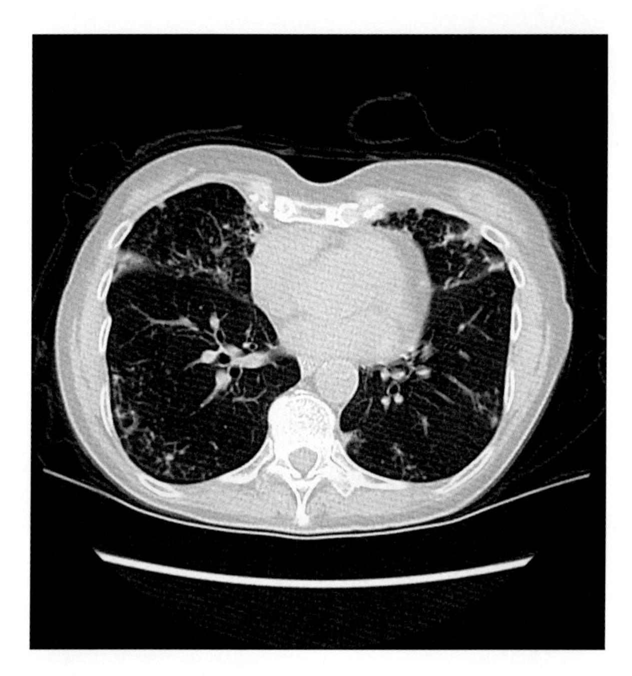

Fig. 2 Chest computed tomography image of a 68-year-old woman with *M. intracellulare* lung disease. Note the mild pectus excavatum, right middle lobe and lingula predominant bronchiectasis, and scattered tree-in-bud opacities in the lower lobes

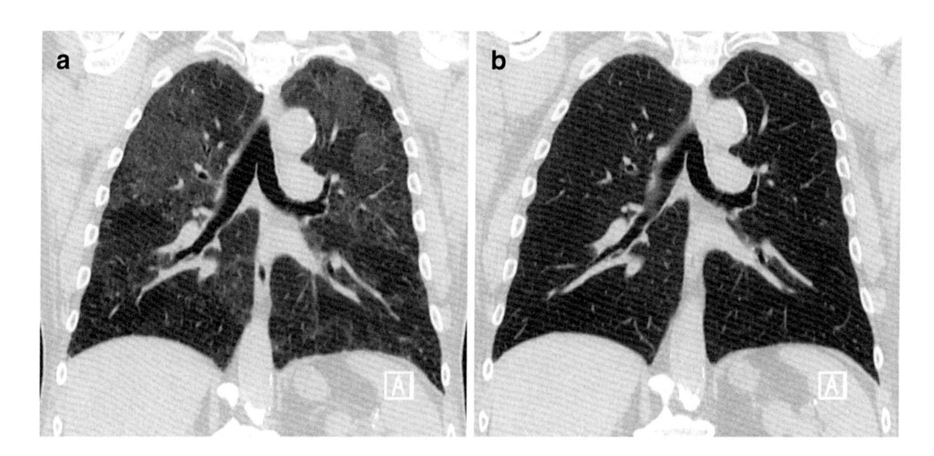

Fig. 3 (**a**) Chest computed tomography (CT) image of an 80-year-old patient who developed hypersensitivity pneumonitis due to *M. avium* after exposure to his indoor hot tub (**b**) Follow-up CT image after 4 months of corticosteroid therapy and drainage of the hot tub

patients with severe disease or respiratory failure [49]. Anti-mycobacterial therapy, administered for 3–6 months according to clinical response, is recommended for patients with immune incompetence, bronchiectasis, or persistent disease despite antigen avoidance irrespective of corticosteroid use [49].

Standard Three-Drug Therapy

The treatment of MAC pulmonary disease involves prolonged administration of multiple antibiotics with significant adverse effect profiles. Thus, the decision to treat should be informed by the patient's symptoms, radiographic progression, and comorbidities. The cornerstone of treatment is a macrolide-containing regimen, administered for a duration sufficient to include 12 months of negative cultures (Table 2) [49]. Daily administration is recommended for patients with fibrocavitary disease or severe nodular bronchiectatic disease.

The efficacy of clarithromycin was demonstrated by a preliminary, open-label, non-comparative trial designed to rapidly assess short-term outcomes of clarithromycin monotherapy [69]. After 4 months of therapy, sputum cultures converted to negative in 58% patients with clarithromycin-susceptible isolates and demonstrated significant microbiologic improvement in a further 21%. Subsequent prospective and retrospective studies supported the superiority of regimens that contained clarithromycin to those that did not [70–72]. Poor microbiologic responses were associated with clarithromycin-resistant isolates and prior anti-MAC treatment [72]. The daily dose of clarithromycin is 500–1000 mg which may be administered in divided doses to improve tolerability. Daily doses greater than 1000 mg are poorly tolerated in elderly patients due to adverse effects, including gastrointestinal symptoms, dysgeusia, central nervous system symptoms, and elevation of serum liver enzymes [73]. Azithromycin-based regimens produce similar microbiologic outcomes to clarithromycin-based regimens [74]. The recommended daily dose of azithromycin is 250–300 mg. A dose of 600 mg per day was efficacious in a study of short-term responses to azithromycin monotherapy, but gastrointestinal adverse effects and

Table 2 Medical treatment regimens for macrolide-susceptible MAC pulmonary disease

Disease category	Macrolide (PO)	Ethambutol (PO)	Rifamycin (PO)	Aminoglycoside (IV)
Treatment-naïve nodular bronchiectatic disease	Clarithromycin 1000 mg or azithromycin 500–600 mg thrice weekly	25 mg/kg thrice weekly	Rifampin 600 mg thrice weekly	None
Treatment-naïve fibrocavitary disease	Clarithromycin 500–1000 mg[a] or azithromycin 250–300 mg daily	15 mg/kg daily	Rifampin 450–600 mg[a] daily	Streptomycin or amikacin thrice weekly for the first 2–3 months
Severe, extensive (multilobar), or previously treated disease	Clarithromycin 500–1000 mg[a] or azithromycin 250–300 mg daily	15 mg/kg daily	Rifampin 450–600 mg[a] or rifabutin 150–300 mg[a, b] daily	Streptomycin or amikacin thrice weekly for the first 2–3 months

Abbreviations: *PO* administered by mouth, *IV* administered intravenously
Source: Griffith et al. [49]
[a]Use lower dose for patients with weight < 50 kg
[b]Use lower dose when co-administered with clarithromycin

ototoxicity were frequent [75]. Unlike clarithromycin, azithromycin is not metabolized by hepatic cytochrome p450 (CYP450) enzymes and does not inhibit CYP3A4 [76]. As a consequence, it has lower potential for interaction with other drugs.

Companion drugs are necessary to protect against macrolide resistance, which complicates clarithromycin monotherapy in 16% of patients with MAC pulmonary disease treated for 4 months and 46% of HIV-infected patients with disseminated MAC after a median of 16 weeks [69, 77]. Ethambutol, 15 mg/kg daily, is the recommended companion agent [49]. The principal adverse effect of ethambutol is optic neuropathy, manifesting as reduced visual acuity or impaired red/green color discrimination. The inclusion of a rifamycin is predicated on the historical success of three-drug regimens in the treatment of pulmonary disease and evidence that rifabutin may reduce the risk of macrolide resistance in HIV-infected patients with disseminated MAC [74, 78]. Nonetheless, the superiority of three-drug regimens in the treatment of pulmonary disease has not been established. The combination of a macrolide and ethambutol may be appropriate for certain patients, including those intolerant of a three-drug regimen [49]. A preliminary open-label randomized controlled trial (RCT), involving HIV-negative patients with treatment-naïve pulmonary disease, suggested that omission of rifampin may produce non-inferior microbiologic outcomes and superior treatment tolerability [79].

Rifabutin is poorly tolerated in elderly patients without HIV infection [80]. Adverse effects include gastrointestinal symptoms, arthralgia, leukopenia, and uveitis. Rifampin, which is regarded as having similar efficacy to rifabutin, is therefore preferred for the treatment of MAC pulmonary disease [49]. A drawback of rifampin is its considerable induction of CYP450 enzyme systems and consequent multiple-drug interactions. Rifampin reduces serum levels of macrolides, especially clarithromycin, but the clinical significance of this remains unclear [81]. The daily dose of rifampin is 10 mg/kg, usually 450 mg or 600 mg. Rifabutin affects serum macrolide levels to a lesser extent than rifampin but is a CYP3A4 substrate. As a consequence, co-administration of clarithromycin warrants reduction of the daily dose of rifabutin from 150–300 mg to 150 mg and consideration of serum drug concentration monitoring.

The Role of Intermittent Administration

A number of prospective and retrospective studies, predominantly involving patients with nodular bronchiectasis, support the notion that intermittent therapy is of similar efficacy and superior tolerability to daily therapy [80, 82, 83]. Intermittent administration permits dosing of azithromycin at 500–600 mg per day, which produces a higher peak plasma concentration (Cmax) than the daily dose. Although the consequences of this higher Cmax are unclear, there is the possibility of improved microbiologic outcomes [84]. A prospective, non-comparative trial suggested that the efficacy of intermittent therapy is reduced in association with cavitary disease, previous treatment for MAC pulmonary disease, chronic obstructive pulmonary disease (COPD), or bronchiectasis [85]. Thrice-weekly therapy is recommended for patients with treatment-naïve nodular bronchiectatic disease and those who are unable to tolerate daily therapy [49]. In the event that sputum cultures remain positive despite 12 months of intermittent therapy, switching to daily therapy may achieve treatment success in 30% of cases [86].

Use of an Aminoglycoside

Many of the studies that established the efficacy of three-drug therapy for MAC pulmonary disease included an initial period of intravenous aminoglycoside treatment. A double-blind RCT investigated the addition of intramuscular streptomycin to the first 3 months of a 24-month three-drug regimen [87]. The frequency of sputum culture conversion was significantly improved, but mortality, microbiologic relapse, clinical trajectory, and radiographic findings were not significantly affected. Adverse effects of aminoglycosides include ototoxicity, vestibulotoxicity, and nephrotoxicity. Ototoxicity has been reported in over a third of patients and is frequently irreversible [88]. In order to mitigate practical encumbrances and toxicity risk, aminoglycoside therapy is reserved for patients with severe, multilobar or fibrocavitary disease. A typical dosing regimen for MAC pulmonary disease is intravenous amikacin, administered thrice weekly for the initial 2–3 months of therapy [49]. The usual starting dose is 15–25 mg/kg, and subsequent doses are adjusted according to serum drug concentration monitoring. Surveillance for toxicity should include serial audiograms and serum creatinine measurements.

Alternative or Adjunctive Antibiotics

Clofazimine may be a suitable alternative to the rifamycins for treatment of MAC pulmonary disease. A retrospective study of patients with predominantly nodular bronchiectatic NTM disease suggested that clofazimine- and rifampin-containing three-drug regimens produce similar microbiologic outcomes and relapse rates [89]. Several retrospective studies have supported the long-term safety and tolerability of clofazimine in patients with NTM pulmonary disease [89–91]. Usual dosing is 100 mg daily. Significant adverse effects include skin pigmentation, gastrointestinal complaints, and prolongation of the electrocardiographic QTc interval.

Moxifloxacin has a limited role in the management of MAC lung disease. Due to the association of macrolide resistance with the use of a fluoroquinolone in lieu of ethambutol, moxifloxacin is not an appropriate choice in the setting of ethambutol intolerance [92]. Expert opinion favors parenteral or inhaled amikacin in this scenario. Adjunctive moxifloxacin, 400 mg daily, may improve outcomes for patients with pulmonary disease refractory to 6 months of macrolide-based multiple-drug therapy. A retrospective study demonstrated treatment success in approximately one third of cases of macrolide-susceptible MAC, but there was no benefit in cases of macrolide-resistant MAC [93]. Serious adverse effects include tendinitis, tendon rupture, and prolongation of the QTc interval.

Bedaquiline, a diarylquinoline antibiotic developed for the treatment of drug-resistant tuberculosis, has excellent in vitro activity against MAC [94]. Bedaquiline causes QTc prolongation and was associated with increased mortality of uncertain etiology in a phase 2b clinical trial [95, 96]. A small case series has provided preliminary data that bedaquiline may be an efficacious salvage therapy for treatment-refractory MAC pulmonary disease [96]. Rifamycins significantly increase the clearance of bedaquiline and concomitant administration of these agents should be avoided [97].

The utility of oxazolidinone antibiotics for MAC pulmonary disease is uncertain. The minimum inhibitory concentration of linezolid is high, and withdrawal from therapy is common due to adverse effects such as bone marrow suppression and peripheral neuropathy [98, 99]. Tedizolid may have superior efficacy and tolerability by virtue of its favorable pharmacokinetic profile and lower MIC [100].

Inhaled amikacin holds promise for patients with treatment-refractory MAC disease. Nebulization of the parenteral formulation is most commonly tolerated at a maximum dose of 250 mg daily. Although this formulation is associated with microbiologic and/or symptomatic improvement in some patients, systemic and local toxicities are common [101]. Amikacin liposomal inhalation suspension (ALIS), a formulation specifically intended for use in NTM pulmonary disease, is designed for uptake by macrophages and retention in the lung. The efficacy and safety of ALIS, 590 mg daily, were recently appraised in a phase 2 study with the primary endpoint of change from baseline on a semiquantitative mycobacterial growth scale [102]. Although the primary endpoint was not met, the addition of ALIS to standard therapy was associated with increased frequency of enduring culture conversion and significant improvement in 6-minute walk test results. ALIS was well-tolerated, the principal adverse effects being respiratory symptoms and infectious complications. During the double-blind phase, 2 of 44 patients who received ALIS developed allergic alveolitis. A randomized, controlled, phase 3 study confirmed the safety and efficacy reported in the earlier trial [102a]. Patients with treatment refractory MAC pulmonary disease who received standard therapy plus LAI were more likely to be culture negative at six months than controls who received standard therapy alone. Respiratory treatment emergent adverse events occurred in 87% of the patients on ALIS compared with 50% on standard treatment. The most common adverse events were dysponia and cough which were generally transient. Serious treatment emergent adverse events were similar in both groups.

Surgery

The indications for surgical treatment of NTM lung disease are focal disease refractory to medical therapy, intractable bothersome or life-threatening symptoms, and parenchymal disease amenable to "debulking" in order to slow progression [103].

Preoperative medical therapy is administered for 8–12 weeks. A number of studies have reported excellent microbiologic outcomes after surgery although morbidity and mortality vary according to the institution and surgical technique [104–106]. Open procedures are associated with higher rates of mortality and major morbidity than thoracoscopic surgery. Ideally, operators should be experienced in the surgical management of mycobacterial lung disease. A review of thoracoscopic lobectomies, performed at an institution with a high degree of expertise, revealed low likelihood of conversion to an open procedure (3%), low morbidity (7%), no mortality, and a short length of stay (mean 3.3 days) [107].

Macrolide Resistance

The likelihood of emergent macrolide resistance despite treatment with recommended three-drug therapy is less than 5% [92]. Macrolide resistance is usually conferred by a point mutation at position 2058 or 2059 in the 23S rRNA gene. Prescribing patterns that predispose to macrolide resistance have differed between retrospective studies but tend to include macrolide monotherapy, macrolide and fluoroquinolone two-drug therapy, and omission of ethambutol [92, 108, 109]. Treatment outcomes are poor but may be significantly improved with the combination of aminoglycoside therapy for at least 6 months and surgical resection [108]. This approach may increase the proportion of patients achieving culture conversion from 5% to 79% [92]. The optimal drug regimen to support aminoglycoside therapy and surgery has not been determined. Although experimental data suggest that clarithromycin may be efficacious despite in vitro resistance, use of a macrolide has not been associated with improved clinical outcomes and is not recommended [49, 108, 110]. Ideally, a multiple-drug regimen should be formulated with reference to drug susceptibility testing and expert consultation (Fig. 4). Candidate agents include ethambutol, the rifamycins, and clofazimine. Bedaquiline is likely to be an efficacious substitute for rifamycins but may be difficult to obtain [96].

Outcomes of MAC Pulmonary Disease

Untreated patients with predominantly nodular bronchiectatic MAC disease in a Korean retrospective study exhibited radiographic progression or requirement for medical therapy in 48% of cases during a mean observation period of 32 months [111]. Predictors of progression or need for therapy were pulmonary cavities and consolidation. Treatment with a recommended macrolide-containing regimen may achieve culture conversion in close to 90% of cases, usually within 6 months. Unfortunately, microbiologic recurrences occur in 48% of cases after completion of therapy. The majority of recurrences are due to a genotypically distinct strain, consistent with reinfection [74]. Clarithromycin susceptibility testing is predictive of

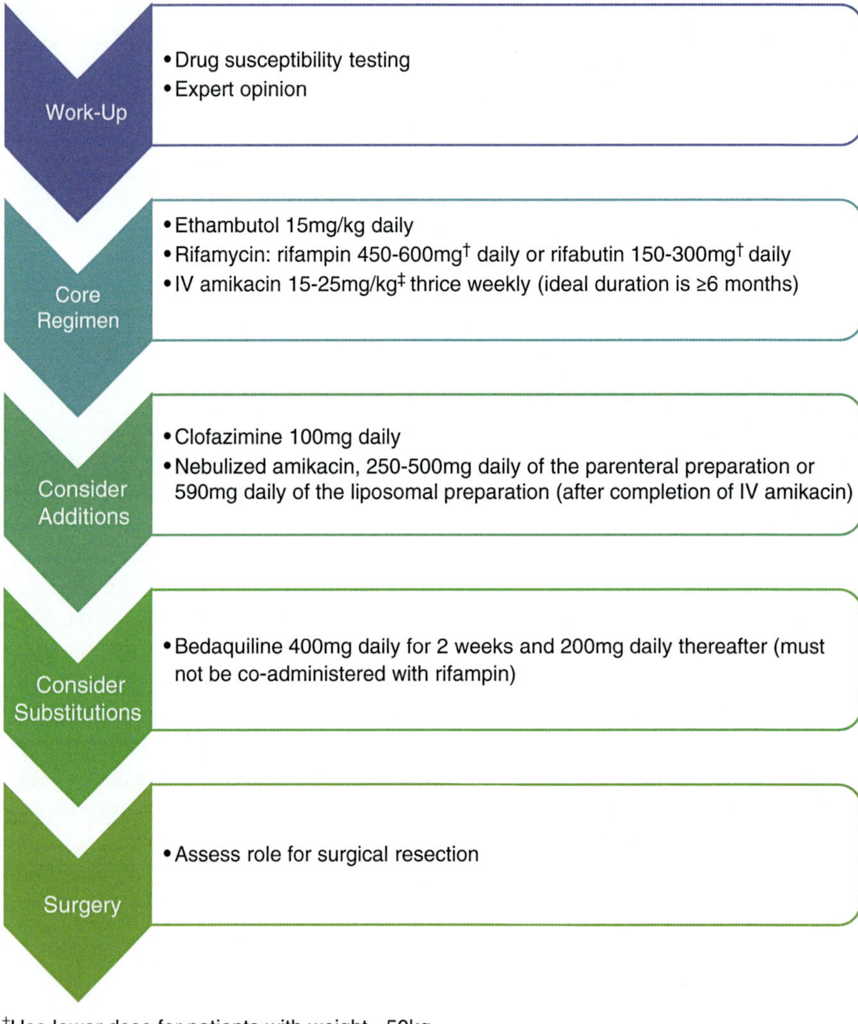

Fig. 4 Suggested algorithm for management of macrolide-resistant MAC pulmonary disease. (†Use lower dose for patients with weight < 50 kg; ‡Adjust dose according to serum amikacin levels)

treatment success. A prospective study conducted in Japan demonstrated culture conversion in 83.9% of patients with clarithromycin-susceptible MAC and in only 25% of patients with isolates exhibiting intermediate susceptibility or resistance [72]. Prior therapy and smear positivity were also associated with reduced frequency of culture conversion.

The prognostic value of the ATS/IDSA definition of disease is questionable but may be greater for MAC than other species of NTM [112, 113]. Five-year all-cause mortality varies from 13% to 40% according to the population being studied [114–116]. Lower mortality rates tend to be associated with female sex, non-cavitary disease, and paucity of comorbidities. The mortality rate of macrolide-resistant MAC pulmonary disease has been likened to that of multiple-drug-resistant tuberculosis and reported to be 47% at 5 years [108, 109].

Extrapulmonary MAC Disease

Disseminated MAC and HIV-AIDS

The HIV-AIDS epidemic ushered in new awareness of the pathogenicity of MAC, particularly *M. avium*. Prior to this time, the MAC entities were considered to be rare pulmonary pathogens of little clinical significance. Reports of disseminated *M. avium* disease increased dramatically after the condition was recognized in 1982 [117]. The diagnosis was usually made after culture of blood or tissue, especially bone marrow, lymph node, or liver. The CD4 count of afflicted patients averaged less than 60cells/mm^3 and was rarely greater than 100cells/mm^3. Median survival was 4 months and autopsies demonstrated wide-ranging organ involvement [118, 119].

Recommended treatment for disseminated MAC in HIV-infected patients is daily administration of a macrolide and ethambutol with consideration of a rifamycin. Although an open-label RCT reported superior microbiologic outcomes with clarithromycin, a double-blind RCT found clarithromycin and azithromycin to have similar efficacy in terms of microbiologic outcomes, mortality, and relapse rates [120, 121]. Doses of clarithromycin exceeding 1000 mg per day are avoided as these have been associated with increased mortality [122]. Ethambutol significantly reduces the risks of relapse and development of clarithromycin resistance [123]. The inclusion of a rifamycin, usually rifabutin, has been associated with improved survival and reduced risk of clarithromycin resistance in RCTs [78, 124]. Rifabutin dosing should be adjusted according to the choice of macrolide in order to manage the risk of adverse effects. Daily doses of rifabutin are 300–450 mg in combination with azithromycin and 300 mg in combination with clarithromycin. Dose adjustment is also necessary to manage interactions with protease inhibitors and non-nucleoside reverse transcriptase inhibitors. Use of clofazimine has not been recommended due to its association with increased mortality in a RCT [49, 125, 126]. However, a subsequent RCT did not associate the substitution of clofazimine for rifabutin with significantly altered mortality or bacteriologic outcomes [122]. Moreover, clofazimine has been used widely to treat HIV-infected patients, including those with drug-resistant tuberculosis.

The incidence of disseminated *M. avium* declined dramatically after the development of highly active antiretroviral therapy (HAART) [127]. HAART also allowed for unprecedented enduring cure of afflicted individuals [128]. Anti-MAC treatment may be ceased for asymptomatic individuals who are receiving HAART and have achieved CD4 counts in excess of 100cells/mm^3 for at least 12 months [49].

Prophylaxis against disseminated *M. avium* is highly effective and recommended for patients with CD4 counts less than 50cells/mm^3 [125]. Azithromycin, 1200 mg once weekly, is the preferred regimen. Clarithromycin, 500 mg twice daily, is efficacious but associated with increased likelihood of breakthrough with macrolide-resistant strains [129, 130]. Rifabutin is less efficacious than either macrolide [130, 131]. Superior efficacy is achieved by the combination of rifabutin and azithromycin, but this regimen is poorly tolerated. Prophylaxis may be withdrawn if CD4 counts exceed 100cells/mm^3 for at least 3 months [125, 132].

Other Extrapulmonary Diseases

Common extrapulmonary manifestations of MAC infection include cutaneous or musculoskeletal infections in adults and cervicofacial adenitis in children. Recommended treatment for adults is surgical excision or debridement in combination with 6–12 months of three-drug therapy [49]. Pediatric adenitis is managed with complete excision which produces superior outcomes to medical therapy and negligible morbidity [133]. Surgical excision without accompanying drug therapy is associated with a high rate of cure and low likelihood of recurrence [134, 135]. Medical therapy may be warranted as an adjunct to surgery for recurrent disease [49].

M. chimaera infections, arising from contamination of heater-cooler units, cause extrapulmonary disease that is challenging to treat and likely to become increasingly prevalent. Common symptoms include fever, shortness of breath, fatigue, and weight loss [26]. Disease manifestations include prosthetic valve endocarditis, prosthetic vein graft infection, myocarditis, bone infection, bacteremia, and dissemination to multiple organs. Symptoms and signs occur at a median of 18 months (range, 11–40 months) after cardiac surgery. Due to this lag time and the non-specific clinical presentation, diagnosis and initiation of treatment may be delayed by several years. Despite surgical intervention and targeted multiple-drug therapy, mortality rates of 50% have been reported [2, 26].

Summary

MAC comprises a number of species of environmental mycobacteria that exhibit geographical diversity. MAC disease is often a consequence of immune dysfunction or structural lung disease. The etiological basis for pulmonary disease in healthy people, especially postmenopausal women, remains unclear. Manifestations of MAC disease include conditions that are clinically and prognostically disparate. Cervicofacial lymphadenitis in immunocompetent children is readily cured with surgical excision. Destructive pulmonary disease in adults is associated with high rates of recurrence despite lengthy combination medical therapy and sometimes surgical lung resection. Disseminated disease in individuals with immune dysfunction is associated with high mortality rates. The treatment of macrolide-resistant disease is particularly challenging. Disseminated MAC disease in HIV-infected patients is rare due to the efficacy of HAART, but prophylaxis remains important in patients with low CD4 counts. Contamination of heater-cooler units by *M. chimaera* is an emerging cause of catastrophic extrapulmonary disease that complicates open cardiac surgery. Research priorities include development of a diagnostic biomarker and expansion of the therapeutic armamentarium.

References

1. Tortoli E, Rindi L, Garcia MJ, Chiaradonna P, Dei R, Garzelli C, et al. Proposal to elevate the genetic variant MAC-A, included in the Mycobacterium avium complex, to species rank as Mycobacterium chimaera sp. nov. Int J Syst Evol Microbiol. 2004;54(Pt 4):1277–85. https://doi.org/10.1099/ijs.0.02777-0.
2. Chand M, Lamagni T, Kranzer K, Hedge J, Moore G, Parks S, et al. Insidious risk of severe Mycobacterium chimaera infection in cardiac surgery patients. Clin Infect Dis. 2017;64(3):335–42. https://doi.org/10.1093/cid/ciw754.
3. Murcia MI, Tortoli E, Menendez MC, Palenque E, Garcia MJ. Mycobacterium colombiense sp. nov., a novel member of the Mycobacterium avium complex and description of MAC-X as a new ITS genetic variant. Int J Syst Evol Microbiol. 2006;56(Pt 9):2049–54. https://doi.org/10.1099/ijs.0.64190-0.
4. Bang D, Herlin T, Stegger M, Andersen AB, Torkko P, Tortoli E, et al. Mycobacterium arosiense sp. nov., a slowly growing, scotochromogenic species causing osteomyelitis in an immunocompromised child. Int J Syst Evol Microbiol. 2008;58(Pt 10):2398–402. https://doi.org/10.1099/ijs.0.65503-0.
5. Ben Salah I, Cayrou C, Raoult D, Drancourt M. Mycobacterium marseillense sp. nov., Mycobacterium timonense sp. nov. and Mycobacterium bouchedurhonense sp. nov., members of the Mycobacterium avium complex. Int J Syst Evol Microbiol. 2009;59(Pt 11):2803–8. https://doi.org/10.1099/ijs.0.010637-0.
6. van Ingen J, Boeree MJ, Kosters K, Wieland A, Tortoli E, Dekhuijzen PN, et al. Proposal to elevate Mycobacterium avium complex ITS sequevar MAC-Q to Mycobacterium vulneris sp. nov. Int J Syst Evol Microbiol. 2009;59(Pt 9):2277–82. https://doi.org/10.1099/ijs.0.008854-0.
7. Kim BJ, Math RK, Jeon CO, Yu HK, Park YG, Kook YH, et al. Mycobacterium yongonense sp. nov., a slow-growing non-chromogenic species closely related to Mycobacterium intracellulare. Int J Syst Evol Microbiol. 2013;63(Pt 1):192–9. https://doi.org/10.1099/ijs.0.037465-0.
8. Rindi L, Garzelli C. Genetic diversity and phylogeny of Mycobacterium avium. Infect Genet Evol. 2014;21:375–83. https://doi.org/10.1016/j.meegid.2013.12.007.
9. Tran QT, Han XY. Subspecies identification and significance of 257 clinical strains of Mycobacterium avium. J Clin Microbiol. 2014;52(4):1201–6. https://doi.org/10.1128/JCM.03399-13.
10. Koh WJ, Jeong BH, Jeon K, Lee NY, Lee KS, Woo SY, et al. Clinical significance of the differentiation between Mycobacterium avium and Mycobacterium intracellulare in M avium complex lung disease. Chest. 2012;142(6):1482–8. https://doi.org/10.1378/chest.12-0494.
11. Kim SY, Shin SH, Moon SM, Yang B, Kim H, Kwon OJ, et al. Distribution and clinical significance of Mycobacterium avium complex species isolated from respiratory

specimens. Diagn Microbiol Infect Dis. 2017;88(2):125–37. https://doi.org/10.1016/j.diagmicrobio.2017.02.017.
12. Prevots DR, Marras TK. Epidemiology of human pulmonary infection with nontuberculous mycobacteria: a review. Clin Chest Med. 2015;36(1):13–34. https://doi.org/10.1016/j.ccm.2014.10.002.
13. Hoefsloot W, van Ingen J, Andrejak C, Angeby K, Bauriaud R, Bemer P, et al. The geographic diversity of nontuberculous mycobacteria isolated from pulmonary samples: an NTM-NET collaborative study. Eur Respir J. 2013;42(6):1604–13. https://doi.org/10.1183/09031936.00149212.
14. Bryant JM, Grogono DM, Greaves D, Foweraker J, Roddick I, Inns T, et al. Whole-genome sequencing to identify transmission of Mycobacterium abscessus between patients with cystic fibrosis: a retrospective cohort study. Lancet. 2013;381(9877):1551–60. https://doi.org/10.1016/S0140-6736(13)60632-7.
15. Telles MA, Yates MD, Curcio M, Ueki SY, Palaci M, Hadad DJ, et al. Molecular epidemiology of Mycobacterium avium complex isolated from patients with and without AIDS in Brazil and England. Epidemiol Infect. 1999;122(3):435–40.
16. Taylor RH, Falkinham JO 3rd, Norton CD, LeChevallier MW. Chlorine, chloramine, chlorine dioxide, and ozone susceptibility of Mycobacterium avium. Appl Environ Microbiol. 2000;66(4):1702–5.
17. Falkinham JO 3rd, Norton CD, LeChevallier MW. Factors influencing numbers of Mycobacterium avium, Mycobacterium intracellulare, and other mycobacteria in drinking water distribution systems. Appl Environ Microbiol. 2001;67(3):1225–31. https://doi.org/10.1128/AEM.67.3.1225-1231.2001.
18. Feazel LM, Baumgartner LK, Peterson KL, Frank DN, Harris JK, Pace NR. Opportunistic pathogens enriched in showerhead biofilms. Proc Natl Acad Sci U S A. 2009;106(38):16393–9. https://doi.org/10.1073/pnas.0908446106.
19. De Groote MA, Pace NR, Fulton K, Falkinham JO 3rd. Relationships between Mycobacterium isolates from patients with pulmonary mycobacterial infection and potting soils. Appl Environ Microbiol. 2006;72(12):7602–6. https://doi.org/10.1128/AEM.00930-06.
20. Nishiuchi Y, Tamura A, Kitada S, Taguri T, Matsumoto S, Tateishi Y, et al. Mycobacterium avium complex organisms predominantly colonize in the bathtub inlets of patients' bathrooms. Jpn J Infect Dis. 2009;62(3):182–6.
21. Falkinham JO 3rd, Iseman MD, de Haas P, van Soolingen D. Mycobacterium avium in a shower linked to pulmonary disease. J Water Health. 2008;6(2):209–13. https://doi.org/10.2166/wh.2008.032.
22. Falkinham JO 3rd. Nontuberculous mycobacteria from household plumbing of patients with nontuberculous mycobacteria disease. Emerg Infect Dis. 2011;17(3):419–24. https://doi.org/10.3201/eid1703.101510.
23. Wallace RJ Jr, Iakhiaeva E, Williams MD, Brown-Elliott BA, Vasireddy S, Vasireddy R, et al. Absence of Mycobacterium intracellulare and presence of Mycobacterium chimaera in household water and biofilm samples of patients in the United States with Mycobacterium avium complex respiratory disease. J Clin Microbiol. 2013;51(6):1747–52. https://doi.org/10.1128/JCM.00186-13.
24. Chou MP, Clements AC, Thomson RM. A spatial epidemiological analysis of nontuberculous mycobacterial infections in Queensland, Australia. BMC Infect Dis. 2014;14:279. https://doi.org/10.1186/1471-2334-14-279.
25. Honda JR, Hasan NA, Davidson RM, Williams MD, Epperson LE, Reynolds PR, et al. Environmental Nontuberculous mycobacteria in the Hawaiian islands. PLoS Negl Trop Dis. 2016;10(10):e0005068. https://doi.org/10.1371/journal.pntd.0005068.
26. Kohler P, Kuster SP, Bloemberg G, Schulthess B, Frank M, Tanner FC, et al. Healthcare-associated prosthetic heart valve, aortic vascular graft, and disseminated Mycobacterium chimaera infections subsequent to open heart surgery. Eur Heart J. 2015;36(40):2745–53. https://doi.org/10.1093/eurheartj/ehv342.

27. Sax H, Bloemberg G, Hasse B, Sommerstein R, Kohler P, Achermann Y, et al. Prolonged outbreak of Mycobacterium chimaera infection after open-chest heart surgery. Clin Infect Dis. 2015;61(1):67–75. https://doi.org/10.1093/cid/civ198.
28. Svensson E, Jensen ET, Rasmussen EM, Folkvardsen DB, Norman A, Lillebaek T. Mycobacterium chimaera in heater-cooler units in Denmark related to isolates from the United States and United Kingdom. Emerg Infect Dis. 2017;23(3):507–9. https://doi.org/10.3201/eid2303.161941.
29. McGarvey JA, Bermudez LE. Phenotypic and genomic analyses of the Mycobacterium avium complex reveal differences in gastrointestinal invasion and genomic composition. Infect Immun. 2001;69(12):7242–9. https://doi.org/10.1128/IAI.69.12.7242-7249.2001.
30. Stout JE, Hopkins GW, McDonald JR, Quinn A, Hamilton CD, Reller LB, et al. Association between 16S-23S internal transcribed spacer sequence groups of Mycobacterium avium complex and pulmonary disease. J Clin Microbiol. 2008;46(8):2790–3. https://doi.org/10.1128/JCM.00719-08.
31. Tateishi Y, Hirayama Y, Ozeki Y, Nishiuchi Y, Yoshimura M, Kang J, et al. Virulence of Mycobacterium avium complex strains isolated from immunocompetent patients. Microb Pathog. 2009;46(1):6–12. https://doi.org/10.1016/j.micpath.2008.10.007.
32. Bruffaerts N, Vluggen C, Roupie V, Duytschaever L, Van den Poel C, Denoel J, et al. Virulence and immunogenicity of genetically defined human and porcine isolates of M. Avium subsp. hominissuis in an experimental mouse infection. PLoS One. 2017;12(2):e0171895. https://doi.org/10.1371/journal.pone.0171895.
33. Kikuchi T, Watanabe A, Gomi K, Sakakibara T, Nishimori K, Daito H, et al. Association between mycobacterial genotypes and disease progression in Mycobacterium avium pulmonary infection. Thorax. 2009;64(10):901–7. https://doi.org/10.1136/thx.2009.114603.
34. Takegaki Y. Effect of serotype specific glycopeptidolipid (GPL) isolated from Mycobacterium avium complex (MAC) on phagocytosis and phagosome-lysosome fusion of human peripheral blood monocytes. Kekkaku. 2000;75(1):9–18.
35. Turenne CY, Wallace R Jr, Behr MA. Mycobacterium avium in the postgenomic era. Clin Microbiol Rev. 2007;20(2):205–29. https://doi.org/10.1128/CMR.00036-06.
36. Sweet L, Singh PP, Azad AK, Rajaram MV, Schlesinger LS, Schorey JS. Mannose receptor-dependent delay in phagosome maturation by Mycobacterium avium glycopeptidolipids. Infect Immun. 2010;78(1):518–26. https://doi.org/10.1128/IAI.00257-09.
37. Bhatnagar S, Schorey JS. Elevated mitogen-activated protein kinase signalling and increased macrophage activation in cells infected with a glycopeptidolipid-deficient Mycobacterium avium. Cell Microbiol. 2006;8(1):85–96. https://doi.org/10.1111/j.1462-5822.2005.00602.x.
38. Belisle JT, Klaczkiewicz K, Brennan PJ, Jacobs WR Jr, Inamine JM. Rough morphological variants of Mycobacterium avium. Characterization of genomic deletions resulting in the loss of glycopeptidolipid expression. J Biol Chem. 1993;268(14):10517–23.
39. Yamazaki Y, Danelishvili L, Wu M, Macnab M, Bermudez LE. Mycobacterium avium genes associated with the ability to form a biofilm. Appl Environ Microbiol. 2006;72(1):819–25. https://doi.org/10.1128/AEM.72.1.819-825.2006.
40. Rindi L, Bonanni D, Lari N, Garzelli C. Most human isolates of Mycobacterium avium Mav-a and Mav-B are strong producers of hemolysin, a putative virulence factor. J Clin Microbiol. 2003;41(12):5738–40.
41. Ghassemi M, Asadi FK, Andersen BR, Novak RM. Mycobacterium avium induces HIV upregulation through mechanisms independent of cytokine induction. AIDS Res Hum Retrovir. 2000;16(5):435–40. https://doi.org/10.1089/088922200309098.
42. Vazquez N, Greenwell-Wild T, Rekka S, Orenstein JM, Wahl SM. Mycobacterium avium-induced SOCS contributes to resistance to IFN-gamma-mediated mycobactericidal activity in human macrophages. J Leukoc Biol. 2006;80(5):1136–44. https://doi.org/10.1189/jlb.0306206.
43. Honda JR, Knight V, Chan ED. Pathogenesis and risk factors for nontuberculous mycobacterial lung disease. Clin Chest Med. 2015;36(1):1–11. https://doi.org/10.1016/j.ccm.2014.10.001.

44. Thomson RM, Armstrong JG, Looke DF. Gastroesophageal reflux disease, acid suppression, and Mycobacterium avium complex pulmonary disease. Chest. 2007;131(4):1166–72. https://doi.org/10.1378/chest.06-1906.
45. Koh WJ, Lee JH, Kwon YS, Lee KS, Suh GY, Chung MP, et al. Prevalence of gastroesophageal reflux disease in patients with nontuberculous mycobacterial lung disease. Chest. 2007;131(6):1825–30. https://doi.org/10.1378/chest.06-2280.
46. Andrejak C, Nielsen R, Thomsen VO, Duhaut P, Sorensen HT, Thomsen RW. Chronic respiratory disease, inhaled corticosteroids and risk of non-tuberculous mycobacteriosis. Thorax. 2013;68(3):256–62. https://doi.org/10.1136/thoraxjnl-2012-201772.
47. van Ingen J, Boeree MJ, Dekhuijzen PN, van Soolingen D. Mycobacterial disease in patients with rheumatic disease. Nat Clin Pract Rheumatol. 2008;4(12):649–56. https://doi.org/10.1038/ncprheum0949.
48. Dorman SE, Picard C, Lammas D, Heyne K, van Dissel JT, Baretto R, et al. Clinical features of dominant and recessive interferon gamma receptor 1 deficiencies. Lancet. 2004;364(9451):2113–21. https://doi.org/10.1016/S0140-6736(04)17552-1.
49. Griffith DE, Aksamit T, Brown-Elliott BA, Catanzaro A, Daley C, Gordin F, et al. An official ATS/IDSA statement: diagnosis, treatment, and prevention of nontuberculous mycobacterial diseases. Am J Respir Crit Care Med. 2007;175(4):367–416. https://doi.org/10.1164/rccm.200604-571ST.
50. Tsukamura M. Diagnosis of disease caused by Mycobacterium avium complex. Chest. 1991;99(3):667–9.
51. Koh WJ, Chang B, Ko Y, Jeong BH, Hong G, Park HY, et al. Clinical significance of a single isolation of pathogenic nontuberculous mycobacteria from sputum specimens. Diagn Microbiol Infect Dis. 2013;75(2):225–6. https://doi.org/10.1016/j.diagmicrobio.2012.09.021.
52. Sugihara E, Hirota N, Niizeki T, Tanaka R, Nagafuchi M, Koyanagi T, et al. Usefulness of bronchial lavage for the diagnosis of pulmonary disease caused by Mycobacterium avium-intracellulare complex (MAC) infection. J Infect Chemother. 2003;9(4):328–32. https://doi.org/10.1007/s10156-003-0267-1.
53. Ikedo Y. The significance of bronchoscopy for the diagnosis of Mycobacterium avium complex (MAC) pulmonary disease. Kurume Med J. 2001;48(1):15–9.
54. Koh WJ, Kwon OJ, Lee KS. Nontuberculous mycobacterial pulmonary diseases in immunocompetent patients. Korean J Radiol. 2002;3(3):145–57. doi:2002v3n3p145 [pii]
55. Reich JM, Johnson RE. Mycobacterium avium complex pulmonary disease presenting as an isolated lingular or middle lobe pattern. The Lady Windermere syndrome. Chest. 1992;101(6):1605–9.
56. Thomson RM. NTM working group at Queensland TB control Centre and Queensland mycobacterial reference laboratory. Emerg Infect Dis. 2010;16(10):1576–83. https://doi.org/10.3201/eid1610.091201.
57. Kim RD, Greenberg DE, Ehrmantraut ME, Guide SV, Ding L, Shea Y, et al. Pulmonary nontuberculous mycobacterial disease: prospective study of a distinct preexisting syndrome. Am J Respir Crit Care Med. 2008;178(10):1066–74. https://doi.org/10.1164/rccm.200805-686OC.
58. Chan ED, Iseman MD. Slender, older women appear to be more susceptible to nontuberculous mycobacterial lung disease. Gend Med. 2010;7(1):5–18. https://doi.org/10.1016/j.genm.2010.01.005.
59. Kartalija M, Ovrutsky AR, Bryan CL, Pott GB, Fantuzzi G, Thomas J, et al. Patients with nontuberculous mycobacterial lung disease exhibit unique body and immune phenotypes. Am J Respir Crit Care Med. 2013;187(2):197–205. https://doi.org/10.1164/rccm.201206-1035OC.
60. Szymanski EP, Leung JM, Fowler CJ, Haney C, Hsu AP, Chen F, et al. Pulmonary Nontuberculous mycobacterial infection. A multisystem, multigenic disease. Am J Respir Crit Care Med. 2015;192(5):618–28. https://doi.org/10.1164/rccm.201502-0387OC.
61. Kitada S, Maekura R, Toyoshima N, Naka T, Fujiwara N, Kobayashi M, et al. Use of glycopeptidolipid core antigen for serodiagnosis of mycobacterium avium complex pulmonary disease in immunocompetent patients. Clin Diagn Lab Immunol. 2005;12(1):44–51. https://doi.org/10.1128/CDLI.12.1.44-51.2005.

62. Kitada S, Kobayashi K, Ichiyama S, Takakura S, Sakatani M, Suzuki K, et al. Serodiagnosis of Mycobacterium avium-complex pulmonary disease using an enzyme immunoassay kit. Am J Respir Crit Care Med. 2008;177(7):793–7. https://doi.org/10.1164/rccm.200705-771OC.
63. Kitada S, Levin A, Hiserote M, Harbeck RJ, Czaja CA, Huitt G, et al. Serodiagnosis of Mycobacterium avium complex pulmonary disease in the USA. Eur Respir J. 2013;42(2):454–60. https://doi.org/10.1183/09031936.00098212.
64. Embil J, Warren P, Yakrus M, Stark R, Corne S, Forrest D, et al. Pulmonary illness associated with exposure to Mycobacterium-avium complex in hot tub water. Hypersensitivity pneumonitis or infection? Chest. 1997;111(3):813–6.
65. Marras TK, Wallace RJ Jr, Koth LL, Stulbarg MS, Cowl CT, Daley CL. Hypersensitivity pneumonitis reaction to Mycobacterium avium in household water. Chest. 2005;127(2):664–71. https://doi.org/10.1378/chest.127.2.664.
66. Centers for Disease Control and Prevention. Respiratory illness in workers exposed to metalworking fluid contaminated with nontuberculous mycobacteria—Ohio, 2001. MMWR. 2002;51(16):349–52.
67. Hartman TE, Jensen E, Tazelaar HD, Hanak V, Ryu JH. CT findings of granulomatous pneumonitis secondary to Mycobacterium avium-intracellulare inhalation: “hot tub lung”. AJR. 2007;188(4):1050–3. https://doi.org/10.2214/AJR.06.0546.
68. Hanak V, Kalra S, Aksamit TR, Hartman TE, Tazelaar HD, Ryu JH. Hot tub lung: presenting features and clinical course of 21 patients. Respir Med. 2006;100(4):610–5. https://doi.org/10.1016/j.rmed.2005.08.005.
69. Wallace RJ Jr, Brown BA, Griffith DE, Girard WM, Murphy DT, Onyi GO, et al. Initial clarithromycin monotherapy for Mycobacterium avium-intracellulare complex lung disease. Am J Respir Crit Care Med. 1994;149(5):1335–41. https://doi.org/10.1164/ajrccm.149.5.8173775.
70. Dautzenberg B, Piperno D, Diot P, Truffot-Pernot C, Chauvin JP. Clarithromycin in the treatment of Mycobacterium avium lung infections in patients without AIDS. Clarithromycin study group of France. Chest. 1995;107(4):1035–40.
71. Wallace RJ Jr, Brown BA, Griffith DE, Girard WM, Murphy DT. Clarithromycin regimens for pulmonary Mycobacterium avium complex. The first 50 patients. Am J Respir Crit Care Med. 1996;153(6 Pt 1):1766–72. https://doi.org/10.1164/ajrccm.153.6.8665032.
72. Tanaka E, Kimoto T, Tsuyuguchi K, Watanabe I, Matsumoto H, Niimi A, et al. Effect of clarithromycin regimen for Mycobacterium avium complex pulmonary disease. Am J Respir Crit Care Med. 1999;160(3):866–72. https://doi.org/10.1164/ajrccm.160.3.9811086.
73. Wallace RJ Jr, Brown BA, Griffith DE. Drug intolerance to high-dose clarithromycin among elderly patients. Diagn Microbiol Infect Dis. 1993;16(3):215–21.
74. Wallace RJ Jr, Brown-Elliott BA, McNulty S, Philley JV, Killingley J, Wilson RW, et al. Macrolide/Azalide therapy for nodular/bronchiectatic mycobacterium avium complex lung disease. Chest. 2014;146(2):276–82. https://doi.org/10.1378/chest.13-2538.
75. Griffith DE, Brown BA, Girard WM, Murphy DT, Wallace RJ Jr. Azithromycin activity against Mycobacterium avium complex lung disease in patients who were not infected with human immunodeficiency virus. Clin Infect Dis. 1996;23(5):983–9.
76. Zuckerman JM, Qamar F, Bono BR. Review of macrolides (azithromycin, clarithromycin), ketolids (telithromycin) and glycylcyclines (tigecycline). Med Clin North Am. 2011;95(4):761–91. https://doi.org/10.1016/j.mcna.2011.03.012.
77. Chaisson RE, Benson CA, Dube MP, Heifets LB, Korvick JA, Elkin S, et al. Clarithromycin therapy for bacteremic Mycobacterium avium complex disease. A randomized, double-blind, dose-ranging study in patients with AIDS. AIDS Clinical Trials Group protocol 157 study team. Ann Intern Med. 1994;121(12):905–11.
78. Gordin FM, Sullam PM, Shafran SD, Cohn DL, Wynne B, Paxton L, et al. A randomized, placebo-controlled study of rifabutin added to a regimen of clarithromycin and ethambutol for treatment of disseminated infection with Mycobacterium avium complex. Clin Infect Dis. 1999;28(5):1080–5. https://doi.org/10.1086/514748.
79. Miwa S, Shirai M, Toyoshima M, Shirai T, Yasuda K, Yokomura K, et al. Efficacy of clarithromycin and ethambutol for Mycobacterium avium complex pulmonary disease.

A preliminary study. Ann Am Thorac Soc. 2014;11(1):23–9. https://doi.org/10.1513/AnnalsATS.201308-266OC.
80. Griffith DE, Brown BA, Girard WM, Griffith BE, Couch LA, Wallace RJ Jr. Azithromycin-containing regimens for treatment of Mycobacterium avium complex lung disease. Clin Infect Dis. 2001;32(11):1547–53. https://doi.org/10.1086/320512.
81. Koh WJ, Jeong BH, Jeon K, Lee SY, Shin SJ. Therapeutic drug monitoring in the treatment of Mycobacterium avium complex lung disease. Am J Respir Crit Care Med. 2012;186(8):797–802. https://doi.org/10.1164/rccm.201206-1088OC.
82. Griffith DE, Brown BA, Murphy DT, Girard WM, Couch L, Wallace RJ Jr. Initial (6-month) results of three-times-weekly azithromycin in treatment regimens for Mycobacterium avium complex lung disease in human immunodeficiency virus-negative patients. J Infect Dis. 1998;178(1):121–6.
83. Jeong BH, Jeon K, Park HY, Kim SY, Lee KS, Huh HJ, et al. Intermittent antibiotic therapy for nodular bronchiectatic Mycobacterium avium complex lung disease. Am J Respir Crit Care Med. 2015;191(1):96–103. https://doi.org/10.1164/rccm.201408-1545OC.
84. Jeong BH, Jeon K, Park HY, Moon SM, Kim SY, Lee SY, et al. Peak plasma concentration of azithromycin and treatment responses in Mycobacterium avium complex lung disease. Antimicrob Agents Chemother. 2016;60(10):6076–83. https://doi.org/10.1128/AAC.00770-16.
85. Lam PK, Griffith DE, Aksamit TR, Ruoss SJ, Garay SM, Daley CL, et al. Factors related to response to intermittent treatment of Mycobacterium avium complex lung disease. Am J Respir Crit Care Med. 2006;173(11):1283–9. https://doi.org/10.1164/rccm.200509-1531OC.
86. Koh WJ, Jeong BH, Jeon K, Park HY, Kim SY, Huh HJ, et al. Response to switch from intermittent therapy to daily therapy for refractory nodular Bronchiectatic Mycobacterium avium complex lung disease. Antimicrob Agents Chemother. 2015;59(8):4994–6. https://doi.org/10.1128/AAC.00648-15.
87. Kobashi Y, Matsushima T, Oka M. A double-blind randomized study of aminoglycoside infusion with combined therapy for pulmonary Mycobacterium avium complex disease. Respir Med. 2007;101(1):130–8. https://doi.org/10.1016/j.rmed.2006.04.002.
88. Peloquin CA, Berning SE, Nitta AT, Simone PM, Goble M, Huitt GA, et al. Aminoglycoside toxicity: daily versus thrice-weekly dosing for treatment of mycobacterial diseases. Clin Infect Dis. 2004;38(11):1538–44. https://doi.org/10.1086/420742.
89. Jarand J, Davis JP, Cowie RL, Field SK, Fisher DA. Long-term follow-up of Mycobacterium avium complex lung disease in patients treated with regimens including Clofazimine and/or rifampin. Chest. 2016;149(5):1285–93. https://doi.org/10.1378/chest.15-0543.
90. Martiniano SL, Wagner BD, Levin A, Nick JA, Sagel SD, Daley CL. Safety and effectiveness of Clofazimine for primary and refractory Nontuberculous mycobacterial infection. Chest. 2017. https://doi.org/10.1016/j.chest.2017.04.175.
91. Yang B, Jhun BW, Moon SM, Lee H, Park HY, Jeon K, et al. Clofazimine-containing regimen for the treatment of Mycobacterium abscessus lung disease. Antimicrob Agents Chemother. 2017;61(6). https://doi.org/10.1128/AAC.02052-16.
92. Griffith DE, Brown-Elliott BA, Langsjoen B, Zhang Y, Pan X, Girard W, et al. Clinical and molecular analysis of macrolide resistance in Mycobacterium avium complex lung disease. Am J Respir Crit Care Med. 2006;174(8):928–34. https://doi.org/10.1164/rccm.200603-450OC.
93. Koh WJ, Hong G, Kim SY, Jeong BH, Park HY, Jeon K, et al. Treatment of refractory Mycobacterium avium complex lung disease with a moxifloxacin-containing regimen. Antimicrob Agents Chemother. 2013;57(5):2281–5. https://doi.org/10.1128/AAC.02281-12.
94. Brown-Elliott BA, Philley JV, Griffith DE, Thakkar F, Wallace RJ Jr. In vitro susceptibility testing of Bedaquiline against Mycobacterium avium complex. Antimicrob Agents Chemother. 2017;61(2). https://doi.org/10.1128/AAC.01798-16.
95. Diacon AH, Pym A, Grobusch MP, de los Rios JM, Gotuzzo E, Vasilyeva I, et al. Multidrug-resistant tuberculosis and culture conversion with bedaquiline. N Engl J Med. 2014;371(8):723–32. https://doi.org/10.1056/NEJMoa1313865.

96. Philley JV, Wallace RJ Jr, Benwill JL, Taskar V, Brown-Elliott BA, Thakkar F, et al. Preliminary results of Bedaquiline as salvage therapy for patients with Nontuberculous mycobacterial lung disease. Chest. 2015;148(2):499–506. https://doi.org/10.1378/chest.14-2764.
97. Svensson EM, Murray S, Karlsson MO, Dooley KE. Rifampicin and rifapentine significantly reduce concentrations of bedaquiline, a new anti-TB drug. J Antimicrob Chemother. 2015;70(4):1106–14. https://doi.org/10.1093/jac/dku504.
98. Brown-Elliott BA, Crist CJ, Mann LB, Wilson RW, Wallace RJ Jr. In vitro activity of linezolid against slowly growing nontuberculous mycobacteria. Antimicrob Agents Chemother. 2003;47(5):1736–8.
99. Winthrop KL, Ku JH, Marras TK, Griffith DE, Daley CL, Olivier KN, et al. The tolerability of linezolid in the treatment of nontuberculous mycobacterial disease. Eur Respir J. 2015;45(4):1177–9. https://doi.org/10.1183/09031936.00169114.
100. Brown-Elliott BA, Wallace RJ Jr. In vitro susceptibility testing of Tedizolid against Nontuberculous mycobacteria. J Clin Microbiol. 2017;55(6):1747–54. https://doi.org/10.1128/JCM.00274-17.
101. Olivier KN, Shaw PA, Glaser TS, Bhattacharyya D, Fleshner M, Brewer CC, et al. Inhaled amikacin for treatment of refractory pulmonary nontuberculous mycobacterial disease. Ann Am Thorac Soc. 2014;11(1):30–5. https://doi.org/10.1513/AnnalsATS.201307-231OC.
102. Griffith DE, Eagle G, Thomson R, Aksamit TR, Hasegawa N, Morimoto K et al. Amikacin Liposome Inhalation Suspension for Treatment-Refractory Lung Disease Caused by *Mycobacterium avium* Complex. Am J Respir and Crit Care Med, In Press, 2018.
103. Mitchell JD. Surgical approach to pulmonary nontuberculous mycobacterial infections. Clin Chest Med. 2015;36(1):117–22. https://doi.org/10.1016/j.ccm.2014.11.004.
104. Watanabe M, Hasegawa N, Ishizaka A, Asakura K, Izumi Y, Eguchi K, et al. Early pulmonary resection for Mycobacterium avium complex lung disease treated with macrolides and quinolones. Ann Thorac Surg. 2006;81(6):2026–30.
105. Koh WJ, Kim YH, Kwon OJ, Choi YS, Kim K, Shim YM, et al. Surgical treatment of pulmonary diseases due to nontuberculous mycobacteria. J Korean Med Sci. 2008;23(3):397–401. https://doi.org/10.3346/jkms.2008.23.3.397.
106. Mitchell JD, Bishop A, Cafaro A, Weyant MJ, Pomerantz M. Anatomic lung resection for nontuberculous mycobacterial disease. Ann Thorac Surg. 2008;85(6):1887–92.; discussion 1892-3. https://doi.org/10.1016/j.athoracsur.2008.02.041.
107. Yu JA, Pomerantz M, Bishop A, Weyant MJ, Mitchell JD. Lady Windermere revisited: treatment with thoracoscopic lobectomy/segmentectomy for right middle lobe and lingular bronchiectasis associated with non-tuberculous mycobacterial disease. Eur J Cardiothorac Surg. 2011;40(3):671–5. https://doi.org/10.1016/j.ejcts.2010.12.028.
108. Morimoto K, Namkoong H, Hasegawa N, Nakagawa T, Morino E, Shiraishi Y, et al. Macrolide-resistant Mycobacterium avium complex lung disease: analysis of 102 consecutive cases. Ann Am Thorac Soc. 2016;13(11):1904–11. https://doi.org/10.1513/AnnalsATS.201604-246OC.
109. Moon SM, Park HY, Kim SY, Jhun BW, Lee H, Jeon K, et al. Clinical characteristics, treatment outcomes, and resistance mutations associated with macrolide-resistant Mycobacterium avium complex lung disease. Antimicrob Agents Chemother. 2016;60(11):6758–65. https://doi.org/10.1128/AAC.01240-16.
110. Bermudez LE, Nash K, Petrofsky M, Young LS, Inderlied CB. Clarithromycin-resistant mycobacterium avium is still susceptible to treatment with clarithromycin and is virulent in mice. Antimicrob Agents Chemother. 2000;44(10):2619–22.
111. Lee G, Lee KS, Moon JW, Koh WJ, Jeong BH, Jeong YJ, et al. Nodular bronchiectatic Mycobacterium avium complex pulmonary disease. Natural course on serial computed tomographic scans. Ann Am Thorac Soc. 2013;10(4):299–306. https://doi.org/10.1513/AnnalsATS.201303-062OC.
112. Kotilainen H, Valtonen V, Tukiainen P, Poussa T, Eskola J, Jarvinen A. Prognostic value of American Thoracic Society criteria for non-tuberculous mycobacterial disease: a retrospec-

tive analysis of 120 cases with four years of follow-up. Scand J Infect Dis. 2013;45(3):194–202. https://doi.org/10.3109/00365548.2012.722227.
113. Marras TK, Campitelli MA, Lu H, Chung H, Brode SK, Marchand-Austin A, et al. Pulmonary Nontuberculous mycobacteria-associated deaths, Ontario, Canada, 2001-2013. Emerg Infect Dis. 2017;23(3):468–76. https://doi.org/10.3201/eid2303.161927.
114. Andrejak C, Thomsen VO, Johansen IS, Riis A, Benfield TL, Duhaut P, et al. Nontuberculous pulmonary mycobacteriosis in Denmark: incidence and prognostic factors. Am J Respir Crit Care Med. 2010;181(5):514–21. https://doi.org/10.1164/rccm.200905-0778OC.
115. Gochi M, Takayanagi N, Kanauchi T, Ishiguro T, Yanagisawa T, Sugita Y. Retrospective study of the predictors of mortality and radiographic deterioration in 782 patients with nodular/bronchiectatic Mycobacterium avium complex lung disease. BMJ Open. 2015;5(8):e008058. https://doi.org/10.1136/bmjopen-2015-008058.
116. Fleshner M, Olivier KN, Shaw PA, Adjemian J, Strollo S, Claypool RJ, et al. Mortality among patients with pulmonary non-tuberculous mycobacteria disease. Int J Tuberc Lung Dis. 2016;20(5):582–7. https://doi.org/10.5588/ijtld.15.0807.
117. Horsburgh CR Jr. Mycobacterium avium complex infection in the acquired immunodeficiency syndrome. N Engl J Med. 1991;324(19):1332–8. https://doi.org/10.1056/NEJM199105093241906.
118. Horsburgh CR Jr, Havlik JA, Ellis DA, Kennedy E, Fann SA, Dubois RE, et al. Survival of patients with acquired immune deficiency syndrome and disseminated Mycobacterium avium complex infection with and without antimycobacterial chemotherapy. Am Rev Respir Dis. 1991;144(3 Pt 1):557–9. https://doi.org/10.1164/ajrccm/144.3_Pt_1.557.
119. Torriani FJ, McCutchan JA, Bozzette SA, Grafe MR, Havlir DV. Autopsy findings in AIDS patients with Mycobacterium avium complex bacteremia. J Infect Dis. 1994;170(6):1601–5.
120. Ward TT, Rimland D, Kauffman C, Huycke M, Evans TG, Heifets L. Randomized, open-label trial of azithromycin plus ethambutol vs. clarithromycin plus ethambutol as therapy for Mycobacterium avium complex bacteremia in patients with human immunodeficiency virus infection. Veterans affairs HIV research consortium. Clin Infect Dis. 1998;27(5):1278–85.
121. Dunne M, Fessel J, Kumar P, Dickenson G, Keiser P, Boulos M, et al. A randomized, double-blind trial comparing azithromycin and clarithromycin in the treatment of disseminated Mycobacterium avium infection in patients with human immunodeficiency virus. Clin Infect Dis. 2000;31(5):1245–52. https://doi.org/10.1086/317468.
122. Cohn DL, Fisher EJ, Peng GT, Hodges JS, Chesnut J, Child CC, et al. A prospective randomized trial of four three-drug regimens in the treatment of disseminated Mycobacterium avium complex disease in AIDS patients: excess mortality associated with high-dose clarithromycin. Terry Beirn community programs for clinical research on AIDS. Clin Infect Dis. 1999;29(1):125–33. https://doi.org/10.1086/520141.
123. Dube MP, Sattler FR, Torriani FJ, See D, Havlir DV, Kemper CA, et al. A randomized evaluation of ethambutol for prevention of relapse and drug resistance during treatment of Mycobacterium avium complex bacteremia with clarithromycin-based combination therapy. California collaborative treatment group. J Infect Dis. 1997;176(5):1225–32.
124. Benson CA, Williams PL, Currier JS, Holland F, Mahon LF, MacGregor RR, et al. A prospective, randomized trial examining the efficacy and safety of clarithromycin in combination with ethambutol, rifabutin, or both for the treatment of disseminated Mycobacterium avium complex disease in persons with acquired immunodeficiency syndrome. Clin Infect Dis. 2003;37(9):1234–43. https://doi.org/10.1086/378807.
125. Panel on Opportunistic Infections in HIV-Infected Adults and Adolescents. Disseminated Mycobacterium avium complex disease. In: Guidelines for the prevention and treatment of opportunistic infections in HIV-infected adults and adolescents: recommendations from the Centers for Disease Control and Prevention, the National Institutes of Health, and the HIV Medicine Association of the Infectious Diseases Society of America. AIDSinfo. 2013. https://aidsinfo.nih.gov/contentfiles/lvguidelines/adult_oi.pdf. Accessed 19 June 2017.

126. Chaisson RE, Keiser P, Pierce M, Fessel WJ, Ruskin J, Lahart C, et al. Clarithromycin and ethambutol with or without clofazimine for the treatment of bacteremic Mycobacterium avium complex disease in patients with HIV infection. AIDS. 1997;11(3):311–7.
127. Karakousis PC, Moore RD, Chaisson RE. Mycobacterium avium complex in patients with HIV infection in the era of highly active antiretroviral therapy. Lancet Infect Dis. 2004;4(9):557–65. https://doi.org/10.1016/S1473-3099(04)01130-2.
128. Aberg JA, Yajko DM, Jacobson MA. Eradication of AIDS-related disseminated mycobacterium avium complex infection after 12 months of antimycobacterial therapy combined with highly active antiretroviral therapy. J Infect Dis. 1998;178(5):1446–9.
129. Pierce M, Crampton S, Henry D, Heifets L, LaMarca A, Montecalvo M, et al. A randomized trial of clarithromycin as prophylaxis against disseminated Mycobacterium avium complex infection in patients with advanced acquired immunodeficiency syndrome. N Engl J Med. 1996;335(6):384–91. https://doi.org/10.1056/NEJM199608083350603.
130. Havlir DV, Dube MP, Sattler FR, Forthal DN, Kemper CA, Dunne MW, et al. Prophylaxis against disseminated Mycobacterium avium complex with weekly azithromycin, daily rifabutin, or both. California collaborative treatment group. N Engl J Med. 1996;335(6):392–8. https://doi.org/10.1056/NEJM199608083350604.
131. Benson CA, Williams PL, Cohn DL, Becker S, Hojczyk P, Nevin T, et al. Clarithromycin or rifabutin alone or in combination for primary prophylaxis of Mycobacterium avium complex disease in patients with AIDS: a randomized, double-blind, placebo-controlled trial. The AIDS Clinical Trials Group 196/Terry Beirn community programs for clinical research on AIDS 009 protocol team. J Infect Dis. 2000;181(4):1289–97. https://doi.org/10.1086/315380.
132. El-Sadr WM, Burman WJ, Grant LB, Matts JP, Hafner R, Crane L, et al. Discontinuation of prophylaxis against Mycobacterium avium complex disease in HIV-infected patients who have a response to antiretroviral therapy. Terry Beirn community programs for clinical research on AIDS. N Engl J Med. 2000;342(15):1085–92. https://doi.org/10.1056/NEJM200004133421503.
133. Lindeboom JA, Kuijper EJ, Bruijnesteijn van Coppenraet ES, Lindeboom R, Prins JM. Surgical excision versus antibiotic treatment for nontuberculous mycobacterial cervicofacial lymphadenitis in children: a multicenter, randomized, controlled trial. Clin Infect Dis. 2007;44(8):1057–64. https://doi.org/10.1086/512675.
134. Rahal A, Abela A, Arcand PH, Quintal MC, Lebel MH, Tapiero BF. Nontuberculous mycobacterial adenitis of the head and neck in children: experience from a tertiary care pediatric center. Laryngoscope. 2001;111(10):1791–6. https://doi.org/10.1097/00005537-200110000-00024.
135. Panesar J, Higgins K, Daya H, Forte V, Allen U. Nontuberculous mycobacterial cervical adenitis: a ten-year retrospective review. Laryngoscope. 2003;113(1):149–54. https://doi.org/10.1097/00005537-200301000-00028.

NTM Disease Caused by *M. kansasii*, *M. xenopi*, *M. malmoense*, and Other Slowly Growing NTM

Theodore K. Marras and Sarah K. Brode

Introduction

Slowly growing NTM species comprise some of the most common species encountered in the management of NTM pulmonary disease (NTM-PD) and non-pulmonary infections as well. This chapter will discuss the more commonly encountered species of this group (Table 1) with the exception of *M. avium* complex, which is discussed in a dedicated chapter. As common causes of NTM-PD, *M. kansasii*, *M. xenopi*, and *M. malmoense* will be discussed in greater detail, but additional species will be included to provide information and guidance to further resources to assist with the clinical management of these infections. For most species, discussion will include information about the organism (environmental sources) and the host (common patient phenotypes), typical anatomic sites of infection (pulmonary, non-pulmonary, disseminated) and clinical presentations, and antimicrobial treatment. Most slowly growing NTM species considered are likely found primarily in waters, and pulmonary infection is believed to often result from inhalation of contaminated water aerosols. This critical but inadequately explored area, especially with respect

T. K. Marras (✉)
Division of Respirology, Department of Medicine, Toronto Western Hospital,
University Health Network, University of Toronto, Toronto, Canada
e-mail: Ted.Marras@uhn.ca

S. K. Brode
Division of Respirology, Department of Medicine, Toronto Western Hospital,
University Health Network and West Park Healthcare Centre, University of Toronto,
Toronto, ON, Canada

D. E. Griffith (ed.), *Nontuberculous Mycobacterial Disease*, Respiratory Medicine,
https://doi.org/10.1007/978-3-319-93473-0_12

Table 1 Epidemiology and clinical aspects of disease caused by slow-growing NTM species excluding MAC

Species	Epidemiology	Typical clinical relevance of an isolate	Lung	Skin/soft tissue/bone	Lymph node	Disseminated
M. kansasii	USA (Central, South), Brazil, Europe, Israel	Among most pathogenic NTM, 40–88% of pulmonary isolates associated with disease	**Common** Upper lobe cavitary disease, often associated with emphysema Fibrocavitary 44–72% Nodular bronchiectasis 28–32%	**Rare** in absence of dissemination	**Rare** in absence of dissemination	**Common** in advanced HIV and some additional immune suppressed states, often with underlying pulmonary infection
M. xenopi	Canada (Ontario), USA (Northeast) British Isles, Croatia, France, Israel	25–30% of pulmonary isolates associated with disease Institutional water systems implicated in pulmonary pseudo-outbreaks	**Common** Strongly associated with emphysema and very often cavitary	**Rare** Postoperative (often delayed) surgical site infections due to contaminated surgical devices and wounds	**Rare**	**Rare**
M. malmoense	Northern Europe, the UK Less commonly Southern Europe, South Africa, North America	Variable, but most series report >70% of pulmonary isolates associated with disease	**Common** Upper lobe cavitary disease, airspace disease, approx. half have underlying COPD	**Uncommon**	**Common** cause of childhood cervical lymphadenitis	**Uncommon**, occurs in immune suppressed patients
M. genavense	Europe	Uncommon	**Rarely** primary lung infection; may involve lung in disseminated disease	**Rare**	**Uncommon**, occurs in HIV and immune suppressed	**Uncommon**, occurs in HIV and other immune suppressed

M. gordonae	Worldwide	Very common contaminant Most pulmonary isolates are contaminants (1–3.5% associated with disease) Implicated in pulmonary pseudo-outbreaks	**Rare**	**Rare** Immune suppressed	**Rare** Immune suppressed	**Rare** Immune suppressed
M. haemophilum	Worldwide	Few data Likely uncommonly a contaminant	**Rare** Immune suppressed, may be associated with disseminated disease	**Uncommon** immune suppressed Often on extremities	**Uncommon** cause of childhood cervical lymphadenitis	**Uncommon** Immune suppressed
M. lentiflavum	USA, Europe, Africa, Asia, Australia	Uncommon isolate Very small proportion of respiratory isolates associated with disease	**Very uncommon**	**Uncommon** Immune suppressed	**Uncommon** overall Lymphadenitis is the most common form of *M. lentiflavum* disease Occurs in immune competent hosts	**Uncommon** Immune suppressed
M. marinum	Worldwide	Uncommon contaminant	**Rare**	**Common** Water or fish exposure Hands most common	**Rare**	**Rare** Immune suppressed
M. scrofulaceum	Worldwide, rarely isolated	Rare	**Uncommon** Fibrocavitary predominant Underlying COPD, silicosis, prior TB	**Rare**	**Rare** (previously common) cause of childhood cervical lymphadenitis	**Rare** Immune suppressed

(continued)

Table 1 (continued)

Species	Epidemiology	Typical clinical relevance of an isolate	Lung	Skin/soft tissue/bone	Lymph node	Disseminated
M. shimoidei	Canada, Europe, Africa, Asia, Australia	Very uncommon isolate Proportion of isolates associated with disease is unknown	**Very uncommon** cause of lung disease Usually preexisting structural lung disease	Not reported	Not reported	Single case report in HIV infection
M. simiae	USA (South), Israel Sporadic elsewhere	Usually contaminant Pulmonary pseudo-outbreaks described from environmental contamination of specimens	**Uncommon**	**Rare**	**Rare**	**Uncommon**
M. szulgai	Rarely identified Nearly worldwide case distribution	Uncommon environmental isolate Rarely thought to contaminate specimens, but a substantial proportion of patients with a respiratory isolate do not have disease	**Rare**	**Rare**	**Rare**	**Rare**
M. terrae complex	Worldwide, temperate climates	Common Pulmonary isolates more likely represent contamination than disease	**Rare**	**Uncommon** Tenosynovitis of hands Immune competent	**Rare**	**Rare** Immune suppressed
M. ulcerans	Tropical regions; Africa, South America, Southeast Asia, Australia, Japan	Not reported	Not reported	**Common** “Buruli ulcer” Nodule or plaque progressing to painless ulcer Immunocompetent	Not reported	**Rare**

to prevention of infection/reinfection, is explored further in chapter "Environmental Niches for NTM and Their Impact on NTM Disease". The role of antimicrobial drug susceptibility testing for most agents in most NTM species remains unclear. Drug susceptibility testing (DST) is discussed in several instances below, but data are generally lacking regarding the definition of clinically meaningful minimal inhibitory concentrations (MICs). The utility of DST is therefore limited, and outside of the cases where relevant MICs have been defined, expert consultation should be considered in the interpretation. Although the importance of long-term follow-up may not be specified throughout, because of the unpredictable progression of what may appear to be indolent disease, and very high recurrence rates after successful treatment, most patients with NTM-PD should have indefinite follow-up. Non-antibiotic therapies are generally not discussed in this chapter but are well-described in chapters "Non-tuberculous Mycobacterial Disease Management Principles, *Mycobacterium avium* Complex Disease, Nontuberculous Mycobacterial Disease in Pediatric Populations, and Non-Tuberculous Mycobacteria in Cystic Fibrosis". Species-specific information regarding epidemiology, diagnosis, and monitoring of treatment may be provided, and additional information can be found in the respective chapters. In general, regular sputum cultures should be obtained to assess microbiologic response to therapy, identify the time of culture conversion, and determine duration of therapy. Periodic clinical and radiologic assessments should be performed to assess response to therapy. Assessments for drug toxicity should include periodic clinical assessments, as well as serum and hematologic testing, which is particularly relevant to rifamycin therapy. Summary tables present information regarding frequency and anatomic site of disease (Table 1) and antimicrobial treatment recommendations for the more commonly encountered causes of pulmonary (Table 2) and non-pulmonary (Table 3) disease. For all species considered in this chapter, the ATS/IDSA guidelines [1] criteria should generally be considered as the starting point for making a diagnosis of NTM-PD. However, some species are more likely than others to be identified in clinical specimens in the absence of disease [2, 3]. *Mycobacterium gordonae* is an example of a species that rarely causes disease, and some investigators have proposed using more stringent diagnostic criteria for this species [4]. At the other end of the spectrum, *M. kansasii* is generally thought to be highly pathogenic, with the disease considered to be present in the majority of patients with a respiratory isolate. However, the apparent pathogenicity, or likelihood of a respiratory isolate to signify disease, may vary by geographic region [3, 5, 6]. Not only is it important to consider what has been described about the pathogenicity of different NTM species, the clinician must also carefully consider all clinical information to make appropriate diagnostic and treatment decisions. Treatment recommendations reflect current guidelines [1] as well as recent literature.

Table 2 Management of lung disease caused by commonly encountered slow-growing NTM species excluding MAC

Species	Diagnostic considerations	Antimicrobial therapies[a]
M. kansasii	ATS/IDSA guidelines Carefully consider any positive specimens due to high pathogenicity	*Rifampin-susceptible*[b] Rifampin 600 mg (or rifabutin 150–300 mg)/d plus Ethambutol 15 mg/kg/d plus, one of Azithromycin 250 mg/d, or Clarithromycin 500 mg/d bid, or Isoniazid[c] 300 mg/d, or Moxifloxacin 400 mg/d *Rifampin-resistant* Three drugs based on drug susceptibility testing Injectable amikacin or streptomycin may be considered Duration of therapy should be 12 months of culture-negative sputum
M. xenopi	ATS/IDSA guidelines Consider possibility of specimen contamination/ institutional pseudo-outbreak depending on local factors	Rifampin 600 mg (or rifabutin 150–300 mg/d) plus Ethambutol 15 mg/kg/d plus at least one of Azithromycin 250 mg/d/clarithromycin 500 mg bid or moxifloxacin 400 mg/d (combining four agents may be considered[d]) Consider addition of injectable amikacin or streptomycin in severe disease Duration of therapy should be 12 months of culture-negative sputum
M. malmoense	ATS/IDSA guidelines	Rifampin 600 mg/d plus Ethambutol 15 mg/kg/d plus Azithromycin 250 mg/d/clarithromycin 500 mg/d bid plus/minus Moxifloxacin 400 mg/d or possibly Isoniazid 300 mg/d Addition of amikacin should be considered for severe disease Optimal duration of therapy unknown; determine according to clinical response and available objective findings during treatment; consider 12 months of culture-negative sputum

Table 2 (continued)

Species	Diagnostic considerations	Antimicrobial therapies[a]
M. simiae	ATS/IDSA guidelines Respiratory isolates usually not indicative of disease	Few data; consider Amikacin 5–15 mg/kg IV tiw plus Trimethoprim-sulfamethoxazole (double strength, bid) Plus at least one of Azithromycin 250 mg/d or clarithromycin 500 mg bid Moxifloxacin 400 mg/d Consider additionally Linezolid 600 mg/d or clofazimine 100 mg/d Optimal duration of therapy unknown; determine according to clinical response and available objective findings during treatment; Consider 12 months of culture-negative sputum
M. szulgai	ATS/IDSA guidelines Respiratory isolates frequently indicative of disease	Few data; Three or four drugs selected from: Azithromycin 250 mg/d or clarithromycin 500 mg bid Rifampin 600 mg/d or Rifabutin 150–300 mg/d Ethambutol daily +/− Fluoroquinolone daily Consider adding parenteral amikacin for severe disease Optimal duration of therapy unknown; determine according to clinical response and available objective findings during treatment; Consider 12 months of culture-negative sputum

Recommendations are not presented for species that very rarely cause lung disease and/or with very little published data, although discussion regarding drug treatment is provided in the text
If moxifloxacin is not tolerated, other fluoroquinolones may provide a similar contribution to regimens listed above, although they are generally considered to be less active agents
[a]Listed treatment regimens are daily for the oral agents; injectable agents generally administered intermittently
[b]Thrice weekly rifampin 600 mg, ethambutol 25 mg/kg, and clarithromycin 500–1000 mg for 12 months of culture-negative sputum demonstrated excellent results in one small study [20]
[c]Isoniazid is likely the weakest drug among the list that also includes macrolides and moxifloxacin and likely should be considered when other agents cannot be used
[d]Given the high mortality associated with *M. xenopi* disease, some believe an aggressive approach including four oral agents (rifampin, ethambutol, macrolide, moxifloxacin) may often be warranted, carefully considering the increased and potentially severe toxicity burden

Table 3 Management of non-pulmonary disease caused by commonly encountered slow-growing NTM species excluding MAC[a]

Species	Organ involved	Antimicrobial therapies
M. genavense	Intestinal, hepatic, splenic, lymph node, bone marrow, bloodstream	Very few data; Consider therapy with at least two of: Azithromycin 250 mg/clarithromycin 500 mg bid Moxifloxacin 400 mg/ciprofloxacin 500–750 mg bid Rifampin 600 mg (or rifabutin 150–300 mg) Ethambutol Amikacin/streptomycin Optimal duration of therapy unknown
M. haemophilum	Childhood cervical lymphadenitis Skin/soft tissue, disseminated disease	Complete surgical excision alone is curative in immune competent children Multidrug therapy including clarithromycin plus Rifampin or rifabutin plus Ciprofloxacin +/– amikacin in severe disease Duration 12–24 months Consider surgical debridement
M. kansasii	Disseminated, usually in context of lung disease	Generally recommended to treat similar to lung disease (Table 1)
M. lentiflavum	Most commonly lymphadenopathy in immune competent children Focal non-pulmonary and disseminated disease less common	Very few data Surgical resection alone may be curative in lymphadenopathy in immune competent patients Medical therapy (adjuvant or primary – see below) may also be useful Consider at least two or three drugs selected from: Azithromycin 250 mg or clarithromycin 500 mg bid Moxifloxacin 400 mg or ciprofloxacin 500–750 mg bid Rifampin 600 mg (or rifabutin 150–300 mg) Ethambutol Amikacin or streptomycin Optimal duration of therapy unknown
M. malmoense	Childhood cervical lymphadenitis Skin/soft tissue	Complete surgical excision alone is curative in immune competent children Same as for pulmonary disease, consider debridement
M. marinum	Skin/soft tissue	Clarithromycin 500 mg bid plus Ethambutol 15 mg/kg daily Other drugs may be used; see text Duration 1–2 months after symptom resolution (at least 3 months total)

Table 3 (continued)

Species	Organ involved	Antimicrobial therapies
M. xenopi	Contaminated wound or surgical device, tenosynovitis	Generally recommended to treat similar to lung disease (Table 1) plus debridement of infected tissues
M. ulcerans	Skin/soft tissue	Rifampin 10 mg/kg daily plus Streptomycin 15 mg/kg I.M. daily, or Clarithromycin 7.5 mg/kg BID, or Moxifloxacin 400 mg daily Duration of 8 weeks Consider adjunctive surgery See WHO treatment guidelines [184]

In the above recommendations, unless otherwise indicated, oral agents are daily, and injectable agents are intermittent

[a]Treatment of non-pulmonary infection with species not presented in Table 3 is generally suggested to employ the same regimens as presented in Table 2 (pulmonary disease), with additional details provided in the text

Mycobacterium kansasii

The Organism and the Host

It is likely that *M. kansasii* infections are contracted from potable waters, in that the organism has often been isolated from water distribution systems [7]. Five to seven strain types have been classified using DNA-based analyses, and although the large majority of human isolates from the USA, Europe, and Japan are of subtype I [8], this subtype has not been readily identified in environmental samples [9]. *Mycobacterium kansasii* commonly causes pulmonary disease, and in some reports this species is among the most commonly encountered causes of NTM-PD, though this is highly variable by region. High rates of *M. kansasii* lung disease have been reported in some areas in Europe, the Southern and Midwestern USA, and some regions in the Middle East, East Asia, and Australia [10, 11]. Risk factors include COPD, prior TB, bronchiectasis [11], and silicosis, the latter described in South African gold miners identified as having an extremely high incidence, regardless of HIV status [12]. Disseminated *M. kansasii* infection occurs in immune-compromised patients, most commonly with advanced HIV, as well as other immune-deficient states [13]. In the respiratory tract, *M. kansasii* has generally been assumed to be relatively pathogenic compared with most other NTM species, and so isolation of this organism was believed to be indicative of disease more often than specimen contamination or patient colonization. It is therefore perhaps not surprising that phylogenetic analyses suggest that, among all NTM species, *M. kansasii* is particularly closely related to *M. tuberculosis* [14]. Early North American studies reported that 45–88% of patients with pulmonary *M. kansasii* isolates had disease, compared with 25–47% of patients with *M. avium* complex (MAC) isolates [11]. The proportion of disease among patients with pulmonary *M. kansasii* isolation in more recent studies from New York City (70%) [15], the Netherlands (71%), [2] Germany (68%), [16] Croatia (40%) [17], Israel (≥50%) [5], and South Korea (52%) [6] have suggested relatively high levels of pathogenicity. Data suggest that pulmonary isolation of *M. kansasii* must be carefully considered regarding pathogenicity, but in a significant proportion of patients, disease will be absent.

Clinical Presentation of M. kansasii *Lung Disease*

Mycobacterium kansasii infects mostly males, with an average reported patient age of 45–62 years [18, 19]. Preexisting pulmonary disease, including COPD, prior TB, bronchiectasis, and silicosis, appears to be common. Radiologically, fibrocavitary disease is the most common, reported in 46–72% [6, 19, 20], with the nodular bronchiectatic type reported in 28–32% (Fig. 1a, b) [6, 20]. The disease presentation is very often similar to TB, with upper lobe cavitary disease and productive cough seen in the majority [6, 19]. The diagnosis of pulmonary *M. kansasii* disease should generally be made according to guideline definitions [1], but the relatively high pathogenicity mandates very careful scrutiny and follow-up of patients with even a single sputum isolate.

Antimicrobial Treatment of M. kansasii *Lung Disease*

Although a single preferred regimen has not been identified, highly successful treatments for *M. kansasii* lung disease have been reported with several multidrug regimens. Rifampin, isoniazid, ethambutol, streptomycin, amikacin, clarithromycin,

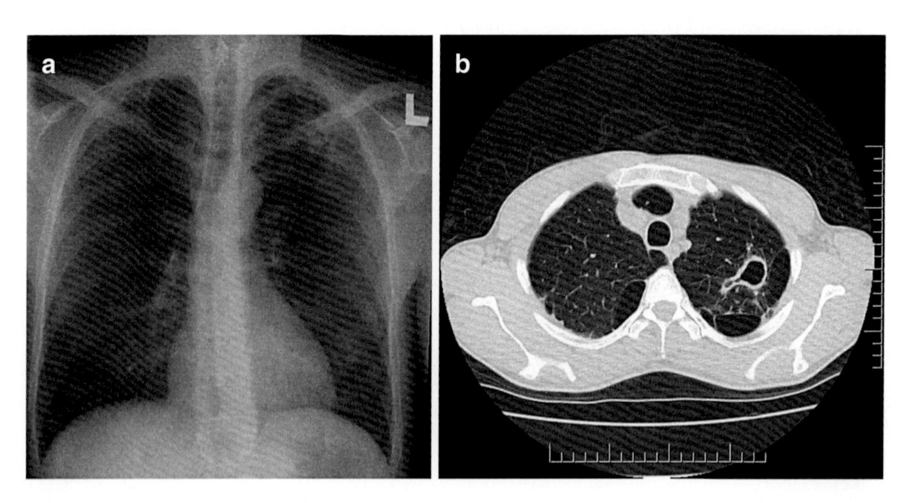

Fig. 1 *M. kansasii.* Chest radiograph (**a**) and chest CT (**b**) images from a 44 yo man, heavy smoker, emphysema (centrilobular and paraseptal) with bullae, and cavitary LUL M. kansasii disease

azithromycin, and sulfamethoxazole have all been reported as components of successful regimens [1, 6]. In vitro data suggest that newer moxifloxacin and linezolid are active against *M. kansasii* [21], and expert opinion holds that these agents may be useful components of therapy [22].

In contrast to most NTM species, DST for *M. kansasii* can generally be interpreted in the clinical context, and testing should be routinely performed for both rifampin and clarithromycin [22]. Rifampin resistance (MIC >1.0 mcg/mL) is associated with treatment failure in the absence of adequate additional effective agents. A description of low-level rifampin resistance (MIC 2.0–8.0 mcg/mL) has been reported as usually susceptible to rifabutin (MIC ≤0.5 mcg/mL), while higher-level rifampin resistance (>8.0 mcg/mL) was resistant to rifabutin [23]. Clarithromycin susceptibility is also considered to be clinically relevant, and testing should be considered. Susceptibility testing for additional agents has been recommended in the setting of rifampin resistance (MIC >1.0 mcg/mL) or drug intolerance [1]. It is stressed that there are not recognized clinically relevant isoniazid or ethambutol MICs and that MICs to these antimycobacterial drugs are not clearly relevant in the context of an otherwise strong regimen (e.g., rifampin containing with demonstrated rifampin susceptibility) [1].

Although there are few controlled data regarding *M. kansasii* treatment, consideration of numerous studies illustrates the importance of rifampin. Pre-rifampin 6-month sputum conversion rates were 52–81%, compared with 100% in several studies containing rifampin [1]. The combination of rifampin (600 mg/d), isoniazid (300 mg/d), and ethambutol (25 mg/kg/d for 2 months followed by 15 mg/kg/d) for 18 months [24–26], or for a shorter period of 12 months of culture-negative sputum (with [27, 28] or without [29] intermittent streptomycin for 3 months), has been highly successful. In addition, a small study demonstrated excellent results using thrice weekly rifampin 600 mg, ethambutol 25 mg/kg, and clarithromycin 500–1000 mg for 12 months of culture-negative sputum [20]. Sputum conversion was nearly universal, and recurrence rates are extremely low in studies using the above regimens. Some experts, citing the questionable value of isoniazid and the apparent efficacy of macrolides and fluoroquinolones, favor treating drug-susceptible *M. kansasii* with the following regimen: rifampin 600 mg/d (or rifabutin 150–300 mg/d) plus ethambutol 15 mg/kg/d plus one of either azithromycin 250 mg/d or clarithromycin 500 mg bid or moxifloxacin 400 mg/d [22].

Recommendations for the treatment of *M. kansasii* lung disease are presented in Table 2. Current guidelines' first-line recommendation for treating rifampin-susceptible *M. kansasii* PD comprises rifampin (600 mg/d), isoniazid (300 mg/d), and ethambutol (15 mg/kg/d) for 12 months of negative sputum cultures [1]. The recommendation for 15 mg/kg/day of ethambutol is based on the lack of data proving that a higher dose is required in the initial months of therapy, the reduced risk of ocular toxicity with the lower dose, and the recognition of the primary importance of rifampin in this combination. Some experts favor rifampin 600 mg/d (or rifabutin 150–300 mg/d) plus ethambutol 15 mg/kg/d plus one of either azithromycin 250 mg/d or clarithromycin 500 mg bid or moxifloxacin 400 mg/d [22], effective in two small studies [6, 20]. Given the excellent response to antimicrobial therapy, adjuvant surgical intervention is rarely indicated in *M. kansasii* lung disease [1].

Pulmonary disease with rifampin-resistant *M. kansasii*, usually resulting from prior therapy, requires special consideration. One demonstrated successful treatment regimen includes high-dose isoniazid (900 mg/d), pyridoxine (50 mg/d), ethambutol (25 mg/kg/d), sulfamethoxazole (1.0 g thrice daily), and streptomycin or amikacin (daily or five times weekly for 2–3 months followed by intermittent 3–4 months), continued for 12–15 months of sputum negativity [30]. The 2007 guidelines also suggest that a three-drug regimen, based on susceptibility testing, combining a macrolide (clarithromycin or azithromycin), moxifloxacin, ethambutol, sulfamethoxazole, or streptomycin, is likely to comprise an effective regimen for rifampin-resistant *M. kansasii* disease [1].

Non-pulmonary M. kansasii *Disease*

M. kansasii is the second most common cause of disseminated NTM disease, most commonly occurring in patients with advanced HIV, as well as other immune-deficient states [13]. Disseminated *M. kansasii* is usually seen in the context of underlying pulmonary infection. Disease limited to non-pulmonary tissues is rare. The treatment of disseminated *M. kansasii* is generally the same as for pulmonary disease. The importance of rifamycins should be borne in mind with respect to compatibility with antiviral regimens among HIV-infected patients receiving antiviral therapy. Updated guidelines for rifamycin compatibility with antivirals are available (http://www.cdc.gov/tb/publications/guidelines/tb_hiv_drugs/default.htm). As described above, in the absence of a rifamycin, a macrolide or moxifloxacin could be substituted.

Mycobacterium xenopi

The Organism and the Host

Potable water systems are extensively described potential sources of human infection with M. *xenopi* [31–34]. The organism has been isolated from household water among patients with *M. xenopi* lung disease [31], from hospital water causing disease outbreaks of postoperative spinal infections when used to rinse surgical devices [34], and pseudo-outbreaks when tap water contaminated clinical specimens in numerous ways [33]. *Mycobacterium xenopi* commonly causes pulmonary disease, and in some reports this species is among the most commonly encountered causes of NTM-PD [10, 11], though this is highly variable by region. Highest rates have been reported in several regions in Europe [17, 35, 36], an institution in Israel [5], and the province of Ontario, Canada [37]. Although uncommonly encountered in the USA overall, appreciable numbers of cases may be encountered in the Northeast [15, 38]. Disseminated *M. xenopi* disease is rare, but localized non-pulmonary disease is described, including cases apparently acquired via wound or surgical device contamination with tap water [34] as well as some cases of tenosynovitis [39]. There are conflicting data regarding the pathogenicity of *M. xenopi.* Some studies suggest the vast majority of isolates represent contamination [40], while more recent investigations suggest that 28–36% of patients with pulmonary isolates have true infection [3, 41]. Although it is important to consider the possibility of contamination or colonization, the high frequency of *M. xenopi* lung disease in many regions dictates that respiratory isolates of this species should carefully be considered as possible pathogens.

Clinical Presentation of M. xenopi *Lung Disease*

Mycobacterium xenopi lung disease is strongly associated with COPD [42] and presents with cavitation in 36–46% [43, 44]. Given this association, it is not surprising that *M. xenopi* has an associated mortality that is among the highest of all NTM species [45, 46]. The typical patient with *M. xenopi* is male, with COPD and fibrocavitary disease. However, perhaps a third of patients have been identified with a pattern of random, small to large nodules, usually in the setting of COPD [43, 47], possibly representing a precursor to fibrocavitation (Fig. 2a, b). Some patients with *M. xenopi* lung disease present with a nodular bronchiectatic pattern that is typical of MAC lung disease, although the frequency of this phenotype is inadequately studied. A nodular bronchiectatic pattern was clearly described in 4% (1/24 patients) in one study [43]. Inferences from two studies where a pattern of nodular bronchiectasis was not explicitly sought include a frequency of 44% (4/9 patients with coexisting bronchiectasis and nodules but no cavitation) [48] and 4.4% (6/136 patients with a nodular form that included bronchiectasis) [47]. We believe that

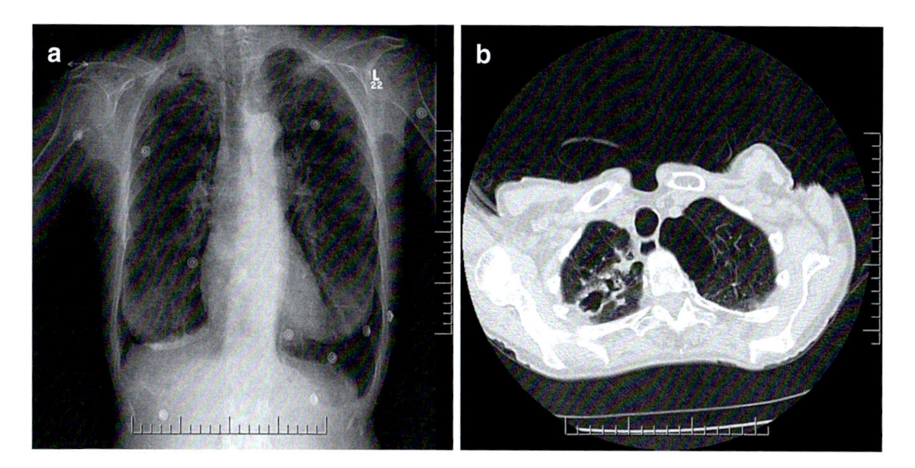

Fig. 2 *M. xenopi.* Chest radiograph (**a**) and chest CT (**b**) images from 73 yo woman, heavy smoker with emphysema and bullae presents with RUL cavitation from M. xenopi

some minority of patients with *M. xenopi* lung disease has the nodular bronchiectatic radiologic pattern and that clinicians should consider this possibility. The diagnosis of *M. xenopi* lung disease should be made using the ATS/IDSA diagnostic criteria [1], bearing in mind the possibility of contamination and pseudo-outbreaks.

Treatment of M. xenopi *Lung Disease*

Controlled data regarding the treatment of *M. xenopi* lung disease are scarce. Drugs thought to have utility have included clarithromycin/azithromycin, ethambutol, rifampin, rifabutin, fluoroquinolones, isoniazid, and streptomycin/amikacin [1]. Susceptibility testing may be difficult to interpret, and although recommendations have been made that susceptibility should be tested according to the recommendations for rifampin-resistant *M. kansasii* [49], data regarding clinically relevant MIC thresholds are scarce.

At least two randomized trials have been published in full [50, 51], and one randomized trial has interim results available [52]. The British Thoracic Society (BTS) sponsored a randomized trial, including 42 patients with *M. xenopi* disease, comparing rifampin 450–600 mg/day plus ethambutol 15 mg/kg/day, with or without isoniazid 300 mg/d [50]. Cavitation was present in 85% of cases, and 67% had underlying lung disease, likely contributing to the high mortality of 57%. Outcomes otherwise included treatment failure or relapse of 12% and a very low rate of 5-year infection-free survival of 17%. There were no statistically significant differences between treatment arms, but patients receiving isoniazid had somewhat lower treatment failure or relapse (5% versus 10%) but also reduced 5-year infection-free survival (10% versus 23%). A second BTS study randomized 34 patients with *M. xenopi* disease to receive rifampin 450–600 mg/day plus ethambutol 15 mg/kg/day plus clarithromycin 500 mg bid or rifampin plus ethambutol plus ciprofloxacin 750 mg bid [51]. Cavitation was present in 62%, and underlying lung disease on radiography was apparent in 62%. Outcomes were poor with 35% in each arm alive and cured at 5 years. It is noteworthy that both of the above studies also included separate sub-studies of patients with MAC and *M. malmoense* and that the survival was by far the worst among patients with *M. xenopi*. A study randomizing patients to rifampin plus ethambutol plus either clarithromycin or moxifloxacin has reported on a preliminary analysis of 36 patients, with an 83% 6-month sputum culture conversion rate overall, with no difference between groups [52]. The extremely high mortality rates in the BTS-sponsored studies make interpretation of drug treatment data somewhat difficult, although the second study might suggest that fluoroquinolones may play a useful role in the treatment of *M. xenopi*. This notion is supported by the ongoing randomized trial, wherein interim 6-month culture results suggest similar efficacy of clarithromycin- and moxifloxacin-based regimens.

Antimicrobial treatment data from uncontrolled studies were summarized in a systematic review, combining 188 patients from 23 reports, including 34 different drug combinations [44]. The two statistically significant findings comprised slightly lower "end-of-treatment" success rates among patients who receive isoniazid (73% versus 87%, $p = 0.028$) and also among patients who received aminoglycosides (56% versus 82%, $p = 0.019$). The authors stressed that firm conclusions cannot be made due to the heterogeneity of the studies and patients and the high likelihood of bias in drug treatment selection. A large retrospective study, published after the systematic review, including 13 centers in France and 136 patients, found that drug treatment and the use of rifamycin-containing regimens were both associated with better survival [47]. The retrospective methods leave this study vulnerable to the inherent biases of treatment assignment and drug selection however. Although we have inadequate human data regarding aminoglycosides in *M. xenopi*, data from murine models may be useful. Two studies in mice infected with *M. xenopi* have shown reduced colony-forming units among mice treated with amikacin in addition to comparator regimens [53, 54]. One study used intravenously infected mice treated with clarithromycin and ofloxacin plus/minus amikacin [53], and the other study used an inhalational infection and treatment with either clarithromycin/ethambutol/rifampin or moxifloxacin/ethambutol/rifampin plus/minus amikacin [54], and both studies identified microbiologic benefit.

Resectional surgery has been reported as adjunctive therapy for *M. xenopi* lung disease. The largest series studied 57 patients treated between 1964 and 1980, including 40 individuals who had been treated with various unsuccessful courses of isoniazid, rifampin, and ethambutol and 17 patients with surgery for presumed lung cancer but were found to have *M. xenopi* on surgical resection [55]. Postoperative complications were frequent, including 19 patients requiring reoperation (indication not described) with complex procedures and 2 deaths (1 pulmonary embolism and 1 advanced cancer). Patients with pleural lesions, bullae, and respiratory impairment had higher rates of complications. It is difficult to know how the results of this study should be translated into practice today, with the advent of additional drugs employed for *M. xenopi* (i.e., macrolides, fluoroquinolones, aminoglycosides). We think it is likely that the observed risk factors for postoperative complications likely remain important, but the magnitude of postoperative risk may be lower in selected patients. Other studies have presented only very small numbers of patients operated for known *M. xenopi*, and so minimal additional data are available.

Recommendations for the treatment of *M. xenopi* lung disease are presented in Table 2. The 2007 ATS/IDSA guidelines express uncertainty regarding the optimal regimen for *M. xenopi* lung disease but propose isoniazid, rifamycin, ethambutol, and clarithromycin ± an initial course of streptomycin and that a fluoroquinolone (preferably moxifloxacin) might be substituted for one of the antituberculous drugs. Based on the available data, we recommend a regimen containing at least three oral agents, comprising rifampin (or rifabutin), plus ethambutol plus either a macrolide (clarithromycin is most studied) or a fluoroquinolone (preferably moxifloxacin). Although it is unclear what proportion of the high mortality is attributable to underlying disease versus *M. xenopi* infection per se, we think an aggressive approach is often warranted. Accordingly, we carefully consider a four-drug oral combination of macrolide + fluoroquinolone + rifampin + ethambutol in patients with severe disease, mindful of the increased adverse effects associated with additional agents. In addition, an injectable aminoglycoside (i.e., amikacin or streptomycin) should be considered in advanced disease. In the absence of data regarding the optimal duration of antimicrobial therapy for *M. xenopi* PD, we suggest continuing therapy until sputum specimens have been persistently culture negative for 12 months. In our experience, adjunctive surgery for localized residual disease, despite intensive therapy, is safe and effective for carefully selected patients. Patients with inadequately controlled infection, multifocal destructive infective lesions, and extensive underlying lung disease are at substantially increased risk of postoperative morbidity and mortality. Surgical management of NTM disease is described in detail in a separate chapter of this monograph.

Non-pulmonary M. xenopi *Disease*

Therapy for non-pulmonary *M. xenopi* infection has not been well-studied. We recommend considering regimens similar to those listed for treating pulmonary disease. Surgical debridement is often important for soft tissue infections [39].

Mycobacterium malmoense

The Organism and the Host

M. malmoense has rarely been isolated from the environment, perhaps because of difficulties in culturing it [56], although it has been found in natural water in Finland and soil in Zaire and Japan [1]. *M. malmoense* most commonly causes lung disease but can also cause cervical lymphadenitis, tenosynovitis, skin infections, and rarely disseminated disease in the severely immunocompromised [57]. *M. malmoense* is one of the most common species causing NTM lung disease in regions of Northern Europe [10, 58]. It has also been reported in laboratory specimens in Southern Europe, South Africa, and North America [58].

Clinical Presentation and Treatment of M. malmoense *Lung Disease*

The clinical presentation of *M. malmoense* pulmonary disease often mimics tuberculosis, with cavities and airspace disease being the most common radiographic findings (Fig. 3a, b) [57]. Approximately 60% of affected patients are male, and almost half have underlying COPD or a history of pulmonary TB [57]. When *M. malmoense* is isolated from respiratory secretions, it is usually of clinical significance; in most studies, greater than 70% of individuals with isolation had clinically significant disease [57]. One notable exception came from the USA, where only 10% of isolates were clinically significant [59]; this suggests that pathogenicity may vary by region.

Drugs with some evidence of clinical efficacy against *M. malmoense* include rifampin, ethambutol, isoniazid, macrolides, and fluoroquinolones. Results of in vitro drug susceptibility testing have been inconsistent, which may be due to differences in laboratory techniques or regional variability of the organism. A lack of correlation between in vitro drug susceptibility test results and clinical response has been described [50, 60, 61], although this has been incompletely investigated.

There have been very few studies of treatment for *M. malmoense* pulmonary disease. The British Thoracic Society performed a randomized trial involving 106 patients with *M. malmoense* pulmonary disease [50, 61], in which patients were treated for 2 years with either rifampin and ethambutol or rifampin, ethambutol, and isoniazid. Ten percent of patients had a poor clinical outcome (death due to NTM, treatment failure, relapse), with no significant difference between groups. Fifty-nine percent of patients were alive at 5 years, and 42% of the original 106 patients were considered cured. A second BTS trial randomized 167 patients to 2 years of treatment with rifampin, ethambutol, and either clarithromycin or ciprofloxacin [51]. Again, there was no significant difference in rates of poor outcomes between groups (overall, 7%). However, significantly more patients receiving the clarithromycin containing regimen completed treatment and were alive and cured at 5 years (38.4% versus 19.8%). That said, there were more side effects in the group that received clarithromycin, and the proportion alive and cured at 5 years was the same as that seen with only rifampin and ethambutol in the prior trial. It is also of note that compared to patients with MAC and *M. xenopi* pulmonary disease, who were also included in the trial, patients with *M. malmoense* were significantly less likely to have a poor outcome.

A retrospective review of 30 patients with *M. malmoense* pulmonary disease from the Netherlands found that treatment regimens varied considerably [62]. In

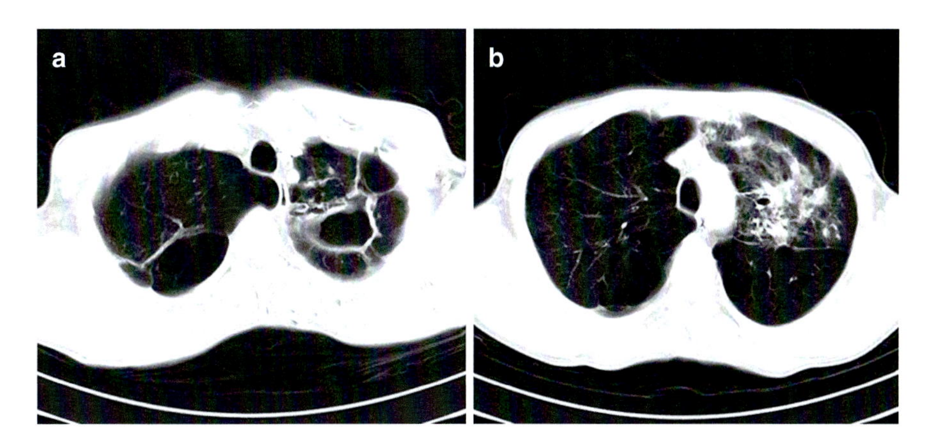

Fig. 3 *M. malmoense.* Chest CT (**a** and **b**) images from 68 year old man with emphysema and bullae with left upper lobe fibrocavitary M. malmoense and adjacent airspace disease. (Images courtesy of Dr. Jakko van Ingen)

that study, 21 patients (70%) had a good clinical response to treatment, 5 patients (17%) had failure or relapse, and 4 (13%) died. In a series of 14 consecutive patients with fibrocavitary *M. malmoense* lung disease in Edinburgh, Scotland, investigators reported 100% culture conversion and symptom reduction after 24 months of daily rifampin (450–600 mg), ethambutol (15 mg/kg), and clarithromycin (500 mg bid) [63]. In this study, all isolates were judged to be susceptible to rifampin and resistant to isoniazid. Resistance to ethambutol was observed in 23%, clarithromycin in 8%, and ciprofloxacin in 46%.

Recommendations for the treatment of *M. malmoense* pulmonary disease are presented in Table 2. Given the minimal data, there is no consensus on the optimal therapy for *M. malmoense* pulmonary disease. The ATS/IDSA guidelines have suggested the use of a regimen including isoniazid, rifampin, and ethambutol with or without a macrolide or fluoroquinolone [1]. Based on the findings across studies described above, we think it is likely that clarithromycin (or azithromycin) is preferable to isoniazid and would probably consider rifampin, ethambutol, and clarithromycin (or azithromycin) as a first-line regimen. The addition of a fluoroquinolone or isoniazid may be additionally helpful. In severe disease, we think that adding amikacin is a reasonable consideration despite an absence of published experience for this organism. This opinion is based on the utility of amikacin for most NTM species, often favorable DST profiles, and expert opinion.

M. malmoense *Non-pulmonary Disease*

M. malmoense cervical lymphadenitis presents similarly to that caused by MAC; it most commonly affects children and causes painless unilateral swelling of the cervical or submandibular lymph nodes, without other symptoms. A presumptive diagnosis of NTM lymphadenitis is based on lymph node biopsy material showing caseating granuloma with or without AFB and a negative tuberculin skin test, while a definite diagnosis requires recovery of the causative organism from the lymph node cultures, obtained either by fine needle aspiration, incision and draining, or excisional biopsy [1]. Complete surgical excision alone is typically curative in immunocompetent children [62]. *M. malmoense* is an uncommon cause of tenosynovitis, skin infections, and rarely disseminated disease in the severely immunocompromised [57]. There is inadequate evidence to guide treatment of *M. malmoense* infection of these body sites, but use of a similar antibiotic regimen to that recommended for *M. malmoense* pulmonary disease, plus surgical debridement when possible, is recommended.

Mycobacterium simiae

Mycobacterium simiae has been isolated from various freshwater sources [64–66], numerous types of animals (originally monkeys [1]), and milk from dairy animals [67]. Its presence in potable water supplies and thus a cause of pseudo-outbreaks implicates water as a likely source of *M. simiae* causing human infections [66]. Uncommon in most regions, *M. simiae* has been most extensively described in Israel [68, 69], Arizona [70], and Texas [71]. Respiratory *M. simiae* isolation is usually not indicative of disease and is reported to be associated with disease in various series from 4% to 21% of cases [70–72]. Recognizing that respiratory isolates of *M. simiae* are most often contaminants, clinicians should maintain a relatively high threshold to diagnose lung disease with this NTM species, looking carefully for other causes of worsening symptoms or imaging abnormalities. Regarding anatomic sites of infection, pulmonary isolation and disease are most common (Fig. 4a–c). Disseminated disease in immune-compromised patients [71] and lymph node disease in normal hosts [73] are also both described, although the latter appears to be quite rare.

Reports regarding DST of *M. simiae* paint a bleak picture, with resistance to almost all tested drugs [71, 72, 74], and most authors describe this infection as extremely difficult to treat. Very few data support a particular regimen. Susceptibility to clarithromycin, trimethoprim-sulfamethoxazole, moxifloxacin, clofazimine and aminoglycosides may variably be observed, and anecdotal support exists for the use of linezolid. The use of DST has been recommended to guide drug selection, with the caveat that results may not correlate with clinical outcomes [22].

Expert opinion recommends the use of a multidrug regimen selecting several agents from a list that includes amikacin, azithromycin or clarithromycin, moxifloxacin, trimethoprim-sulfamethoxazole, linezolid and perhaps clofazimine [1, 22]. The optimal duration of therapy is unknown. Should durable sputum conversion occur, treatment for 12 months after the first negative culture is likely appropriate. Otherwise treatment duration should be guided according to clinical circumstance.

Mycobacterium szulgai

Mycobacterium szulgai has been only rarely isolated from the environment, including a case from water and fish from a tropical fish aquarium owned by a patient with pulmonary *M. szulgai* infection [75]. Because of the scarcity of reports of environmental isolation, respiratory isolates had generally been regarded as indicative of disease, assuming that environmental contamination was unlikely [1]. This has been challenged by findings of a comprehensive national survey from the Netherlands [76] and a consecutive series from a South Korean institution [77], wherein the proportion of patients with pulmonary *M. szulgai* isolation who fulfilled disease criteria was 76% and 43%, respectively. Given that a significant proportion of patients with *M. szulgai* respiratory isolates do not have disease, clinicians should carefully apply current NTM diagnostic criteria while recognizing that the likelihood of disease is substantial. Infections with *M. szulgai* are uncommon but have been reported widely across the globe [78]. Over a 13-year period in Ontario, Canada, among 9658 patients who fulfilled microbiological criteria for NTM lung disease, only 10 (0.1%) had *M. szulgai* (unpublished data). The majority of

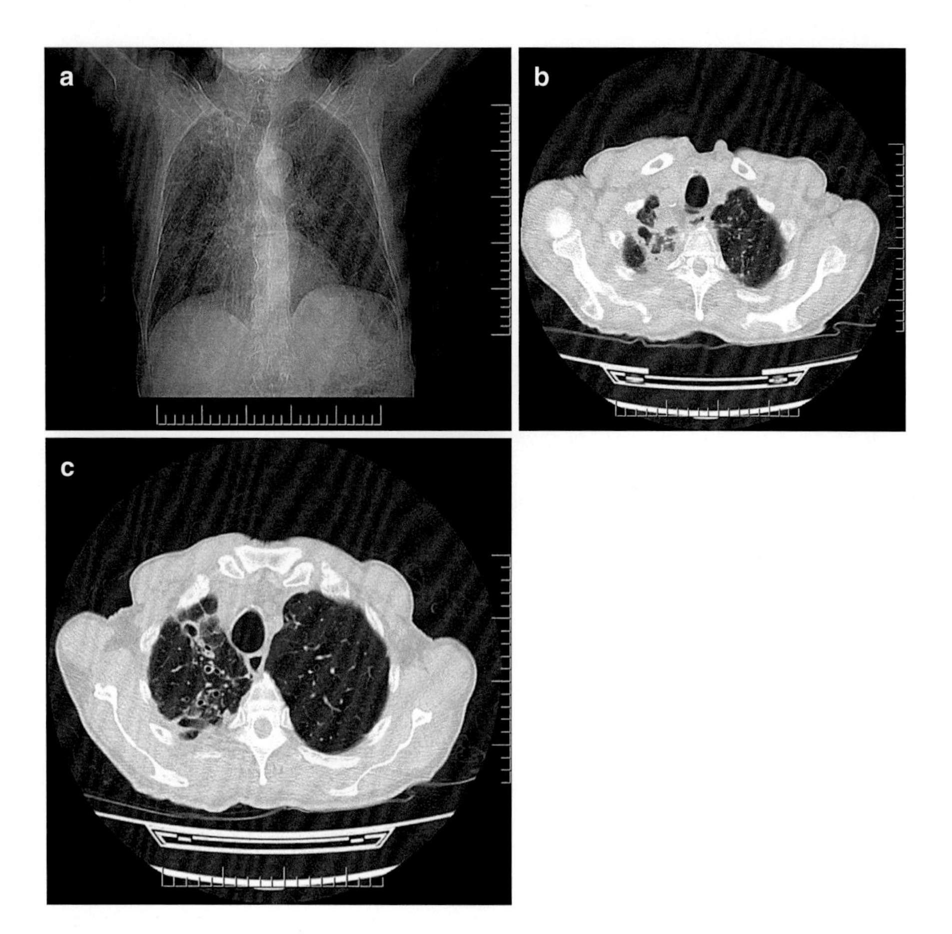

Fig. 4 *M. simiae*. Chest radiograph (**a**) and chest CT (**b** and **c**) images from 82 year old man, 45 pack-year former smoker, with RUL bronchiectasis and cavitation with *M. simiae* disease

published cases comprise lung disease, although lymphadenitis, skin and soft tissue, disseminated, and other focal infections have been described [1, 78, 79]. *M. szulgai* lung disease is frequently described as a TB-like illness, usually in men, with cavitation and systemic symptoms, often in the presence of COPD, and with a prior history of pulmonary TB (Fig. 5a, b) [76, 77].

Drug susceptibility testing for *M. szulgai* has been reported in several reviews, suggesting frequent susceptibility to antituberculous agents, including rifampin, ethambutol and amikacin [49, 78, 79], and more recently clarithromycin [76]. However, as for most NTM species, there are no data regarding the appropriate clinical correlates for in vitro MIC values for this organism.

Current guidelines suggest treating pulmonary *M. szulgai* with a three- or four-drug regimen and that macrolides, fluoroquinolones, and standard TB drugs often have activity. In addition to numerous case reports, two recent series report that clarithromycin, azithromycin, rifampin, ethambutol, isoniazid, and less often fluo-

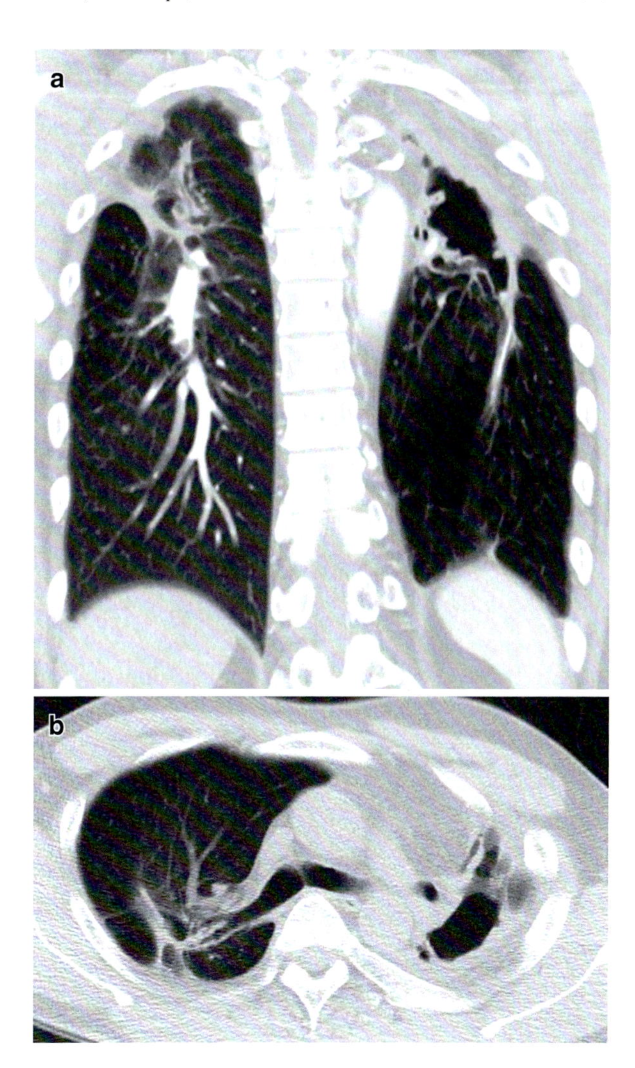

Fig. 5 *M. szulgai*. Chest CT (**a** and **b**) from 47 year old man with *M. szulgai* lung disease with associated bilateral upper lobe disease and extensive fibrocavitation in the left upper lobe. (Images courtesy of Dr. Won-Jung Koh)

roquinolones have been components of successful treatment regimens in pulmonary disease [76, 77]. Treatment duration of 8–18 months is most frequently reported, and response to treatment is usually very good, with prognosis often dependent on comorbid conditions. Based on the available observational retrospective data, we think that a reasonable treatment regimen comprises a three- or four-drug combination selected from clarithromycin or azithromycin, a rifamycin (usually rifampin), ethambutol, and possibly a fluoroquinolone, continued until 12 months after achieving negative sputum cultures. In severe disease, we think that adding amikacin is a reasonable consideration despite very limited published experience for this organism. This opinion is based on the utility of amikacin for most NTM species, often favorable DST profiles, and expert opinion. Although interpretation is not clear, DST should be considered. Non-pulmonary *M. szulgai* infection, when limited to a single focus, is likely to be adequately treated with 4–6 months of treatment with similar agents [1], though in the setting of immune deficiency, dissemination, or multifocal disease, longer duration therapy is undoubtedly preferred [76]. Treatment response is expected to be good in the absence of immune deficiency or severe underlying comorbidities.

Mycobacterium genavense

The Organism and the Host

Mycobacterium genavense is an uncommon cause of NTM disease that was first described in patients with advanced HIV infection and extensive disseminated mycobacterial disease [80]. The organism has been detected by molecular methods in hospital tap water [81] and unchlorinated water from the distribution system of water treatment plants [82], as well as by culture from infected pets and zoo animals, predominantly birds [83, 84]. *Mycobacterium genavense* is a fastidious organism requiring prolonged incubation in liquid media or supplemented solid media for growth and detection. Human infection with *M. genavense* has been described almost exclusively in immunocompromised individuals and usually with disseminated infection [85, 86]. Pulmonary involvement occurs in few patients, and predominantly pulmonary disease appears to be very rare [87]. *Mycobacterium genavense* is not a common cause of NTM disease in any region. Most reported cases have been in Europe, and it appears to be very rare in North America.

Non-pulmonary M. genavense *Disease*

As noted above, almost all infections with *M. genavense* are non-pulmonary and involve immunocompromised patients [85, 86]. Sites of involvement include intestinal, hepatic, splenic, lymph node, bone marrow, and bloodstream [85, 86]. Although susceptibility has been reported to macrolides, rifamycins, fluoroquinolones, amikacin, and streptomycin, and resistance has been reported to ethambutol and isoniazid [88, 89], clinically significant MIC thresholds have not been defined. The treatment of non-pulmonary *M. genavense* disease is probably best guided by published series. In a multicenter series of 25 patients in France, including 20 HIV-infected patients, treatment included clarithromycin (84%), rifamycins (72%), ethambutol (88%), fluoroquinolones (24%), and amikacin (28%) [85]. Overall, cure was achieved in 32%, while an additional 25% had chronic controlled disease. In 13 patients from the Netherlands, including four HIV-infected patients, treatment included clarithromycin (92%), rifamycins (77%), ethambutol (84%), and fluoroquinolones (15%) [86]. Cure was achieved in 31%, while 38% had chronic controlled disease.

The ATS/IDSA guidelines did not provide a recommendation on treatment but suggested that macrolide-containing regimens may be preferred [1]. On balance, it appears that a combination of drugs including at least two agents selected from a macrolide, rifamycin, ethambutol, fluoroquinolone, and amikacin or streptomycin may comprise an appropriate regimen. Clinicians should be mindful of interactions between rifamycins and antiviral drugs used for treating HIV infection and consult appropriate resources to develop compatible regimens. The optimal duration of therapy is unknown and will therefore be determined according to clinical response and available objective findings during treatment.

Pulmonary M. genavense *Disease*

Successful treatment in two cases of predominantly pulmonary disease has been described [87]. One patient had HIV coinfection and was treated with clarithromycin, ethambutol, amikacin, as well as combination anti-retroviral therapy. The addition of subcutaneous thrice-weekly interferon-gamma injections was followed by improvement. Antimycobacterial therapy was administered for 2 years. The other patient, post renal transplant, was treated with clarithromycin, rifabutin, moxifloxacin, and ethambutol. Despite major regression of lesions, therapy was continued beyond 2 years because of ongoing immunosuppressive therapy to prevent rejection of the renal graft. There are inadequate data to distinguish treatment of pulmonary from non-pulmonary *M. genavense*, either for drug selection or duration of therapy.

Mycobacterium gordonae

The Organism and the Host

M. gordonae has been isolated from freshwater and water distribution systems. Many clusters of positive *M. gordonae* cultures, or pseudo-outbreaks, have been reported and have implicated contaminated drinking water [90, 91], ice machines [91, 92], laboratory water [93], an antibiotic solution [94], a topical anesthetic [95], and bronchoscope cleaning fluid [96]. *M. gordonae* is a very commonly isolated contaminant in respiratory specimens [1]. It has been hypothesized that *M. gordonae* in tap water or ice water is deposited in the mouth or oropharynx when ingested and then subsequently contaminates respiratory specimens during sputum expectoration, tracheal suctioning, or bronchoscopy [91, 92]. Therefore, it has been recommended that patients avoid mouth rinsing or drinking beverages containing tap water for several hours before collection of respiratory specimens [1, 97].

Although weakly pathogenic, *M. gordonae* has rarely been reported to cause pulmonary [4, 98–100], peritoneal [101], skin and soft tissue [102, 103], lymph node [2], and disseminated disease [104–106] with extrapulmonary disease typically occurring in immunocompromised patients.

True pulmonary disease caused by *M. gordonae* is very uncommon, with the vast majority of respiratory isolates representing contamination. The proportion of patients with *M. gordonae* isolated in respiratory specimens in the laboratory who meet ATS/IDSA microbiologic criteria (i.e., ≥2 positive sputum samples or 1 positive bronchoalveolar lavage [BAL] or lung biopsy) for pulmonary disease has been reported to be as high as 10.3% in one study from Croatia but only 1.7% when stricter microbiologic criteria are applied (> 2 positive sputum samples or 1 positive BAL and ≥ 1 positive sputum samples) [3]. The proportion of patients with *M. gordonae* respiratory isolates who meet the full ATS/IDSA pulmonary disease criteria is very small; 3.5% [3], 2% [2], and 1% [4] have been reported (the latter study used stricter microbiologic criteria, including ≥3 positive sputum cultures, with ≥1 smear positive) [4]. Therefore, possible pulmonary disease due to *M. gordonae* should be considered with healthy skepticism, including diligent efforts to identify other cause(s) for a patient's illness. Some authors have proposed stricter diagnostic criteria be applied for *M. gordonae* pulmonary disease than for pulmonary disease caused by other mycobacteria, given that it is frequently isolated in respiratory specimens as a contaminant or colonizer and only very rarely associated with true disease [3, 4]. Morimoto et al. proposed that time to positive culture detection in liquid broth could be useful, as could rpoB gene sequencing, as all of their definite cases of *M. gordonae* pulmonary disease had relatively short median times to culture detection, and rpoB subtype C [4]. However, these criteria require further evaluation. Although anecdotal, among nearly 800 NTM patients we have assessed, we have treated only a single case of *M. gordonae* lung disease, despite its frequency of isolation as the third most common pulmonary NTM isolate in Ontario, Canada. We wish to stress that making a diagnosis of *M. gordonae* lung disease requires a much higher level of evidence than other NTM species discussed in this chapter and likely throughout this monograph.

M. gordonae *Pulmonary Disease*

Given the rarity of true *M. gordonae* pulmonary disease, there is little data on clinical presentation or treatment. Morimoto et al. describe five patients with *M. gordonae* pulmonary disease who presented similar to patients with MAC lung disease: two males with smoking histories and fibrocavitary disease and three female nonsmokers with nodular bronchiectatic disease [4]. Interstitial fibrosis [99, 107], organizing pneumonia [107], and mediastinal lymphadenitis [99, 108] have also been histologically described as part of the possible reaction to *M. gordonae* infection.

In vitro, antimicrobial agents most consistently active against *M. gordonae* include ethambutol, rifabutin, clarithromycin, linezolid, and the fluoroquinolones [109, 110]. The optimal antimicrobial regimen is unknown. Several reports have described successful treatment outcomes for *M. gordonae* lung disease with a combination of rifampin and ethambutol [98], plus macrolide [4, 100].

Mycobacterium haemophilum

The Organism and the Host

Water appears to be the most likely source of *M. haemophilum* in the environment [56, 111], and the organism has been isolated from water distribution systems [112]. *M. haemophilum* has unique culture requirements; it requires special medium with hemin- or iron-containing compounds for growth and grows optimally at lower temperatures of 28–30 °C [113, 114]. In immunocompromised patients, it can cause skin and soft tissue infections, septic arthritis, and osteomyelitis [111]. It can also cause cervicofacial lymphadenitis in immunocompetent children, and rarely in immunocompetent adults, while lung disease is very rare [111]. The organism has been isolated from patients on most continents [111].

Clinical Presentation of M. haemophilum *Disease*

M. haemophilum has been reported to cause a variety of infections in immunocompromised patients, particularly those with HIV infection and post-solid organ or bone marrow transplant, and less commonly in patients with rheumatic diseases or malignancy receiving corticosteroids, chemotherapy, or other immune-suppressing medications [111]. Skin lesions are the most common manifestation and can present as erythematous papules, plaques, nodules, necrotic abscesses, or chronic ulcers. They are most frequently found on the extremities. Septic arthritis, osteomyelitis, eye infections, catheter infections, and disseminated disease (with positive blood cultures) have been described [111].

In immunocompetent children, M. *haemophilum* can cause cervicofacial lymphadenitis [111]. In Israel and the Netherlands, *M. haemophilum* was the second most commonly recognized pathogen in children with cervical NTM lymphadenitis, after *M. avium* [115, 116]. The disease presents similarly to that caused by *M. avium*, except that children tend to be older. In immunocompetent adults, *M. haemophilum* rarely causes skin or lymph node infection [111]. An outbreak of *M. haemophilum* skin and lymph node infection due to permanent eyebrow makeup in healthy adults has been reported [117].

M. haemophilum very rarely causes pulmonary disease. Most reports of lung infection due to *M. haemophilum* have occurred in immunocompromised hosts, in whom it may cause isolated lung infection, manifesting most commonly as airspace disease or pulmonary nodules, or as one of several organs infected in patients with disseminated disease [111].

Given its unique culture requirements, clinical suspicion of *M. haemophilum* should prompt supplementation of culture media to recover the organism. *M. haemophilum* should be considered with an AFB smear positive, draining skin lesion that has no growth on ordinary (routine) AFB media. Specimens from immunocompromised patients, such as skin lesions or ulcerations, lymph node aspirates, joint fluid, or other undiagnosed lesions, with a positive AFB smear result, should be cultured for *M. haemophilum*. Additionally, specimens obtained from adenitis in immunocompetent children should be cultured for *M. haemophilum* [1].

Antimicrobial Treatment of M. haemophilum

There are no standardized susceptibility methods for *M. haemophilum*, but drugs that appear to be active in vitro include clarithromycin, ciprofloxacin, rifampin, rifabutin, and amikacin [1]. Discrepant results have been found with streptomycin [111, 118], doxycycline, and sulfonamides [1]. All isolates are resistant to ethambutol [1] and isoniazid [111, 118].

Optimal therapy for *M. haemophilum* infection is unknown. In immunocompromised hosts, successful outcomes have been reported with multidrug regimens including clarithromycin, rifampin or rifabutin, and ciprofloxacin [111, 118–120] with the addition of amikacin in severe disease [118]. Treatment for 12–24 months has been recommended by some authors [111, 118]. Surgical excision/debridement may be helpful for localized disease [111, 121]. Surgical excision alone is usually adequate treatment for lymphadenitis in immunocompetent hosts [111].

Mycobacterium lentiflavum

The Organism and the Host

It is likely that human infection with *M. lentiflavum* is contracted through exposure to contaminated waters. The organism has been isolated from drinking water and water-related facilities in Australia [65], the Czech Republic [64], Finland [122], and the USA. In a clinical environmental study in Brisbane, Australia, environmental sites that yielded *M. lentiflavum* overlapped geographically with home addresses of patients who had clinically significant disease, and automated repetitive sequence-based PCR genotyping showed a dominant environmental clone closely related to clinical strains [123].

Mycobacterium lentiflavum disease has been described most commonly as cervical adenopathy, especially in immunocompetent children, likely comprising approximately half of reported cases, followed by disseminated, focal non-pulmonary, and pulmonary disease [124]. In a large sample of patients with pulmonary *M. lentiflavum* isolates in Australia, only 7.4% (2/27) patients were judged to have a significant disease [123]. A study from Crete, Greece, identified *M. lentiflavum* as the most common NTM isolate, but clinical significance was questioned [125]. Among 4417 immunosuppressed cancer patients with mycobacterial isolates in the USA, *M. lentiflavum* was identified in respiratory specimens of 10 patients and was judged to be insignificant in all [126]. It may therefore be concluded that this generally uncommon isolate is of relatively low pathogenicity and most often a colonizer or contaminant when isolated from pulmonary specimens. Accordingly, care should be taken before deciding that patients with respiratory *M. lentiflavum* isolates have clinically important disease. However, non-pulmonary isolates with objective evidence of infection, especially in the context of immune suppression, should be carefully considered as important pathogens [124].

Pulmonary **M. lentiflavum** *Disease*

Although *M. lentiflavum* pulmonary isolates have been reported to be most often clinically insignificant, disease that appears similar to more common NTM species seems to occur. Pulmonary *M. lentiflavum* disease has been described in nodular bronchiectatic form typical for *M. avium* [123, 127], and at least one case of acute necrotizing pneumonia has been described [128]. Among 354 cystic fibrosis patients in France, 6 (1.7%) had *M. lentiflavum* isolated from respiratory tract secretions, and in 2/6 the organism was judged to be causing disease [129].

There are few data regarding treatment of *M. lentiflavum* lung disease. Case reports generally describe treatment responses with the use of three or more of clarithromycin, rifampin (or rifabutin), ethambutol, or ciprofloxacin [123, 128, 130]. In a study including *M. lentiflavum* isolates from 36 patients in the USA, all isolates were deemed susceptible to clarithromycin and moxifloxacin, while resistance was judged to be variably present to rifampin (79%), rifabutin (46%), ethambutol (83%), ciprofloxacin (4%), amikacin (33%), kanamycin (71%), streptomycin (8%), clofazimine (50%), and linezolid (66%) [74]. Judging by these data, macrolides and fluoroquinolones would be expected to be particularly useful agents, but recommendations advise susceptibility testing to guide therapy of clinically significant disease [49].

There are inadequate data upon which to make firm treatment recommendations. It seems prudent to obtain DST on clinically significant respiratory isolates, though data to assist with interpretation are lacking. A combination of agents including clarithromycin and a fluoroquinolone and possibly one or more of a rifamycin or ethambutol may comprise an effective regimen. The optimal duration of treatment is unknown. Treatment duration might reasonably be guided by clinical and objective testing response as well as 12 months of negative sputum cultures.

Non-pulmonary M. lentiflavum *Disease*

Non-pulmonary *M. lentiflavum* disease may be divided into cervicofacial adenopathy or other non-pulmonary sites that may include disseminated disease. In a study of 17 children with *M. lentiflavum* cervicofacial adenitis, treatment included surgery alone (5/17, 29.5%), antimycobacterials plus surgery (11/17, 64.7%), and antimicrobials alone (1/17, 5.9%) and all experienced resolution [131]. The antimycobacterials included clarithromycin in 4/12 (33%), rifampin in 3/12 (25%), isoniazid in 2/12 (17%), and ciprofloxacin, ethambutol, and pyrazinamide in 1/12 (8%). Based on this information, and the often high rate of apparent drug resistance, perhaps *M. lentiflavum* cervicofacial adenopathy is best treated with a combination of surgery plus antimycobacterials to which the isolate appears susceptible. Disease in other body sites might be treated in a fashion similar to that described for pulmonary disease, with drug susceptibility test results employed to guide selection of drugs. One might consider the use of a combination of agents including clarithromycin and a fluoroquinolone and possibly one or more of a rifamycin or ethambutol as part of an effective regimen. The optimal duration of treatment is unknown and should probably be guided by clinical and objective evidence of response and the host's immune status.

Mycobacterium marinum

The Organism and the Host

The environmental source of *M. marinum* infection in humans is water, and *M. marinum* has been found in a variety of aquatic environments, including fish tanks, unchlorinated swimming pools, and natural bodies of fresh and salt water [56]. *M. marinum* typically causes infection of the extremities, which may be cutaneous or may invade to deeper structures. Skin and soft tissue infections occur in both immunocompetent and immunocompromised hosts. Infection usually occurs following a break in the skin in contaminated water or from direct contact with fish or shellfish [56, 132]. Rare cases of pulmonary disease, cervical lymphadenitis, and disseminated disease in immunocompromised patients have been reported. *M. marinum* infections have been reported worldwide [132].

Clinical Presentation of M. marinum *Disease*

M. marinum skin and soft tissue infection typically occurs in the extremities, most commonly the hands. Cutaneous infections present as nodular or ulcerating skin lesions [1, 133]. Most lesions are solitary, but "ascending" lesions that resemble sporotrichosis can occur [1, 134]. The infection can invade to involve the deeper structures of the extremities, resulting in tenosynovitis, septic arthritis, or osteomyelitis (i.e., invasive disease) [132, 133]. Most cases have a history of aquatic exposure, including fish tank exposure, handling fish or seafood, or boating/fishing [132]. Most case series have described deeper/invasive infection in less than half of cases, with the exception of a recent series from a tertiary referral center that reported 68% of cases were invasive [132]. The diagnosis is made when the tissue from a surgical biopsy or aspiration from an involved site cultures *M. marinum*; histologic examination of involved tissue demonstrating granulomatous inflammation and acid fast bacilli is supportive. *M. marinum* has also rarely been reported to cause cervical lymphadenitis in healthy children [135] and to cause disseminated disease in immunocompromised patients, which typically involves the skin [136–139]. Although case reports of pulmonary *M. marinum* infection have been published [140, 141], this entity appears to be very rare in comparison with non-pulmonary disease.

Antimicrobial Treatment of **M. marinum** *Infection*

Routine susceptibility testing of *M. marinum* is not recommended by the ATS/IDSA guidelines, because there are no reports of mutational resistance, and there is little variability in susceptibility patterns to clinically useful agents [1]. However, it should be done in cases of treatment failure. In vitro, *M. marinum* isolates are susceptible to rifampin, rifabutin, clarithromycin, ethambutol, sulfonamides, and trimethoprim-sulfamethoxazole; intermediately susceptible to streptomycin, doxycycline, and minocycline; and resistant to isoniazid, pyrazinamide [1], and most fluoroquinolones [133].

There are no robust studies evaluating the treatment of *M. marinum* skin and soft tissue infections. The ATS/IDSA guidelines advise treatment with at least two active agents for 1–2 months after resolution of symptoms, typically for at least 3 months total duration [1]. Longer durations may be needed for invasive disease [132, 133]. However, some studies have reported good outcomes with a single drug (clarithromycin or tetracyclines) for a limited cutaneous disease [142–144]. In the largest case series reported to date, 63 patients were treated for a median of 3.5 months, with monotherapy in 37% and at least two drugs in 63%. The combination of clarithromycin and rifampin was most commonly used. Eighty-seven percent of patients were cured and 13% failed; failure was associated with infection of deeper structures, but not related to antibiotic regimen. Excellent outcomes have also been reported with the combination of clarithromycin and ethambutol [145], which may be an ideal combination because of tolerability [1]. Addition of a third drug may be beneficial for deep structure infection [1]. Surgical debridement(s), sometimes multiple, may be indicated for invasive infection or cases of treatment failure [1, 132, 133].

Mycobacterium scrofulaceum

The Organism and the Host

M. scrofulaceum has been found in natural bodies of freshwater [56]. Water distribution systems do not appear to be a source of human *M. scrofulaceum* exposure currently [56, 146]. In the early 1980s, *M. scrofulaceum* was a fairly common mycobacterial isolate found in clinical samples in the USA, and most clinical cases were due to cervical lymphadenitis in children. However, cases of *M. scrofulaceum* cervical adenitis have significantly decreased over time [147], and the organism is now rarely seen in the laboratory. Some have hypothesized that tap water was previously the source of human infection, and changes in chlorination have removed it from water distribution systems [1], although there is data to suggest that the organism has become less prevalent in the natural environment [56]. In addition to childhood cervical lymphadenitis, *M. scrofulaceum* has been rarely reported to cause pulmonary disease, skin and soft tissue, and disseminated infections in immunocompromised hosts. *M. scrofulaceum* has been found in clinical laboratories worldwide [1].

M. scrofulaceum *Pulmonary Disease*

M. scrofulaceum is an uncommon cause of NTM lung disease. Case series suggest that most *M. scrofulaceum* lung infections present in middle-aged or elderly men, with underlying lung disease such as COPD, previous pulmonary TB, or silicosis [12, 148]. A high incidence of *M. scrofulaceum* lung disease has been reported in HIV-negative South African gold miners [12]. Fibrocavitary disease predominates, but nodular bronchiectatic disease has also been reported [148].

There is little data on the in vitro susceptibility of *M. scrofulaceum* nor on the treatment of these infections. Susceptibility testing for clinically significant isolates is recommended, with multidrug antibiotic regimens chosen based on susceptibility results [1]. Drugs that have been used previously and may be considered based on drug susceptibility test results include macrolides, fluoroquinolones, rifamycins, ethambutol, isoniazid, and aminoglycosides [12, 148]. Daily therapy with at least three drugs based on drug susceptibility testing, selected from azithromycin or clarithromycin, moxifloxacin, rifampin or rifabutin, ethambutol, isoniazid, and an aminoglycoside (streptomycin, amikacin, or kanamycin), may comprise a useful regimen.

M. scrofulaceum *Non-pulmonary Disease*

Childhood cervical lymphadenitis due to *M. scrofulaceum* is rarely seen today; excisional surgery without chemotherapy is the recommended treatment. *M. scrofulaceum* has been reported to cause skin [149, 150] and soft tissue infection [151, 152] and disseminated disease [153] usually in immunocompromised hosts. While skin infections have been reported to respond to clarithromycin monotherapy [150, 154], multidrug antibiotic therapy is recommended for deeper infections and disseminated disease.

Mycobacterium shimoidei

Mycobacterium shimoidei is a rarely isolated species [1]. Although data regarding its frequency are lacking, in Ontario, Canada, we observed *M. shimoidei* in only 43/35,556 (0.12%) NTM isolations (unpublished data). It has been isolated almost exclusively from human respiratory specimens, and environmental sources have not been identified. Most reports identified lung infection in the setting of underlying destructive lung disease and have emanated from Canada, Europe, Australia, Madagascar, and Japan [155, 156]. Reports of lung disease presenting with bronchiectasis and nodules, typical of MAC lung disease [156], as well as disseminated disease with positive blood cultures in an HIV-infected patient [157] have both been described. In a recent study, Baird and colleagues reported on all isolates of *M. shimoidei* in Queensland, Australia, during 2000–2014 (contemporary population 3.6–4.7 million) [158]. They identified 23 patients (16 (69.6%) male), of mean age 66 years, 43.5% (10/23) with obstructive airways and 26.1% (6/23) with bronchiectasis. Cavitation, nodules, and consolidation were observed in nine (39.1%), eight (34.8%), and two (8.7%) patients, respectively. Within their cohort, ten (43.5%) patients fulfilled ATS/IDSA diagnostic criteria, and clinically significant disease was felt to be "likely" in nine (39.1%) patients and "possible" in seven (30.4%). Based on this study, it appears that a respiratory isolate of *M. shimoidei* is often indicative of disease and so should be carefully considered. We recommend that clinicians consider the ATS/IDSA criteria for diagnosing *M. shimoidei* lung disease.

Drug susceptibility test results have been reported in some studies, with results of most published cases summarized in two reviews [155, 156]. Moxifloxacin, clarithromycin, ethambutol, rifabutin, streptomycin, kanamycin, linezolid, and sulfamethoxazole seem to be drugs to which the organism is most often susceptible. Because these data reflect only a very small number of cases, drug susceptibility testing should be sought when antibiotic treatment is being considered, but as is usually the case, data guiding interpretation of DST are lacking. Reports on clinical outcomes are variable. Good outcomes have been reported in some cases with the use of clarithromycin, rifampin or rifabutin, ethambutol, and sometimes additional agents [156, 159–161]. Other reports suggest a high mortality with death frequently attributed to underlying lung disease [155]. In a recent study, Baird and colleagues reported on all isolates of M. shimoidei in Queensland, Australia during 2000-2014 (contemporary population 3.6-4.7 million) [158]. In the recent Australian study, ten (43.4%) patients improved or remained stable, five (27.1%) died, while eight (34.7%) did not have follow-up data. Antimicrobial treatment was administered in six patients, including clarithromycin/azithromycin in 5/6 (83.3%), a rifamycin in 3/6 (50%), ethambutol in 3/6 (50%), and miscellaneous other agents. Five of six treated patients (83.3%) stabilized or improved, and 1/6 (16.7%) died of lung disease [158].

Based on very little evidence, antibiotic treatment of *M. shimoidei* lung disease should probably include at least three drugs, the selection of which may be guided by drug susceptibility test results. Daily administrations of macrolide, moxifloxacin, rifabutin (or rifampin), and ethambutol appear to be reasonable considerations for first-line therapy and might be supplemented with an injectable agent (streptomycin and kanamycin most commonly described) in severe disease. The optimal duration of therapy is unknown. In some recent cases with good outcomes, treatment was continued for 12 months after sputum culture conversion [156, 159], which we would support.

Mycobacterium terrae Complex

The Organism and the Host

M. terrae complex is composed of multiple species, including *M. terrae*, *M. nonchromogenicum*, *M. arupense*, *M. heraklionense*, *M. virginiense*, and others [162, 163]. The individual species are difficult to distinguish by biochemical and culture methods, and molecular methods are required. The number of species in the complex has grown considerably in recent years with the greater availability of DNA sequencing. However, molecular identification of these organisms to the species level is not performed in most clinical laboratories. The environmental source of *M. terrae* complex human exposure is thought to be natural water, soil, and/or water distribution systems [164, 165]. *M. terrae* complex is weakly pathogenic and often felt to be a contaminant when isolated in clinical laboratories [1]. However, it is known to cause skin, soft tissue, and/or bone infection (typically tenosynovitis and/or osteomyelitis) and has also been rarely reported to cause infection of the lungs [166–172], genitourinary system [173], gastrointestinal tract [174], lymph nodes [175], and disseminated disease [176].

Clinical Presentation of M. terrae *Complex Disease*

Numerous cases of chronic tenosynovitis and/or osteomyelitis due to *M. terrae* complex have been described [162, 165]. They typically occur in immunocompetent hosts and involve the hand. Many are associated with an antecedent wound or trauma to the involved area [165]. Although most cases were previously believed due to *M. nonchromogenicum*, a recent study using molecular techniques has shown this to be false and has attributed these infections to other species in the complex [162]. *M. terrae* complex pulmonary disease is rare, and most cases of *M. terrae* complex isolation in respiratory specimens are likely due to contamination or colonization [1]. However, cases of true lung disease have been reported [166–172].

Antimicrobial Treatment of M. terrae *Complex Disease*

In vitro, *M. terrae* complex isolates are susceptible to clarithromycin. Most isolates are susceptible to ethambutol and trimethoprim-sulfamethoxazole, and most are resistant to rifampin and fluoroquinolones [162, 165]. Findings regarding rifabutin are mixed [162, 165]. In one series, half of the isolates were susceptible aminoglycosides [165].

The optimal antimicrobial treatment for *M. terrae* complex infections has not been established. The use of a macrolide plus ethambutol and/or another drug based on antimicrobial susceptibility testing seems reasonable [1]. Surgical debridement(s) for soft tissue infections may also be indicated. Clinical outcomes of *M. terrae* complex tenosynovitis cases have not been ideal, with one review of 31 cases reporting 29% with persistent disease that required repeated debridement, tendon extirpation, or amputation [165].

Mycobacterium ulcerans

The Organism and the Host

Mycobacterium ulcerans typically requires 6–12 weeks to grow at temperatures of 25–33 °C [1]. Conventional decontamination methods may prevent growth of the organism, and supplementation of media with egg yolk or reduction of oxygen tension may enhance recovery [1]. Molecular techniques are often used to more rapidly identify the organism. *M. ulcerans* is unique in its ability to produce mycolactone, a cytotoxin that induces necrosis and ulceration [177].

M. ulcerans is the third most common cause of mycobacterial infection worldwide, after *M. tuberculosis* and *M. leprae*. The organism causes "Buruli ulcer" in individuals residing in humid, rural, tropical regions, most commonly in Africa. The infection has also been reported in South and Central America, Southeast Asia, Australia, and Japan [178, 179].

Clinical Presentation and Management of M. ulcerans *Skin and Soft Tissue Disease*

Buruli ulcer presents initially as a painless nodule, papule, plaque, or area of edema, which progresses to a painless ulcer with undermined edges within days to weeks [178]. Ulcerations slowly progress, and can become quite extensive, and involve deep tissues such as tendons, joints, and bones [178]. Involvement of other organs is very rare. Extremities are the most commonly involved sites, but other sites can be involved. The infection can affect any age group but most commonly affects immunocompetent children and young adults [180]. Infection is believed to occur when compromised skin comes in contact with contaminated water, typically stagnant or slow-moving water [181, 182].

The diagnosis of Buruli ulcer is often made based on clinical manifestations in an endemic region and not confirmed microbiologically because of limited access to laboratory services. Histologic findings may be supportive. Specimens obtained by swabs of undermined edges of ulcers, punch or surgical biopsies, or fine need aspiration may demonstrate AFB or be culture positive, although the sensitivity is low (up to 60%) [183]. Molecular techniques have improved sensitivity (approximately 85%) [183].

The World Health Organization has published treatment guidelines that divide Buruli ulcer infections into three categories based on lesion size and stage [184]. Antibiotic therapy for 8 weeks is recommended for all infections, with surgical therapy (excision for small lesions, debridement for larger lesions) sometimes indicated as an adjunct, after the first 4 weeks of antibiotics [184]. Several antibiotic regimens have been evaluated, but the best outcomes have been observed with the combination of rifampin plus one other agent, including an aminoglycoside, fluoroquinolone, or macrolide. The combination of rifampin and streptomycin is most widely accepted and can be given for the entire 8 weeks, or for the initial 4 weeks, followed by 4 weeks of rifampin and clarithromycin [185]. A regimen consisting of rifampin plus clarithromycin or moxifloxacin for 8 weeks is also effective and is preferred in some countries [184, 186]. Cure rates are high [185, 187], but patients with more extensive lesions are often left with significant deformity and functional disability [188].

References

1. Griffith DE, Aksamit T, Brown-Elliott BA, Catanzaro A, Daley C, Gordin F, et al. An official ATS/IDSA statement: diagnosis, treatment, and prevention of nontuberculous mycobacterial diseases. Am J Respir Crit Care Med. 2007;175(4):367–416.
2. van Ingen J, Bendien SA, de Lange WC, Hoefsloot W, Dekhuijzen PN, Boeree MJ, et al. Clinical relevance of non-tuberculous mycobacteria isolated in the Nijmegen-Arnhem region, The Netherlands. Thorax. 2009;64(6):502–6.
3. Jankovic M, Sabol I, Zmak L, Jankovic VK, Jakopovic M, Obrovac M, et al. Microbiological criteria in non-tuberculous mycobacteria pulmonary disease: a tool for diagnosis and epidemiology. The international journal of tuberculosis and lung disease : the official journal of the International Union against. Tuberc Lung Dis. 2016;20(7):934–40.
4. Morimoto K, Kazumi Y, Shiraishi Y, Yoshiyama T, Murase Y, Ikushima S, et al. Clinical and microbiological features of definite Mycobacterium gordonae pulmonary disease: the establishment of diagnostic criteria for low-virulence mycobacteria. Trans R Soc Trop Med Hyg. 2015;109(9):589–93.
5. Braun E, Sprecher H, Davidson S, Kassis I. Epidemiology and clinical significance of non-tuberculous mycobacteria isolated from pulmonary specimens. Int J Tuberc Lung Dis. 2012;17(1):96–9.
6. Moon SM, Park HY, Jeon K, Kim SY, Chung MJ, Huh HJ, Ki CS, Lee NY, Shin SJ, Koh WJ. Clinical Significance of Mycobacterium kansasii Isolates from Respiratory Specimens. PLoS One. 2015;10(10):12.
7. Collins CH, Grange J, Yates MD. Mycobacteria in water. J Appl Bacteriol. 1984;57(2):193–211.
8. Zhang Y, Mann L, Wilson RW, Brown-Elliott BA, Vincent V, Iinuma Y, Wallace RJ Jr. Molecular analysis of Mycobacterium kansasii isolates from the United States. J Clin Microbiol. 2004;42(1):119–25.

9. Alcaide F, Richter I, Bernasconi C, Springer B, Hagenau C, Schulze-Röbbecke R, Tortoli E, Martín R, Böttger EC, Telenti A. Heterogeneity and clonality among isolates of Mycobacterium kansasii: implications for epidemiological and pathogenicity studies. J Clin Microbiol. 1997;35(8):6.
10. Prevots DR, Marras TK. Epidemiology of human pulmonary infection with nontuberculous mycobacteria: a review. Clin Chest Med. 2015;36(1):13–34.
11. Marras TK, Daley CL. Epidemiology of human pulmonary infection with nontuberculous mycobacteria. Clin Chest Med. 2002;23(3):553–67.
12. Corbett EL, Hay M, Churchyard GJ, Herselman P, Clayton T, Williams BG, et al. Mycobacterium kansasii and M. scrofulaceum isolates from HIV-negative South African gold miners: incidence, clinical significance and radiology. The international journal of tuberculosis and lung disease : the official journal of the International Union against. Tuberc Lung Dis. 1999;3(6):501–7.
13. Lovell JP, Zerbe C, Olivier KN, Claypool RJ, Frein C, Anderson VL, Freeman AF, Holland SM. Mediastinal and Disseminated Mycobacterium kansasii Disease in GATA2 Deficiency. Ann Am Thorac Soc. 2016;13(12):2169–73.
14. Veyrier F, Pletzer D, Turenne C, Behr MA. Phylogenetic detection of horizontal gene transfer during the step-wise genesis of Mycobacterium tuberculosis. BMC Evol Biol. 2009;9:196.
15. Bodle EE, Cunningham JA, Della-Latta P, Schluger NW, Saiman L, Bodle EE, et al. Epidemiology of nontuberculous mycobacteria in patients without HIV infection, New York City. Emerg Infect Dis. 2008;14(3):390–6.
16. Vesenbeckh S, Wagner S, Mauch H, Roth A, Streubel A, Russmann H, Bauer TT, Matthiessen W, Schonfeld N. Pathogenicity of Mycobacterium kansasii. Pneumologie. 2014;68(8):526–31.
17. Jankovic M, Samarzija M, Sabol I, Jakopovic M, Katalinic Jankovic V, Zmak L, Ticac B, Marusic A, Obrovac M, van Ingen J. Geographical distribution and clinical relevance of nontuberculous mycobacteria in Croatia. Int J Tuberc Lung Dis. 2013;17(6):836–41.
18. Ahn CH, Lowell JR, Onstad GD, Shuford EH, Hurst GA. A demographic study of disease due to Mycobacterium kansasii or M intracellulare-avium in Texas. Chest. 1979;75:120–5.
19. Shitrit D, Baum G, Priess R, Lavy A, Shitrit AB, Raz M, Shlomi D, Daniele B, Kramer MR. Pulmonary Mycobacterium kansasii infection in Israel, 1999-2004: clinical features, drug susceptibility, and outcome. Chest. 2006;129(3):771–6.
20. Griffith DE, Brown-Elliott B, Wallace RJ Jr. Thrice-weekly clarithromycin-containing regimen for treatment of Mycobacterium kansasii lung disease: results of a preliminary study. Clin Infect Dis. 2003;37(9):1178–82.
21. Rodriguez Diaz JC, López M, Ruiz M, Royo G. In vitro activity of new fluoroquinolones and linezolid against non-tuberculous mycobacteria. Int J Antimicrob Agents. 2003;21(6):4.
22. Philley JV, Griffith DE. Treatment of slowly growing mycobacteria. Clin Chest Med. 2015;36(1):79–90.
23. Griffith DE. Management of disease due to Mycobacterium kansasii. Clin Chest Med. 2002;23:9.
24. Pezzia W, Raleigh J, Bailey MC, Toth EA, Silverblatt J. Treatment of pulmonary disease due to Mycobacterium kansasii: recent experience with rifampin. Rev Infect Dis. 1981;3(5):1035–9.
25. Ahn CH, Lowell J, Ahn SS, Ahn S, Hurst GA. Chemotherapy for pulmonary disease due to Mycobacterium kansasii: efficacies of some individual drugs. Rev Infect Dis. 1981;3(5):1028–34.
26. Banks J, Hunter A, Campbell IA, Jenkins PA, Smith AP. Pulmonary infection with Mycobacterium kansasii in Wales, 1970-9: review of treatment and response. Thorax. 1983;38(4):271–4.
27. Ahn CH, Lowell J, Ahn SS, Ahn SI, Hurst GA. Short-course chemotherapy for pulmonary disease caused by Mycobacterium kansasii. Am Rev Respir Dis. 1983;128(6):1048–50.
28. Santin M, Dorca J, Alcaide F, Gonzalez L, Casas S, Lopez M, Guerra MR. Long-term relapses after 12-month treatment for Mycobacterium kansasii lung disease. Eur Respir J. 2009;33(1):148–52.

29. Sauret J, Hernández-Flix S, Castro E, Hernandez L, Ausina V, Coll P. Treatment of pulmonary disease caused by Mycobacterium kansasii: results of 18 vs 12 months' chemotherapy. Tuber Lung Dis. 1995;76(2):104–8.
30. Wallace RJ Jr, Dunbar D, Brown BA, Onyi G, Dunlap R, Ahn CH, Murphy DT. Rifampin-resistant Mycobacterium kansasii. Clin Infect Dis. 1994;18(5):736–43.
31. Slosarek M, Kubín M, Jaresova M. Water-borne household infections due to Mycobacterium xenopi. Cent Eur J Public Health. 1993;1(2):78–80.
32. Slosarek M, Kubín M, Pokorny J. Water as a possible factor of transmission in mycobacterial infections. Cent Eur J Public Health. 1994;2(2):103–5.
33. Sniadack DH, Ostroff S, Karlix MA, Smithwick RW, Schwartz B, Sprauer MA, Silcox VA, Good RC. A nosocomial pseudo-outbreak of Mycobacterium xenopi due to a contaminated potable water supply: lessons in prevention. Infect Control Hosp Epidemiol. 1993;14(11):636–41.
34. Astagneau P, Desplaces N, Vincent V, Chicheportiche V, Botherel A, Maugat S, Lebascle K, Leonard P, Desenclos J, Grosset J, Ziza J, Brucker G. Mycobacterium xenopi spinal infections after discovertebral surgery: investigation and screening of a large outbreak. Lancet. 2001;358(9283):747–51.
35. Dailloux M, Abalain ML, Laurain C, Lebrun L, Loos-Ayav C, Lozniewski A, et al. Respiratory infections associated with nontuberculous mycobacteria in non-HIV patients. Eur Respir J. 2006;28:1211–5.
36. Del Giudice G, Iadevaia C, Santoro G, Moscariello E, Smeraglia R, Marzo C, et al. Nontuberculous mycobacterial lung disease in patients without HIV infection: a retrospective analysis over 3 years. Clin Respir J. 2011;5(4):203–10.
37. Marras TK, Mendelson D, Marchand-Austin A, May K, Jamieson FB. Pulmonary Nontuberculous Mycobacterial Disease, Ontario, Canada, 1998-2010. Emerg Infect Dis. 2013;19(11):1889–91.
38. Costrini AM, Mahler D, Gross WM, Hawkins JE, Yesner R, D'Esopo ND. Clinical and roentgenographic features of nosocomial pulmonary disease due to Mycobacterium xenopi. Am Rev Respir Dis. 1981;123(1):104–9.
39. Zenone T, Boibieux A, Tigaud S, Fredenucci JF, Vincent V, Chidiac C, Peyramond D. Nontuberculous mycobacterial tenosynovitis: a review. Scand J Infect Dis. 1999;31(3):221–8.
40. Donnabella V, Salazar-Schicchi J, Bonk S, Hanna B, Rom WN. Increasing incidence of Mycobacterium xenopi at Bellevue Hospital: An emerging pathogen or a product of improved lasboratory methods? Chest. 2000;118:1365–70.
41. Marras TK, Chedore P, Jamieson F. Gender and age characteristics of patients with pulmonary nontuberculous mycobateria ata Canadian tertiary-care institution. Am J Respir Crit Care Med. 2008;177:583.
42. Marras TK, Campitelli M, Kwong JC, Lu H, Brode SK, Marchand-Austin A, Gershon AS, Jamieson FB. Risk of nontuberculous mycobacterial pulmonary disease with obstructive lung disease. Eur Respir J. 2016;48:928–31.
43. Carrillo MC, Patsios D, Wagnetz U, Jamieson F, Marras TK. Comparison of the Spectrum of Radiologic and Clinical Manifestations of Pulmonary Disease Caused by Mycobacterium avium Complex and Mycobacterium xenopi. Can Assoc Radiol J. 2014;65(3):207–13.
44. Varadi RG, Marras T. Pulmonary Mycobacterium xenopi infection in non-HIV-infected patients: a systematic review. Int J Tuberc Lung Dis. 2009;13(10):1210–8.
45. Andrejak C, Thomsen VO, Johansen IS, Riis A, Benfield TL, Duhaut P, et al. Nontuberculous Pulmonary Mycobacteriosis in Denmark: Incidence and Prognostic Factors. Am J Respir Crit Care Med. 2010;181:514–21.
46. Marras TK, Campitelli MA, Lu H, Chung H, Brode SK, Marchand-Austin A, et al. Pulmonary nontuberculous mycobacteria–associated deaths, Ontario, Canada, 2001–2013. Emerg Infect Dis [Internet]. 2017. Available from: https://doi.org/10.3201/eid2303.161927.
47. Andrejak C, Lescure F, Pukenyte E, Douadi Y, Yazdanpanah Y, Laurans G, Schmit JL, Jounieaux V. Mycobacterium xenopi pulmonary infections: a multicentric retrospective study of 136 cases in north-east France. Thorax. 2009;64(4):291–6.

48. Hollings NP, Wells AU, Wilson R, Hansell DM. Comparative appearances of non-tuberculous mycobacteria species: a CT study. Eur Radiol. 2002;12(9):2211–7.
49. Brown-Elliott BA, Nash K, Wallace RJ Jr. Antimicrobial susceptibility testing, drug resistance mechanisms, and therapy of infections with nontuberculous mycobacteria. Clin Microbiol Rev. 2012;25(3):545–82.
50. Research Committee of the British Thoracic Society. First randomised trial of treatments for pulmonary disease caused by M avium intracellulare, M malmoense, and M xenopi in HIV negative patients: rifampicin, ethambutol and isoniazid versus rifampicin and ethambutol. Thorax. 2001;56(3):167–72.
51. Jenkins PA, Campbell IA, Banks J, Gelder CM, Prescott RJ, Smith AP. Clarithromycin vs ciprofloxacin as adjuncts to rifampicin and ethambutol in treating opportunist mycobacterial lung diseases and an assessment of Mycobacterium vaccae immunotherapy. Thorax. 2008;63(7):627–34.
52. Andrejak C, Véziris N, Lescure F-X, Mal H, Bouvry D, Bassinet L, Blanc F-X, Camuset J, Couturaud F, Bervar J-F, Marquette C-H, Thiberville L, Vallerand H, Dalphin J-C, Morel H, Andrejak J, Cadranel J, Jounieaux V. CaMoMy Trial: A Prospective Randomized Clinical Trial to Compare Six-Months Sputum Conversion Rate with a Clarithromycin or Moxifloxacin Containing Regimen in Patients with a M. Xenopi Pulmonary Infection:Â Intermediate Analysis. B49 NON-TUBERCULOUS MYCOBACTERIAL DISEASE AND CASE REPORTS. American Thoracic Society International Conference Abstracts: American Thoracic Society; 2016. p. A3733.
53. Lounis N, Truffot-Pernot C, Bentoucha A, Robert J, Ji B, Grosset J. Efficacies of clarithromycin regimens against Mycobacterium xenopi in mice. Antimicrob Agents Chemother. 2001;45(11):3229–30.
54. Andrejak C, Almeida DV, Tyagi S, Converse PJ, Ammerman NC, Grosset JH. Improving existing tools for Mycobacterium xenopi treatment: assessment of drug combinations and characterization of mouse models of infection and chemotherapy. J Antimicrob Chemother. 2016;68(3):659–65.
55. Parrot RG, Grosset J. Post-surgical outcome of 57 patients with Mycobacterium xenopi pulmonary infection. Tubercle. 1988;69(1):47–55.
56. Falkinham JO 3rd. Epidemiology of infection by nontuberculous mycobacteria. Clin Microbiol Rev. 1996;9(2):177–215.
57. Hoefsloot W, Boeree MJ, van Ingen J, Bendien S, Magis C, de Lange W, et al. The rising incidence and clinical relevance of Mycobacterium malmoense: a review of the literature. Int J Tuberc Lung Dis. 2008;12(9):987–93.
58. Hoefsloot W, van Ingen J, Andrejak C, Angeby K, Bauriaud R, Bemer P, et al. The geographic diversity of nontuberculous mycobacteria isolated from pulmonary samples: an NTM-NET collaborative study. Eur Respir J. 2013;42(6):1604–13.
59. Buchholz UT, McNeil MM, Keyes LE, Good RC. Mycobacterium malmoense infections in the United States, January 1993 through June 1995. Clin Infect Dis. 1998;27(3):551–8.
60. Chocarra A, Gonzalez Lopez A, Breznes MF, Canut A, Rodriguez J, Diego JM. Disseminated infection due to Mycobacterium malmoense in a patient infected with human immunodeficiency virus. Clin Infect Dis. 1994;19(1):203–4.
61. Pulmonary disease caused by M. malmoense in HIV negative patients: 5-yr follow-up of patients receiving standardised treatment. Eur Respir J. 2003;21(3):478–82.
62. Hoefsloot W, van Ingen J, de Lange WC, Dekhuijzen PN, Boeree MJ, van Soolingen D. Clinical relevance of Mycobacterium malmoense isolation in The Netherlands. Eur Respir J. 2009;34(4):926–31.
63. Murray MP, Laurenson IF, Hill AT. Outcomes of a standardized triple-drug regimen for the treatment of nontuberculous mycobacterial pulmonary infection. Clin Infect Dis. 2008;47(2):222–4.
64. Makovcova J, Slany M, Babak V, Slana I, Kralik P. The water environment as a source of potentially pathogenic mycobacteria. J Water Health. 2014;12(2):254–63.

65. Thomson RM, Carter R, Tolson C, Coulter C, Huygens F, Hargreaves M. Factors associated with the isolation of Nontuberculous mycobacteria (NTM) from a large municipal water system in Brisbane, Australia. BMC Microbiol. 2013;13:89.
66. El Sahly HM, Septimus E, Soini H, Septimus J, Wallace RJ, Pan X, Williams-Bouyer N, Musser JM, Graviss EA. Mycobacterium simiae pseudo-outbreak resulting from a contaminated hospital water supply in Houston, Texas. Clin Infect Dis. 2002;35(7):802–7.
67. Jordao Junior CM, Lopes F, David S, Farache Filho A, Leite CQ. Detection of nontuberculous mycobacteria from water buffalo raw milk in Brazil. Food Microbiol. 2009;26(6):658–61.
68. Shitrit D, Peled N, Bishara J, Priess R, Pitlik S, Samra Z, Kramer MR. Clinical and radiological features of Mycobacterium kansasii infection and Mycobacterium simiae infection. Respir Med. 2008;102(11):1598–603.
69. Lavy A, Yoshpe-Purer Y. Isolation of Mycobacterium simiae from clinical specimens in Israel. Tubercle. 1982;63(4):279–85.
70. Rynkiewicz DL, Cage G, Butler WR, Ampel NM. Clinical and microbiological assessment of Mycobacterium simiae isolates from a single laboratory in southern Arizona. Clin Infect Dis. 1998;26(3):625–30.
71. Valero G, Peters J, Jorgensen JH, Graybill JR. Clinical isolates of Mycobacterium simiae in San Antonio, Texas. An 11-yr review. Am J Respir Crit Care Med. 1995;152(5 Pt 1):1555–7.
72. van Ingen J, Boeree M, Dekhuijzen PN, van Soolingen D. Clinical relevance of Mycobacterium simiae in pulmonary samples. Eur Respir J. 2008;31(1):106–9.
73. Patel NC, Minifee P, Dishop MK, Munoz FM. Mycobacterium simiae cervical lymphadenitis. Pediatr Infect Dis J. 2007;26(4):362–3.
74. van Ingen J, Totten SE, Heifets LB, Boeree MJ, Daley CL. Drug susceptibility testing and pharmacokinetics question current treatment regimens in Mycobacterium simiae complex disease. Int J Antimicrob Agents. 2012;39(2):173–6.
75. Abalain-Colloc ML, Guillerm D, Salaun M, Gouriou S, Vincent V, Picard B. Mycobacterium szulgai isolated from a patient, a tropical fish and aquarium water. Eur J Clin Microbiol Infect Dis. 2003;22(12):768–9.
76. van Ingen J, Boeree M, de Lange WC, de Haas PE, Dekhuijzen PN, van Soolingen D. Clinical relevance of Mycobacterium szulgai in The Netherlands. Clin Infect Dis. 2008;46(8):1200–5.
77. Yoo H, Jeon K, Kim SY, Jeong BH, Park HY, Ki CS, Lee NY, Shin SJ, Koh WJ. Clinical significance of Mycobacterium szulgai isolates from respiratory specimens. Scandinavica. 2014;46(3):169–74.
78. Benator DA, Kan V, Gordin FM. Mycobacterium szulgai infection of the lung: case report and review of an unusual pathogen. Am J Med Sci. 1997;313(6):346–51.
79. Tortoli E, Besozzi G, Lacchini C, Penati V, Simonetti MT, Emler S. Pulmonary infection due to Mycobacterium szulgai, case report and review of the literature. Eur Respir J. 1998;11(4):975–7.
80. Bottger EC, Teske A, Kirschner P, Bost S, Chang HR, Beer V, Hirschel B. Disseminated "Mycobacterium genavense" infection in patients with AIDS. Lancet. 1992;340(8811):76–80.
81. Hillebrand-Haverkort ME, Kolk A, Kox LF, Ten Velden JJ, Ten Veen JH. Generalized mycobacterium genavense infection in HIV-infected patients: detection of the mycobacterium in hospital tap water. Scand J Infect Dis. 1999;31(1):63–8.
82. van der Wielen PW, Heijnen L, van der Kooij D. Pyrosequence analysis of the hsp65 genes of nontuberculous mycobacterium communities in unchlorinated drinking water in the Netherlands. Appl Environ Microbiol. 2013;79(19):6160–6.
83. Kiehn TE, Hoefer H, Bottger EC, Ross R, Wong M, Edwards F, Antinoff N, Armstrong D. Mycobacterium genavense infections in pet animals. J Clin Microbiol. 1996;34(7):1840–2.
84. Portaels F, Realini L, Bauwens L, Hirschel B, Meyers WM, de Meurichy W. Mycobacteriosis caused by Mycobacterium genavense in birds kept in a zoo: 11-year survey. J Clin Microbiol. 1996;34(2):319–23.
85. Charles P, Lortholary O, Dechartres A, Doustdar F, Viard JP, Lecuit M, Gutierrez MC. Mycobacterium genavense infections: a retrospective multicenter study in France, 1996-2007. Medicine (Baltimore). 2011;90(4):223–30.

86. Hoefsloot W, van Ingen J, Peters EJ, Magis-Escurra C, Dekhuijzen PN, Boeree MJ, van Soolingen D. Mycobacterium genavense in the Netherlands: an opportunistic pathogen in HIV and non-HIV immunocompromised patients. An observational study in 14 cases. Clin Microbiol Infect. 2012;19(5):432–7.
87. Rammaert B, Couderc L, Rivaud E, Honderlick P, Zucman D, Mamzer MF, Cahen P, Bille E, Lecuit M, Lortholary O, Catherinot E. Mycobacterium genavense as a cause of subacute pneumonia in patients with severe cellular immunodeficiency. BMC Infect Dis. 2011;11:311.
88. Thomsen VO, Dragsted U, Bauer J, Fuurste K, Lundgren J. Disseminated infection with Mycobacterium genavense: a challenge to physicians and mycobacteriologists. J Clin Microbiol. 1999;37(12):3901–5.
89. Siddiqi SH, Laszlo A, Butler WR, Kilburn JO. Bacteriologic investigations of unusual mycobacteria isolated from immunocompromised patients. Diagn Microbiol Infect Dis. 1993;16(4):321–3.
90. Prabaker K, Muthiah C, Hayden MK, Weinstein RA, Cheerala J, Scorza ML, et al. Pseudo-outbreak of Mycobacterium gordonae Following the Opening of a newly constructed hospital at a Chicago Medical Center. Infect Control Hosp Epidemiol. 2015;36(2):198–203.
91. Arnow PM, Bakir M, Thompson K, Bova JL. Endemic contamination of clinical specimens by Mycobacterium gordonae. Clin Infect Dis. 2000;31(2):472–6.
92. Panwalker AP, Fuhse E. Nosocomial Mycobacterium gordonae pseudoinfection from contaminated ice machines. Infect Control. 1986;7(2):67–70.
93. Stine TM, Harris AA, Levin S, Rivera N, Kaplan RL. A pseudoepidemic due to atypical mycobacteria in a hospital water supply. JAMA. 1987;258(6):809–11.
94. Tokars JI, McNeil MM, Tablan OC, Chapin-Robertson K, Patterson JE, Edberg SC, et al. Mycobacterium gordonae pseudoinfection associated with a contaminated antimicrobial solution. J Clin Microbiol. 1990;28(12):2765–9.
95. Steere AC, Corrales J, von Graevenitz A. A cluster of Mycobacterium gordonae isolates from bronchoscopy specimens. Am Rev Respir Dis. 1979;120(1):214–6.
96. Gubler JG, Salfinger M, von Graevenitz A. Pseudoepidemic of nontuberculous mycobacteria due to a contaminated bronchoscope cleaning machine. Report of an outbreak and review of the literature. Chest. 1992;101(5):1245–9.
97. Metchock B, Nolte F, Wallace RJ Jr. In: Murray P, editor. Mycobacterium. Washington, D.C.: ASM Press; 1999.
98. Aguado JM, Gomez-Garces JL, Manrique A, Soriano F. Pulmonary infection by Mycobacterium gordonae in an immunocompromised patient. Diagn Microbiol Infect Dis. 1987;7(4):261–3.
99. Youssef D, Shams WE, Elshenawy Y, El-Abbassi A, Moorman JP. Pulmonary infection with caseating mediastinal lymphadenitis caused by Mycobacterium gordonae. Int J Mycobacteriol. 2014;3(3):220–3.
100. Nanda Kumar U, Varkey B. Pulmonary infection caused by Mycobacterium gordonae. Br J Dis Chest. 1980;74(2):189–92.
101. Harro C, Braden GL, Morris AB, Lipkowitz GS, Madden RL. Failure to cure Mycobacterium gordonae peritonitis associated with continuous ambulatory peritoneal dialysis. Clin Infect Dis. 1997;24(5):955–7.
102. Rusconi S, Gori A, Vago L, Marchetti G, Franzetti F. Cutaneous infection caused by Mycobacterium gordonae in a human immunodeficiency virus-infected patient receiving antimycobacterial treatment. Clin Infect Dis. 1997;25(6):1490–1.
103. Hirohama D, Ishibashi Y, Kawarazaki H, Kume H, Fujita T. Successful treatment of Mycobacterium gordonae exit-site and tunnel infection by partial catheter reimplantation of the Tenckhoff catheter. Perit Dial Int. 2011;31(3):368–70.
104. Weinberger M, Berg SL, Feuerstein IM, Pizzo PA, Witebsky FG. Disseminated infection with Mycobacterium gordonae: report of a case and critical review of the literature. Clin Infect Dis. 1992;14(6):1229–39.

105. Bonnet E, Massip P, Bauriaud R, Alric L, Auvergnat JC. Disseminated Mycobacterium gordonae infection in a patient infected with human immunodeficiency virus. Clin Infect Dis. 1996;23(3):644–5.
106. Asija A, Prasad A, Eskridge E. Disseminated Mycobacterium gordonae infection in an immunocompetent host. Am J Ther. 2011;18(3):e75–7.
107. Marchevsky A, Damsker B, Gribetz A, Tepper S, Geller SA. The spectrum of pathology of nontuberculous mycobacterial infections in open-lung biopsy specimens. Am J Clin Pathol. 1982;78(5):695–700.
108. Mazumder SA, Hicks A, Norwood J. Mycobacterium gordonae pulmonary infection in an immunocompetent adult. N Am J Med Sci. 2010;2(4):205–7.
109. Brown BA, Wallace RJ Jr, Onyi GO. Activities of clarithromycin against eight slowly growing species of nontuberculous mycobacteria, determined by using a broth microdilution MIC system. Antimicrob Agents Chemother. 1992;36(9):1987–90.
110. Rastogi N, Goh KS, Guillou N, Labrousse V. Spectrum of drugs against atypical mycobacteria: how valid is the current practice of drug susceptibility testing and the choice of drugs? Zentralbl Bakteriol. 1992;277(4):474–84.
111. Lindeboom JA, Bruijnesteijn van Coppenraet LE, van Soolingen D, Prins JM, Kuijper EJ. Clinical manifestations, diagnosis, and treatment of Mycobacterium haemophilum infections. Clin Microbiol Rev. 2011;24(4):701–17.
112. Falkinham JO 3rd, Norton CD, LeChevallier MW. Factors influencing numbers of Mycobacterium avium, Mycobacterium intracellulare, and other Mycobacteria in drinking water distribution systems. Appl Environ Microbiol. 2001;67(3):1225–31.
113. Dawson DJ, Jennis F. Mycobacteria with a growth requirement for ferric ammonium citrate, identified as Mycobacterium haemophilum. J Clin Microbiol. 1980;11(2):190–2.
114. Samra Z, Kaufmann L, Zeharia A, Ashkenazi S, Amir J, Bahar J, et al. Optimal detection and identification of Mycobacterium haemophilum in specimens from pediatric patients with cervical lymphadenopathy. J Clin Microbiol. 1999;37(3):832–4.
115. Lindeboom JA, Prins JM, Bruijnesteijn van Coppenraet ES, Lindeboom R, Kuijper EJ. Cervicofacial lymphadenitis in children caused by Mycobacterium haemophilum. Clin Infect Dis. 2005;41(11):1569–75.
116. Zeharia A, Eidlitz-Markus T, Haimi-Cohen Y, Samra Z, Kaufman L, Amir J. Management of nontuberculous mycobacteria-induced cervical lymphadenitis with observation alone. Pediatr Infect Dis J. 2008;27(10):920–2.
117. Giulieri S, Morisod B, Edney T, Odman M, Genne D, Malinverni R, et al. Outbreak of Mycobacterium haemophilum infections after permanent makeup of the eyebrows. Clin Infect Dis. 2011;52(4):488–91.
118. Shah MK, Sebti A, Kiehn TE, Massarella SA, Sepkowitz KA. Mycobacterium haemophilum in immunocompromised patients. Clin Infect Dis. 2001;33(3):330–7.
119. Kiehn TE, White M. Mycobacterium haemophilum: an emerging pathogen. Eur J Clin Microbiol Infect Dis. 1994;13(11):925–31.
120. McBride ME, Rudolph AH, Tschen JA, Cernoch P, Davis J, Brown BA, et al. Diagnostic and therapeutic considerations for cutaneous Mycobacterium haemophilum infections. Arch Dermatol. 1991;127(2):276–7.
121. Cross GB, Le Q, Webb B, Jenkin GA, Korman TM, Francis M, et al. Mycobacterium haemophilum bone and joint infection in HIV/AIDS: case report and literature review. Int J STD AIDS. 2015;26(13):974–81.
122. Torvinen E, Suomalainen S, Paulin L, Kusnetsov J. Mycobacteria in Finnish cooling tower waters. APMIS. 2013;122(4):353–8.
123. Marshall HM, Carter R, Torbey MJ, Minion S, Tolson C, Sidjabat HE, Huygens F, Hargreaves M, Thomson RM. Mycobacterium lentiflavum in drinking water supplies, Australia. Emerg Infect Dis. 2011;17(3):395–402.
124. Tortoli E, Bartoloni A, Erba ML, Levre E, Lombardi N, Mantella A, Mecocci L. Human infections due to Mycobacterium lentiflavum. J Clin Microbiol. 2002;40(2):728–9.
125. Neonakis IK, Gitti Z, Kourbeti IS, Michelaki H, Baritaki M, Alevraki G, Papadomanolaki E, Tsafaraki E, Tsouri A, Baritaki S, Krambovitis E, Spandidos DA. Mycobacterial species

diversity at a general hospital on the island of Crete: first detection of Mycobacterium lentiflavum in Greece. Scand J Infect Dis. 2007;39(10):875–9.
126. Safdar A, Han X. Mycobacterium lentiflavum, a recently identified slow-growing mycobacterial species: clinical significance in immunosuppressed cancer patients and summary of reported cases of infection. Eur J Clin Microbiol Infect Dis. 2005;24(8):554–8.
127. Jeong BH, Song J, Kim W, Han SG, Ko Y, Song J, Chang B, Hong G, Kim SY, Choi GE, Shin SJ, Koh WJ. Nontuberculous Mycobacterial Lung Disease Caused by Mycobacterium lentiflavum in a Patient with Bronchiectasis. Tuberc Respir Dis (Seoul). 2013;74(4):187–90.
128. Lee YC, Kim S, Gang SJ, Park SY, Kim SR. Acute necrotizing pneumonia combined with parapneumonic effusion caused by Mycobacterium lentiflavum: a case report. BMC Infect Dis. 2015;15:354.
129. Phelippeau M, Dubus J, Reynaud-Gaubert M, Gomez C, Stremler le Bel N, Bedotto M, Prudent E, Drancourt M. Prevalence of Mycobacterium lentiflavum in cystic fibrosis patients, France. BMC Pulm Med. 2015;15:131.
130. Molteni C, Gazzola L, Cesari M, Lombardi A, Salerno F, Tortoli E, Codecasa L, Penati V, Franzetti F, Gori A. Mycobacterium lentiflavum infection in immunocompetent patient. Emerg Infect Dis. 2005;11(1):119–22.
131. Jimenez-Montero B, Baquero-Artigao F, Saavedra-Lozano J, Tagarro-Garcia A, Blazquez-Gamero D, Cilleruelo-Ortega MJ, Ramos-Amador JT, Gale-Anso I, Marin N, Gomez-Garcia R, Santiago-Garcia B, Garrido J, Lopez G. Comparison of Mycobacterium lentiflavum and Mycobacterium avium-intracellulare complex lymphadenitis. Pediatr Infect Dis J. 2014;33(1):28–34.
132. Johnson MG, Stout JE. Twenty-eight cases of Mycobacterium marinum infection: retrospective case series and literature review. Infection. 2015;43(6):655–62.
133. Aubry A, Chosidow O, Caumes E, Robert J, Cambau E. Sixty-three cases of Mycobacterium marinum infection: clinical features, treatment, and antibiotic susceptibility of causative isolates. Arch Intern Med. 2002;162(15):1746–52.
134. Lewis FM, Marsh BJ, von Reyn CF. Fish tank exposure and cutaneous infections due to Mycobacterium marinum: tuberculin skin testing, treatment, and prevention. Clin Infect Dis. 2003;37(3):390–7.
135. Colville A. Retrospective review of culture-positive mycobacterial lymphadenitis cases in children in Nottingham, 1979-1990. Eur J Clin Microbiol Infect Dis. 1993;12(3):192–5.
136. Jacobs S, George A, Papanicolaou GA, Lacouture ME, Tan BH, Jakubowski AA, et al. Disseminated Mycobacterium marinum infection in a hematopoietic stem cell transplant recipient. Transpl Infect Dis. 2012;14(4):410–4.
137. Tchornobay AM, Claudy AL, Perrot JL, Levigne V, Denis M. Fatal disseminated Mycobacterium marinum infection. Int J Dermatol. 1992;31(4):286–7.
138. Asakura T, Ishii M, Kikuchi T, Kameyama K, Namkoong H, Nakata N, et al. Disseminated Mycobacterium marinum Infection With a Destructive Nasal Lesion Mimicking Extranodal NK/T Cell Lymphoma: A Case Report. Medicine. 2016;95(11):e3131.
139. Danko JR, Gilliland WR, Miller RS, Decker CF. Disseminated Mycobacterium marinum infection in a patient with rheumatoid arthritis receiving infliximab therapy. Scand J Infect Dis. 2009;41(4):252–5.
140. Lai CC, Lee LN, Chang YL, Lee YC, Ding LW, Hsueh PR. Pulmonary infection due to Mycobacterium marinum in an immunocompetent patient. Clin Infect Dis. 2005;40(1):206–8.
141. Velu PP, Fernandes SE, Laurenson IF, Noble DD. Pulmonary Mycobacterium marinum infection: 'fish tank granuloma' of the lung. Scott Med J. 2016;61(4):203–6.
142. Feng Y, Xu H, Wang H, Zhang C, Zong W, Wu Q. Outbreak of a cutaneous Mycobacterium marinum infection in Jiangsu Haian, China. Diagn Microbiol Infect Dis. 2011;71(3):267–72.
143. Eberst E, Dereure O, Guillot B, Trento C, Terru D, van de Perre P, et al. Epidemiological, clinical, and therapeutic pattern of Mycobacterium marinum infection: a retrospective series of 35 cases from southern France. J Am Acad Dermatol. 2012;66(1):e15–6.

144. Abbas O, Marrouch N, Kattar MM, Zeynoun S, Kibbi AG, Rached RA, et al. Cutaneous nontuberculous Mycobacterial infections: a clinical and histopathological study of 17 cases from Lebanon. J Eur Acad Dermatol Venereol. 2011;25(1):33–42.
145. Wolinsky E, Gomez F, Zimpfer F. Sporotrichoid Mycobacterium marinum infection treated with rifampin-ethambutol. Am Rev Respir Dis. 1972;105(6):964–7.
146. von Reyn CF, Waddell RD, Eaton T, Arbeit RD, Maslow JN, Barber TW, et al. Isolation of Mycobacterium avium complex from water in the United States, Finland, Zaire and Kenya. J Clin Microbiol. 1993;31(12):3227–30.
147. Wolinsky E. Mycobacterial lymphadenitis in children: a prospective study of 105 nontuberculous cases with long-term follow-up. Clin Infect Dis. 1995;20(4):954–63.
148. Suzuki S, Morino E, Ishii M, Namkoong H, Yagi K, Asakura T, et al. Clinical characteristics of pulmonary Mycobacterium scrofulaceum disease in 2001-2011: A case series and literature review. J Infect Chemother. 2016;22(9):611–6.
149. Ito A, Yoshida Y, Higaki-Mori H, Watanabe T, Nakanaga K, Ishii N, et al. Multiple skin lesions caused by Mycobacterium scrofulaceum infection. Eur J Dermatol. 2011;21(6):1014–5.
150. Lai J, Abbey BV, Jakubovic HR. Epithelioid histiocytic infiltrate caused by Mycobacterium scrofulaceum infection: a potential mimic of various neoplastic entities. Am J Dermatopathol. 2013;35(2):266–9.
151. Carter TI, Frelinghuysen P, Daluiski A, Brause BD, Wolfe SW. Flexor tenosynovitis caused by Mycobacterium scrofulaceum: case report. J Hand Surg. 2006;31(8):1292–5.
152. Phoa LL, Khong KS, Thamboo TP, Lam KN. A case of Mycobacterium scrofulaceum osteomyelitis of the right wrist. Ann Acad Med Singap. 2000;29(5):678–81.
153. Sanders JW, Walsh AD, Snider RL, Sahn EE. Disseminated Mycobacterium scrofulaceum infection: a potentially treatable complication of AIDS. Clin Infect Dis. 1995;20(3):549.
154. Jang HS, Jo JH, Oh CK, Kim MB, Lee JB, Chang CL, et al. Successful treatment of localized cutaneous infection caused by Mycobacterium scrofulaceum with clarithromycin. Pediatr Dermatol. 2005;22(5):476–9.
155. Mayall B, Gurtler V, Irving L, Marzec A, Leslie D. Identification of Mycobacterium shimoidei by molecular techniques: case report and summary of the literature. Int J Tuberc Lung Dis. 1999;3(2):169–73.
156. Galizzi N, Tortoli E, Gori A, Morini F, Lapadula G. A case of mild pulmonary disease due to Mycobacterium shimoidei with a favorable outcome. J Clin Microbiol. 2013;51(10):3467–8.
157. Furrer H, Bodmer T, von Overbeck J. Disseminated nontuberculous mycobacterial infections in AIDS patients. Schweiz Med Wochenschr. 1994;124(3):89–96.
158. Baird TM, Carter R, Eather G, Thomson R. Mycobacterium shimoidei, a Rare Pulmonary Pathogen, Queensland, Australia. Emerg Infect Dis. 2017;23(11):1919–1922.
159. Kanaji N, Kushida Y, Bandoh S, Ishii T, Haba R, Tadokoro A, Watanabe N, Takahama T, Kita N, Dobashi H, Matsunaga T. Membranous glomerulonephritis associated with Mycobacterium shimoidei pulmonary infection. Am J Case Rep. 2013;14:543–7.
160. Saito H, Zayasu K, Shigeto E, Iwamoto T, Nakanaga K, Kodama A, Ishii N. Two cases of lung infection due to Mycobacterium shimoidei, with special reference to bacteriological investigation. Kansenshogaku Zasshi. 2007;81(1):12–9.
161. Takayama S, Tominaga S, Tsukada Y, Ohkochi M, Inase N. A case of pulmonary Mycobacterium shimoidei infection. Kekkaku. 2006;81(8):537–41.
162. Vasireddy R, Vasireddy S, Brown-Elliott BA, Wengenack NL, Eke UA, Benwill JL, et al. Mycobacterium arupense, Mycobacterium heraklionense, and a Newly Proposed Species, "Mycobacterium virginiense" sp. nov., but Not Mycobacterium nonchromogenicum, as Species of the Mycobacterium terrae Complex Causing Tenosynovitis and Osteomyelitis. J Clin Microbiol. 2016;54(5):1340–51.
163. Tortoli E, Gitti Z, Klenk HP, Lauria S, Mannino R, Mantegani P, et al. Survey of 150 strains belonging to the Mycobacterium terrae complex and description of Mycobacterium engbaekii sp. nov., Mycobacterium heraklionense sp. nov. and Mycobacterium longobardum sp. nov. Int J Syst Evol Microbiol. 2013;63(Pt 2):401–11.

164. Lockwood WW, Friedman C, Bus N, Pierson C, Gaynes R. An outbreak of Mycobacterium terrae in clinical specimens associated with a hospital potable water supply. Am Rev Respir Dis. 1989;140(6):1614–7.
165. Smith DS, Lindholm-Levy P, Huitt GA, Heifets LB, Cook JL. Mycobacterium terrae: case reports, literature review, and in vitro antibiotic susceptibility testing. Clin Infect Dis. 2000;30(3):444–53.
166. Spence TH, Ferris VM. Spontaneous resolution of a lung mass due to infection with Mycobacterium terrae. South Med J. 1996;89(4):414–6.
167. Peters EJ, Morice R. Miliary pulmonary infection caused by Mycobacterium terrae in an autologous bone marrow transplant patient. Chest. 1991;100(5):1449–50.
168. Tonner JA, Hammond MD. Pulmonary disease caused by Mycobacterium terrae complex. South Med J. 1989;82(10):1279–82.
169. Palmero DJ, Teres RI, Eiguchi K. Pulmonary disease due to Mycobacterium terrae. Tubercle. 1989;70(4):301–3.
170. Krisher KK, Kallay MC, Nolte FS. Primary pulmonary infection caused by Mycobacterium terrae complex. Diagn Microbiol Infect Dis. 1988;11(3):171–5.
171. Kuze F, Mitsuoka A, Chiba W, Shimizu Y, Ito M, Teramatsu T, et al. Chronic pulmonary infection caused by Mycobacterium terrae complex: a resected case. Am Rev Respir Dis. 1983;128(3):561–5.
172. Tsukamura M, Kita N, Otsuka W, Shimoide H. A study of the taxonomy of the Mycobacterium nonchromogenicum complex and report of six cases of lung infection due to Mycobacterium nonchromogenicum. Microbiol Immunol. 1983;27(3):219–36.
173. Chan TH, Ng KC, Ho A, Scheel O, Lai CK, Leung R. Urinary tract infection caused by Mycobacterium terrae complex. Tuber Lung Dis. 1996;77(6):555–7.
174. Fujisawa K, Watanabe H, Yamamoto K, Nasu T, Kitahara Y, Nakano M. Primary atypical mycobacteriosis of the intestine: a report of three cases. Gut. 1989;30(4):541–5.
175. Shimizu T, Furumoto H, Takahashi T, Yasuno H, Muto M. Lymphadenitis due to Mycobacterium terrae in an immunocompetent patient. Dermatology (Basel, Switz). 1999;198(1):97–8.
176. Carbonara S, Tortoli E, Costa D, Monno L, Fiorentino G, Grimaldi A, et al. Disseminated Mycobacterium terrae infection in a patient with advanced human immunodeficiency virus disease. Clin Infect Dis. 2000;30(5):831–5.
177. van der Werf TS, Stinear T, Stienstra Y, van der Graaf WT, Small PL. Mycolactones and Mycobacterium ulcerans disease. Lancet (Lond Engl). 2003;362(9389):1062–4.
178. van der Werf TS, van der Graaf WT, Tappero JW, Asiedu K. Mycobacterium ulcerans infection. Lancet (Lond Engl). 1999;354(9183):1013–8.
179. Sugawara M, Ishii N, Nakanaga K, Suzuki K, Umebayashi Y, Makigami K, et al. Exploration of a standard treatment for Buruli ulcer through a comprehensive analysis of all cases diagnosed in Japan. J Dermatol. 2015;42(6):588–95.
180. Debacker M, Aguiar J, Steunou C, Zinsou C, Meyers WM, Scott JT, et al. Mycobacterium ulcerans disease: role of age and gender in incidence and morbidity. Tropical Med Int Health. 2004;9(12):1297–304.
181. Raghunathan PL, Whitney EA, Asamoa K, Stienstra Y, Taylor TH Jr, Amofah GK, et al. Risk factors for Buruli ulcer disease (Mycobacterium ulcerans Infection): results from a case-control study in Ghana. Clin Infect Dis. 2005;40(10):1445–53.
182. Bratschi MW, Ruf MT, Andreoli A, Minyem JC, Kerber S, Wantong FG, et al. Mycobacterium ulcerans persistence at a village water source of Buruli ulcer patients. PLoS Negl Trop Dis. 2014;8(3):e2756.
183. Herbinger KH, Adjei O, Awua-Boateng NY, Nienhuis WA, Kunaa L, Siegmund V, et al. Comparative study of the sensitivity of different diagnostic methods for the laboratory diagnosis of Buruli ulcer disease. Clin Infect Dis. 2009;48(8):1055–64.
184. World Health Organization. Treatment of Mycobacterium ulcerans disease (Buruli ulcer): guidance for health workers. WHO Press. Geneva, Switzerland. 2012.

185. Nienhuis WA, Stienstra Y, Thompson WA, Awuah PC, Abass KM, Tuah W, et al. Antimicrobial treatment for early, limited Mycobacterium ulcerans infection: a randomised controlled trial. Lancet (Lond, Engl). 2010;375(9715):664–72.
186. Chauty A, Ardant MF, Marsollier L, Pluschke G, Landier J, Adeye A, et al. Oral treatment for Mycobacterium ulcerans infection: results from a pilot study in Benin. Clin Infect Dis. 2011;52(1):94–6.
187. Converse PJ, Nuermberger EL, Almeida DV, Grosset JH. Treating Mycobacterium ulcerans disease (Buruli ulcer): from surgery to antibiotics, is the pill mightier than the knife? Future Microbiol. 2011;6(10):1185–98.
188. Stienstra Y, van Roest MH, van Wezel MJ, Wiersma IC, Hospers IC, Dijkstra PU, et al. Factors associated with functional limitations and subsequent employment or schooling in Buruli ulcer patients. Tropical Med Int Health. 2005;10(12):1251–7.

Disease Caused by *Mycobacterium Abscessus* and Other Rapidly Growing Mycobacteria (RGM)

Julie V. Philley and David E. Griffith

Taxonomy

Historical Background and Current RGM Classification

The term rapidly growing mycobacteria (RGM) was originally used to describe mycobacterial organisms with growth on subculture in less than 7 days. The first rapidly growing mycobacterium was described in the early twentieth century when *Mycobacterium chelonae* was recovered from the lungs of sea turtles [1]. *Mycobacterium fortuitum* was originally recovered from frogs in 1905 and dubbed *Mycobacterium rana*e later to become *M. fortuitum* at the suggestion of Ernest Runyon [2]. In 1953 *Mycobacterium abscessus* was first reported as a cause of human skin and soft tissue infection in a patient with multiple soft tissue abscesses [3].

There are now more than 75 recognized RGM species, representing approximately 50% of all mycobacterial species. (Table 1) [4–11]. The three most clinically important pathogenic species that represent more than 80% of clinical RGM isolates are *M. fortuitum*, *M. chelonae*, and *M. abscessus*. *Mycobacterium abscessus* was separated from *M. chelonae* more than 20 years ago, but some mycobacterial laboratories still utilize anachronistic nomenclature labeling *M. abscessus* isolates with a group label, "*M. chelonae/abscessus* complex," rather than species or subspecies identification. This approach is not adequate or acceptable for contemporary RGM disease management.

Unfortunately, even the advent of molecular organism identification methods has not eliminated nomenclature confusion as illustrated by the recent controversies surrounding *M. abscessus* species and subspecies taxonomic designations.

J. V. Philley (✉) · D. E. Griffith
University of Texas Health Science Center, Tyler, TX, USA
e-mail: Julie.philley@uthct.edu; David.griffith@uthct.edu

D. E. Griffith (ed.), *Nontuberculous Mycobacterial Disease*, Respiratory Medicine,
https://doi.org/10.1007/978-3-319-93473-0_13

Table 1 RGM species associated with human disease

Common	*M. abscessus* subsp. *abscessus*, *M. abscessus* subsp. *massiliense*
Infrequent	*M. chelonae*, *M. fortuitum*, *M. porcinum*, *M. abscessus* subsp. *bolletii*
Rare but proven pathogens	*M. fortuitum* group (*M. boenickei*, *M. houstonense*, *M. peregrinum*, *M. senegalense*), *M. franklinii*, *M. immunogenum*, *M. mageritense*, *M. mucogenicum*[a], *M. wolinskyi*, *M. bacteremicum*, *M. canariasense*[c], *M. celeriflavum*, *M. cosmeticum*, *M. goodii*[c], *M. iranicum*, *M. neoaurum*, *M. smegmatis*

[a]*M. mucogenicum* group is composed of *M. mucogenicum*, *M. aubagnense*, and *M. phocaicum*
[c]Late pigmentation

Mycobacterium bolletii and *M massiliense* were originally described as unique species distinct from *M abscessus* in 2004 and 2006, respectively, but were subsequently found to be indistinguishable based on 16S rRNA sequence analysis [5, 12–15]. This latter finding suggested that the putatively unique species were in fact the same species [12–15]. In 2011, a proposal was made that the two organisms should be combined as one species and reclassified as *M. bolletii* [14]. Subsequently, with multigene and whole genome sequencing, it became apparent that the two organisms are in fact distinct species with an extremely important genetic difference [15, 16]. One organism has an active erythromycin ribosomal resistance methylase (*erm*) gene and one does not. The *erm* gene results in methylation of the 23S rRNA macrolide binding site and inducible loss of macrolide activity/function, discussed in detail below.

Adoption of the 2011 proposal would have created significant confusion because the taxonomic designation *M. massiliense* had already been widely accepted in the medical literature to describe the "*M. abscessus*" organism without an active *erm* gene, while the term *M. bolletii* was used to describe the other similar organism with an active *erm* gene. There was a clear need to standardize the nomenclature for the three closely related *M. abscessus* organisms, two with active *erm* genes and one with an inactive *erm* gene.

In an effort to clarify this situation, Tortoli et al. recently proposed that the three organisms in the *M. abscessus* complex should be emended to three subspecies: *M. abscessus* subsp. *abscessus* and *M. abscessus* subsp. *bolletii* with active *erm* genes and *M. abscessus* subsp. *massiliense* with an inactive *erm* gene [16, 17]. The Tortoli proposal is not without controversy as universally accepted criteria for species and subspecies designations does not exist. A counter proposal has recently been advocated to grant species designations for the three *M. abscessus* organisms [18]. It should be emphasized that the M *abscessus* subsp. *massiliense*, or *M. massiliense* as sometimes cited, remains the generally accepted name for the *M. abscessus* organism with an inactive *erm* gene which remains the most clinically important aspect of the taxonomy debate [16, 17, 19, 20]. From a nihilistic standpoint, the clinician may not care what an organism is named but absolutely must know the status of the *erm* gene and in vitro macrolide susceptibility of the organism.

For this chapter we have adopted the *M abscessus* nomenclature suggested by Tortoli et al. with *M. abscessus* subspecies designations [16]. Where possible, an *M. abscessus* subspecies is identified in the text. When the term *M. abscessus* without a subspecies designation is used, it was not possible to tease out a particular subspecies designation. The term *M. abscessus* without a subspecies designation is not, therefore, synonymous with *M. abscessus* subspecies *abscessus* but indicates a lack of subspecies differentiation in the cited references.

Identifying RGM

Laboratory techniques for identifying RGM are briefly summarized here. A detailed discussion is beyond the scope of this chapter. For a more in-depth discussion, please see chapter "Laboratory Diagnosis and Antimicrobial Susceptibility Testing of Nontuberculous Mycobacteria".

Phenotypic laboratory identification of RGM was previously based on growth in subculture in less than 7 days, colony morphology, and biochemical growth requirements. While the group of organisms categorized as RGM can be identified in this manner, definitive RGM identification is not possible. High-performance liquid chromatography (HPLC) of mycolic acids can identify only a few RGM species. HPLC may be useful for separating organisms into complexes or groups but lacks the specificity needed for full species-level identification [11, 21].

Lipid and ribosomal protein analyses by matrix-assisted laser desorption ionization-time of flight mass spectrometry (MALDI-TOF MS) is a method currently being used to identify the NTM including RGM [21–23]. Although MALDI-TOF MS has been successful in the identification of many species of NTM, several clinically significant species and subspecies of RGM, including *M. abscessus* subsp. *abscessus*, *M. abscessus* subsp. *massiliense, and M. abscessus* subsp. *bolletii*, have been difficult to differentiate by MALDI-TOF.

Currently, the only nucleic acid probe available for identification of the RGM is the INNO-LiPA multiplex probe assay (Innogenetics, Ghent, Belgium). The major advantage is that a large variety of species can be identified by a single probe without necessitating the selection of a specific probe for each species. Disadvantages include cross-reactivity among closely related *M. fortuitum* groups and the inability to differentiate isolates of *M. chelonae* from *M. abscessus* [24, 25].

The primary gene target of molecular taxonomic studies has been the 16S rRNA gene which is a highly conserved gene within mycobacterial species [26]. Differentiation of *M. chelonae* and *M. abscessus* and some species within the *M. fortuitum* group requires complete 16S rRNA sequence analysis for species identification unless other gene targets are sequenced. The limitations of complete 16S rRNA sequencing for RGM species and subspecies differentiation were exposed by the recent *M. abscessus* subspecies controversy described above.

The 65-kDa heat shock protein gene (*hsp*65) is also useful for species-level RGM isolates such as *M. abscessus* and *M. chelonae* and for most of the common RGM species [27, 28]. Some RGM species, including *M. fortuitum*, are more readily discriminated with the *hsp*65 gene analysis than by 16S rRNA gene analysis [29]. The *rpo*β gene has been used recently in the identification of RGM, including the identification of several new species [5, 12, 30–33].

Multiple different *erm* genes have been recognized in RGM species, including *erm* (*M. fortuitum*) and *erm* (*M. abscessus* subsp. *abscessus*). Some RGM species such as *M. chelonae* have no detectable *erm* gene which allows reliable identification of this organism.

Whole genome sequencing (WGS) and phylogenomic analysis are the most recently utilized methods for studying genetic variations and population studies in mycobacteria. Whole genome sequencing enables the study of multiple genetic regions which may be associated with pathogenicity, antibiotic resistance, virulence, and/or host relationships to the NTM. [34–38]. As previously noted, there are no universally accepted WGS criteria for defining NTM species and subspecies. Recent publications in cystic fibrosis literature have revealed *M. abscessus* subsp. *massiliense* isolates from patients in different countries including the United Kingdom, Brazil, and the United States, to have high levels of genetic relatedness by WGS and thus provided the first suggestion of possible person-to-person RGM transmission [17, 39, 40]. The future of NTM identification and epidemiological studies will likely be based upon WGS findings.

Clinicians must be familiar with the laboratory methods used for RGM identification including the limitations of the specific methods available to them. Clinicians should clearly communicate to laboratorians that optimal management of RGM disease patients requires timely and accurate organism identification. Clinicians face many challenges impeding successful treatment of these patients, and the laboratory should not be one of them.

Clinical Disease

The RGM are frequently isolated in the environment and have been found in 30% to 78% of soil samples from various geographical regions in the United States [11, 41, 42]. Although it is a much less common occurrence than with *Mycobacterium avium* complex (MAC), *M. abscessus* has also been isolated from municipal water [43]. Until recently, the majority of reported cases were from the United States, with a strong disease geographic localization in the southern United States [11, 44]. Human infections have now been reported from most areas of the developed world (see chapters "Epidemiology of Nontuberculous Mycobacterial Pulmonary Disease (NTM PD) in the USA" and "Global Epidemiology of NTM Disease (Except Northern America)"). The route of organism acquisition for RGM pulmonary disease is presumably inhalation of contaminated aerosols, similar to MAC [44]. Community-acquired localized skin, soft tissue, and/or bone disease usually follows a traumatic injury with potential soil contamination. Nosocomial infections are strongly associated with tap water exposure or contamination.

Chronic Pulmonary Infections

Chronic RGM pulmonary infections are most often associated with *M. abscessus* subspecies [11, 42, 44] (Tables 2 and 3). The majority are due to either *M. abscessus* subsp. *abscessus* or *M. abscessus* subsp. *massiliense*. These infections are typically

Table 2 Rapidly growing mycobacterial (RGM) species and their common clinical diseases

M. fortuitum group	*M. chelonae*	*M. abscessus*
Localized posttraumatic infections	Disseminated skin infections	Chronic lung infections
Catheter infections	Localized posttraumatic wound infections	Localized posttraumatic wound infections
Surgical wound infections	Catheter infections	Catheter infections
Augmentation mammaplasty	Sinusitis	Disseminated skin infections
Cardiac surgery	Corneal infections	Corneal infections

Table 3 Pulmonary syndromes associated with positive respiratory cultures for RGM

Finding	Interpretation	Most common RGM involved
Single AFB smear-negative culture-positive specimen	Transient infection or specimen contamination	*M. abscessus*
Multiple culture-positive specimens		
Elderly nonsmoking patients, usually female, with nonspecific radiographic findings consistent with bronchiectasis and multiple sputum AFB-positive cultures	Probable nodular bronchiectasis disease: confirm by HRCT[a]	*M. abscessus*
Cavitary radiographic changes either apical typical of fibrocavitary mycobacterial disease, usually male, or mid and lower lung field associated with bronchiectasis, usually female	Possible association with prior granulomatous disease (TB) or due to progressive nodular/bronchiectatic disease	*M. abscessus*
Achalasia with chronic vomiting and bilateral interstitial/alveolar infiltrates or known lipoid pneumonia	Chronic pneumonitis, reversal of underlying GI problem essential for adequate therapy	*M. fortuitum any RGM*
Cystic fibrosis	Focal pneumonitis, transient infection or progressive disease (HRCT may be helpful in determination)	*M. abscessus*

[a]HRCT, high-resolution computerized tomography

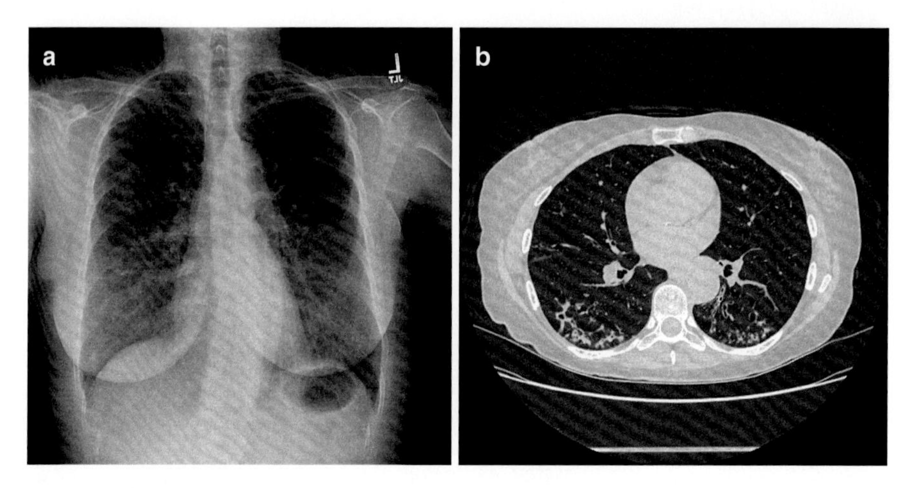

Fig. 1 *M. abscessus* lung disease with nodular bronchiectasis

found in postmenopausal nonsmoking females who present with chronic cough, weight loss, fatigue, and sometimes hemoptysis. By high-resolution computerized tomography (HRCT) of the chest, most patients have patchy cylindrical bronchiectasis and small nodules involving the right middle lobe and lingua (Fig. 1a, b). This radiographic pattern is referred to as nodular bronchiectatic (NB) disease and is also typical of patients with *Mycobacterium avium* complex (MAC) lung disease. Chronic NB lung disease caused by MAC and *M. abscessus* are radiographically and clinically indistinguishable. RGM lung disease tends to be slowly progressive, with indolent symptomatic and radiographic progression [42, 44].

Patients undergoing therapy for MAC lung disease sometimes have co-isolation of *M. abscessus* from respiratory specimens. The clinical significance of these isolates is variable and must be determined with longitudinal evaluation [45]. In one recent series of MAC patients with co-isolation of *M. abscessus*, most *M. abscessus* isolates did not appear to be clinically significant when they occurred only once or twice without clinical and radiographic impact [45]. For some patients, however, *M. abscessus* was isolated from multiple specimens over time and was associated with radiographic progression, especially new or expanding lung cavitation. In these patients *M. abscessus* appeared to be a significant pathogen requiring therapeutic intervention. This latter consideration is not inconsequential as there are few antibiotic agents with activity against both MAC and *M. abscessus* subspecies with an active *erm* gene and therefore little overlap in treatment regimens for *M. abscessus* and MAC.

Cavitary lung disease that occurs with *M. abscessus* subspecies is a more aggressive disease than the NB form. Patients with long-standing NB RGM lung disease, either untreated or refractory to treatment, can evolve from NB lung disease to a mixed NB and cavitary lung disease (Fig. 2). Patients with primary cavitary disease are similar clinically to cavitary MAC lung disease patients, usually male with a history of cigarette smoking and COPD. *Mycobacterium abscessus* infection is also

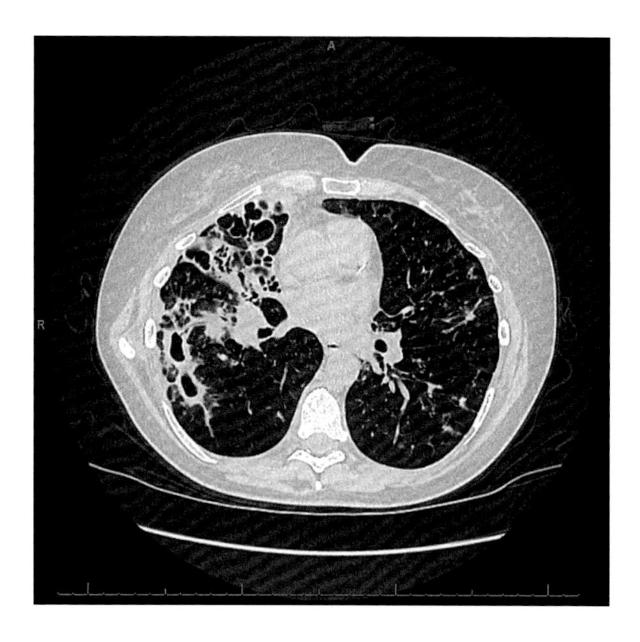

Fig. 2 Bronchiectasis with cavitary lesions in a patient with M. *abscessus* spp. *abscessus*

sometimes associated with other lung diseases and infections that leave residual areas of bronchiectasis and scarring such as tuberculosis.

Pulmonary infections with the *M. fortuitum* group are rare and most often seen in patients with achalasia, chronic vomiting, and other forms of gastrointestinal disturbances associated with chronic aspiration [42]. *Mycobacterium fortuitum* is a relatively non-virulent pathogen and a rare cause of lung disease outside of these conditions [42, 46]. *Mycobacterium mucogenicum* is sometimes a significant pathogen in the setting of chronic obstructive lung disease but overall is rarely associated with progressive chronic pulmonary infection [42]. *Mycobacterium fortuitum* and *M. mucogenicum* are examples of the poor specificity of the current NTM diagnostic guidelines in that patients with multiple sputum AFB cultures positive for *M. fortuitum* or *M. mucogenicum* are unlikely to develop progressive disease and require therapy. In the absence of a specific predisposition, such as chronic aspiration, clinicians are urged to use more rigorous diagnostic criteria for *M. fortuitum* and *M. mucogenicum* isolates than for other more common NTM respiratory pathogens.

Chronic pulmonary infections with *M. abscessus* also are seen in patients with cystic fibrosis (CF) (chapter "Non-tuberculous Mycobacteria in Cystic Fibrosis") [47–49]. Patients with CF have aggressive and rapidly progressive bronchiectasis in addition to chronic recurrent airway and parenchymal infections due to *Pseudomonas aeruginosa* and other bacterial pathogens [48, 49]. Nontuberculous mycobacteria were previously thought not transmissible between humans. Recently, however, geographically widely dispersed cases associated with *M. abscessus* subsp. *massiliense* isolates that are indistinguishable by DNA analysis have been identified, suggesting direct or indirect spread of identical infections that can occur at least in this highly vulnerable population [17, 50].

The most common pathogens associated with CF lung disease in some series have been members of the *M. abscessus* group. One-half of 104 NTM isolates recovered in a multicenter study involving 1582 CF patients in France were *M. abscessus* [51]. A 3-year longitudinal clinical study from Brazil recovered NTM from 8% of 129 children with the majority of RGM isolates *M. abscessus* [52]. An early large study from US CF centers found that *M. abscessus* isolates were second in prevalence to those of the MAC [49]. Although some patients appear to have transient carriage, other patients remain culture positive, with significant symptoms and high morbidity and mortality [49].

The difficulty effectively treating *M. abscessus* lung infections in CF patients adds an additional layer of complexity to their management. Because of resistance to antibiotic therapy and the specter of postoperative infections with poor wound healing, isolation of *M. abscessus* can preclude lung transplant for CF patients in some centers.

Localized Posttraumatic Wound Infections

Wound infections are typically associated with accidental penetrating trauma with soft tissue infection sometimes followed by osteomyelitis [11] (Fig. 3). Patients with this type of infection are usually healthy without systemic immune suppression. After an incubation period of 3–6 weeks, local redness and swelling with spontaneous drainage typically occurs. Systemic symptoms such as fever, chills, malaise, and fatigue are infrequent. The drainage is usually thin and clear but occasionally can be thick and purulent. Sinus tract formations with intermittent drainage are common.

The most common pathogens in these settings are the *M. fortuitum* group including *M. fortuitum*, *M. porcinum*, and *M. houstonense*, but almost any of the pathogenic

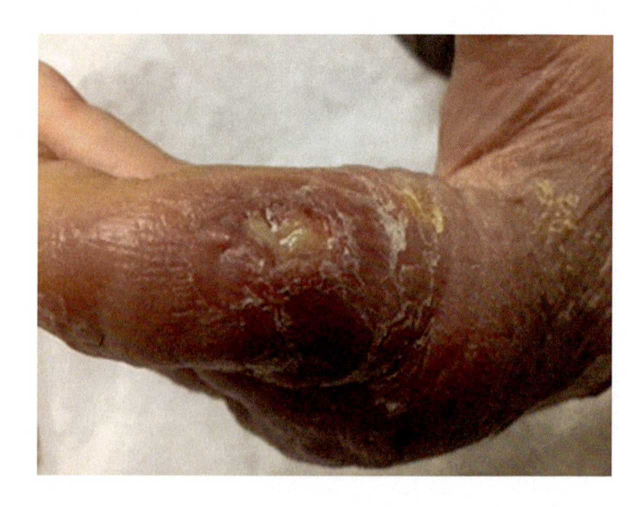

Fig. 3 *M. abscessus* subsp. *abscessus* skin and soft tissue infection following penetrating trauma to the finger demonstrating characteristic purple discoloration associated with purulent drainage

RGM species can cause disease in patients with infected open fractures [4, 8, 11]. Furthermore, these infections may be polymicrobial, reflecting environmental contamination with more than one species of mycobacteria or a combination of bacteria and mycobacteria. Timely diagnosis requires clinical suspicion in a patient who is not responding to or improving with standard antibacterial antibiotics.

Surgical Wound Infections

Most outbreaks of healthcare-associated RGM infections or pseudo-infections have been associated epidemiologically with various water sources, including water-based solutions, distilled water, tap water, ice, and ice water [11, 52, 53]. The utilization of pulsed-field gel electrophoresis and randomly amplified polymorphic DNA PCR methods for analyzing genomic DNA has improved the investigation of nosocomial outbreaks [11, 54, 55]. DNA fingerprinting for some of these outbreaks has confirmed molecular identity between water and human isolates. One study of hemodialysis centers in the United States showed that 55% of incoming city water contained mycobacteria, of which RGM species were the most common [56]. Biofilms, which are the lipid-rich layers that form at water-solid interfaces, are present in most water transport pipes. Up to 90% of these from community-piped water systems contain mycobacteria [57]. Compared with free-living mycobacteria, mycobacteria associated with biofilms are more resistant to water treatment [58].

Surgical wound infections, including cataract excision, corneal graft, laser surgery, extremity amputations, dacryocystorhinostomy, plastic surgery of the face, prosthetic hip or knee insertions, coronary artery bypass, excision of basal cell carcinoma, augmentation mammaplasty, and cosmetic surgeries including liposuction and liposculpture, clinically present in a similar fashion to accidental trauma [11, 59–64]. After an incubation period of 2–8 weeks, the healing wound will develop serous drainage and redness. Localized nodular areas adjacent to the incision may develop which are often painful and may require incision and drainage. Isolates of the *M. fortuitum* group are most commonly recovered in these settings, but other species may also be involved [11, 64]. Rare cases of *M. wolinskyi* infection have been reported, mainly skin and soft tissue infection following surgery including cosmetic procedures [65–68]. Additionally, pseudo-outbreaks of *M. abscessus* or *M. immunogenum* related to contaminated automated bronchoscope disinfection machines, contaminated gastric endoscopes, and laboratory contamination have been described [11, 69, 70].

A recent large nosocomial outbreak of *M. abscessus* subspecies in lung transplant patients was associated with positive cultures for *M. abscessus* from patients, biofilms, and water sources obtained from hospital water outlets [71]. Using pulsed-field gel electrophoresis (PFGE), 4 of 10 patients and 8 of 12 environmental cultures showed the same strain of *M. abscessus*. Multiple interventions were made, and the incidence rate of *M. abscessus* decreased from 3.9 cases/month during the outbreak period to 1 case/month during the intervention period. This decrease in incidence showed that the outbreak of clonally related pulmonary *M. abscessus* was epidemiologically linked to the water sources and amenable to targeted infection control efforts [71]. Please see chapter "Environmental Niches for NTM and Their Impact on NTM Disease" for a more extensive discussion of nosocomial RGM disease.

Catheter-Related Infections

The most common healthcare-associated NTM disease since the 1990s has been central venous catheter infections. These may manifest as occult bacteremia, granulomatous hepatitis, septic lung infiltrates, tunnel infections, or exit site infections [11]. The timing of these infections usually involves catheters that have been in place at least several months. The most frequent etiologic agent is *M. fortuitum*, although pigmented species such as *M. neoaurum* and *M. bacteremicum* have also been associated with catheter sepsis [72–75]. Other long-term catheters, including chronic peritoneal dialysis catheters, hemodialysis catheters, nasolacrimal duct catheters, and ventriculoperitoneal shunts also have been associated with RGM infection [4, 76, 77]. Removal of catheters is necessary for successful treatment.

Disseminated Cutaneous Infections

One type of RGM skin involvement occurs in patients who require chronic steroid therapy and is characterized by the presence of multiple noncontiguous nodules with spontaneous drainage on one or more extremities. These multiple skin lesions develop without life-threatening sequelae and almost always involve the lower extremity and are typically due to *M. chelonae* [11]. Steroid doses may be as low as 5–10 mg of prednisone daily. The most common underlying disease is rheumatoid arthritis but may also include organ transplants and chronic autoimmune disorders [11]. The patients are frequently asymptomatic except for the local discomfort of the lesions.

A second type of disseminated skin disease is seen in immune-compromised patients, some with rapidly fatal disorders such as poorly controlled hematologic malignancies [11, 77, 78]. This type of infection is usually systemic, with positive cultures of the blood and bone marrow. A portal of entry for the organism is rarely identified, although central catheters may be involved. It is usually caused by *M. abscessus* and, combined with the underlying disease, was often fatal in the era before current antimicrobial therapy was available. Interestingly, members of the *M. fortuitum* and *M. mucogenicum* groups are rarely associated with either type of disseminated disease [11, 79]. As with most infections associated with immune suppression, reversal of the immune suppression is necessary for adequate treatment response for the RGM pathogen.

Ophthalmic Infections

A recent study of 100 patients with nontuberculous mycobacterial ophthalmic infections showed that 95% of the infections were due to RGM, commonly *M. chelonae* and *M. abscessus* [80]. Ophthalmic infections due to RGM are often associated with poor visual outcomes despite aggressive treatment [81–83].

Nail Salon/Footbath-Associated Folliculitis

RGM lower extremity skin infections involving *M. fortuitum*, *M. mageritense*, and a newly described species, *M. cosmeticum*, is associated with the use of contaminated nail salon whirlpool footbaths [81–83]. Patients were salon customers with persistent skin infections below the knee [83]. Most often the infections involved a furunculosis of the lower leg hair follicles [82]. The disease pathogenesis likely results from microtrauma caused by shaving the legs prior to pedicures and footbath water that is heavily contaminated with RGM due to failure to routinely clear the footbath filters [82–84].

Anti-TNF-α Therapy-Associated Infections

NTM infections associated with the use of biologic therapies that inhibit tumor necrosis factor alpha (TNF-α) have been reported and are discussed in detail in chapters "Vulnerability to Nontuberculous Mycobacterial Lung Disease or Systemic Infection Due to Genetic/Heritable Disorders and Immune Dysfunction and Nontuberculous Mycobacterial Disease". Patients receiving anti-TNF-α therapy are at high risk for activation of tuberculosis and appear to have some increase in disease risk or difference in clinical manifestation for NTM as well [85]. A recent review of the US Food and Drug Administration MedWatch database reports identified 239 possible cases of NTM disease associated with TNF-α inhibitor use from 1999 to 2006 of which 105 cases (44%) met NTM disease criteria. NTM infections were associated with immunosuppressive therapies including infliximab, etanercept, and adalimumab, while 65% and 55% of patients were also taking prednisone and methotrexate, respectively. Infections with MAC were most commonly reported, while 20/105 cases (20%) involved RGM species including *M. abscessus*, *M. chelonae*, and *M. fortuitum* [85].

Interferon-Gamma/Interleukin 12 (IFN- γ/IL-12)-Associated Infections

IFN-γ/IL12 is an immunological pathway designed for intracellular killing of mycobacteria and is discussed in more detail in chapter "Vulnerability to Nontuberculous Mycobacterial Lung Disease or Systemic Infection Due to Genetic/Heritable Disorders". An apparently acquired autoantibody-mediated immunodeficiency was recently described and found to be almost exclusively among Asian-born women [86, 87]. Most cases related to NTM were associated with MAC; however, recent reports suggest that approximately 45% of the NTM cases are due to RGM including *M. abscessus* (32%) and *M. fortuitum* (12%). Infections were typically multifocal affecting lymph nodes, osteoarticular tissue, lungs, skin, and/or soft tissues. IFN-γ autoantibodies should be considered in cases of unexplained disseminated NTM infection, especially in the Asian-born population. [86, 87].

Drug Therapy/Drug Resistance

Antimicrobial Susceptibility

Antimicrobial regimens for disease caused by the RGM are usually based upon their in vitro susceptibility patterns. While this approach is appealing from a traditional infectious diseases perspective, it is frequently frustrating with limited practical applicability [11, 42, 88–94]. Recognizing those limitations is a prerequisite for successful RGM patient treatment. RGM isolates are not susceptible to the first-line antituberculous drugs and require in vitro susceptibility testing in specialized laboratories that are experienced in processing RGM. The current drugs that should be tested for susceptibility include amikacin, cefoxitin, imipenem, moxifloxacin, meropenem, sulfamethoxazole or trimethoprim-sulfamethoxazole, clarithromycin, ciprofloxacin, doxycycline/minocycline, linezolid, and tobramycin (the latter only for isolates of *M. chelonae*) [90]. Tigecycline is frequently included in routine testing, but no minimum inhibitory concentration (MIC) breakpoints have been determined [88, 89].

Clarithromycin inhibits all RGM isolates with no functional *erm* gene, including *M. chelonae* and *M. abscessus* subsp. *massiliense* at a concentration of 4 μg/ml at 3 days [90]. The MICs for several species, including *M. abscessus* subsp. *abscessus*, *M abscessus* subsp. *bolletii*, and *M. fortuitum* may be in the susceptible range with only 3 days of incubation but due to the presence of the *erm* gene become resistant with a 14-day macrolide incubation. Most RGM species that have late intrinsic resistance to clarithromycin contain an inducible *erm* gene [91–94]. The finding of the *erm* gene in the RGM helps explain the lack of effectiveness for macrolide therapy despite ostensibly "susceptible" routine in vitro MICs with standard (3 days) incubation times.

It cannot be overemphasized that routine in vitro macrolide susceptibility testing that does not include macrolide preincubation of the RGM isolate for 14 days is not reliable and cannot be used for basing treatment decisions [92]. In 2011, the Clinical and Laboratory Standards Institute (CLSI) recommended the final reading of the clarithromycin MICs for RGM at 14 days, to detect inducible resistance [88]. Alternatively, the need for extended incubation could be eliminated for isolates where *erm* gene functionality has been molecularly determined by *erm* gene sequencing [89].

Multiple studies have shown that patients with RGM lung disease due to an *erm* gene inactive isolate respond significantly more favorably to therapy than patients with an *erm* gene active RGM [95, 96]. Because of the impact on treatment response, determining the *erm* gene functionality of clinically significant RGM isolates is absolutely essential, either with preincubation of the RGM isolate in the presence of macrolide or by molecular methods. Clinicians can also be confident that RGM isolates accurately identified as *M. chelonae* or *M. abscessus* subsp. *massiliense* do not have an active *erm* gene and will respond to macrolide-containing regimens.

For antibiotics other than the macrolides, the terms "susceptible" and "resistant" have limited meaning as treatment regimens based on these designations are not predictably effective. In general, isolates of *M. abscessus* (all subspecies) and *M. chelonae* are more resistant to antibiotics than other RGM species and are usually susceptible or intermediate only to amikacin, imipenem, and clarithromycin, in the absence of an active *erm* gene [11]. Isolates of *M. abscessus* (all subspecies) are moderately susceptible (intermediate) to cefoxitin (MIC ≤64 μg/mL), whereas isolates of *M. chelonae* are highly resistant (MIC ≥256 μg/mL). Additionally, MICs of tobramycin for *M. chelonae* are lower than those of amikacin, so that

Table 4 Antimicrobials used for treatment of commonly encountered species of RGM

Species	Drugs[a]
M. fortuitum	Oral: ciprofloxacin, levofloxacin, trimethoprim-sulfamethoxazole, moxifloxacin, clarithromycin[b] (80%), doxycycline (50%), linezolid (86%)
	Parenteral: amikacin, cefoxitin, imipenem,[c] tigecycline[d]
M. abscessus subspecies	Oral: clarithromycin or azithromycin in isolates with nonfunctional *erm* genes (i.e., approximately 20% *M. abscessus* subsp. *abscessus*, 100% *M. abscessus* subsp. *massiliense*), doxycycline (<5%), ciprofloxacin (<5%), moxifloxacin (<15%), linezolid (23%)
	Parenteral: amikacin, tigecycline,[d] cefoxitin (70%), imipenem
M. chelonae	Oral: doxycycline (25%), ciprofloxacin (25%), linezolid (54%), clarithromycin or azithromycin
	Parenteral: tobramycin, amikacin (70%), imipenem, tigecycline[d]

[a]Untreated strains are 100% susceptible unless otherwise noted
[b]Does not include inducible (*erm*) resistance determination. *M. fortuitum* isolates must be assumed to be macrolide resistant. No studies have yet addressed specific percentage of inducible *erm* genes present in the *M. fortuitum* group
[c]Susceptible or intermediate
[d]No breakpoints are currently available for tigecycline with RGM, but most MICs for *M. fortuitum*, *M. abscessus*, and *M. chelonae* have been ≤1 μg/mL

M. chelonae is the only species of RGM for which amikacin is not the preferred aminoglycoside. Approximately 20% of the strains of *M. chelonae* are also susceptible to achievable serum levels of ciprofloxacin and/or doxycycline or minocycline [11, 97] (Table 4).

Amikacin is an important antibiotic in RGM treatment regimens. An amikacin resistance breakpoint MIC of >64 μg/mL (compared to ≥64 μg/mL for IV amikacin) has been proposed to the CLSI for inhaled amikacin and has been correlated with the finding of a mutation in the 23S rRNA gene [89, 98]. A recent randomized placebo controlled trial of inhaled liposomal amikacin in patients with MAC and *M. abscessus* lung disease demonstrated the efficacy of inhaled amikacin in *Mycobacterium avium* complex (MAC), but only a small number of conversions occurred in patients with *M. abscessus* [99].

Among the oxazolidinones, linezolid has in vitro activity against the *M. fortuitum* group and *M. chelonae* [100]. Linezolid has been used subsequently in the treatment of infections due to RGM, including disseminated *M. chelonae* with acquired mutational resistance to clarithromycin [101]. Isolates of *M. abscessus* have variable susceptibility to linezolid. Recently a new oxazolidinone, tedizolid, has shown early in vitro activity among isolates of RGM [102]. No clinical experience in the treatment of RGM disease has been reported so far with this antibiotic.

Tigecycline, a glycylcycline derivative of minocycline, has shown excellent in vitro activity against all species of RGM, including *M. chelonae*, *M. abscessus*, and *M. fortuitum* group with MICs of <1 μg/mL [97]. A clinical trial of patients treated with tigecycline showed limited efficacy of the drug for salvage therapy of patients with respiratory *M. abscessus* infections [103]. However, MIC breakpoints for this agent with the RGM have not been established so that tigecycline MICs are reported without interpretation [89].

Clofazimine, a riminophenazine antibiotic, has been used in the treatment of *M. leprae* and multidrug-resistant (MDR) *M. tuberculosis*. Recently there has been a revival of interest in its efficacy against RGM, especially *M. chelonae* and *M. abscessus.* Synergism between clofazimine and amikacin against *M. abscessus* was demonstrated in vitro [104, 105]. Clofazimine may also act to increase exposure to other important antimicrobials including macrolides. There are no CLSI breakpoints available yet for the interpretation of clofazimine MICs, and large prospective clinical studies have not been done. A recent retrospective study of 42 patients treated with clofazimine as part of multidrug treatment regimens for *M. abscessus* subsp. *abscessus* found a 24% sputum conversion rate and 81% treatment response rate based on symptoms with 31% radiographic response [106]. Although clofazimine MICs are frequently reported and the drug appears to be widely used, the clinical efficacy of clofazimine for treating *M. abscessus* subsp. *abscessus* remains unclear and its role in this context not yet established. As with most aspects of this difficult process, further studies are necessary.

The new diarylquinolone antibiotic, bedaquiline, was recently approved by the FDA to treat multidrug-resistant (MDR) tuberculosis. It causes disruption of ATP synthesis and has impressive in vitro activity against isolates of NTM including *M. abscessus* subspecies. In a study of four patients with *M. abscessus* lung infection, two patients had microbiologic improvement (i.e., decrease in semiquantitative colony counts), and all but one patient showed symptomatic improvement [107]. Large clinical studies along with in vitro susceptibility studies, however, remain to be performed.

A 2014 in vitro study showed the lack of bactericidal antibiotics, including amikacin, in the currently recommended treatment regimens for *M. abscessus* [108]. This finding was hypothesized to be due to the presence of functional aminoglycoside-modifying enzymes encoded by specific genes in the *M. abscessus* chromosome [108]. In vitro studies, such as this, emphasize the need for novel antibiotics and/or therapeutic options to improve the therapeutic outcome of patients with RGM, especially *M. abscessus* [108, 109].

Isolates of the *M. fortuitum* group, *M. smegmatis* group, and *M. mucogenicum* group are generally the most susceptible of the commonly encountered RGM species [6, 11]. In vitro, they are usually susceptible or intermediate to amikacin, cefoxitin, imipenem, ciprofloxacin, sulfonamides, and moxifloxacin, with about 50% of the isolates of *M. fortuitum* susceptible to doxycycline (Table 4). Both minocycline and doxycycline are preferred over tetracycline because of greater in vitro activity of the former two antimicrobials in previous studies of the RGM [97].

Treatment of RGM Disease

Therapy for most RGM infections has not been established from clinical trials. Current recommendations are generally based on uncontrolled case series and clinical experience. There is a clear dichotomy in the anticipated response of RGM pathogens to therapy. Infections caused by *M. fortuitum* group pathogens are usually responsive to antibiotic therapy, while infections due to *M. abscessus* subsp. are usually less responsive to antibiotic therapy.

Effective therapy for many RGM, especially *M. abscessus* subsp. *abscessus*, is thwarted by two major mechanisms of drug resistance. The first is innate or natural drug resistance best illustrated by the *erm* gene whereby initial, routine, MICs for macrolides appear susceptible, but subsequent MICs after macrolide exposure show resistance consistent with the activation of the inducible macrolide resistance gene [91–94, 109]. While the description of the *erm* gene has opened windows into the perplexing and complicated realm of innate antibiotic resistance, it is by no means the end of the story. *M. abscessus* subsp. *abscessus* possesses multiple innate drug resistance mechanisms that frustrate antibiotic therapy [109].

Patients infected with *M. abscessus* with a nonfunctional *erm*(41) gene, primarily those due to *M. abscessus* subsp. *massiliense*, have the best prognosis as they are macrolide susceptible [96, 109]. The prognostic difference between an RGM pathogen with an active versus an inactive *erm* gene is so critical that it necessitates a pause at this point in the narrative to explicitly emphasize this concept. When a clinician is faced with an RGM infection, especially one due to an organism identified as "*M. abscessus*," the clinician must know the functional status of the *erm* gene from that organism. The *erm* gene status can usually be deduced from the identification of the organism but must be corroborated after incubation of the organism in the presence of macrolide for 14 days. Identification of an organism as *M. abscessus* subsp. *massiliense* will inevitably mean the organism does not have an active *erm* gene. Identification of an organism as *M. abscessus* subsp. *abscessus* does not inevitably mean that the organism will have an active *erm* gene as approximately 20% of these isolates will have an *erm* gene mutation that renders it nonfunctional [110]. The clinician must have both taxonomic and *erm* gene functional status for all clinically significant RGM isolates. It is imperative that the clinician is familiar with the method used to determine in vitro macrolide susceptibility.

The second mechanism of resistance is through acquired mutational drug resistance in isolates that do not have an active *erm* gene such as *Mycobacterium abscessus* subsp. *massiliense*, *M. chelonae*, and approximately 20% of *M. abscessus* subsp. *abscessus* isolates [109]. Mutational drug resistance can develop on therapy and is a concern for ribosomal active drugs, such as clarithromycin (23S rRNA gene) and amikacin (16S rRNA gene). The mechanism of acquired mutational resistance for these organisms is the same as for all mycobacteria, inadequate companion medications to prevent the emergence of isolates with naturally occurring ribosomal gene mutations. The recognition that clinically significant RGM isolates without *erm* gene activity can subsequently develop acquired mutational macrolide resistance dictates that these isolates must be treated with adequate companion medications for macrolide to prevent the emergence of acquired mutational macrolide resistance.

The difference in clinical response for *M. abscessus* subspecies with and without active *erm* genes is profound [95, 111–114]. It would be extraordinarily disappointing for a patient with *M. abscessus* subsp. *abscessus* disease who was fortunate enough to have an inactive *erm* gene to become macrolide resistant because of inappropriate therapy resulting in acquired mutational macrolide resistance. Protection against the emergence of acquired mutational amikacin resistance is equally as important. Physicians must treat these patients with adequate aggressiveness to prevent acquired mutational resistance. This is, yet again, another reminder that there are no shortcuts in the management of these infections.

Mutational resistance is also a concern for the quinolones. Hence, therapy with these agents for organisms such as *M. chelonae* and *M. fortuitum* should include combination therapy, especially for extensive disease with large numbers of organisms. Acquired mutational resistance with the tetracyclines and sulfonamide monotherapy has not been described.

Treatment of *M. abscessus* lung disease is impeded not only by the antibiotic resistance of the organisms but by the complexity, expense, and toxicity of the required antibiotics [42]. See Tables 4 and 5 for a summary of drugs and suggested therapeutic approaches.

A critical evaluation, comparison, and summary of the literature reporting treatment outcomes for "*M. abscessus*" are impossible for studies done prior to knowledge and demonstration of *erm* gene activity. Prior to *erm* gene analysis, published *M. abscessus* case series included variable numbers of patients with *erm* gene active

Table 5 General principles of therapy of RGM disease

Clinical setting	Drug treatment[a]
Pulmonary disease	
M. fortuitum.	Short-term parenteral treatment (4–6 wks.) and then multiple (at least 2) oral antibiotics with in vitro activity for minimum of 6 mos.
M. abscessus subsp. *abscessus,* and *M. abscessus* subsp. *bolletii*	Isolates with functional *erm*(41) gene will be difficult to treat effectively, relatively low-dose amikacin (single daily dose with peaks in low- to mid-20 range), cefoxitin, and/or tigecycline or imipenem. Consider linezolid and/or clofazimine. We recommend a minimum of three antibiotics initially. Consider inhaled amikacin after microbiologic conversion. Initial therapeutic attempt should be to attain sputum culture negativity for 12 months
	For isolates with nonfunctional *erm* genes (i.e., approximately 20% *M. abscessus* subsp. *abscessus*, 100% *M. abscessus* subsp. *massiliense*), best oral antimicrobials are clarithromycin or azithromycin and linezolid with addition of pyridoxine (to potentially reduce risk of peripheral neuropathy). In addition to oral drugs, begin with relatively low-dose amikacin (single daily dose with peaks in low- to mid-20 range) and consider inhaled amikacin after microbiologic conversion
Localized skin/soft tissue/bone disease	
M. fortuitum, M. chelonae, M. abscessus	Initial (4–8 weeks) parenteral antibiotics for extensive disease followed by oral medicines as guided by in vitro susceptibility; for minor disease, oral antibiotics only. Essential to remove catheter or foreign body. Treat 6 mos. total for significant disease, including all cases with osteomyelitis
	Linezolid (oral or parenteral) with pyridoxine has also been successful for treatment of *M. chelonae*
Disseminated (cutaneous) disease	
M. chelonae	Once-daily low-dose tobramycin or intermittent daily imipenem plus clarithromycin for first 2–4 wks and then clarithromycin only to complete 6 mos.; linezolid with pyridoxine may also be effective
M. abscessus subsp. *abscessus*	Same as for *M. chelonae* except use of amikacin in place of tobramycin and cefoxitin may replace imipenem; 80% of the isolates will be macrolide resistant, 20% macrolide susceptible
M. abscessus subsp. *massiliense*	Best oral antimicrobials are clarithromycin or azithromycin and linezolid with pyridoxine

[a]Guided by in vitro susceptibility results

(*M. abscessus* subsp. *abscessus*) and *erm* gene inactive (*M abscessus* subsp. *massiliense*) patients which explains the wide variability in reported treatment responses. For instance, in one study from the United States, 69 *M. abscessus* lung disease patients without subspecies identification or reported *erm* gene activity were aggressively treated for 52 antibiotic months, including 6 months parenteral antibiotic therapy and surgery in 33% of cases [115]. Forty-eight percent of patients were AFB culture negative by the end of the study including 57% of the patients who had surgery [115]. In a retrospective study from the Republic of Korea, 65 patients with *M. abscessus* lung disease were treated with 1 month parenteral and oral antibiotics followed by 24 months oral antibiotics usually including a macrolide [116]. Sputum conversion and maintenance of negative sputum cultures for ≥12 months were achieved in 58%. In another study from the Republic of Korea 41 patients, 41% received macrolide and one parenteral agent while 59% received a macrolide and 2 parenteral agents [117]. Sputum conversion was achieved in 81% of patients without a difference between the two groups. It is clear that the improved outcomes in the Korean studies cited above were due to a much higher prevalence of *M. abscessus* subsp. *massiliense* patients in the study cohorts compared to the US cohorts.

In a subsequent study from the Republic of South Korea with subspecies organism identification, sputum conversion and maintenance of culture negativity occurred in 88% of 33 patients with *M. massiliense* lung disease compared with only 25% of 24 patients with *M. abscessus* subsp. *abscessus* infection [95]. Similar results were reported in a Japanese study that evaluated 62 patients and found higher sputum conversion rates (50% vs 31%) and lower relapse rates (30% vs 65%) in patients with *M. abscessus* subsp. *massiliense* lung disease compared with *M. abscessus* subsp. *abscessus* lung disease [118]. In a prospective cohort of 16 *M. abscessus* subsp. *massiliense* and 27 *M. abscessus* subsp. *abscessus* lung infection cases with CF, clarithromycin-based combination therapies led to mycobacterial eradication in 100% of *M. massiliense* cases but only in 27% of *M. abscessus* cases [119].

The preponderance of patients with *erm* gene inactive *M. abscessus* subsp. *massiliense* not only explains the improved treatment outcomes reported in the Republic of Korea compared with the United States but also the success of less aggressive treatment regimens with only brief exposure to parenteral medications and long-term oral antibiotic therapy including macrolide monotherapy [120]. While attractive from the standpoint of patient convenience and avoidance of drug toxicity, this approach has been reported to result in acquired macrolide resistance and is not recommended [95, 111–114].

It cannot be emphasized too strongly that patients with macrolide susceptible *M. abscessus* subsp. *abscessus* isolates because of an inactive *erm* gene are vulnerable to acquired mutational macrolide resistance due to mutations in the 23S rRNA gene. These mutations emerge during therapy with inadequate companion medications for the macrolides. This development is especially pernicious because it signals a significant decline in the chances for successful therapy of the patient [95, 111–114]. We are concerned by the recommendations for oral drugs of uncertain, even dubious, activity as companions for macrolides against *M. abscessus* isolates that do not have an active *erm* gene, and we strongly disagree with this approach [121]. It perhaps doesn't matter what oral drugs are used with *erm* gene active *M. abscessus* subsp. *abscessus*, but the creation of acquired mutational macrolide resistance for these difficult to treat organisms is an incalculable disservice to the patient. Expert consultation for management of RGM lung disease patients should occur prior to the advent of acquired mutational macrolide resistance.

So far, there is no consensus on the optimal composition of multidrug therapy for *M. abscessus* subsp. *abscessus* with the exception of the necessity for including a macrolide for patients with an inactive *erm* gene. The therapy of the patient is directed by the *erm* gene activity analysis and in vitro macrolide susceptibility status. Patients who are macrolide susceptible unquestionably have a much greater chance of completing successful therapy than those patients with macrolide-resistant isolates. For *erm* gene active isolates, there is also consensus that amikacin is an important element in the therapeutic regimen.

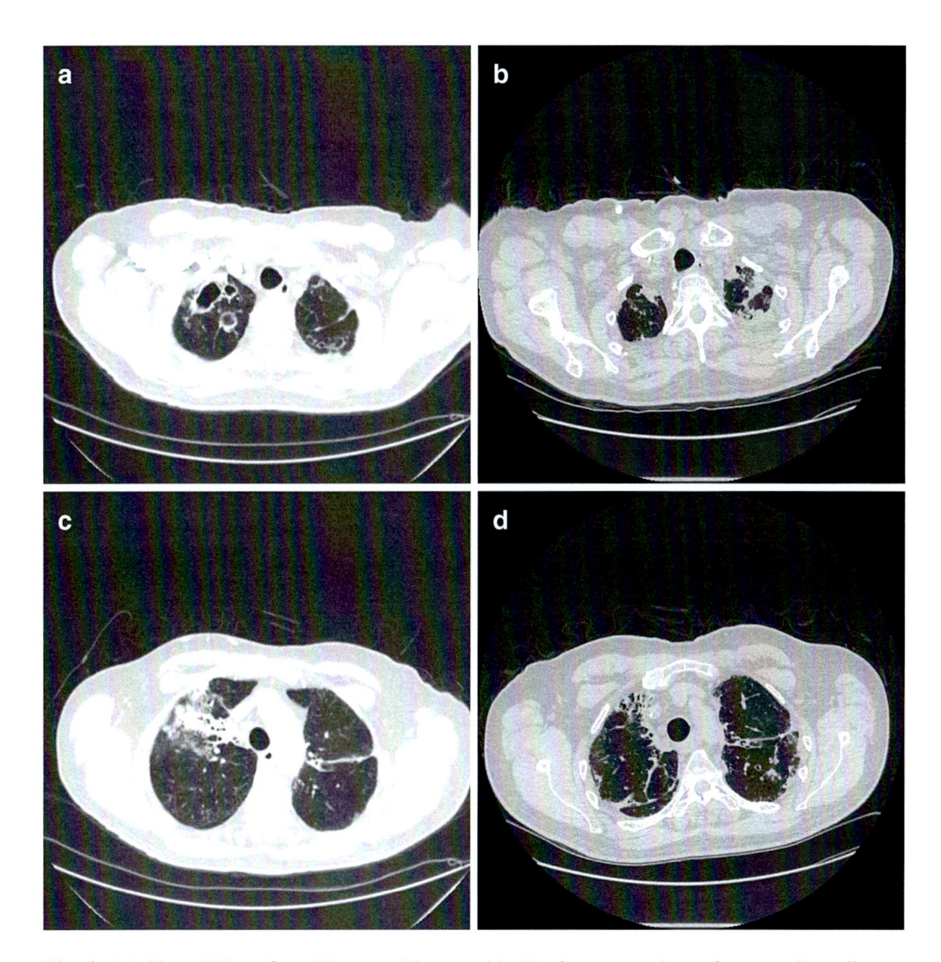

Fig. 4 (**a**) Chest CT cut from 71-year-old man with *M. abscessus* subsp. *abscessus* lung disease and right apical cavitation. (**b**) Chest CT cut from the same patient showing extensive right upper lobe inflammatory density. (**c**) Chest CT cut at same level as Fig. 4a showing improvement in cavitary lesions after extended therapy with tigecycline, tedizolid switched to linezolid, and amikacin switched to imipenem for a duration including 12 months sputum culture negativity. (**d**) Chest CT cut a same level as Fig. 4b showing improvement in extensive inflammatory densities after therapy. (**e**) Chest CT cut from 62-year-old patient with *M. abscessus* subsp. *abscessus* lung disease. Her *M. abscessus* subsp. *abscessus* isolate was macrolide susceptible in vitro because the isolate had an *erm* gene mutation rendering it inactive. (**f**) Chest CT cut from the same patient at the same level showing closure of the cavity associated with 12 months sputum culture negativity while on therapy. Her treatment consisted of 6 months oral azithromycin and linezolid with intravenous amikacin followed by 8 months oral azithromycin and linezolid with inhaled amikacin

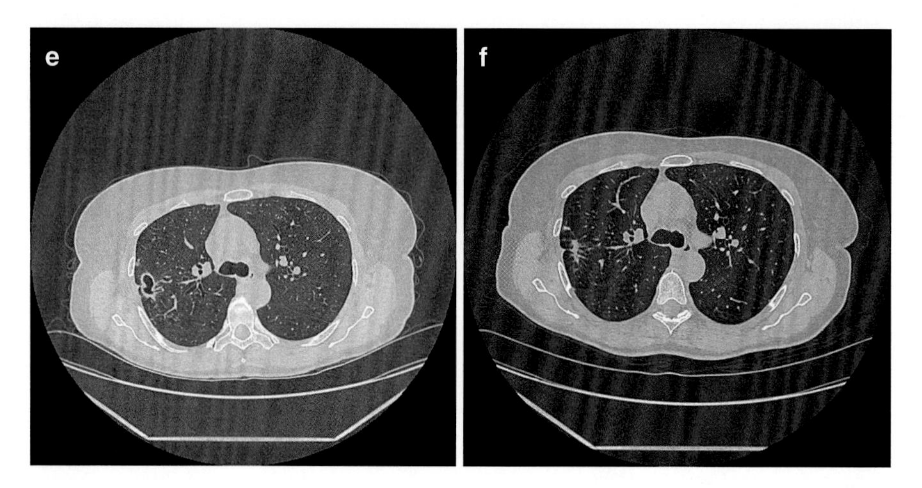

Fig. 4 (continued)

Our approach to patients with RGM diseases including lung disease is to treat with the intent to eradicate the infection (Fig. 4a–d). We recognize that this approach is not universally accepted due to extensive RBM drug resistance and the need for potentially toxic parenteral medications in a prolonged multidrug treatment regimen. We also do not promote the concept of "induction" and "maintenance" therapy for RGM in general and *M. abscessus* subsp. *abscessus* in particular. First, those concepts make pathophysiologic sense for tuberculosis, but we are not aware of information that supports similar pathophysiology for RGM lung disease. Second, also in contrast to TB, there are not a plethora of oral medications with activity against *M. abscessus* subsp. *abscessus* to choose from which brings into play the possibility that an *erm* gene inactive macrolide susceptible *M. abscessus* subsp. *abscessus* isolate might undergo macrolide therapy with oral drugs of questionable activity resulting in acquired mutational macrolide resistance.

We are also aware that some clinicians, as a result of multiple obstacles, toxicity, and complexities presented by parenteral medication administration, choose to treat patients with relatively short and intermittent regimens. Given the lack of a predictably favorable response to current antibiotic choices, it is hard to argue vehemently against this approach which should be left to the discretion of the patient and treating physician.

We base initial therapy on in vitro susceptibility results, which is clearly not optimal but is the best default approach available in our view. Obviously, an inactive *erm* gene isolate should be treated with a macrolide (Fig. 4e, f). For *M. abscessus* subsp. *abscessus* and *M. massiliense*, we would then choose at least one parenteral agent, usually amikacin. The next choice would depend on disease severity. For cavitary disease we would choose a second parenteral agent, such as tigecycline, imipenem, or cefoxitin. If the isolate was susceptible in vitro to linezolid, that would be a reasonable third drug choice as well. Some experts would use clofazimine in this situation. As noted, our goal initially is eradication of the organism. We do not a priori assume that

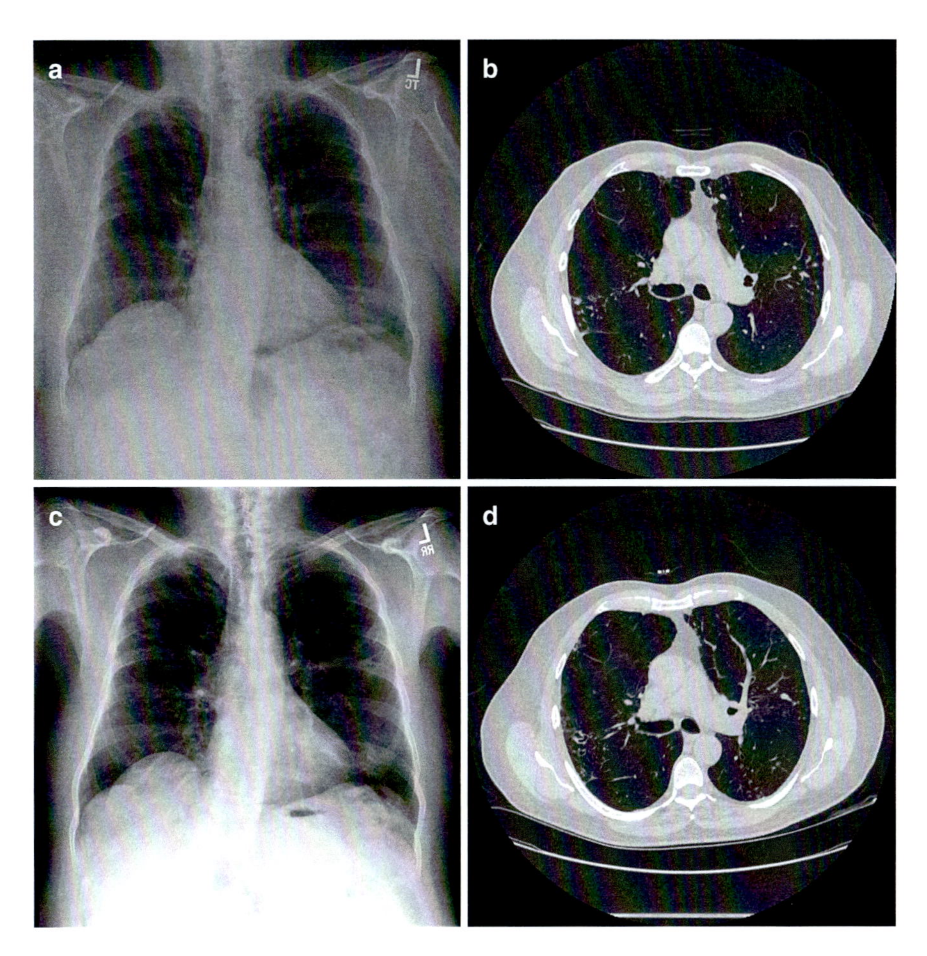

Fig. 5 (**a**) Chest radiograph from 76-year-old patient with *M. abscessus* subsp. *abscessus* lung disease showing nonspecific bilateral densities consistent with bronchiectasis and nontuberculous mycobacterial disease. (**b**) Chest CT cut from the same patient showing bronchiectasis and nodular densities with tree-in-bud pattern. (**c**) Chest radiograph showing progressive right lower lung field densities associated with persistently positive sputum cultures for *M. abscessus* subsp. *abscessus* in spite of treatment with amikacin, imipenem, and tedizolid switched to linezolid and azithromycin. (**d**) Chest CT cut at same level as Fig. 5b at same time as chest radiograph in Fig. 5c showing bronchiectasis with increased peribronchial inflammation and early cavitation

any patient is untreatable, although regrettably, a significant proportion of these patients will turn out to be refractory to even aggressive therapeutic efforts (Fig. 5a–d). Our recommendation for a specific treatment strategy is outlined in Table 5.

For *erm* gene active *M. abscessus* subsp. *abscessus* isolates, we recommend at least two parenteral agents including amikacin plus oral linezolid or clofazimine (Table 5). We do not believe that the inclusion of a macrolide favorably influences the outcome of therapy for these organisms, but the macrolide may have a beneficial effect on bronchiectasis as an immune modulating agent. *Mycobacterium bolletii*

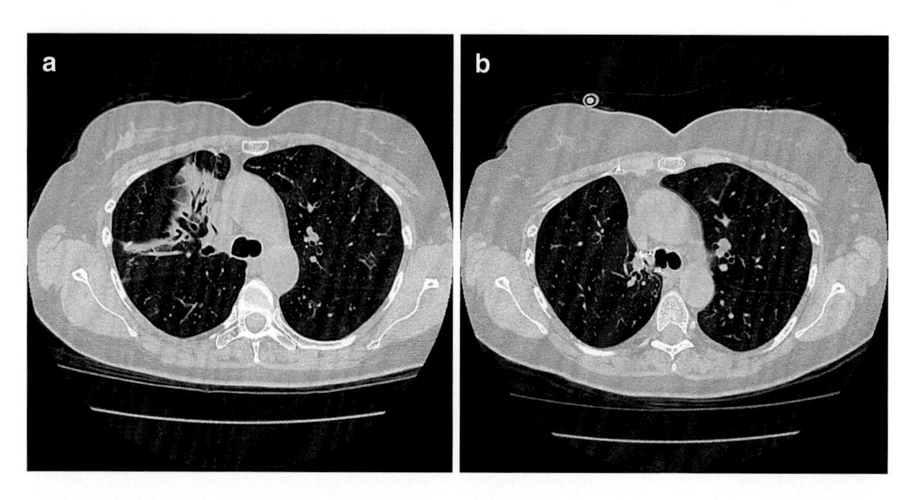

Fig. 6 (**a**) Chest CT cut from 64-year-old female with *M. abscessus* subsp. *abscessus* lung disease and persistently positive sputum cultures in spite of aggressive therapy including parenteral antibiotics. (**b**) Chest CT cut from the same patient at a comparable level to Fig. 6b after right middle lobe lobectomy and 12 months sputum culture negativity following surgery and antibiotic therapy including tigecycline and amikacin

isolates would be treated identically to *M. abscessus* subsp. abscessus isolates, guided by *erm* gene activity. As with other NTM respiratory pathogens, the target duration of therapy is 12 months of sputum culture negativity while on therapy.

In the (hopefully) rare situation where a clinician is faced with an *M. abscessus* isolate of unknown *erm* gene activity, then we believe it would be reasonable to include a macrolide in the regimen on the small chance that the isolate had an inactive *erm* gene.

Surgery is an important option for selected patients with *M. abscessus* disease, especially those with macrolide resistance from any mechanism, and is covered in detail in chapter "Surgical Management of NTM Diseases". For an organism that is so difficult to eradicate with medication alone, surgery is associated with improved clinical outcomes [115] (Fig. 6a, b). While a minority of patients will be appropriate candidates for adjunctive surgery, for various reasons, we feel that the surgical option should be considered for all *M. abscessus* patients, even if only briefly to confirm a patient's non-suitability.

Fortunately, *M. fortuitum* is usually susceptible in vitro to multiple oral antibiotics with the exception of macrolides due to the presence of an active *erm* gene. We recommend treatment of *M. fortuitum* disease with at least two agents to which the organism is susceptible in vitro. Parenteral agents may be necessary for severe or refractory disease. Removal of any foreign body is also essential for successful therapy of extrapulmonary or disseminated *M. fortuitum* infection associated with foreign material.

M. chelonae does not have an active *erm* gene and is also frequently more susceptible to oral antibiotics than *M. abscessus* subsp. *abscessus*, including macrolide, which offers opportunities for therapy without parenteral agents. It must be empha-

Table 6 Essentials for evaluation of an RGM clinical isolate

Accurate Organism Identification, Do Not Accept "Group" Designation or Nomenclature	
For in vitrosusceptibility testing, all clinically significant RGM isolates[a] must be preincubated with macrolide to determine *erm* gene activity;do not accept any in vitrosusceptibility result without this step	
Preincubation step must be done even if the isolate is identified as *M. abscessus* subsp. *abscessus* because 20% of these isolates will have mutations inactivating the *erm* gene	
erm Gene Active ↙ ↘	
YES	NO
Therapy without macrolide	Therapy with macrolide-based regimen and adequate companion drugs to prevent acquired macrolide resistance

[a]With the exceptions of *M. chelonae* and *M. abscessus* subsp. *massiliense* because they do not have an active *erm* gene

sized again that monotherapy with macrolide can result in acquired macrolide resistance and worse prognosis.

The general recommendation for serious wound infections is combination therapy with initial parenteral therapy for *M. chelonae* and *M. abscessus* until clinically improved, followed by oral therapy for a total treatment duration of at least 6 months. Development of new skin lesions on therapy is not necessarily indicative of treatment failure. Cultures should be obtained, and antibiotic therapy should be continued. In most cases, cultures are sterile following approximately 6–8 weeks of proper treatment. The new skin lesions likely represent a paradoxical immunological response rather than microbiological persistence or relapse. New lesions on therapy are not unexpected but should be aggressively evaluated including new AFB cultures. As long as cultures of these new lesions remain negative, treatment should be continued for a minimum of 4 months for less serious infections and 6 months for more serious infections. Abscess drainage and surgical debridement are essential. Surgical debridement may be necessary more than once. Cultures for AFB must be sent with each biopsy and debridement procedure. Sadly, this simple requirement is not always met.

In summary, the management of RGM infections in general and *M. abscessus* infections, specifically, remains challenging. There are many confounding aspects to RGM disease from nomenclature to drug resistance mechanisms that confuse or intimidate clinicians caring for these patients. Given those unfavorable circumstances, there is no other group of NTM organism that require as much knowledge by the clinician for optimal patient management. The clinician must have accurate organism identification and must know the significance of in vitro susceptibility results. With that knowledge, the clinician must craft drug treatment regimens that, at a minimum, do not make the patient's status worse. There is simply no substitute for an in-depth understanding by the clinician of the nuances and idiosyncrasies of RGM disease and no short cuts in the management of these patients (Table 6).

Acknowledgments We humbly acknowledge and thank our patients with *M. abscessus* lung disease whose patience with their physicians, courage and strength are both boundless and inspirational.

References

1. Cobbett L. An acid-fast bacillus obtained from a pustular eruption. Br Med J. 1918;2:158.
2. Runyon H. Conservation of the specific epithet fortuitum in the name of the organism known as *Mycobacterium fortuitum* da Costa Cruz. Int J Syst Bacteriol. 1972;22:50–1.
3. Moore M, Frerichs JB. An unusual acid fast infection of the knee with subcutaneous, abscess-like lesions of the gluteal region: report of a case study with a study of the organism, *Mycobacterium abscessus*. J Investig Dermatol. 1953;20:133–69.
4. Schinsky MF, Morey RE, Steigerwalt AG, Douglas MP, Wilson RW, Floyd MM, Butler WR, Daneshvar MI, Brown-Elliott BA, Wallace RJ Jr, McNeil MM, Brenner DJ, Brown JM. Taxonomic variation in the *Mycobacterium fortuitum* third-biovariant complex: description of *Mycobacterium boenickei* sp. nov., *Mycobacterium houstonense* sp. nov., *Mycobacterium neworleansense* sp. nov., *Mycobacterium brisbanense* sp. nov., and recognition of *Mycobacterium porcinum* from human clinical isolates. Int J Syst Evol Microbiol. 2004;54:1653–67.
5. Adékambi T, Berger P, Raoult D, Drancourt M. *rpoB* gene sequence-based characterization of emerging non-tuberculous mycobacteria with descriptions of *Mycobacterium bolletii* sp. nov., *M. phocaicum* sp. nov. and *Mycobacterium aubagnense* sp. nov. Int J Syst Evol Microbiol. 2006;56:133–43.
6. Brown-Elliott BA, Wallace RJ Jr. *Mycobacterium*: clinical and laboratory characteristics of rapidly growing mycobacteria. In: Manual of clinical microbiology, vol. 1. 11th ed. Washington, D.C: ASM Press; 2015.
7. Brown BA, Springer B, Steingrube VA, Wilson RW, Pfyffer GE, Garcia MJ, Menendez MC, Rodriguez-Salgado B, Jost KC Jr, Chiu SH, Onyi GO, Bottger EC, Wallace RJ Jr. *Mycobacterium wolinskyi* sp. nov. and *Mycobacterium goodii* sp. nov., two new rapidly growing species related to *Mycobacterium smegmatis* and associated with human wound infections: a cooperative study from the international working group on mycobacterial taxonomy. Int J Syst Bacteriol. 1999;49:1493–511.
8. Wallace RJ Jr, Brown-Elliott BA, Wilson RW, Mann L, Hall L, Zhang Y, Jost KC Jr, Brown JM, Kabani A, Schinsky MF, Steigerwalt AG, Crist CJ, Roberts GD, Blacklock Z, Tsukamura M, Silcox V, Turenne C. Clinical and laboratory features of *Mycobacterium porcinum*. J Clin Microbiol. 2004;42:5689–97.
9. Jiménez MS, Campos-Herrero MI, García D, Luquin M, Herrera L, García MJ. *Mycobacterium canariasense* sp. nov. Int J Syst Evol Microbiol. 2004;54:1729–34.
10. Whipps CM, Butler WR, Pourahmad F, Watral VG, Kent ML. Molecular systematics support the revival of *Mycobacterium salmoniphilum* (ex Ross 1960) sp. nov., nom. Rev., a species closely related to *Mycobacterium chelonae*. Int J Syst Evol Microbiol. 2007;57:2525–31.
11. Brown-Elliott BA, Wallace RJ Jr. Clinical and taxonomic status of pathogenic nonpigmented or late-pigmenting rapidly growing mycobacteria. Clin Microbiol Rev. 2002;15:716–46.
12. Adékambi T, Drancourt M. Dissection of phylogenetic relationships among nineteen rapidly growing mycobacterium species by 16S r-RNA, *hsp65, sod*A, *rec*A, and *rpoB* gene sequencing. Int J Syst Evol Microbiol. 2004;54:2095–105.
13. Leao SC, Tortoli E, Viana-Niero C, Ueki SYM, Batista Lima KV, Lopes ML, Yubero J, Menendez MC, Garcia MJ. Characterization of mycobacteria from a major Brazilian outbreak suggests a revision of the taxonomic status of members of the *Mycobacterium chelonae-abscessus* group. J Clin Microbiol. 2009;47:2691–8.

14. Leao SC, Tortoli E, Euzeby JP, Garcia MJ. Proposal that *Mycobacterium massiliense* and *Mycobacterium bolletii* be united and reclassified as *Mycobacterium abscessus* subsp. *bolletii* comb. nov., designation of *Mycobacterium abscessus* subsp. *abscessus* subsp. nov. an amended description of *Mycobacterium abscessus*. Int J Syst Evol Microbiol. 2011;61:2311–3.
15. Zelazny AM, Root JM, Shea YR, Colombo RE, Shamputa IC, Stock F, Conlan SS, McNulty S, Brown-Elliott BA, Wallace RJ Jr, Olivier KN, Holland SM, Sampaio EP. Cohort study of molecular identification and typing of *Mycobacterium abscessus, Mycobacterium massiliense* and *Mycobacterium bolletii*. J Clin Microbiol. 2009;47:1985–95.
16. Tortoli E, Kohl TA, Trovato A, Garcia MJ, Leao SC, Baldan R, Campana S, Cariani L, Colombo C, Costa D, Pizzamiglio G, Rancoita PM, Russo MC, Simonetti TM, Sottotetti S, Taccetti G, Teri A, Niemann S, Cirillo DM, Brown-Elliott BA, Wallace Jr. RJ. Emended description of *Mycobacterium abscessus, Mycobacterium abscessus* subsp. *abscessus, Mycobacterium abscessus* subsp. *bolletii* and designation of *Mycobacterium abscessus* subsp. *massiliense* subsp. nov. Int J Syst Evol Microbiol. 2016; In press.
17. Tettelin H, Davidson RM, Agrawal S, Aitken ML, Shallom S, Hasan NA, Strong M, de Moura VCN, De Groote MA, Duarte RS, Hine E, Parankush S, Su Q, Daugherty SC, Fraser CM, Brown-Elliott BA, Wallace RJ Jr, Holland SM, Sampaio EP, Olivier KN, Jackson M, Zelazny AM. High-level relatedness among *Mycobacterium abscessus* subsp. *massiliense* strains from widely separated outbreaks. Emerg Infect Dis. 2014;20:364–71.
18. Adekambi T, Sassi M, van Ingen J, Drancourt M. Reinstating Mycobacterium massiliense and Mycobacterium bolletii as species of the Mycobacterium abscessus complex. Int J Syst Evol Microbiol. 2017;67(8):2726–30.
19. Kim K, Hong S-H, Kim B-J, Kim B-R, Lee S-Y, Kim G-N, Shim TS, Kook Y-H, Kim B-J. Separation of *Mycobacterium abscessus* into subspecies or genotype level by direct application of peptide nucleic acid multi-probe-real-time PCR method into sputa samples. BMC Infect Dis. 2015;15:325–31.
20. Tan JL, Ngeow YF, Choo SW. Support from phylogenomic networks and subspecies signatures for separation of *Mycobacterium massiliense* from *Mycobacterium bolletii*. J Clin Microbiol. 2015;53:3042–6.
21. Tortoli E. Microbiological features and clinical relevance of new species of the genus *Mycobacterium*. Clin Microbiol Rev. 2014;27:727–52.
22. Buckwalter SP, Olson SL, Connelly BJ, Lucas BC, Rodning AA, Walchak RC, Deml SM, Wohlfiel SL, Wengenack NL. Evaluation of matrix-assisted laser desorption ionization-time of flight mass spectrometry for identification of *Mycobacterium* species, *Nocardia* species, and other aerobic actinomycetes. J Clin Microbiol. 2016;54:376–84.
23. Saleeb PG, Drake SK, Murray PR, Zelazny AM. Identification of mycobacteria in solid-culture media by matrix-assisted laser desorption ionization-time of flight mass spectrometry. J Clin Microbiol. 2011;49:1790–4.
24. Tortoli E, Nanetti A, Piersimoni C, Cichero P, Farina C, Mucignat G, Scarparo C, Bartolini L, Valentini R, Nista D, Gesu G, Passerini Tosi C, Crovatto M, Brusarosco G. Performance assessment of new multiplex probe assay for identification of mycobacteria. J Clin Microbiol. 2001;39:1079–84.
25. Tortoli E, Pecorari M, Fabio G, Messinò M, Fabio A. Commercial DNA probes for mycobacteria incorrectly identify a number of less frequently encountered species. J Clin Microbiol. 2010;48:307–10.
26. Tortoli E. Impact of genotypic studies on mycobacterial taxonomy: the new mycobacteria of the 1990s. Clin Microbiol Rev. 2003;16:319–54.
27. Steingrube VA, Gibson JL, Brown BA, Zhang Y, Wilson RW, Rajagopalan M, Wallace RJ Jr. PCR amplification and restriction endonuclease analysis of a 65-kilodalton heat shock protein gene sequence for taxonomic separation of rapidly growing mycobacteria [ERRATUM 1995;33:1686]. J Clin Microbiol. 1995;33:149–53.

28. Telenti A, Marchesi F, Balz M, Bally F, Böttger EC, Bodmer T. Rapid identification of mycobacteria to the species level by polymerase chain reaction and restriction enzyme analysis. J Clin Microbiol. 1993;31:175–8.
29. McNabb A, Eisler D, Adie K, Amos M, Rodrigues M, Stephens G, Black WA, Isaac-Renton J. Assessment of partial sequencing of the 65-kilodalton heat shock protein gene (hsp65) for routine identification of *Mycobacterium* species isolated from clinical sources. J Clin Microbiol. 2004;42:3000–11.
30. Adékambi T, Reynaud-Gaubert M, Greub G, Gevaudan MJ, La Scola B, Raoult D, Drancourt M. Amoebal coculture of "*Mycobacterium massiliense*" sp. nov. from the sputum of a patient with hemoptoic pneumonia. J Clin Microbiol. 2004;42:5493–501.
31. Adékambi T, Colson P, Drancourt M. *rpo B*-based identification of nonpigmented and late pigmented rapidly growing mycobacteria. J Clin Microbiol. 2003;41:5699–708.
32. Kim H-Y, Kook Y, Yun Y-J, Park CG, Lee NY, Shim TS, Kim B-J, Kook Y-H. Proportion of *Mycobacterium massiliense* and *Mycobacterium bolletii* in strains among Korean *Mycobacterium chelonae-Mycobacterium abscessus* group isolates. J Clin Microbiol. 2008;46:3384–90.
33. Forbes BA, Banaiee N, Beavis KG, Brown-Elliott BA, Della Latta P, Elliott LB, Hall GS, Hanna B, Perkins MD, Siddiqi SH, Wallace Jr. RJ, Warren NG. Laboratory detection and identification of mycobacteria; approved guideline. CLSI document M48-A. 2008.
34. Ngeow YF, Wee WY, Wong YL, Tan JL, Ongi CS, Ng KP, Choo SW. Genomic analysis of *Mycobacterium abscessus* strain M139, which has an ambiguous subspecies taxonomic position. J Bacteriol. 2012;194:6002–3.
35. Ngeow YF, Wong YL, Lokanathan N, Wong GJ, Ong CS, Ng KP, Choo SW. Genomic analysis of *Mycobacterium massiliense* strain M115, an isolate from human sputum. J Bacteriol. 2012;194:4786.
36. Ngeow YF, Wong YL, Tan JL, Arumugam R, Wong GJ, Ong CS, Ng KP, Choo SW. Genome sequence of *Mycobacterium massiliense* M18, isolated from a lymph node biopsy specimen. J Bacteriol. 2012;194:4125.
37. Tettelin H, Sampaio EP, Daugherty SC, Hine E, Riley DR, Sadzewicz L, Sengamalay N, Shefchek K, Su Q, Tallon LJ, Conville P, Olivier KN, Holland SM, Fraser CM, Zelazny AM. Genomic insights into the emerging human pathogen *Mycobacterium massiliense*. J Bacteriol. 2012;194:5450.
38. Chan J, Halachev M, Yates E, Smith G, Pallen M. Whole-genome sequence of the emerging pathogen *Mycobacterium abscessus* strain 47J26. J Bacteriol. 2012;194:549.
39. Bryant JM, Grogono DM, Greaves D, Foweraker J, Roddick I, Inns T, Reacher M, Haworth CS, Curran MD, Harris SR, Peacock SJ, Parkhill J, Floto RA. Whole-genome sequencing to identify transmission of *Mycobacterium abscessus* between patients with cystic fibrosis: a retrospective cohort study. Lancet. 2013;381:1551–60.
40. Davidson RM, Hasan N, Reynolds PR, Totten S, Garcia B, Levin A, Ramamoorthy P, Heifets L, Daley CL, Strong M. Genome sequencing of *Mycobacterium abscessus* isolates from patients in the United States and comparisons to globally diverse clinical strains. J Clin Microbiol. 2014;52:3573–82.
41. Wolinsky E. State of the art: nontuberculous mycobacterial and associated disease. Am Rev Respir Dis. 1979;119:107–59.
42. Griffith DE, Aksamit T, Brown-Elliott BA, Catanzaro A, Daley C, Gordin F, Holland SM, Horsburgh R, Huitt G, Iademarco MF, Iseman M, Olivier K, Ruoss S, von Reyn CF, Wallace RJ Jr, Winthrop K. An official ATS/IDSA statement: diagnosis, treatment and prevention of nontuberculous mycobacterial diseases. Am J Respir Crit Care Med. 2007;175:367–416.
43. Thomson R, Tolson C, Sidjabat H, Huygens F, Hargreaves M. Mycobacterium abscessus isolated from municipal water - a potential source of human infection. BMC Infect Dis. 2013;13:241.
44. Falkinham JO. The changing pattern of nontuberculous mycobacterial disease. Can. J Infect Dis. 2003;14:281–6.

45. Griffith DE, Philley JV, Brown-Elliott BA, Benwill JL, Shepherd S, York D, Wallace RJ Jr. The significance of Mycobacterium abscessus subspecies abscessus isolation during Mycobacterium avium complex lung disease therapy. Chest. 2015;147(5):1369–75.
46. Jun HJ, Jeon K, Um SW, Kwon OJ, Lee NY, Koh WJ. Nontuberculous mycobacteria isolated during the treatment of pulmonary tuberculosis. Respir Med. 2009;103(12):1936–40.
47. Cullen AR, Cannon CL, Mark EJ, Colin AA. *Mycobacterium abscessus* infection in cystic fibrosis. Am J Respir Crit Care Med. 2000;161:641–5.
48. Fauroux B, Delaisi B, Clément A, Saizou C, Moissenet D, Truffot-Pernot C, Tournier G, Vu Thien H. Mycobacterial lung disease in cystic fibrosis: a prospective study. Pediatr Infect Dis J. 1997;16:354–8.
49. Olivier KN, Weber DJ, Wallace RJ Jr, Faiz AR, Lee J-H, Zhang Y, Brown-Elliott BA, Handler A, Wilson RW, Schechter MS, Edwards LJ, Chakraborti S, Knowles MR, Group ftNMiCFS. Nontuberculous mycobacteria. I. Multicenter prevalence study in cystic fibrosis. Am J Respir Crit Care Med. 2003;167:828–34.
50. Aitken ML, Limaye A, Pottinger P, Whimbey E, Goss GH, Tonelli MR, Cangelosi GA, Ashworth M, Olivier KN, Brown-Elliott BA, Wallace RJ Jr. Respiratory outbreak of *Mycobacterium abscessus* subspecies *massiliense* in a lung transplant and cystic fibrosis center. Letter to the Editor Am J Respir Crit Care Med. 2012;185:231–3.
51. Roux A-L, Catherinot E, Ripoll F, Soismier N, Macheras E, Ravilly S, Bellis G, Vibet M-A, Le Roux E, Lemonnier L, Gutierrez C, Vincent V, Fauroux B, Rottman M, Guillemot D, Gaillard J-L, Herrman J-L, Group. ftO. Multicenter study of prevalence of nontuberculous mycobacteria in patients with cystic fibrosis in France. J Clin Microbiol. 2009;47:4124–8.
52. Cândido PHC, De Souza Nunes L, Marques EA, Folescu TW, Coelho FS, Nogueira de Moura VC, da Silva MG, Gomes KM, da Silva Lourenço MC, Aguiar FS, Chitolina F, Armstrong DT, Leão SC, Neves FPG, de Queiroz Mello FC, Duarte RS. 2014. Multidrug-resistant nontuberculous mycobacteria isolated from cystic fibrosis patients. J Clin Microbiol 58:2990–2997 [52] Gubler JGH, Salfinger M, von Graevenitz A. 1992. Pseudoepidemic of nontuberculous mycobacteria due to a contaminated bronchoscope cleaning machine: report of an outbreak and review of the literature. Chest. 101:1245–1249.
53. Tiwari TSP, Ray B, Jost KC Jr, Rathod MK, Zhang Y, Brown-Elliott BA, Hendricks K, Wallace RJ Jr. Forty years of disinfectant failure: outbreak of postinjection *Mycobacterium abscessus* infection caused by contamination of benzalkonium chloride. Clin Infect Dis. 2003;36:954–62.
54. Hector JSR, Pang Y, Mazurek GH, Zhang Y, Brown BA, Wallace RJ Jr. Large restriction fragment patterns of genomic *Mycobacterium fortuitum* DNA as strain-specific markers and their use in epidemiologic investigation of four nosocomial outbreaks. J Clin Microbiol. 1992;30:1250–5.
55. Zhang Y, Rajagopalan M, Brown BA, Wallace RJ Jr. Randomly amplified polymorphic DNA PCR for comparison of *Mycobacterium abscessus* strains from nosocomial outbreaks. J Clin Microbiol. 1997;35:3132–9.
56. Carson LA, Bland LA, Cusick LB, Favero MS, Bolan GA, Reingold AL, Good RC. Prevalence of nontuberculous mycobacteria in water supplies of hemodialysis centers. App Environ Microbiol. 1988;54:3122–5.
57. Schulze-Röbbecke R, Janning B, Fischeder R. Occurrence of mycobacteria in biofilm samples. Tuberc Lung Dis. 1992;73:141–4.
58. Galassi L, Donato R, Tortoli E, Burrini D, Santianni D, Dei R. Nontuberculous mycobacteria in hospital water systems: application of HPLC for identification of environmental mycobacteria. J Water Health. 2003;1:133–9.
59. Sudesh S, Cohen EJ, Schwartz LW, Myers JS. *Mycobacterium chelonae* infection in a corneal graft. Arch Ophthalmol. 2000;118:294–5.
60. Reviglio V, Rodriguez ML, Picotti GS, Paradello M, Luna JD, Juárez CP. *Mycobacterium chelonae* keratitis following laser in situ keratomileusis. J Refract Surg. 1998;14:357–60.
61. Saluja A, Peters NT, Lowe L, Johnson TM. A surgical wound infection due to *Mycobacterium chelonae* successfully treated with clarithromycin. Dermatol Surg. 1997;23:539–43.

62. Friedman ND, Sexton DJ. Bursitis due to *Mycobacterium goodii*, a recently described, rapidly growing mycobacterium. J Clin Microbiol. 2001;39:404–5.
63. Meyers H, Brown-Elliott BA, Moore D, Curry J, Truong C, Zhang Y, Wallace RJ Jr. An outbreak of *Mycobacterium chelonae* infection following liposuction. Clin Infect Dis. 2002;34:1500–7.
64. Centers for Disease Control and Prevention. Rapidly growing mycobacterial infection following liposuction and liposculpture—Caracas, Venezuela, 1996-1998. Morb Mortal Wkly Rep. 1998;47:1065.
65. Nagpal A, Wentink JE, Berbari EF, Aronhalt KC, Wright AJ, Krageschmidt DA, Wengenack NL, Thompson RI, Tosh PK. A cluster of *Mycobacterium wolinskyi* surgical site infection at an academic medical center. Infect Control Hosp Epidemiol. 2014;35:1169–75.
66. Dupont C, Terru D, Aguilhon S, Frapier J-M, Paquis M-P, Morquin D, Lamy B, Godreuil S, Parer S, Lotthé A, Jumas-Bilak E, Romano-Bertrand S. Source-case investigation of *Mycobacterium wolinskyi* cardiac surgical site infection. J Hosp Infect. 2016; In press.
67. Ariza-Heredia EJ, Databneh AS, Wilhelm MP, Wengenack NL, Razonable RR, Wilson JW. *Mycobacterium wolinskyi*: a case series and review of the literature. Diagn Microbiol Infect Dis. 2011;71:421–7.
68. Bossart S, Schnell B, Kerl K, Urosevic-Maiwald M. Ulcers as a sign of skin infection with *Mycobacterium wolinskyi*: report of a case and review of the literature. Case Rep Dermatol. 2016;8:151–5.
69. Lai KK, Brown BA, Westerling JA, Fontecchio SA, Zhang Y, Wallace RJ Jr. Long-term laboratory contamination by *Mycobacterium abscessus* resulting in two pseudo-outbreaks: recognition with use of random amplified polymorphic DNA (RAPD) polymerase chain reaction. Clin Infect Dis. 1998;27:169–75.
70. Wilson RW, Steingrube VA, Böttger EC, Springer B, Brown-Elliott BA, Vincent V, Jost KC Jr, Zhang Y, Garcia MJ, Chiu SH, Onyi GO, Rossmoore H, Nash DR, Wallace RJ Jr. *Mycobacterium immunogenum* sp. nov., a novel species related to *Mycobacterium abscessus* and associated with clinical disease, pseudo-outbreaks, and contaminated metalworking fluids: an international cooperative study on mycobacterial taxonomy. Int J Syst Evol Microbiol. 2001;51:1751–64.
71. Baker AW, Lewis SS, Alexander BD, Chen LF, Wall S, Wallace RJ Jr, Brown-Elliott BA, Isaacs PJ, Pickett LC, Patel CB, Smith PK, Reynolds JM, Engel J, Wolfe CR, Milano CA, Schroder JN, Davis RD, Hartwig MG, Stout JE, Strittholt N, Maziarz EK, Saullo JH, Hazen KC, Walczak RJ Jr, Vasireddy R, Vasireddy S, CM MK, Anderson DJ, Sexton DJ. A cluster of *Mycobacterium abscessus* among lung transplant patients: investigation and mitigation. Clin Infect Dis. 2017;64(7):902–11.
72. Brown-Elliott BA, Wallace RJ Jr, Petti CA, Mann LB, McGlasson M, Chihara S, Smith GL, Painter P, Hail D, Wilson R, Simmon KE. *Mycobacterium neoaurum* and *Mycobacterium bacteremicum* sp. nov. as causes of bacteremia. J Clin Microbiol. 2010;48:4377–85.
73. Raad II, Vartivarian S, Khan A, Bodey GP. Catheter-related infections caused by the *Mycobacterium fortuitum* complex: 15 cases and review. Rev Infect Dis. 1991;13: 1120–5.
74. Washer LL, Riddell IVJ, Rider J, Chenoweth CE. *Mycobacterium neoaurum* bloodstream infection: report of 4 cases and review of the literature. Clin Infect Dis. 2007;45:e10–3.
75. Martínez López AB, Álvarez Blanco O, Ruíz Serrano MJ, Morales San-José MD, Luque de Pablos A. *Mycobacterium fortuitum* as a cause of peritoneal dialysis catheter port infection. A clinical case and a review of the literature. Nefrologia. 2015;35:584–6.
76. Al Shaalan M, Law BJ, Israels SJ, Pianosi P, Lacson AG, Higgins R. *Mycobacterium fortuitum* interstitial pneumonia with vasculitis in a child with Wilms tumor. Pediatr Infect Dis J. 1997;16:996–1000.
77. Levendoglu-Tugal O, Munoz J, Brudnicki A, Ozkaynak MF, Sandoval C, Jayabose S. Infections due to nontuberculous mycobacteria in children with leukemia. Clin Infect Dis. 1998;27:1227–30.

78. Tahara M, Yatera K, Yamasaki K, Orihashi T, Hirosawa M, Ogoshi T, Noguchi S, Nishida C, Ishimoto H, Yonezawa A, Tsukada J, Mukae H. Disseminated *Mycobacterium abscessus* complex infection manifesting as multiple areas of lymphadenitis and skin abscess in the preclinical stage of acute lymphocytic leukemia. Intern Med. 2016;55:1787–91.
79. Chetchotisakd P, Mootsikapun P, Anunnatsiri S, Jirarattanapochai K, Choonhakarn C, Chaiprasert A, Ubol PN, Wheat LJ, Davis TE. Disseminated infection due to rapidly growing mycobacteria in immunocompetent hosts presenting with chronic lymphadenopathy: a previously unrecognized clinical entity. Clin Infect Dis. 2000;32:29–34.
80. Brown-Elliott BA, Mann LB, Hail D, Whitney C, Wallace RJ Jr. Antimicrobial susceptibility of nontuberculous mycobacteria from eye infections. Cornea. 2012;31:900–6.
81. Cooksey RC, de Waard JH, Yakrus MA, Rivera I, Chopite M, Toney SR, Morlock GP, Butler WR. *Mycobacterium cosmeticum* sp. nov., a novel rapidly growing species isolated from a cosmetic infection and from a nail salon. Int J Syst Evol Microbiol. 2004;54:2385–91.
82. Gira AK, Reisenauer H, Hammock L, Nadiminti U, Macy JT, Reeves A, Burnett C, Yakrus MA, Toney S, Jensen BJ, Blumberg HM, Caughman SW, Nolte FS. Furunculosis due to *Mycobacterium mageritense* associated with footbaths at a nail salon. J Clin Microbiol. 2004;42:1813–7.
83. Winthrop KL, Albridge K, South D, Albrecht P, Abrams M, Samuel MC, Leonard W, Wagner J, Vugia DJ. The clinical management and outcome of nail salon-acquired *Mycobacterium fortuitum* skin infection. Clin Infect Dis. 2004;38:38–44.
84. Vugia DJ, Jang Y, Zizek C, Ely J, Winthrop KL, Desmond E. Mycobacteria in nail salon whirlpool footbaths. California Emerg Infect Dis. 2005;11:616–8.
85. Winthrop KL, Chang E, Yamashita S, Iademarco MF, LoBue PA. Nontuberculous mycobacteria infections and anti-tumor necrosis factor-alpha therapy. Emerg Infect Dis. 2009;15:1556–61.
86. Czaja CA, Merkel PA, Chan ED, Lenz LL, Wolf ML, Alam R, Franke SK, Fischer A, Gogate S, Perez-Velez CM, Knight V. Rituximab as successful adjunct treatment in a patient with disseminated nontuberculous mycobacterial infection due to acquired anti-interferon-γ autoantibody. Clin. Infect. Dis. In: **58:**e-115-118; 2014.
87. Valour F, Perpoint T, Sénéchal A, Kong X-F, Bustamante J, Ferry T, Chidiac C, Ader A, group obotLTs. Interferon-γ autoantibodies as predisposing factor for nontuberculous mycobacterial infection. Emerg. Infect. Dis. 2016;22:1124–6.
88. Clinical and Laboratory Standards Institute. Susceptibility testing of mycobacteria, nocardiae, and other aerobic actinomycetes: approved standard—second edition. CLSI document M24-A2. 2011.
89. Clinical and Laboratory Standards Institute. Susceptibility testing of mycobacteria, nocardia, and other aerobic actinomycetes. 3rd ed. Wayne, PA: Clinical and Laboratory Standards Institute; 2017. To be submitted
90. Brown BA, Wallace RJ Jr, Onyi GO, De Rosas V, Wallace RJ III. Activities of four macrolides, including clarithromycin, against *Mycobacterium fortuitum*, *Mycobacterium chelonae*, and *M. chelonae*-like organisms. Antimicrob Agents Chemother. 1992;36:180–4.
91. Nash KA. Intrinsic macrolide resistance in *Mycobacterium smegmatis* is conferred by a novel *erm* gene, *erm*(38). Antimicrob Agents Chemother. 2003;47:3053–60.
92. Nash KA, Andini N, Zhang Y, Brown-Elliott BA, Wallace RJ Jr. Intrinsic macrolide resistance in rapidly growing mycobacteria. Antimicrob Agents Chemother. 2006;50:3476–8.
93. Nash KA, Brown-Elliott BA, Wallace RJ Jr. A novel gene, erm(41), confers inducible macrolide resistance to clinical isolates of *Mycobacterium abscessus* but is absent from *Mycobacterium chelonae*. Antimicrob Agents Chemother. 2009;53:1367–76.
94. Nash KA, Zhang Y, Brown-Elliott BA, Wallace RJ Jr. Molecular basis of intrinsic macrolide resistance in clinical isolates of *Mycobacterium fortuitum*. J Antimicrob Chemother. 2005;55:170–7.

95. Koh WJ, Jeon K, Lee NY, Kim B-J, Kook Y-H, Lee S-H, Park Y-K, Kim CK, Shin SJ, Huitt GA, Daley CL, Kwon OJ. Clinical significance of differentiation of *Mycobacterium massiliense* from *Mycobacterium abscessus*. Am J Respir Crit Care Med. 2011;183:405–10.
96. Koh WJ, Stout JE, Yew WW. Advances in the management of pulmonary disease due to Mycobacterium abscessus complex. Int J Tuberc Lung Dis. 2014;18(10):1141–8.
97. Wallace RJ Jr, Brown-Elliott BA, Crist CJ, Mann L, Wilson RW. Comparison of the *in vitro* activity of the glycylcycline tigecycline (formerly GAR-936) with those of tetracycline, minocycline, and doxycycline against isolates of nontuberculous mycobacteria. Antimicrob Agents Chemother. 2002;46:3164–7.
98. Brown-Elliott BA, Iakhiaeva E, Griffith DE, Woods GL, Stout JE, Wolfe CR, Turenne CY, Wallace Jr. RJ. 2013. *In vitro* activity of amikacin against isolates of *Mycobacterium avium* complex with proposed MIC breakpoints and finding of a 16S rRNA gene mutation in treated isolates. J Clin Microbiol 51:3389–3394. ERRATUM JCM 2014; 3352:1311.
99. Olivier KN, Griffith DE, Eagle G, McGinnis II JP, Micioni L, Liu K, Daley CL, Winthrop KL, Ruoss S, Addrizzo-Harris DJ, Flume PA, Dorgan D, Salathe M, Brown-Elliott BA, Gupta R, Wallace Jr. RJ. Randomized trial of liposomal amikacin for inhalation in nontuberculous mycobacterial lung disease. Am J Respir Crit Care Med. 2016; Submitted.
100. Wallace RJ Jr, Brown-Elliott BA, Ward SC, Crist CJ, Mann LB, Wilson RW. Activities of linezolid against rapidly growing mycobacteria. Antimicrob Agents Chemother. 2001;45:764–7.
101. Brown-Elliott BA, Wallace RJJ, Blinkhorn R, Crist CJ, Mann LB. Successful treatment of disseminated *Mycobacterium chelonae* infection with linezolid. Clin Infect Dis. 2001;33:1433–4.
102. Brown-Elliott BA, Philley JV, Griffith DE, Wallace RJ Jr. Comparison of in vitro susceptibility testing of tedizolid (TZD) and linezolid (LZD) against isolates of nontuberculous mycobacteria (NTM). Boston, MA: General Meeting of the American Society of Microbiology 2016; 2016.
103. Wallace RJ Jr, Dukart G, Brown-Elliott BA, Griffith DE, Scerpella EG, Marshall B. Clinical experience in 52 patients with tigecycline-containing regimens for salvage treatment of Mycobacterium abscessus and Mycobacterium chelonae infections. J Antimicrob Chemother. 2014;69(7):1945–53.
104. Shen G-H, Wu B-D, Hu S-T, Lin C-F, Wu K-M, Chen J-H. High efficacy of clofazimine and its synergistic effect with amikacin against rapidly growing mycobacteria. Int J Antimicrob Agents. 2010;35:400–4.
105. van Ingen J, Totten SE, Helstrom NK, Heifets LB, Boeree MJ, Daley CL. In vitro synergy between clofazimine and amikacin in treatment of nontuberculous mycobacterial disease. Antimicrob Agents Chemother. 2012;56(12):6324–7.
106. Yang B, Jhun BW, Moon SM, Lee H, Park HY, Jeon K, Kim DH, Kim SY, Shin SJ, Daley CL, Koh WJ. Clofazimine-containing regimen for the treatment of Mycobacterium abscessus lung disease. Antimicrob Agents Chemother. 2017;61(6):e02052–16.
107. Philley JV, Wallace RJ Jr, Benwill JL, Taskar V, Brown-Elliott BA, Thakkar F, Aksamit TR, Griffith DE. Preliminary results of bedaquiline as salvage therapy for patients with nontuberculous mycobacterial lung disease. Chest. 2015;148:499–506.
108. Maurer FP, Bruderer VL, Ritter C, Castelberg C, Bloemberg GV, Böttger EC. Lack of antimicrobial bactericidal activity in *Mycobacterium abscessus*. Antimicrob Agents Chemother. 2014;58:3828–36.
109. van Ingen J, Boeree MJ, van Soolingen D, Mouton JW. Resistance mechanisms and drug susceptibility testing of nontuberculous mycobacteria. Drug Resist Updat. 2012;15:149–61.
110. Brown-Elliott BA, Vasireddy S, Vasireddy R, Iakhiaeva E, Howard ST, Nash K, Parodi N, Strong A, Gee M, Smith T, Wallace RJ Jr. Utility of sequencing the erm(41) gene in isolates of Mycobacterium abscessus subsp. abscessus with low and intermediate clarithromycin MICs. J Clin Microbiol. 2015;53(4):1211–5.
111. Koh WJ, Jeong BH, Jeon K, Kim SY, Park KU, Park HY, Huh HJ, Ki CS, Lee NY, Le SH, Kim CK, Daley CL, Shin SJ, Kim H, Kwon OJ. Oral macrolide therapy following short-

term combination antibiotic treatment for *Mycobacterium massiliense* lung disease. Chest. 2016;150(6):1211–21.
112. Choi H, Kim S-Y, Kim DH, Huh HJ, Ki C-S, Lee NY, Lee S-H, Shin S, Shin SJ, Daley CL, Koh W-J. Clinical characteristics and treatment outcomes of patients with acquired macrolide-resistant Mycobacterium abscessus lung disease. Antimicrob Agents Chemother. 2017;61:e01146–17, https://doi.org/10.1128/AAC.01146–17.
113. Choi H, Kim SY, Lee H, Jhun BW, Park HY, Jeon K, Kim DH, Huh HJ, Ki CS, Lee NY, Lee SH, Shin SJ, Daley CL, Clinical Characteristics KWJ. Treatment outcomes of patients with macrolide-resistant Mycobacterium massiliense lung disease. Antimicrob Agents Chemother. 2017;61(2):293–5
114. Griffith DE, Aksamit TR. Nontuberculous mycobacterial disease therapy: take it to the limit one more time. Chest. 2016;150(6):1177–8.
115. Jarand J, Levin A, Zhang L, Huitt G, Mitchell JD, Daley CL. Clinical and microbiologic outcomes in patients receiving treatment for Mycobacterium abscessus pulmonary disease. Clin Infect Dis. 2011;52(5):565–71.
116. Jeon K, Kwon OJ, Lee NY, Kim BJ, Kook YH, Lee SH, Park YK, Kim CK, Koh WJ. Antibiotic treatment of Mycobacterium abscessus lung disease: a retrospective analysis of 65 patients. Am J Respir Crit Care Med. 2009;180(9):896–902.
117. Lyu J, Jang HJ, Song JW, Choi CM, Oh YM, Lee SD, Kim WS, Kim DS, Shim TS. Outcomes in patients with Mycobacterium abscessus pulmonary disease treated with long-term injectable drugs. Respir Med. 2011;105(5):781–7.
118. Harada T, Akiyama Y, Kurashima A, Nagai H, Tsuyuguchi K, Fujii T, Yano S, Shigeto E, Kuraoka T, Kajiki A, Kobashi Y, Kokubu F, Sato A, Yoshida S, Iwamoto T, Saito H. Clinical and microbiological differences between Mycobacterium abscessus and Mycobacterium massiliense lung diseases. J Clin Microbiol. 2012;50(11):3556–61.
119. Roux AL, Catherinot E, Soismier N, Heym B, Bellis G, Lemonnier L, Chiron R, Fauroux B, Le Bourgeois M, Munck A, Pin I, Sermet I, Gutierrez C, Véziris N, Jarlier V, Cambau E, Herrmann JL, Guillemot D, Gaillard JL, group O. Comparing *Mycobacterium massiliense* and *Mycobacterium abscessus* lung infections in cystic fibrosis patients. J Cyst Fibros. 2015;14:63–9.
120. Lee SH, Yoo HK, Kim SH, Koh WJ, Kim CK, Park YK, Kim HJ. The drug resistance profile of Mycobacterium abscessus group strains from Korea. Ann Lab Med. 2014;34(1):31–7.
121. Floto RA, Olivier KN, Saiman L, Daley CL, Herrmann JL, Nick JA, Noone PG, Bilton D, Corris P, Gibson RL, Hempstead SE, Koetz K, Sabadosa KA, Ermet-Gaudelus I, Smyth AR, van Ingen J, Wallace RJ, Winthrop KL, Marshall BC, Haworth CS. US Cystic Fibrosis Foundation and European cystic fibrosis society. US Cystic Fibrosis Foundation and European cystic fibrosis society consensus recommendations for the management of non-tuberculous mycobacteria in individuals with cystic fibrosis. Thorax. 2016;71(Suppl 1):i1–22.

Management of Lung Diseases Associated with NTM Infection

Anne E. O'Donnell

Introduction

Nontuberculous mycobacterial (NTM) infections are associated with a variety of structural lung diseases, especially bronchiectasis but also chronic obstructive pulmonary disease (COPD) and fibrotic lung disease. Many patients with NTM lung infections are also coinfected with other bacteria as well as fungi and viruses. In dealing with patients who are infected with NTM organisms, it is important for the clinician to take a holistic approach to the patient. Treatable underlying causes of the lung disease need to be identified, all infections need to be addressed, and treatment plans may need to include multiple therapeutic modalities individualized for the specific patient.

Underlying Diseases

NTM lung infection is a complication of several different lung diseases as well as a cause of structural lung disease. For many years, NTM infections were most commonly seen in older males with postinfectious fibro-cavitary disease or chronic obstructive pulmonary disease (COPD) [1–3]. NTM infections are increasingly being diagnosed in patients with underlying bronchiectasis due to multiple etiologies but especially in cystic fibrosis (CF) bronchiectasis [4–6]. A recently emerging group of NTM-infected patients, primarily female, have NTM-related fibro-nodular bronchiectasis and bronchiolitis, seemingly caused by the NTM infection. This

A. E. O'Donnell
Division of Pulmonary, Critical Care and Sleep Medicine, Georgetown University Medical Center, Washington, DC, USA
e-mail: ODONNELA@gunet.georgetown.edu

D. E. Griffith (ed.), *Nontuberculous Mycobacterial Disease*, Respiratory Medicine,
https://doi.org/10.1007/978-3-319-93473-0_14

group of patients has been recognized with increasing frequency over the past 25 years, particularly in North America and Asia [7, 8]. NTM infection is also seen in patients with fibrotic lung disease and "traction" bronchiectasis, especially in association with underlying autoimmune diseases like rheumatoid arthritis [9]. A much smaller number of NTM patients present with a solitary or dominant pulmonary nodule with or without other obvious structural lung disease [10]. Finally, a rare phenomenon of "hot tub" lung has also been reported where NTM infection due to intensive aerosol exposure causes an inflammatory reaction in the lungs [11].

Addressing the Underlying Disease

Chronic fibro-cavitary lung damage due to various causes predisposes patients to NTM infection. Clinicians need to recognize that patients who have parenchymal lung scarring due to prior infections (including mycobacterium tuberculosis) or other insults may become infected with NTM organisms. Sarcoidosis and prior radiation therapy can cause such scarring; patients with these underlying conditions who present with cough and sputum production need to be assessed, by respiratory secretion culture, for NTM infection. NTM infection in COPD is likely under-recognized; one retrospective series of 142 lung volume reduction surgery specimens showed that 10% had histologic evidence of mycobacterial infection [12]. Bronchiectasis due to any cause also predisposes the patient to NTM infection. As mentioned above, approximately 13 percent of CF patients in the United States are chronically infected with NTM [6]; outbreaks of NTM infection have been reported in CF centers in the United States and United Kingdom [13, 14]. Patients with ciliary disorders are at risk for development of NTM lung infections [15]. A recent publication from the US Bronchiectasis Research Registry showed that 50% of 1826 patients with diverse causes of non-cystic fibrosis bronchiectasis enrolled in the multicenter registry project had at least one culture positive for NTM organism [16]. Eleven percent of patients enrolled in a large retrospective study in Shanghai, China, were found to have at least one positive AFB culture with 5% having NTM lung disease as defined by ATS/IDSA guidelines [17, 18].

Increasingly recognized since the late 1980s is nodular bronchiectatic lung disease that appears to be caused by NTM infection. This disease primary affects older Caucasian women with a tall, lean body habitus and other morphometric abnormalities; there may be genetic or immune predisposing factors that have not yet been clearly identified [19–21]. A study from Japan published in 2003 demonstrated the pathologic changes that occur with NTM airway infection resulted in nodular bronchiectasis [22].

NTM infections appear to be increased in patients with lung disease associated with rheumatoid arthritis [10] as well as in patients on antirheumatic medications [23].

Imaging Findings in Lung Disease Associated with NTM Infection

See Figs. 1, 2, 3, 4, and 5; these are representative images of the five main radiographic presentations of the lung disease associated with NTM infection. None of the radiographic features are pathognomonic for NTM infection. Fibro-cavitary disease (Fig. 1) may or may not have superimposed NTM infections; other infecting organisms, including fungi, may mimic the findings seen in NTM. Traction bronchiectasis seen in interstitial/fibrotic lung diseases (Fig. 2) is a non-specific finding.

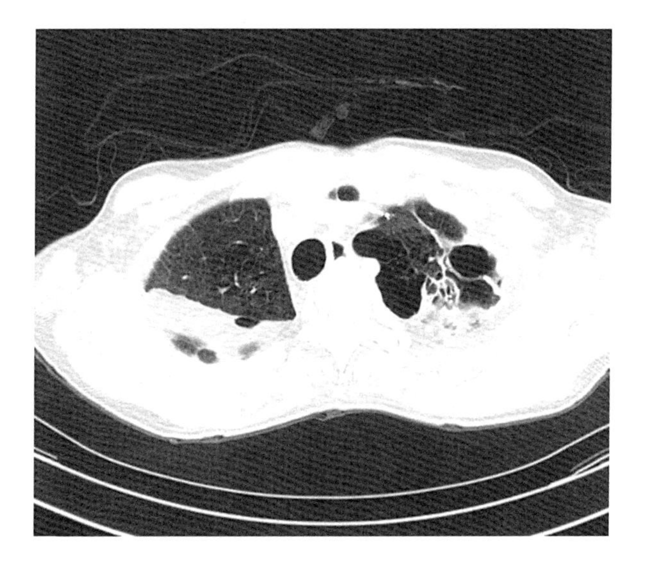

Fig. 1 CT slice demonstrating fibro-cavitary disease; sputum culture from this patient confirmed superimposed MAC infection

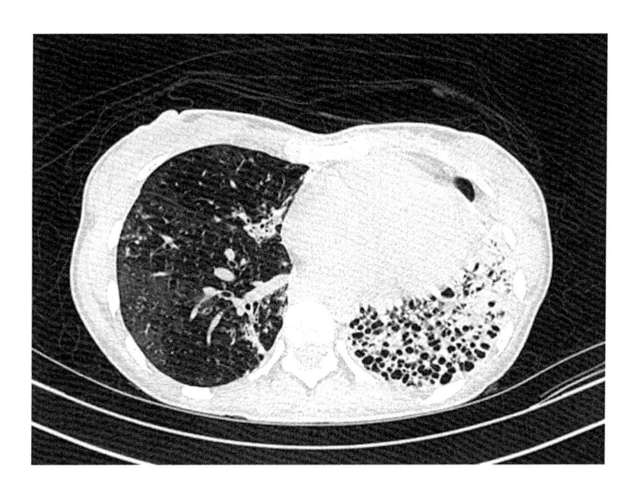

Fig. 2 CT slice demonstrating fibrotic lung disease in the left lower lobe; bronchoalveolar lavage confirmed MAC infection

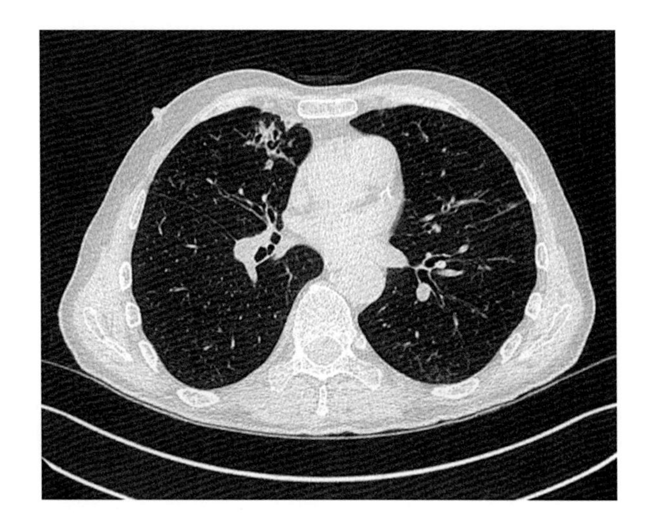

Fig. 3 CT slice demonstrating right middle lobe and lingua bronchiectasis and scattered areas of "tree-in-bud" bronchiolitis; sputum culture confirmed MAC infection

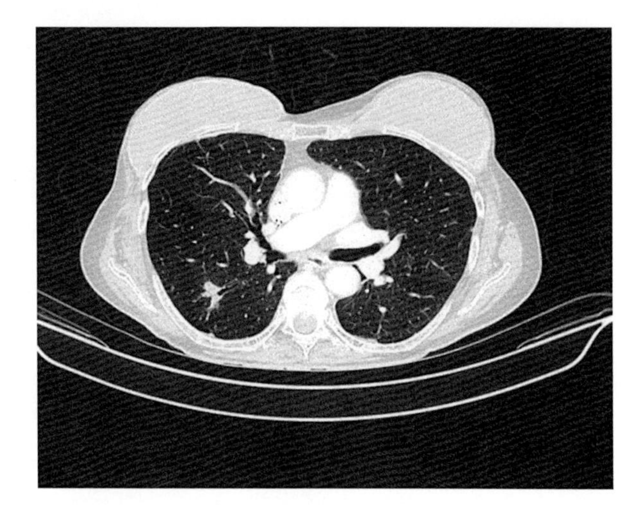

Fig. 4 CT slice demonstrating a solitary pulmonary nodule surrounding an airway; surgical pathology confirmed granulomatous inflammation and tissue culture was positive for MAC

The tree-in-bud nodularity seen in Fig. 3 is thought to be diagnostic of NTM infection; however, two recent studies have shown the tree-in-bud pattern reflects endobronchial inflammation from any cause, including noninfectious etiologies [24, 25]. Though right middle lobe and lingular bronchiectasis (Fig. 3) is often thought to be strongly suggestive of NTM infection, other infecting organisms can result in the same radiographic presentation. A solitary pulmonary nodule (Fig. 4) due to NTM is indistinguishable from a malignant nodule; PET/CT scans cannot definitively distinguish an infectious versus malignant nodule [26]. The radiographic findings in "hot tub" lung (Fig. 5) are no different from any other interstitial/hypersensitivity

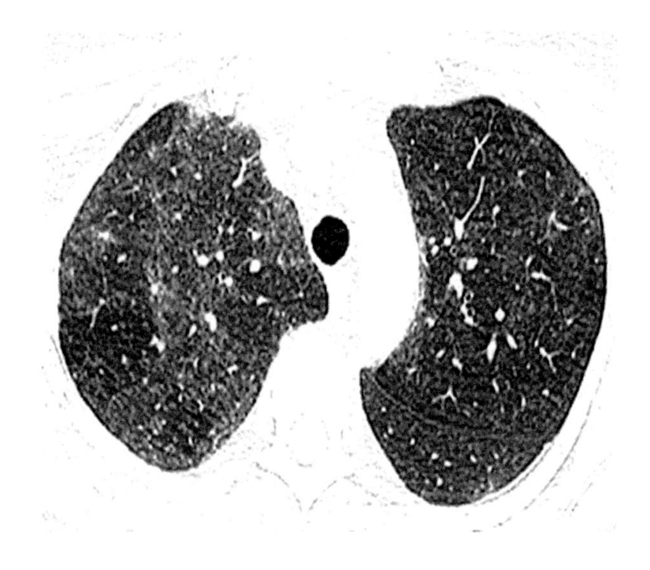

Fig. 5 CT slice demonstrating diffuse patchy ground-glass infiltrates with mosaicism in a patient who used an indoor hot tub; bronchoalveolar lavage culture confirmed MAC infection

Table 1 Lung diseases associated with NTM infection

Lung conditions associated with NTM infection
Fibro-cavitary scarring
COPD/emphysema
Bronchiectasis: multiple etiologies including cystic fibrosis
Bronchiolitis
Solitary pulmonary nodules
Fibrotic lung diseases
Inhalation injury

process. Hence, the clinician and the radiologist need to confirm "suspicious for NTM" findings with culture and/or tissue identification of the organism (Table 1).

Why the Underlying Disease Matters

Identifying and categorizing the underlying lung disease are important metrics in NTM-infected patients. Prognosis may be impacted by the severity of the underlying disease (COPD/bronchiectasis). The finding of overlapping bronchiectasis in patients with COPD portends a worse prognosis [27]. A rising mortality rate from NTM infection has been observed in the United States, particularly when the patient also carries the diagnosis of bronchiectasis [28]. NTM infection, particularly infection with *Mycobacterium abscessus*, subspecies abscessus, is associated with poorer outcomes in CF patients [29]. Because NTM infection is not a reportable disease, it is difficult to assess how much impact it has in patients with underlying complex lung diseases. Hence it is important to recognize both the underlying disease and the impact of the NTM infection when personalizing treatment options.

Microbiology/Coinfection

Many patients with NTM lung disease are infected only with NTM, at least when assessed by routine culture methods. However, the US Bronchiectasis Research Registry data shows that coinfection with other organisms is common. Thirty percent of the NTM-infected patients in that Registry were also culture positive for *Pseudomonas aeruginosa*, and many other coinfecting organisms were also seen, including other gram-negative organisms, *Staphylococcus aureus* and *Streptococcal pneumoniae* [16]. More sophisticated microbiome technology, now only used in research settings, shows that anaerobic organisms may play a role in the pathogenesis of nontuberculous mycobacterial infection [30]. It is also clear that antibiotic treatment aimed at various organisms in bronchiectasis patients may alter the composition and balance of the airway microbiome [31].

Treatment of the Underlying Disease in NTM-Infected Patients

Antibiotic treatment aimed at NTM infection is discussed extensively in the previous chapters. It is important to keep in mind that therapies aimed at the underlying disease and at coinfecting organisms are also major components of the treatment of patients with NTM infection. In fact, optimization of the underlying lung condition and antibiotic treatment aimed at the "other" microorganisms may obviate or at least delay the need to initiate NTM-specific antibiotic therapy. However, there may be conflicts between treatments that might help the structural lung disease (such as corticosteroids or macrolides) yet might result in a proliferation of the NTM organisms. Balancing the various treatment components is crucial in each patient.

Treating Underlying COPD

Though there is little published data regarding NTM-infected COPD patients, these patients probably benefit from standard COPD regimens that include bronchodilators, especially beta-agonist and anticholinergic therapies. However, clinicians need to be cautious about using inhaled corticosteroids (ICS), oral steroids, and chronic oral macrolides in COPD patients with NTM infection. Steroid medications may enhance the growth of NTM organisms (or at least make the airways more susceptible), and monotherapy with macrolide agents might result in development of resistant NTM organisms [3, 32, 33].

Treating Underlying Bronchiectasis

Much more is known about the treatment of bronchiectasis (with or without NTM infection), both due to cystic fibrosis and to other etiologies. Airway clearance is considered to be a basic component of treatment in bronchiectasis [34, 35]. Airway clearance can be done by mechanical means (manual chest physiotherapy, positive expiratory/resistance devices, chest wall oscillating vests). The superiority of one technique over the other has not been clearly documented, so the choice of therapy should be based on what best suits the patient's needs [36]. Hypertonic saline (3–7%) treatments have salutary effects in both CF and non-CF bronchiectasis [37, 38]. Regular exercise and participation in pulmonary rehabilitation also may benefit patients with bronchiectasis.

Other therapies that have become common in the treatment of bronchiectasis must be viewed with extreme caution when the patient is infected with NTM organisms. Chronic macrolide therapy may be beneficial in bronchiectasis patients infected with "routine" organisms like *Pseudomonas aeruginosa* [39], but there is a risk of developing macrolide-resistant NTM organisms which may already be present (though not necessarily identified) in the airways of these patients [40]. Growth of other resistant organisms may also be enhanced by a chronic macrolide strategy [41]. Additionally, adverse effects from macrolides such as hearing loss or cardiac dysfunction may be more prominent in an older bronchiectasis population with underlying comorbidities. Inhaled corticosteroids with or without long-acting bronchodilators may benefit patients with bronchiectasis [42], but there is a risk of propagating NTM infection with ICS therapy [32].

When the bronchiectasis patient is coinfected with NTM and other organisms, it can be difficult to decide which infection to target with antibiotics. *Pseudomonas aeruginosa* is commonly seen in these patients. It must not be assumed that all of the patient's infectious symptoms are due to one organism versus another. There is no consensus on which organism to treat "first"; expert opinion is to use airway clearance treatments as a cornerstone of therapy and then consider shorter-term antibiotic strategies against the "regular" organisms before embarking on full-blown prolonged NTM antibiotic treatment.

Treating Other Underlying Lung Diseases When NTM Is Present

Patients with underlying fibrotic lung disease and traction bronchiectasis may have positive cultures for NTM organisms; there is no evidence that treatments aimed at lung fibrosis have any impact on the secondary infections. Patients with such underlying lung disease may have positive NTM cultures that should only be treated if they fit the treatment threshold recommended in the ATS/IDSA guidelines [18]. Patients with alpha-1 antitrypsin deficiency who have emphysema or bronchiectasis may benefit from replacement therapy [43], but there is no evidence that this treatment impacts the infectious complications. Likewise, patients who have bronchiectasis due to immunoglobulin deficiency may benefit from immunoglobulin replacement [44], but the impact of this therapy on control of NTM infections in these patients is uncertain. An unknown number of patient have structural lung disease and possibly enhanced susceptibility to NTM infections because of gastroesophageal reflux and aspiration; one study from South Korea showed that 26% (15 of 58) of patients with NTM infection had proven GERD even though only 27% of those patients (4 of 15) had typical GERD symptoms [45]. Testing may include modified barium swallow and video stroboscopy for swallowing and barium esophagram/24 h pH monitoring for assessment of esophageal and gastroesophageal sphincter function. It remains unproven that treatment of acid reflux, esophageal dysmotility, or aspiration impacts outcome of NTM infections; however, simple measures such as elevation of the head of the bed at night may be reasonable for patients with these infections.

Treating Solitary Pulmonary Nodules Due to NTM and "Hot Tub" Lung

NTM in solitary nodules is often diagnosed at the time of resectional surgery. Generally, NTM antibiotic treatment is not needed after such surgery, but the patient must be closely followed for subsequent spread of disease. "Hot tub" lung due to NTM exposure is treated by cessation of hot tub use; steroids have also been reported to have benefit.

Summary and Conclusions

There is a spectrum of lung diseases that are either the cause or the consequence of NTM infections. The most common of these diseases are COPD and bronchiectasis. Identifying and treating the underlying lung diseases may have a role in mitigating the impact of the NTM infection. Without question, the quality of life of patients with NTM infection can be improved by addressing the lung disease and other complicating factors that contribute to morbidity and possibly mortality. Antibiotic therapy for NTM lung infections, even when well tolerated and given according to guidelines, is unlikely to "cure" many of these patients. A multimodality approach that includes airway clearance, exercise, attention to nutrition, and treatment of coinfections is needed so that symptoms can be controlled and quality of life can be maximized. A collaborative approach between treating clinicians from various disciplines (pulmonary medicine, respiratory care, infectious diseases) can maximize the benefit to the individual patient. Patient-centered priorities have recently been published that stress the need for additional clinical education to improve screening and diagnosis of NTM infections and development of a geographically distributed network of NTM disease specialists [46].

References

1. Rosenzweig DY. Pulmonary mycobacterial infections due to mycobacterium intracellulare-avium complex. Chest. 1979;75:115–9.
2. Wolinsky E. Nontuberculous mycobacterium and associated diseases. Am Rev Respir Dis. 1979;119:107–59.
3. Andrejak C, Nielsen R, Thomsen VO, Duhaut P, et al. Chronic respiratory disease, inhaled corticosteroids and risk of non-tuberculous mycobacteriosis. Thorax. 2013;68: 256–62.
4. Adjemian J, Olivier KN, Seitz AE, Holland SM, Prevots DR. Prevalence of nontuberculous mycobacterial lung disease in U.S. Medicare beneficiaries. Am J Respir Crit Care Med. 2012;185:881–6.
5. Henkle E, Hedberg K, Schafer S, Novosad S, Winthrop KL. Population-based incidence of pulmonary nontuberculous mycobacterial disease in Oregon 2007 to 2012. Ann Am Thorac Soc. 2015;12:642–7.
6. Olivier KN, Weber DJ, Wallace RJ, Faiz AR, et al. Nontuberculous mycobacteria in Cystic Fibrosis Study Group. Nontuberculous mycobacteria. I: multicenter prevalence study in cystic fibrosis. Am J Respir Crit Care Med. 2003;167:828–34.
7. McShane PJ, Glassroth J. Pulmonary disease due to nontuberculous mycobacterium. Current state and new insights. Chest. 2015;148:1517–27.
8. Ito Y, Hirai T, Fujita K, Maekawa K, et al. Increasing patients with pulmonary mycobacterium avium complex disease and associated underlying disease in Japan. J Infect Chemother. 2015;21:352–6.
9. Brode SK, Jamieson FB, Ng R, et al. Risk of mycobacterial infections associated with rheumatoid arthritis in Ontario Canada. Chest. 2014;146:563–72.
10. Lim J, Lyu J, Choi C-M, Oh Y-M, et al. Non-tuberculous mycobacterial diseases presenting as solitary pulmonary nodules. Int J Tuberc Lung Dis. 2010;14:1635–40.
11. Khoor A, Leslie KO, Tazelaar HD, Helmers RA, Colby TV. Diffuse pulmonary disease caused by nontuberculous mycobacteria in immunocompetent people (hot tub lung). Am J Clin Pathol. 2001;115:755–62.
12. Char A, Hopkinson NS, Hansell DM, Nicholson AG, et al. Evidence of mycobacterial disease in COPD patients with lung volume reduction surgery; the importance of histologic assessment of specimens: a cohort study. BMC Pulm Med. 2014;12:124.
13. Aitken MB, Limaye A, Pottinger P, et al. Respiratory outbreak of Mycobacterium abscessus subspecies massiliense in a lung transplant and cystic fibrosis center. Am J Respir Crit Care Med. 2012;185:231–2.
14. Bryant JM, Grogono DM, Greaves D, Foweraker J, et al. Whole-genome sequencing to identify transmission of Mycobacterium abscessus between patients with cystic fibrosis: a retrospective cohort study. Lancet. 2013;381:1551–60.
15. Fowler CJ, Olivier KN, Leung JM, et al. Abnormal nasal nitric oxide production, ciliary beat frequency and toll-like receptor response in pulmonary nontuberculous mycobacterial disease epithelium. Am J Respir Crit Care Med. 2013;187:1374–81.
16. Aksamit TR, O'Donnell AE, Barker A, Olivier KL, et al. Adult patients with bronchiectasis: a first look at the US Bronchiectasis Research Registry. Chest. 2017;151:982–92.
17. Lin J-L, Xu J-F, Qu J-M. Bronchiectasis in China. Ann Am Thorac Soc. 2016;5:609–16.

18. Griffith DE, Aksamit T, Brown-Elliott BA, Catanzaro A, et al. An official ATS/ISDA statement: diagnosis, treatment and prevention of Nontuberculous mycobacterial disease. Am J Respir Crit Care Med. 2007;175:367–416.
19. Kim RD, Greenberg DE, Ehrmantraut ME, et al. Pulmonary nontuberculous mycobacterial disease: prospective study of a distinct pre-existing syndrome. Am J Respir Crit Care Med. 2008;178:1066–74.
20. Colombo RE, Hill SC, Claypool RJ, Holland SM, Olivier KN. Familial clustering of pulmonary nontuberculous mycobacterial disease. Chest. 2010;137:629–34.
21. O'Connell ML, Birkenkamp KE, Kleiner DE, et al. Lung manifestations in an autopsy-based series of pulmonary or disseminated nontuberculous mycobacterial disease. Chest. 2012;141:1203–9.
22. Fujita J, Ohtsuki Y, Shigeto E, et al. Pathologic findings of bronchiectasis caused by mycobacterium avium intracellulare complex. Respir Med. 2003;97:933–8.
23. Brode SK, Jamieson FB, Ng R, et al. Increased risk of mycobacterial infections associated with anti-rheumatic medications. Thorax. 2015;70:677–82.
24. Shimon G, Yonit WW, Gabriel I, et al. The "tree in bud" pattern on chest CT: radiologic and microbiologic correlation. Lung. 2015;193:823–9.
25. Miller WT, Panosian JS. Causes and imaging patterns of tree-in-bud opacities. Chest. 2013;144:1883–92.
26. DelGuidice G, Bianco A, Cennamo A, et al. Lung and nodal involvement in nontuberculous mycobacterial disease: PET/CT role. Biomed Res Int. 2015;25:8. https://dx.doi.org/10.1155/2105/353202
27. Martinez-Garcia M-A, de la Rosa Carrillo D, Soler-Cataluna J-J, et al. Prognostic value of bronchiectasis in patients with moderate-to severe chronic obstructive pulmonary disease. Am J Respir Crit Care Med. 2013;187:823–31.
28. Vinnard C, Longworth S, Mezochow A, et al. Deaths related to non-tuberculous mycobacterial infections in the United States, 1999–2014. Ann Am Thorac Soc. 2016;13:1951–55.
29. Qvist T, Taylor-Robinson D, Waldmann E, et al. Comparing harmful effects of non-tuberculous mycobacteria and Gram negative bacteria on lung function in patients with cystic fibrosis. J Cyst Fibros. 2016;15:380–5.
30. Yamasaki K, Mukae H, Kawanami T, et al. Possible role of anaerobes in the pathogenesis of nontuberculous mycobacterial infection. Respirology. 2015;20:758–65.
31. Rogers GB, Bruce KD, Martin ML, Burr LD, Serisier DJ. The effect of long-term macrolide treatment on respiratory microbiota composition in non-cystic fibrosis bronchiectasis: an analysis from the randomised, double-blind, placebo-controlled BLESS trial. Lancet Respir Med. 2014;2:988–96.
32. Dirac MA, Horan KL, Doody DR, Meschke JS, et al. Environment or host? A case-control study of risk factors for mycobacterium avium complex lung disease. Am J Respir Crit Care Med. 2012;186:684–91.
33. Griffith DE, Brown-Elliott BA, Langsjoen B, Zhang Y, et al. Clinical and molecular analysis of macrolide resistance in mycobacterium avium complex lung infection. Am J Respir Crit Care Med. 2006;174:928–34.
34. Smith AR, Bell SC, Bojcin S, Bryon M, et al. European cystic fibrosis society standards of care: best practice guidelines. J Cyst Fibros. 2014;13:S23–42.
35. Pasteur MC, Bilton D, Hill AT, et al. British Thoracic Society guideline for non-CF bronchiectasis. Thorax. 2010;65:i1–58.
36. Volsko TA. Airway clearance therapy: finding the evidence. Respir Care. 2013;58:1669–78.
37. Elkins MR, Robinson M, Rose BR, et al. A controlled trial of long term hypertonic saline in patients with cystic fibrosis. N Engl J Med. 2006;354:229–40.
38. Kellett F, Robert NM. Nebulised 7% saline improves lung function and quality of life in bronchiectasis. Respir Med. 2011;105:1831–5.
39. Hill AT. Macrolides for clinically significant bronchiectasis in adults: who should receive this treatment? Chest. 2016;150:1187–93.

40. Binder AM, Adjemian J, Olivier KN, Prevots DR. Epidemiology of nontuberculous mycobacterial infection and associated chronic macrolide use among patients with cystic fibrosis. Am J Respir Crit Care Med. 2013;188:807–12.
41. Elborn JS, Tunney MM. Macrolides and bronchiectasis. Clinical benefit with a resistance price. JAMA. 2013;309:1295–6.
42. Martinez-Garcia MA, Soler-Cataluna JJ, Catalan-Serra P, Roman-Sanchez P, Perpina Tordera M. Clinical efficacy and safety of budesonide-formoterol in non-cystic fibrosis bronchiectasis. Chest. 2012;141:461–8.
43. Mohanka M, Khemasuwan D, Stoller JK. A review of augmentation therapy for alpha-1 antitrypsin deficiency. Expert Opin Biol Ther. 2012;12:685–700.
44. Quinti I, Soresina A, Geurra A, Rondelli R, et al. Effectiveness of immunoglobulin replacement therapy on clinical outcomes in patients with primary antibody deficiencies: results from a multicenter prospective cohort study. J Clin Immunol. 2011;31:315–22.
45. Koh WJ, Lee JH, Kwon YS, Lee KS, et al. Prevalence of gastroesophageal reflux disease in patients with nontuberculous mycobacterial lung disease. Chest. 2007;131:1825–30.
46. Henkle E, Aksamit T, Barker A, Daley CL, et al. Patient-centered research priorities for pulmonary nontuberculous mycobacterial (NTM) infection. An NTM research consortium workshop report. Ann Am Thorac Soc. 2016;13:S379–84.

Surgical Management of NTM Diseases

James A. Caccitolo

History of Thoracic Surgery and Tuberculosis

Hippocrates of ancient Greece is credited with the first reference describing a thoracic surgical procedure [1]. He described treatment of "empyema" first with herbal remedies and chest physical therapy and if unsuccessful then open surgical evacuation of the empyema. For centuries, thoracic surgery was limited to surgical drainage of the chest cavity. Many patients died due to the open pneumothorax that was created to allow drainage of blood or infection. Early thoracic surgeons persisted and developed means to apply suction to drains which could then evacuate air from the chest cavity and re-expand a collapsed lung. These early techniques were mainly employed for treatment of traumatic injuries sustained in battle or complications due to infections. In the early 1900s, *Mycobacterium tuberculosis* was a leading cause of death. With no effective antimicrobial agents, and the knowledge that *M. tuberculosis* was an obligate aerobe, thoracic surgeons began to develop procedures that were intended to treat mycobacterial infections. A range of surgical techniques were developed including collapse therapy, thoracoplasty, plumbage, and even early thoracoscopy [2]. Swiss reports of patients with *Mycobacterium tuberculosis* treated with thoracoplasty and extrapleural pneumothorax provided the first hope that patients could be cured of the disease. At the time, limitations of anesthetic techniques prevented much surgical intervention inside of the chest [3]. Pneumonectomy and early lobectomies were rare and crude, accomplished essentially by tourniquet techniques. With the concurrent developments of endotracheal intubation and positive pressure ventilation, more surgical options became available, and isolated single lung ventilation paved the way for the development of modern anatomic lung resection procedures. The development of antimicrobial agents quickly led to

J. A. Caccitolo
Cardiothoracic Surgery, CHRISTUS Trinity Clinic, University of Texas Health Science Center Tyler, Tyler, TX, USA
e-mail: James.Caccitolo@uthct.edu

D. E. Griffith (ed.), *Nontuberculous Mycobacterial Disease*, Respiratory Medicine,
https://doi.org/10.1007/978-3-319-93473-0_15

medical treatment becoming the primary therapy for the majority of patients with mycobacterial disease and surgical intervention becoming a rarity.

Today, for selected patients with mycobacterial disease, the original indications for thoracic surgical intervention still apply, although the specific surgical techniques available have evolved. Thoracoplasty, rather than a first-line technique, is reserved as a last resort for only the most challenging cases, while minimally invasive video-assisted thoracic surgery (VATS) is often the initial procedure of choice. In general, lower morbidity and faster recovery associated with minimally invasive techniques have helped to make thoracic surgical lung resection an important adjunct to antibiotic therapy in the treatment of mycobacterial disease.

Indications for Surgical Intervention

The primary treatment of patients diagnosed with NTM is a multidrug antibiotic regimen; however, a number of patients will not respond to medical therapy alone. The rate of successful sputum clearance in patients with NTM treated with antibiotics varies widely and, particularly in the case of *Mycobacterium abscessus* infections, can be quite poor. Furthermore, compliance with multidrug regimens over many months can be difficult due to significant side effects and, in some studies, has been less than 80% [4]. Surgical intervention and specifically anatomic resection of areas of gross cavitary disease or severe bronchiectasis can be of significant benefit in treating select patients. The goals of surgery are similar to those of antibiotic therapy and include eradication of the infection to prevent further destruction of the lung and relief of symptoms including cough, sputum production, and hemoptysis. The official ATS/IDSA statement does not list clear indications for which surgical resection is recommended, but does state that the more difficult to treat an infection is, the more likely surgery should be considered. The statement also recommends expert consultation with a multispecialty group experienced with treating NTM disease [5]. While there is no definitive statement regarding indications for surgical resection, there are circumstances in which surgical resection should be considered. Table 1 includes a list of generally accepted indications for surgical resection in patients with NTM disease. The Japanese Society for Tuberculosis (JST) has

Table 1 Indications for consideration of surgical resection

Failed antibiotic therapy
Localized disease amenable to anatomic resection
Patients have cardiopulmonary fitness to tolerate a lung resection
Hemoptysis
Large cavitary disease
Severe bronchiectasis
Progressive disease on imaging
Recurrence of positive sputum off antibiotic therapy
Medication intolerance/non-compliance
Macrolide resistance

Table 2 Indications for NTM lung disease surgery (JST guidelines) [6]

1. When sources of bacterial discharge or major lesions being sources of bacterial discharge are clearly noted and, in addition, one of the following disease conditions is observed:
(a) Chemotherapy has failed to stop bacterial discharge.
(b) Bacteriological relapse is noted.
(c) Radiographically enlarged lesions or tendencies of lesion enlargement are either revealed or predicted.
(d) Even though bacterial discharge has been stopped, cavitary lesions or bronchiectatic lesions remain, suggesting that relapse or reactivation may occur.
(e) Acute exacerbation has repeatedly occurred due to lesions that are sources of massive bacterial discharge, leading to the rapid progression of disease.
2. In patients with hemoptysis, repeated airway infection, or comorbid aspergillosis, responsible lesions are subject to resection irrespective of the status of bacterial discharge.

published guidelines for surgical therapy for pulmonary NTM disease in 2008 [6]. Those guidelines are similar to the generally accepted indications and are listed in Table 2.

Failed Antibiotic Therapy

Treatment failure has been described as no microbiologic, clinical, or radiographic response after 6 months of appropriate therapy or no achieved conversion of sputum to AFB culture negative after 12 months of appropriate therapy [5]. Based on this definition, most patients who are considered for surgery have been on a prolonged antibiotic therapy regimen of usually 1 year or more. After 1 year of unsuccessful sputum clearance, it is reasonable to consider surgical intervention as an option to continued antibiotic therapy alone. Referral of patients for surgical resection earlier in their antibiotic course can also be considered if based on expert opinion, likelihood of sputum clearance with antibiotics alone is low due to radiographic extent of disease, or patients exhibit other confounding sequalae such as unrelenting cough or hemoptysis.

Localized Disease Amenable to Anatomic Resection

Patients who are considered for surgical resection have a distribution of cavitary disease or bronchiectasis that is confined to an anatomically defined pulmonary lobe or segment. Optimally, once that anatomic area is removed, the remaining of the lung is relatively free of gross disease. Patients with diffuse involvement of both lungs are generally not candidates for resection. Depending on preoperative pulmonary function testing, however, patients with diffuse disease of one lung or bilateral disease present may be candidates for pneumonectomy or staged bilateral resections. The latter scenario is commonly seen in patients who have middle lobe and lingular bronchiectasis with sparing of the upper and lower lobes bilaterally. Patents

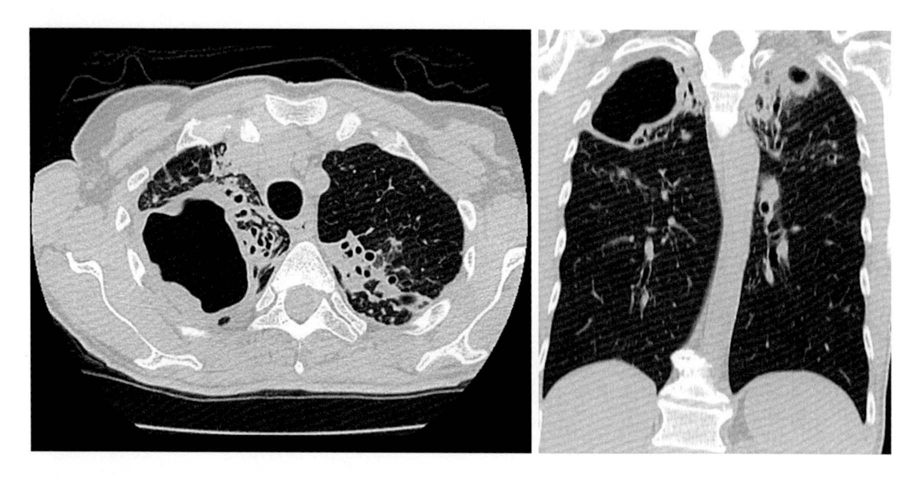

Fig. 1 Axial and coronal CT images demonstrating a large right upper lobe cavity and smaller cavity of the left upper lobe

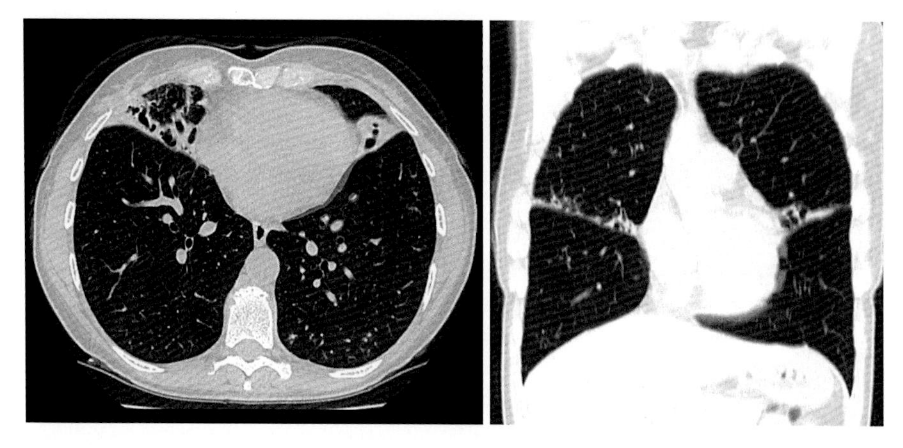

Fig. 2 CT images of a patient with right middle lobe and lingular nodular bronchiectasis due to *Mycobacterium avium* complex disease

with upper lobe involvement may also have concomitant lower lobe disease. When the lower lobe involvement is limited to superior segment, upper lobectomy with lower lobe superior segmentectomy can be accomplished to preserve the majority of the unaffected lower lobe.

Figure 1 depicts a patient with right upper lobe cavitary disease. Typically, with large apical cavities, much of the upper lobe is destroyed, and a lesser resection than lobectomy is not practical. In the lower lobe, however, it is not uncommon to have a cavity in the superior segment with sparing of the basilar segments. Patients with classic Lady Windermere syndrome, (right middle lobe bronchiectasis) or lingular involvement, are good candidates for middle lobectomy or lingulectomy via a minimally invasive or VATS approach [7]. Figure 2 shows a patient with right middle lobe and lingular bronchiectasis who had *Mycobacterium avium* complex (MAC) infection

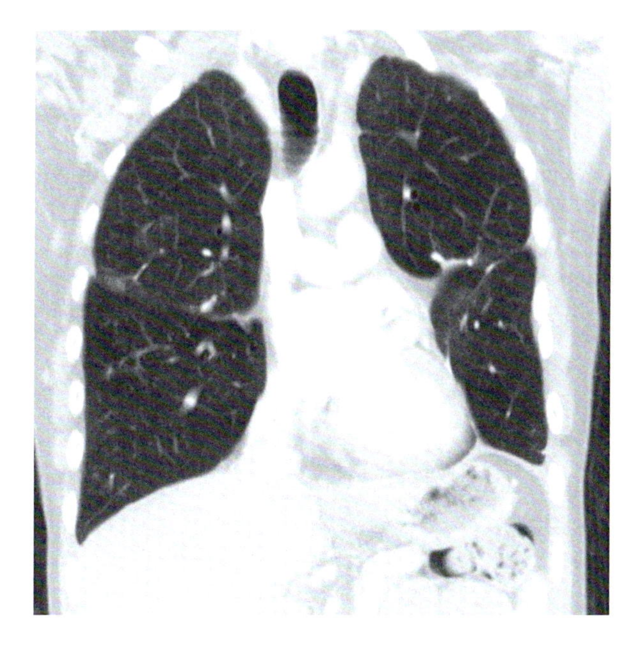

Fig. 3 Coronal CT image following right VATS middle lobectomy and left VATS lingulectomy

and presented with recurrent hemoptysis. Figure 3 shows a coronal CT image of the same patient following bilateral VATS right middle lobectomy and lingulectomy.

Patients Have Cardiopulmonary Fitness to Tolerate a Lung Resection

In order to be considered for lung resection, patients must have adequate pulmonary reserve to tolerate a resection and must not have other uncontrolled medical conditions that put them at a prohibitive risk for a major surgical procedure. CT imaging and pulmonary function testing are needed to determine the extent of resection and the pulmonary reserve of a candidate for resection. Nutritional status and general functional status are also important considerations. The preoperative evaluation of patients is discussed in detail below.

Hemoptysis

Hemoptysis is most commonly due to bronchiectasis. Massive hemoptysis of greater than 500 cc of blood in 24 h or greater than 100 cc of blood loss an hour requires immediate intervention [8]. Prior to the 1970s, the emergency lung resection was performed in this situation but was associated with high complication rates and mortality. Modern series demonstrate a significant reduction in morbidity and mortality with bronchial artery embolization for treatment of hemoptysis [9]. In patients with known NTM disease, who present with hemoptysis, an initial evaluation should include high-resolution CT scan and flexible bronchoscopy. Initial management includes ICU admission, cardiopulmonary stabilization, and airway management including intubation if needed. Approximately 25% of patients with massive hemoptysis for all reasons will not require intervention; half typically undergo interventional radiologic embolization [10]. Bronchoscopy, in addition to localizing the bleeding source, can provide therapeutic options for some control of bleeding. In over half of cases with hemoptysis, a bleeding site is not identified at the time of bronchoscopy [11]. CT imaging in patients with NTM disease can identify the most suspect areas of involvement localizing a potential source of bleeding and helping to guide a bronchoscopy examination.

Bronchial artery embolization is the treatment of choice for management of massive or recurrent hemoptysis [12–14]. Success rates for treatment of massive hemoptysis range from 75% to 99% and in studies with recurrent hemoptysis are in the range of 10–55% [15]. Urgent surgical resection is reserved for patients who have ongoing massive bleeding not controlled with other measures and should be considered as a last resort. Emergency lung resection for massive hemoptysis carries a mortality risk of 20% and morbidity as high as 50% [16]. Elective surgical resection for the primary treatment of recurrent hemoptysis associated with NTM as the primary indication for surgical intervention has been reported for about 10% of patients [17].

Large Cavitary Disease

Cavitary lesions in NTM lung disease patents can remain as reservoirs for large numbers of infectious organisms. In patients with large fibrocavitary lesions, lung destruction may progress more rapidly than in patients with primarily bronchiectatic nodular disease [5]. Antimicrobial agents do not effectively penetrate these areas and thus can continue to spread and cause further lung destruction over time [18]. Further destruction of lung can lead to continuing decline in pulmonary function until a patient no longer has adequate pulmonary reserve to tolerate resection. Resection of large cavities can reduce the mycobacterial burden and help to prevent further lung destruction. Additionally, large cavities can provide refuge for secondary bacterial or fungal infections, which can also lead to hemoptysis. Figure 4 shows CT images of a patient with cavitary disease due to MAC with a secondary mycetoma.

Severe Bronchiectasis

Severe bronchiectasis can be associated with chronic cough, daily sputum production, and hemoptysis. Additionally, patients are susceptible to recurrent secondary bacterial infections, which over time may include multidrug-resistant strains. The CT scanning is more useful that plain chest X-ray in identifying and quantifying the degree of bronchiectasis. Patients with NTM infections, who have continued symptoms and progression of disease on CT imaging, are candidates for possible lung resection.

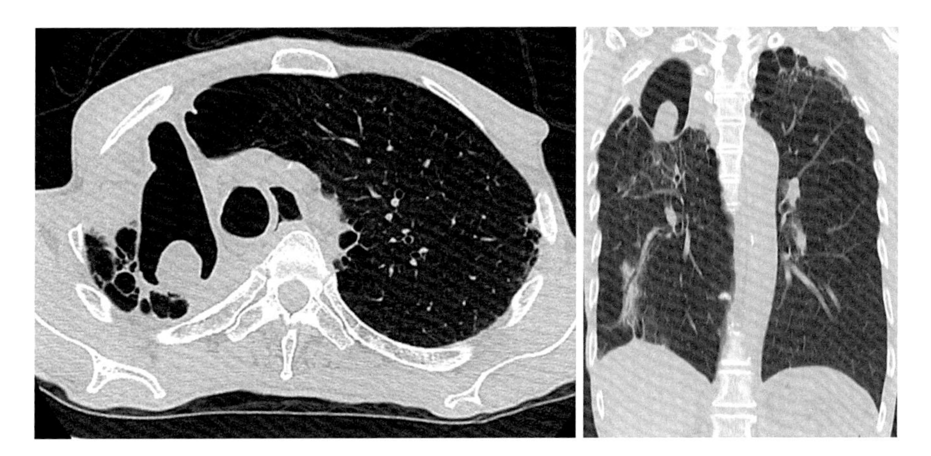

Fig. 4 CT image shows a large right upper lobe cavity with secondary development of a mycetoma

Cough and sputum production continuing over long periods can significantly impact quality of life. Lady Windermere syndrome refers to a pattern of middle lobe and lingular bronchiectasis seen with NTM infections. In a report by Yu et al. [7], a total of 134 patients underwent 172 operations, with 38 patients having staged bilateral resections. Video-assisted thoracic surgery (VATS) approach was used, and 102 middle lobectomies and 70 lingulectomies were performed. There were no operative mortalities and a complication rate of 7%. The reported cure rate with antibiotic therapy alone in similar groups ranges between 55% and 67% [5, 19, 20]. The operative group reported 84% of patients were negative for mycobacterial disease after surgical resection, and 27 patients (29%) who had operative tissue were positive for mycobacteria subsequently converted to a sputum culture-negative status [7].

Progressive Disease on Imaging

Despite appropriate antibiotic therapy, patients who have progressive disease including enlarging cavities or increasing nodular bronchiectasis should be evaluated for surgical resection. Large cavitary disease in particular tends to progress at a more rapid pace than nodular bronchiectasis. Ongoing destruction of the lung without eradication of infection can lead to worsening pulmonary function which will eventually exclude a patient from consideration of lung resection.

Recurrence of Positive Sputum Off Antibiotic Therapy

Reactivation of MAC after a period of negative sputum smear and culture has been reported in patients with nodular bronchiectasis. It has been reported that patients with NTM and bronchiectasis may be infected with multiple genetic strains [21].

When patients with MAC develop recurrent positive cultures after a period of negative sputum cultures, the infection is often due to a different genetic strain. This has led to the implication that the underlying bronchiectasis is the substrate which allows for NTM infection [22]. Lung resection in this circumstance should help to eliminate the infection and reduce the likelihood of recurrent infection, by removing the affected lung tissue.

Medication Intolerance/Non-compliance

Drug treatment regimens for NTM disease require multiple medications over extended periods. Medication side effects, intolerance, and non-compliance have been a significant factor in the treatment of patients with those infections [23]. Side effects from antibiotic regiments range from gastrointestinal upset to hepatitis and hearing loss. Weight loss and malaise can be a significant sequalae of antibiotic therapy. Continuance of antibiotic therapy in this circumstance can lead to a debilitated metabolic state, excluding the patient from possible surgical candidacy. Monitoring for side effects and toxicity is imperative for patients on undergoing long-term therapy. In patients who have difficulty maintaining a prescribed antibiotic regimen, surgical resection should be considered.

Macrolide Resistance

Two risk factors have been associated with the development of macrolide-resistant NTM disease. These include treatment of the disease with macrolide monotherapy or treatment with a macrolide and an inadequate companion medication [24]. The management of macrolide-resistant NTM requires "complex clinical decision making" and, according to 2007 ATS guidelines, should only be undertaken with expert consultation [5]. In regard to treatment of patients with macrolide-resistant MAC lung disease, the overall outcome is poor, but the treatment strategy associated with the most success included both the use of a parenteral aminoglycoside (streptomycin or amikacin) and surgical resection of disease for patients with either cavitary or nodular/bronchiectatic disease [24].

Preoperative Evaluation of Patients

When patients are referred for surgical resection, they are best considered for a procedure in the context of a multidisciplinary evaluation by a team experienced with treatment of NTM disease [5]. At a minimum, patients should have complete culture data with species identification of their infection, pulmonary function testing with FEV1 and DLCO measurements, and recent CT imaging of the chest. History and physical examination should focus on determining if patients are fit for surgery with focus on pulmonary, cardiac, nutritional, and overall physical conditions. Other medical conditions should be under good control. An initial surgical evaluation may identify areas of potential risk that require further testing or even intervention prior to undergoing lung resection.

All patients considered for resection should have undergone at least an initial regimen of multidrug antibiotic therapy prior to consideration of surgical intervention. The duration of antibiotic therapy prior to surgery is dependent on the particular organism identified, response to antibiotics, extent of disease, and the patient's fitness to undergo a resection [25].

Pulmonary Function Testing

Most recommendations on preoperative assessment of patient's fitness for lung resection are developed from literature focusing on patients with malignancies. While such recommendations do not apply in every circumstance, patients with NTM referred for possible lung resection are evaluated preoperatively utilizing similar testing methods. The goal of testing remains the same, to determine if patients who are considered for resection of NTM disease have adequate cardiopulmonary fitness to safely endure a lung resection. As recommended by the American College of Chest Physicians (ACCO) [26] and British Thoracic Society (BTS) [27], the initial step in evaluating patients for potential lung resection candidates is to complete pulmonary function testing that incudes spirometry and measurement of DLCO. Patients with an FEV1 and DLCO>80% can usually tolerate a pneumonectomy. Patients who have an FEV1 > 1.5 liters can usually tolerate a lobectomy [28]. Patients who do fall in these categories may still be candidates for resection; however, further risk stratification is usually recommended. Although still obtained, preoperative pulmonary functions may not be as predictive of outcomes for patients who undergo minimally invasive procedures as opposed to thoracotomy [29].

Due to the frequency of underlying lung disease in many patients with NTM disease, many patients considered for lung resection will have significantly impaired pulmonary function and will not fall into a low-risk group. Patients who undergo lung resection for infectious reasons, however, tend to lose mostly destroyed and nonfunctioning lung. Thus, the loss of cavitated and bronchiectatic lung may not result in as much a detriment to the patient's overall lung function as one would calculate using the standard anatomic method (post-op predicted FEV1 = pre-op FEV1 *(1- # of segments resected /19). A similar phenomenon is seen in patients with severe bullous emphysema who undergo lung resection for treatment of malignancy. The concomitant lung volume reduction that occurs with removal of a severely emphysematous lobe can improve post-op lung function by decompressing the remaining functional lung [30].

When calculated post-op FEV1 and DLCO are both <30%; patients are clearly at increased risk of morbidity and mortality with lung resection for malignancy [31]. In patients with NTM disease, it is in that circumstance, with severe impairment, that eradication of infection becomes paramount to preserve the remaining lung function patients have. Every effort should be made to pursue a course that will lead to cure. Radionuclide ventilation and perfusion scanning can be helpful in more objectively predicting post-op FEV1 and DLCO [32]. In certain circumstances the lung slated for resection will contribute marginally to overall pulmonary function.

To illustrate this scenario, Fig. 5 shows the CT images and VQ scan of a patient with a large left upper lobe cavity due to *Mycobacterium avium* disease. FEV1 and DLCO were 28% and 32%, respectively. A VQ scan however, revealed only minimal perfusion to the left lung (Fig. 6). Since the contribution from the left lung to overall lung function was so little, this patient underwent pneumonectomy without complication and achieved negative sputum cultures.

Cardiopulmonary exercise stress testing can also stratify operative risk for patients who based on pulmonary function are classified as high risk. Patients with a maximal oxygen consumption (VO2 max) less than 1 liter/minute are at prohibitively high risk and should not undergo resection [33]. Expressing VO2 max in terms of body mass is useful in evaluation of patients who may have had significant weight loss as an effect of long-standing NTM disease. Patients who demonstrate a VO2 max <10 mL/kg per min are at very high risk for morbidity and mortality and also should not be considered for lung resection [34]. Although patients with poor FEV1 and DLCO and reduced VO2 max are at high risk, their pulmonary function may improve with optimal medical management of their underlying lung disease and a formal pulmonary rehabilitation program [35]. Patients who are still smoking must stop and then commit to completing a formal pulmonary rehabilitation program. Following completion of the program, repeat testing can be completed to see if the patient has improved enough to be reconsidered for resection.

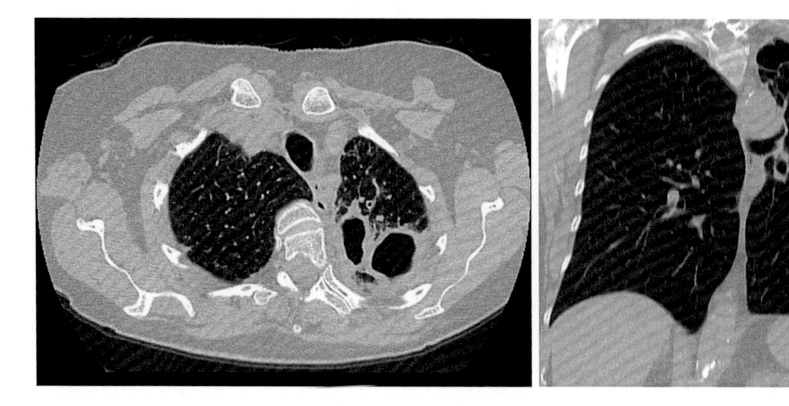

Fig. 5 The CT images show large cavity of the left upper lobe, with an apparently spared left lower lobe

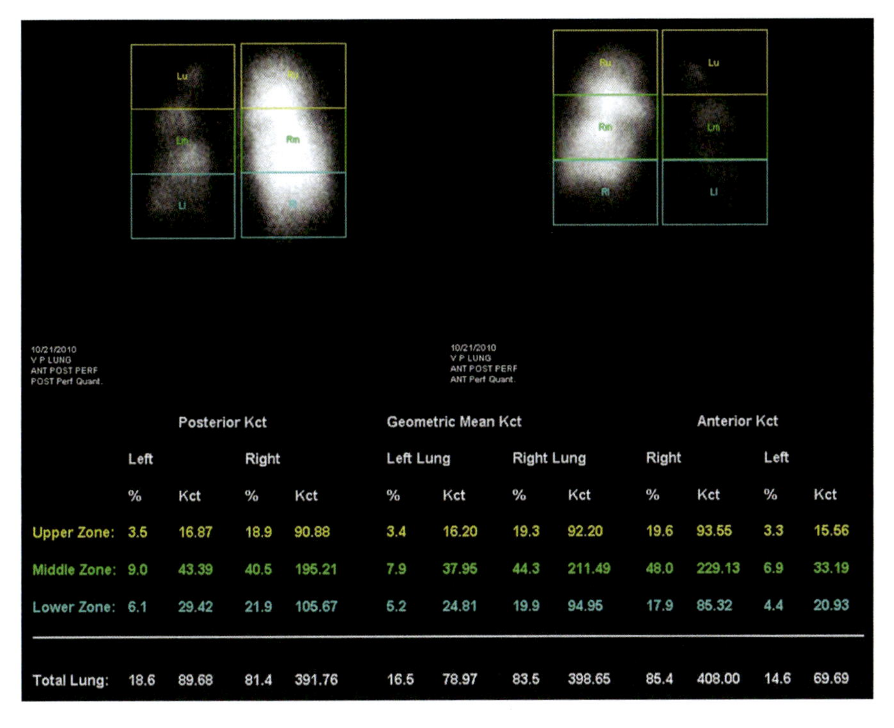

Fig. 6 Pulmonary function testing showed an FEV1 of 9 L (32% predicted) and a DLCO of 35% perfusion images from a quantitative VQ scan show that the left lung accounts for less than 15% of perfusion. The patient underwent a left pneumonectomy and recovered without incident

Cardiac Evaluation

Cardiac evaluation prior to lung surgery follows guidelines established and published by the American Heart Association for patients undergoing surgery for other non-cardiac procedures [36]. Initial evaluation starts with a complete history and physical and includes questions specifically about angina, prior cardiac events, stroke, history of valve disease, arrhythmias, smoking, peripheral vascular disease, cerebrovascular disease, stroke, hypertension, and dyslipidemia. Patients also undergo EKG and a determination of cardiac functional status. For patients with low risk, no further evaluation is needed. In patients with high risk, additional cardiac testing including echocardiography, stress testing, formal cardiology evaluation, or invasive cardiac procedures may be indicated.

Metabolic Status

Second to pulmonary function, overall metabolic status is an important consideration in patients undergoing lung resection. Chronic malaise and weight loss may result from either NTM infection or as a side effect of antibiotic therapy. Weight loss due to malnutrition can lead to increased complication rates including infection and poor wound healing. A history and physical should include questions about weight loss, diet, and a calculation of BMI. Patients should also undergo routine laboratory testing including a serum albumin. In elderly patients undergoing surgical procedure, a negative catabolic state indicated by a serum albumin of less than 2.2 g/dl has been shown predictor of poor outcomes [37]. Patients who are identified to be malnourished should be seen by a dietitian who can help define a program of nutritional supplementation to improve the patients' overall nutritional status. In some case of severe malnutrition, feeding tube placement and enteral supplementation may be necessary prior to undergoing an operation to reverse severe malnutrition. Unlike thoracic surgical patients who undergo lung resection for malignancies, most patients with NTM disease do have ample time to optimize their metabolic state prior to surgery.

Perioperative Antibiotic Coverage

All patients undergoing surgical interventions for NTM disease must also be on optimized antibiotic coverage pre- and postoperatively to minimize the risk of postoperative complications related to poor wound healing. Ideally, patients should have sputum that is acid-fast bacilli (AFB) culture negative prior to surgery, but unfortunately, many patients are deemed surgical candidates due to the inability to accomplish sputum culture conversion with antimicrobials alone. Choosing adequate antimicrobial therapy is further impeded by the general drug resistance of NTM pathogens. Regardless, surgical candidates should be on the best antimicrobial regimen possible for as long as possible, typically at least 3 months, prior to a surgical procedure. Parenteral agents such as amikacin should be included in the antimicrobial regimen during hospitalization in the immediate perioperative period. Appropriate antimicrobials should be continued postoperatively until the patient has met the microbiologic criterion for treatment success, 12 months of sputum culture negativity while on therapy.

Surgical Techniques

The majority of patients who require surgical intervention for NTM disease do so as an adjunct to antibiotic therapy to increase the likelihood of cure. A minority of patients undergo thoracic surgical procedures to treat complications of NTM infections which can include bleeding, pneumothorax, or empyema. Indications for treatments of these complications are considered on a case by case basis. In general, major lung resection is best not performed in an emergency setting, especially when a more conservative option is available such as chest tube drainage or an interventional radiologic procedure. Once the acute issue is addressed, then patients can be evaluated for a definitive procedure.

In planning an operation for a patient with NTM disease, there are three key elements to consider including the virulence of the infection, the pattern of disease present, and the patient's ability to tolerate an operation. An understanding of these elements will help guide decisions on timing of an operation, approach, and extent of resection.

Of the hundreds of NTM species, only a few account for lung disease that may need to be considered for surgical resection. These can be divided into fast- and slow-growing categories. Rapid-growing mycobacteria include *M. abscessus*, *M. fortuitum*, and *M. chelonae*. The slow-growing organisms include MAC (*M. avium* and *M. intracellulare*, *M. Kansasii*, *M. xenopi*, and *M. simiae*) [38]. *Mycobacterium avium* complex is the most commonly found respiratory pathogen. *Mycobacterium abscessus* can be very difficult to eradicate with antibiotics; therefore, surgical resection is typically considered earlier than for other organisms after an initial antibiotic course [5]. Similarly, with macrolide-resistant organisms, earlier surgical resection is favored.

The pattern of disease is also important in deciding the timing of surgery and the specific operation to be planned. Fibrocavitary disease tends to progress rapidly, and the large cavities serve as a reservoir of mycobacteria into which antimicrobial penetration is limited. Patients with large residual cavities after a period of negative sputum samples may tend to become positive again, due to survival of mycobacteria in those cavities. Nodular bronchiectasis pattern may be more diffuse and often involves both the middle lobe and lingula. In patients with bilateral disease, where staged resection is necessary, reassessment of fitness with repeat pulmonary function testing needs to be considered prior to a second operation.

In both disease patterns, the volume of the lung to be resected is determined by the extent of disease. The goal of surgical resection is to remove all gross areas of disease. Typically, this is accomplished with anatomic lung resection such as lobectomy or segmentectomy; however, smaller cavities may sometimes be resectable with wedge excision. All three, lobectomy, segmentectomy, and wedge resection, may be utilized in a single case to remove all areas of disease. Historically, these resections were accomplished through a thoracotomy incision, but application of minimally invasive VATS has demonstrated excellent results [39].

While the surgical approaches utilized today in the treatments of NTM have advanced to include minimally invasive procedures such as VATS, certain principles of treating patients with NTM have not changed. The goal of surgical approaches is gross resection of diseased lung while preserving as much functional lung as possible. Gross resection of any large cavities is necessary, while small areas of nodular disease can be allowed to remain if removing them requires loss of significant "normal" lung.

Thoracotomy

In planning an approach for resection of NTM disease, the preoperative CT scan is invaluable. In patients with large cavities that have destroyed much of the upper lobe and appear to extend to the chest wall, typically a thoracotomy approach is warranted. When patients have had a prior thoracotomy, the previous incision is utilized. If they have no history of prior chest surgery, then a longitudinal muscle-sparing incision is used just anterior to the border of the latissimus dorsi muscle. The latissimus is retracted posteriorly, and the serratus anterior is retracted anteriorly to expose the ribs. This muscle-sparing incision allows for the harvest of full latissimus if muscle transposition is needed. Once exposed, the chest is then entered above the fifth rib if possible. With dense fibrous adhesions, resection of a rib may be required to gain entry. A complete pneumonolysis is then performed by freeing the lung within the pleural space. The use of extrapleural dissection is avoided if possible due to the increased bleeding it causes. Electrocautery and blunt and sharp dissection are used, with meticulous technique to avoid tears in the visceral pleura. If an extrapleural approach is taken, often it is safer to enter an apical cavity to free the lung from the apex of the chest cavity and then with better visualization complete the removal of the attached cavity from the apex of the chest in the area of the subclavian artery. On occasion a second thoracotomy incision is required to completely free the base of the lung. Also, a VATS scope can be used to help visualize the base or apex of the lung when difficult to see through the thoracotomy incision.

Once the lung is free in the chest cavity, a complete assessment is made with careful palpation to identify all areas of disease. Typically, a lobectomy is performed by dividing the hilar structures with an endo-GIA stapler. Dissection in the fissure is avoided if possible, and then it is divided last with serial firings of the stapler. This technique can help to limit air leak from the remaining lung. In some cases, dense reaction in the hilum due to granulomatous nodal inflammation makes initial hilar dissection treacherous, and dissection in the fissure to identify the pulmonary artery is necessary. On rare occasions, dense reaction in the hilum will make an anatomic dissection impossible, with vascular structures being unrecognizable. In that circumstance, a combination of clamping, electrocautery, and oversewing along the clamp can be used to complete a nonanatomic resection of diseased tissue.

Once the main area of disease is resected, the lung is carefully examined, and any additional areas of disease are resected with segmentectomy or wedge resection. Tissues are cultured prior to being sent for pathologic review.

The chest is then irrigated with warm sterile water, and the lung is ventilated, while a careful inspection for air leaks is performed. Special attention is focused on bronchial stumps and other staple lines to ensure that they are pneumostatic. Any air leak from the bronchus is repaired with interrupted sutures. Air leaks from the lung parenchyma are also oversewn. An assessment is then made of any significant residual pleural space. If there is a large space not filled by the remaining lung, then a muscle flap can reinforce the bronchial closure and fill thoracic cavity. The latissimus dorsi can be used. It is usually a large muscle and provides well-vascularized bolstering of the bronchial stump while filling the apical chest cavity. It is freed subcutaneously, and then its inferior and posterior attachments are divided. With careful attention to its axial orientation, the muscle is transposed into the chest through a resected 5 cm portion of the second rib. The muscle is sutured to the bronchial stump. Figure 7 shows a patient with a large left upper lobe cavity. The patient underwent left upper lobectomy, and the remaining left lower lobe did not fill his

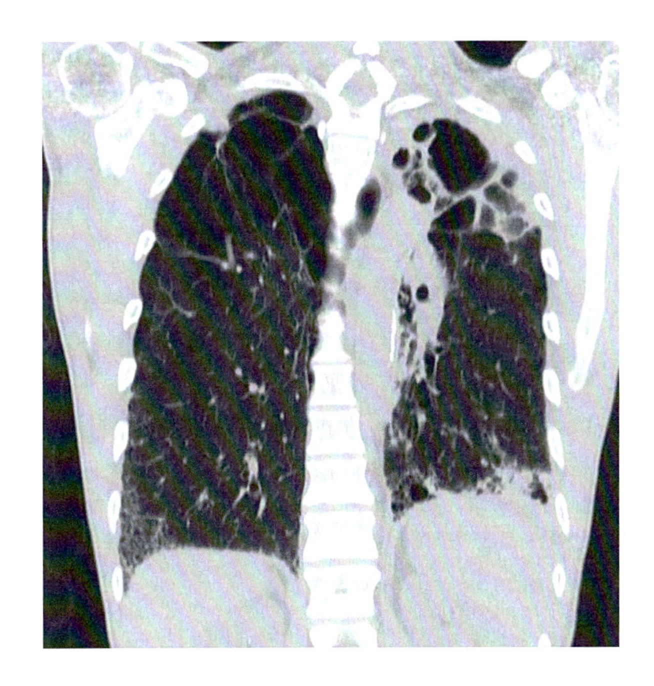

Fig. 7 CT image showing large left upper lobe cavity, due to NTM

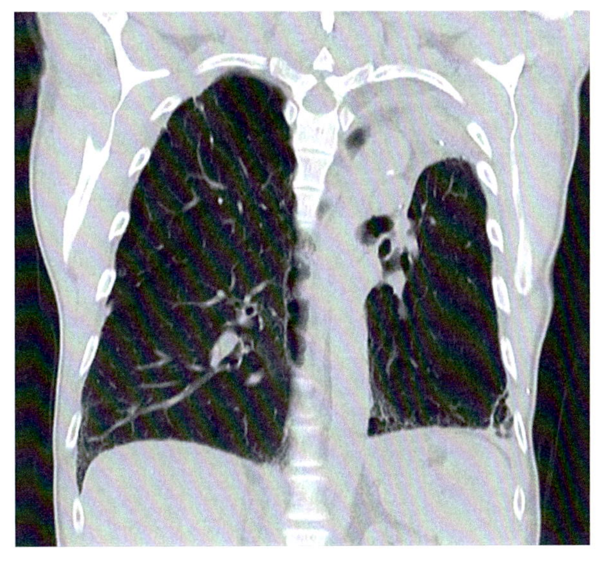

Fig. 8 Post-op coronal CT image following left upper lobectomy and latissimus dorsi muscle flap transposition. The remaining lung is well expanded and the apical space is filled by muscle flap

left chest cavity well at the time of surgery. A latissimus dorsi muscle flap was transposed into the left chest. Figure 8 shows a post-op coronal CT image demonstrating good expansion of the remaining lung and filling of the apical chest cavity space with muscle.

If the lung adequately fills the pleural space, following resection, then a large muscle transposition is not necessary. A pleural flap, pericardial fat pad, or intercostal muscle flap can be used to reinforce the bronchial closure. Two chest tubes are placed, one anterior and one posterior to the lung. For pain control a soaker catheter is tunneled along the rib margins and connected to a pain pump device that runs a continuous infusion of local anesthetic (On-Q pain relief system, Halyard, Irvine, CA). Alternatively, a thoracic epidural may be placed prior to the onset of the procedure.

VATS Resection

For patients with a cavity centrally located (not in contact with the chest wall) or with middle lobe or lingular bronchiectasis, a VATS approach is typically planned. Two 10 mm ports are utilized, one in the eighth interspace at about the midaxillary line and one in the sixth interspace anteriorly. A utility incision of about 4 cm is made in the fourth interspace. A soft tissue retractor is used (Alexis retractor, Applied Medical, Rancho Santa Margarita, CA) at the utility incision site, without rib spreading. A complete mobilization of the lung is performed if adhesions exist. The mediastinal pleura is opened using electrocautery. The pulmonary vein branch to the lobe to be resected is then divided using an endo-GIA stapler. Next the pulmonary artery branches are similarly divided with staplers and finally the bronchus. The remaining fissure is then divided, and the lobe is then placed in a plastic endo-bag and removed though the utility port. Water is instilled and an inspection for air leaks is undertaken. Any leaks are addressed. Pericardial fat pad, pleura, or intercostal muscle flaps can be used to cover the bronchial stump. The port incisions are all thoroughly irrigated to minimize abscess development, especially with M. abscesses infections. Figure 9 shows closed VATS incisions following lingulectomy.

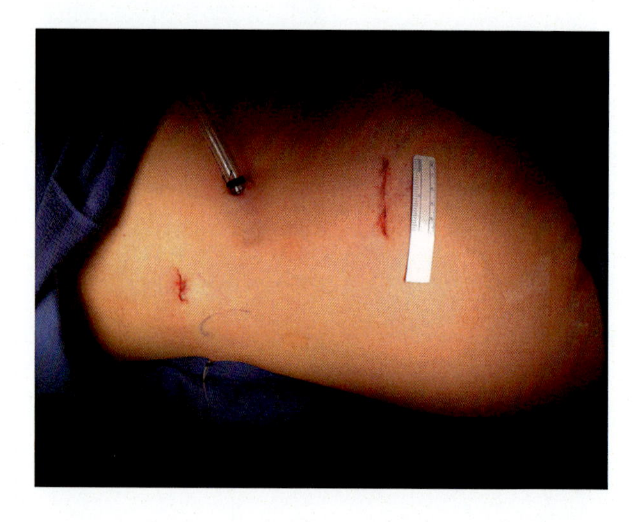

Fig. 9 Operative photo showing closed VATS incisions following lingulectomy. A chest tube is exiting the middle port site. The ruler is near the utility incision. A small pain pump catheter exits posteriorly

Bronchial Coverage

Reoperation and treatment of bronchopleural fistulas is difficult and associated with significant morbidity and mortality. To avoid that complication, routine flap coverage of the bronchial stump can be employed. There are a number of tissue flap options for coverage of bronchial stumps. A pleural flap is relatively easy and does provide vascularized tissue coverage of the bronchial stump. Due to dense pleural adhesions in many patients with NTM disease, a viable pleural flap may not be available following mobilization of the lung. The pericardial fat pad can be used when pleura is not available. Mobilization of the pericardial fat pad is accomplished by elevating the fat from the pericardium starting inferiorly. The upper portion of the pad remains intact and vascularized from branches from the internal mammary artery. It is a good choice for use in covering a middle lobe bronchus, although it can be used to cover any airway closure. Pedicle intercostal muscle flap also provides good vascularized tissue coverage of bronchial closures. The flap can be harvested during the time of a thoracotomy or via a minimally invasive technique [40]. It remains an option in patients who have had a previous thoracotomy involving division of the latissimus dorsi muscle. Latissimus dorsi muscle flap provides well-vascularized muscle coverage as well as helps to fill the pleural space. The muscle flap is created by freeing the latissimus from the skin and then detaching it from the iliac crest inferiorly and spine posteriorly. Creation of a large skin flap in harvesting the muscle can result in the development of post-op seroma. A subcutaneous drain is left in the bed of the latissimus when transposed until drainage remains low.

Thoracoplasty

Thoracoplasty remains as an option for dealing with complicated space issues. The Schede thoracoplasty was originally described as a collapse mechanism in which ribs and intercostal muscles were resected allowing collapse of pectoral muscles and skin [41]. Alexander described extrapleural thoracoplasty in which the ribs were resected, but periosteum, intercostals, and pleura were preserved [42]. While effective in eradicating infected spaces and airway fistulas, thoracoplasty can be disfiguring and over the long-term lead to scoliosis and chronic pain. Today, thoracoplasty is not usually employed as a primary maneuver to treat patients with NTM, but is generally reserved for management of a complication such as empyema or bronchopleural fistula.

Post-op Management

Postoperative management of patients who undergo lung resection for NTM disease is similar to patients who undergo lung resection for other indications. Typically, patients are admitted to the ICU post-op and are monitored closely for adequate respiration and bleeding. Adequate pain control allows early mobilization and participation in physical therapy. Patients are educated on the use of an incentive spirometer and receive regular respiratory therapy evaluation with encouragement to cough and expectorate. Oral mucolytics, nebulized hypertonic saline or mucomyst, and chest physical therapy are used regularly. Chest tubes are kept to −20 cm suction until air leaks resolve and then are placed to water seal. Chest tubes are removed when there are no air leaks, and daily drainage is less than 250 cc per 24 h.

Air Leaks

One of the most common complications with lung resection in NTM patients has been air leak. This often occurs due to the scarring and fibrosis caused by chronic infection. Dense adhesions may be present throughout the pleural space to the point where the entire pleural space is obliterated. Extrapleural dissection can result in significant bleeding, but intrapleural dissection can result in visceral pleural tears which then leak air. Meticulous intra-op technique to avoid tears is important in avoiding prolonged air leak, but a complete pneumonolysis is essential to allow full expansion of the remaining lung. A combination of electrocautery and sharp and blunt dissection is utilized to free the lung. Avoidance of excessive dissection in incomplete fissures also helps to avoid air leaks. Hilar structures are first identified and divided using stapling devices, and then the fissure can be divided with a stapler. Careful attempts at identification and intervention to stop air leaks at the time of surgery is best way to avoid the reoperation.

For small air leaks which do not resolve after 5 days, chest tubes can be placed to Heimlich valve. Serial chest X-rays are done to determine if the patient's lung will remain expanded without continuous suction. If so, the patient may be discharged with a chest tube in place and reevaluated for chest tube removal as an outpatient. Most small air leaks with no large residual space can be managed conservatively. When patients have a large persistent air leak beyond 14 days, develop a large pneumothorax off of suction, or have a large residual space, re-exploration is usually considered. An inspection is performed, and an attempt at direct closure of the leak, additional tissue flap bolstering, or thoracoplasty may be required.

Endobronchial valves (Spiration, Olympus Respiratory, USA) have received an FDA Investigational Device Exemption for use with air leaks after lobectomy or segmentectomy. Figure 10 shows the IBV valves, and Fig. 11 shows a bronchoscopic view following deployment. In patients with persistent air leak, bronchoscopy is performed, and a systematic balloon occlusion of the airways is performed while carefully watching the water seal chamber for a change in air leak, as evidenced by decreased bubbling. Once the airway or airways contributing to the air leak have been identified, IBV valves are deployed. The patient's chest tube can be removed once complete resolution of air leak is confirmed. The valves are left in position for 6 weeks, and then the patient returns for outpatient bronchoscopy and retrieval of the valves.

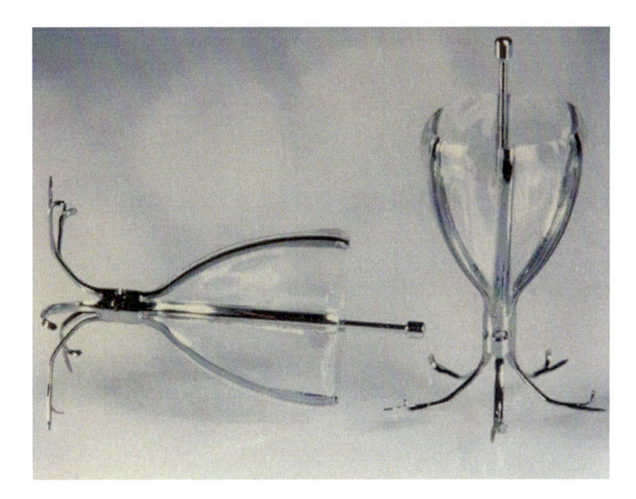

Fig. 10 IBV valves

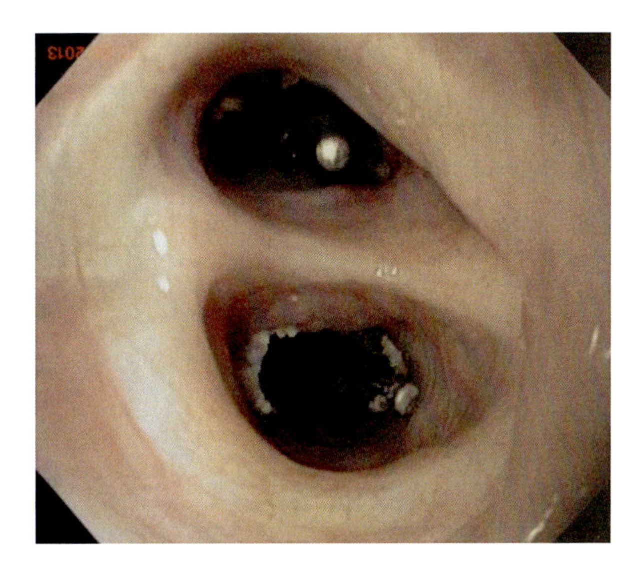

Fig. 11 Two IVB valves following deployment

Outcomes with Pulmonary Resection for NTM Disease

In general, there have been good results with surgical series for NTM disease. Table 3 summarizes some of the largest modern series. The vast majority of patients included in these reports had MAC infections. A smaller number patients had *M. abscessus* infections. Mortality ranges from 0% to 23% with sputum conversion being reported at 54–100%. The morbidity associated with lung resection ranges from 8% to 50%. Centers with more experience tend to have the lowest morbidity and mortality rates. Patients with nodular bronchiectasis more frequently undergo VATS and tend to have fewer complications. Wound complications and infections tend to be higher in patients with *M. abscessus* infections. Overall survival rates are good, but in general morbidity following lung resection for NTM disease is high

Table 3 Surgical outcomes for NTM surgery

Author	Patients	Resections	Organism	Morbidity %	Mortality %	Sputum conversion %
Pomerantz [43]	13	15	MAC (12)	8	0	100
Ono [44]	8	8	MAC (8)	0	0	100
Shiraishi [45]	33	33	MAC (33)	15	0	94
Nelson [25]	28	28	MAC (28)	29	7	79
Shiraishi [46]	21	21	MAC (21)	29	0	100
Shiraishi [47]	11	11	MAC (10) MA (1)	36	18	100
Sherwood [48]	26	26	MAC (15) MA (1)	46	23	82
Watanabe [49]	22	25	MAC (22)	5	0	100
Koh [50]	23	23	MAC (10) MA (12)	35	4	100
Mitchell [38]	236	265	MAC (189) MA (32)	12	3	NA
Van Ingen [51]	8	8	MAC (7)	50	13	88
Yu [7]	134	172	MAC (118) MA (14)	7	0	84
Jarand [52]	24	29	MA (24)	25	17	57
Mitchell [39]	171	212	MAC (147) MA (36)	9	0	NA
Shiraishi [17]	60	65	MAC (55) MA (3)	12	0	100
Kang [53]	70	74	MAC (45) MA (23)	20	1	81
Zvetina [54]	35	38	M. Kansasii (35)	37	3	100

and ranges from 5% to 50%. In comparison, a review of the STS database including 20,657 lobectomy cases from 231 centers, mostly for malignant indications, demonstrated a 1.5% mortality rate and 9.6% major complication rate [55].

Late Complications

Most bronchopleural fistulas are identified early but may present late, even years following the original operation. Patients may present with a range of symptoms ranging from mild to severe sepsis. An enlarging air-filled cavity or air- and fluid-filled cavity is a common radiologic finding. Patients may admit to increased cough, sputum production, or even a bubbling sensation in their chest. CT scanning and bronchoscopy are done to assess the location of the fistula. Ventilation scan can also help determine if a gas-filled space is truly in continuity with the airway. Reoperation is required to drain any infection and close the fistula. Typically, direct closure is quite difficult, and tissue flap buttressing to the area of fistula is necessary.

Wound Infections

Subcutaneous infections due to NTM can occur following surgical resection. The most common occur with *M. abscessus*. Careful irrigation of incisions is important at the time of surgery, and use of tissue protectors and specimen bags for minimally invasive procedures limits contamination. When patients present with a draining wound, incision and drainage with culturing is the initial step. Bacterial infections can usually be treated with drainage, packing, and antibiotics based on culture results. Mycobacterial infections may require wide local excisional debridement and reclosure.

Extrapulmonary Surgical Infections

NTM infections in children may present as painless head and neck lymphadenopathy which involves the skin. The most common NTM pathogen is MAC. Proposed treatments have included antibiotics alone, aspiration, incision and debridement, or surgical excision. The best results in regard to healing favor surgical excision [56]. Excision can be associated with nerve injury, and as an alternative to excision, incision and drainage with a course of antibiotics may be an option when the risk of nerve injury is high.

Mycobacterium marinum is an endemic fish pathogen which is present in aquatic environments. Human infections can occur with skin injuries that are exposed to contaminated water, fish, or shellfish. The American Thoracic Society (ATS) and Infectious Disease Society of America (IDSA) guidelines recommend two active agents (clarithromycin/azithromycin, ethambutol, or rifampin) for 3–4 months with adjunctive surgical debridement for invasive infections [5]. In a recent series and review, only 44% of superficial skin infections required surgical excision, while 95% of invasive infections (tenosynovitis or septic arthritis) required surgery [57].

Infection of cardiac surgical wounds and diffuse systemic infections have also been reported in cardiac surgical patients who were exposed to NTM from the heater/cooler units used with cardiopulmonary bypass. *Mycobacterium chimaera*, a member of the *M. avium* complex, has been cultured from municipal water supplies. That water when used in the reservoir of the heater/cooler equipment can become a source of contamination for patients undergoing cardiac surgery. Prosthetic valve endocarditis, local wound infections, and systemic disease have occurred as a result of exposure. Infections were fatal in four of ten reported cases [58].

NTM infections have also been reported following plastic surgical procedures [59], eye surgery [60], vascular surgery [61], and other surgical procedures. There is variability in the management based on the specific infection and location. When surgical implants are present, they may require surgical removal in addition to antibiotic therapy to resolve the infection [62]. In non-healing and recurrent wounds, a high index of suspicion is needed as standard bacterial wound cultures will not identify NTM organisms. AFB stains and cultures must be specifically obtained. Accurate diagnosis is key to guiding appropriate antibiotic therapy with surgical intervention.

Conclusion

Surgical resection for NTM infections has a reputation of being an option of last resort. This may be partly due to the historical treatment of mycobacterial tuberculosis with collapse therapy and disfigurement that patients had to endure with thoracoplasty. In centers with experienced multidisciplinary teams, modern thoracic surgical techniques can be utilized in patients with NTM disease to achieve a cure and resolve symptoms. Minimally invasive video-assisted techniques have demonstrated excellent outcomes, with fewer complications and faster recovery. Surgical resection is an effective and important adjunct to treating selected patients with NTM infections.

References

1. Hippocrates, (trans: Potter P.) Hippocrates Volume VI. Cambridge: Harvard University Press; 1988. p. 39–43.
2. Bertolaccini L, Viti A, Di Perri G, Terzi A. Surgical treatment of pulmonary tuberculosis: the phoenix of thoracic surgery. J Thorac Dis. 2013;5(2):198–9.
3. Naef AP. The mid-century revolution in thoracic and cardiovascular surgery: part 2 prelude to 20th century cardio-thoracic surgery. Interact Cardiovasc Thorac Surg. 2003;2(4):431–49.
4. Griffith DE. Risk-benefit assessment of therapies for Mycobacterium avium complex infections. Drug Saf. 1999;21(2):137–52.
5. Griffith DE, Aksamit T, Brown-Elliott BA, Catanzaro A, Daley C, Gordin F, Holland SM, Horsburgh R, Huitt G, Iademarco MF, Iseman M, Olivier K, Ruoss S, von Reyn CF, Wallace RJ Jr, Winthrop K. ATS mycobacterial diseases subcommittee, American Thoracic Society, infectious disease Society of America. Am J Respir Crit Care Med. 2007;175(4):367–416.
6. Guideline for surgical therapy of non-tuberculous acid-fast bacterial infection of the lung. Committee on the Management of non-Tuberculous Acid-Fast Bacterial Infections of the lung, the Japanese Society for Tuberculosis. Kekkaku. 2008;83(7):527–8.
7. Yu JA, Pomerantz M, Bishop A, Weyant MJ, Mitchell JD. Lady Windermere revisited: treatment with thoracoscopic lobectomy/segmentectomy for right middle lobe and lingular bronchiectasis associated with non-tuberculous mycobacterial disease. Eur J Cardiothorac Surg. 2011;40:671–5.
8. Jean-Baptiste E. Clinical assessment and management of massive hemoptysis. Crit Care Med. 2000;28:1642–7.
9. Shigemura N, Wan IY, Yu SC, et al. Multidisciplinary management of life-threatening massive hemoptysis: a 10-year experience. Ann Thorac Surg. 2009;87:849–53.
10. Haponik EF, Fein A, Chin R. Managing life-threatening hemoptysis: has anything really changed? Chest. 2000;118(5):1431–5.

11. Yoon W, Kim JK, Kim YH, et al. Bronchial and nonbronchial systemic artery embolization for life-threatening hemoptysis: a comprehensive review. Radiographics. 2002;22:1395–409.
12. Fernando HC, Stein M, Benfield JR, et al. Role of bronchial artery embolization in the management of hemoptysis. Arch Surg. 1998;133:862–6.
13. Uflacker R, Kaemmerer A, Picon PD, et al. Bronchial artery embolization in the management of hemoptysis: technical aspects and long-term results. Radiology. 1985;157:637–44.
14. Poyanli A, Acunas B, Rozanes I, et al. Endovascular therapy in the management of moderate and massive haemoptysis. Br J Radiol. 2007;80:331–6.
15. Ittrich H, Klose H, Adam Radiologic G. Management of Haemoptysis: diagnostic and interventional bronchial arterial embolisation. Fortschr Röntgenstr. 2015;187(04):248–59.
16. Cahill BC. Ingbar DH massive hemoptysis. Assessment and management. Clin Chest Med. 1994;15(1):147.
17. Shiraishi Y, Katsuragi N, Kita H, Hyogotani A, Saito MH, Shimoda K. Adjuvant surgical treatment of nontuberculous mycobacterial lung disease. Ann Thorac Surg. 2013;96:287–91.
18. Shiraishi Y. Surgical treatment of nontuberculous mycobacterial lung disease. Gen Thorac Cardiovasc Surg. 2014;62:475.
19. Field SK, Cowie RL. Treatment of Mycobacterium avium-intracellulare complex lung disease with a macrolide, ethambutol, and clofazimine. Chest. 2003;124:1482–6.
20. Tanaka E, Kimoto T, Tsuyuguchi K, Watanabe I, Matsumoto H, Niimi A, Suzuki K, Murayama T, Amitani R, Kuze F. Effect of clarithromycin regimen for Mycobacterium avium complex pulmonary disease. Am J Respir Crit Care Med. 1999;160:866–72.
21. Wallace RJ Jr, Zhang Y, Brown BA, et al. Polyclonal Mycobacterium avium complex infections in patients with nodular bronchiectasis. Am J Respir Crit Care Med. 1998;158:1235–44.
22. Wallace RJ Jr, Zhang Y, Brown-Elliott BA, Yakrus MA, Wilson RW, Mann L, Couch L, Girard WM, Griffith DE. Repeat positive cultures in Mycobacterium intracellulare lung disease after macrolide therapy represent new infections in patients with nodular bronchiectasis. J Infect Dis. 2002;186:266–73.
23. Maliwan N, Zvetina JR. Clinical features and follow up of 302 patients with Mycobacterium kansasii pulmonary infection: a 50-year experience. Postgrad Med J. 2005;81:530–3.
24. Griffith DE, Brown-Elliott BA, Langsjoen B, Zhang Y, Pan X, Girard W, Nelson K, Caccitolo J, Alvarez J, Shepherd S, et al. Clarithromycin resistant Mycobacterium avium complex lung disease. Am J Respir Crit Care Med. 2006;174:928–34.
25. Nelson KG, Griffith DE, Brown BA, et al. Results of operation in Mycobacterium avium-intracellulare lung disease. Ann Thorac Surg. 1998;66:325–30.
26. Brunelli A, Kim AW, Berger KI, Addrizzo-Harris DJ. Physiologic evaluation of the patient with lung cancer being considered for resectional surgery: diagnosis and management of lung cancer, 3rd ed: American College of Chest Physicians evidence-based clinical practice guidelines. Chest. 2013;143(5 Suppl):e166S–90S.
27. British Thoracic Society. Society of Cardiothoracic Surgeons of great Britain and Ireland working party BTS guidelines: guidelines on the selection of patients with lung cancer for surgery. Thorax. 2001;56(2):89.
28. UMazzone. Preoperative evaluation of the lung resection candidate. Cleve Clin J Med. 2012;79(Electronic Suppl 1):eS17–22.
29. Berry MF, Villamizar-Ortiz NR, Tong BC, Burfeind WR Jr, Harpole DH, D'Amico TA, Onaitis MW. Pulmonary function tests do not predict pulmonary complications after thoracoscopic lobectomy. Ann Thorac Surg. 2010;89(4):1044–51. discussion 1051-2
30. DeRose J Jr, Argenziano M, El-Amir N, Jellen PA, Gorenstein LA, Steinglass KM, Thomashow B, Ginsburg ME. Lung reduction operation and resection of pulmonary nodules in patients with severe emphysema. Ann Thorac Surg. 1998;65(2):314–8.
31. Brunelli A, Charloux A, Bolliger CT, Rocco G, Sculier JP, Varela G, Licker M, Ferguson MK, Faivre-Finn C, Huber RM, Clini EM, Win T, De Ruysscher D, Goldman L, European Respiratory Society and European Society of Thoracic Surgeons joint task force on fitness for radical therapy. ERS/ESTS clinical guidelines on fitness for radical therapy in lung cancer patients (surgery and chemo-radiotherapy). Eur Respir J. 2009;34(1):17–41.

32. Ali MK, Mountain CF, Ewer MS, Johnston D, Haynie TP. Predicting loss of pulmonary function after pulmonary resection for bronchogenic carcinoma. Chest. 1980;77(3):337–42.
33. Eugene H, Brown SE, Light RW, et al. Maximum oxygen consumption: a physiologic guide to pulmonary resection. Surg Forum. 1982;33:260.
34. Colice GL, Shafazand S, Griffin JP, Keenan R, Bolliger CT, American College of Chest Physicians. Physiologic evaluation of the patient with lung cancer being considered for resectional surgery: ACCP evidenced-based clinical practice guidelines (2nd edition). Chest. 2007;132(3 Suppl):161S–77S.
35. Rodriguez-Larrad A, Lascurain-Aguirrebena I, Abecia-Inchaurregui LC, Seco J. Perioperative physiotherapy in patients undergoing lung cancer resection. Interact Cardiovasc Thorac Surg. 2014;19(2):269–81.
36. Fleisher LA, Fleischmann KE, Auerbach AD, Barnason SA, Beckman JA, Bozkurt B, Davila-Roman VG, Gerhard-Herman MD, Holly TA, Kane GC. 2014 ACC/AHA guideline on perioperative cardiovascular evaluation and Management of Patients Undergoing Noncardiac Surgery. J Am Coll Cardiol. 2014;6:e77–e137.
37. Stijn MF, Korkic-Halilovic I, Bakker MS, van der Ploeg T, van Leeuwen PA, Houdijk AP. Preoperative nutrition status and postoperative outcome in elderly general surgery patients: a systematic review. Parenter Enteral Nutr. 2013;37(1):37–43.
38. Mitchell JD, Bishop A, Cafaro A, Weyant MJ, Pomerantz M. Anatomic lung resection for nontuberculous mycobacterial disease. Ann Thorac Surg. 2008;85:1887–92. discussion 92–3
39. Mitchell JD, Yu JA, Bishop A, Weyant MJ, Pomerantz M. Thoracoscopic lobectomy and segmentectomy for infectious lung disease. Ann Thorac Surg. 2012;93:1033–9. discussion 9–40
40. Sagawa M, Sugita M, Takeda Y, Toga H, Sakuma T. Video-assisted bronchial stump reinforcement with an intercostal muscle flap. Ann Thorac Surg. 2004;78(6):2165–6.
41. Schede M. Die Behandlung der Empyeme. Verh Cong Innere Med Wiesbaden. 1890;9:41–141.
42. Alexander J. The collapse therapy of pulmonary tuberculosis. Springfield, IL: Charles C. Thomas; 1937.
43. Pomerantz M, Denton JR, Huitt GA, Brown JM, Powell LA, Iseman MD. Resection of the right middle lobe and lingula for mycobacterial infection. Ann Thorac Surg. 1996;62: 990–3.
44. Ono N, Satoh K, Yokomise H, Tamura K, Horikawa S, Suzuki Y, et al. Surgical management of Mycobacterium avium complex disease. Thorac Cardiovasc Surg. 1997;45:311–3.
45. Shiraishi Y, Fukushima K, Komatsu H, Kurashima A. Early pulmonary resection for localized Mycobacterium avium complex disease. Ann Thorac Surg. 1998;66:183–6.
46. Shiraishi Y, Nakajima Y, Takasuna K, Hanaoka T, Katsuragi N, Konno H. Surgery for Mycobacterium avium complex lung disease in the clarithromycin era. Eur J Cardiothorac Surg. 2002;21:314–8.
47. Shiraishi Y, Nakajima Y, Katsuragi N, Kurai M, Takahashi N. Pneumonectomy for nontuberculous mycobacterial infections. Ann Thorac Surg. 2004;78:399–403.
48. Sherwood JT, Mitchell JD, Pomerantz M. Completion pneumonectomy for chronic mycobacterial disease. J Thorac Cardiovasc Surg. 2005;129:1258–65.
49. Watanabe M, Hasegawa N, Ishizaka A, Asakura K, Izumi Y, Eguchi K, et al. Early pulmonary resection for Mycobacterium avium complex lung disease treated with macrolides and quinolones. Ann Thorac Surg. 2006;81:2026–30.
50. Koh WJ, Kim YH, Kwon OJ, Choi YS, Kim K, Shim YM, et al. Surgical treatment of pulmonary diseases due to nontuberculous mycobacteria. J Korean Med Sci. 2008;23:397–401.
51. van Ingen J, Verhagen AF, Dekhuijzen PN, van Soolingen D, Magis-Escurra C, Boeree MJ, et al. Surgical treatment of non-tuberculous mycobacterial lung disease: strike in time. Int J Tuberc Lung Dis. 2010;14:99–105.
52. Jarand J, Levin A, Zhang L, Huitt G, Mitchell JD, Daley CL. Clinical and microbiologic outcomes in patients receiving treatment for Mycobacterium abscessus pulmonary disease. Clin Infect Dis. 2011;52:565–71.
53. Kang HK, Park HY, Kim D, Jeong BH, Jeon K, Cho JH, Kim HK, Choi YS, Kim J, Koh WJ. Treatment outcomes of adjuvant resectional surgery for nontuberculous mycobacterial lung disease. BMC Infect Dis. 2015;15:76.

54. Zvetina JR, Neville WE, Maben HC, Langston HT, Correll NO. Surgical treatment of pulmonary disease due to Mycobacterium kansasii. Ann Thorac Surg. 1971;11(6):551–6.
55. Kozower BD, O'Brien SM, Kosinski AS, Magee MJ, Dokholyan R, Jacobs JP, Shahian DM, Wright CD, Fernandez FG. The Society of Thoracic Surgeons composite score for rating program performance for lobectomy for lung Cancer. Ann Thorac Surg. 2016;101(4):1379–86. discussion 1386-7
56. Mahadevan M, Neeff M, Van Der Meer G, Baguley C, Wong WK, Gruber M. Non-tuberculous mycobacterial head and neck infections in children: analysis of results and complications for various treatment modalities. Int J Pediatr Otorhinolaryngol. 2016;82:102–6.
57. Johnson MG, Stout JE. Twenty-eight cases of Mycobacterium marinum infection: retrospective case series and literature review. Infection. 2015;43(6):655–62.
58. Kohler P, Kuster SP, Bloemberg G, Schulthess B, Frank M, Tanner FC, Rossle M, Boni C, Falk V, Wilhelm MJ, Sommerstein R, Achermann Y, Ten Oever J, Debast SB, Wolfhagen MJ, Brandon Bravo Bruinsma GJ, Vos MC, Bogers A, Serr A, Beyersdorf F, Sax H, Bottger EC, Weber R, van Ingen J, Wagner D, Hasse B. Healthcare-associated prosthetic heart valve, aortic vascular graft, and disseminated Mycobacterium chimaera infections subsequent to open heart surgery. Eur Heart J. 2015;36(40):2745–53.
59. Bowles P, Miller MC, Cartwright S, Jones M. Presentation of Mycobacterium abscessus infection following rhytidectomy to a UK plastic surgery unit. BMJ Case Rep. 2014.
60. Edens C, Liebich L, Halpin AL, Moulton-Meissner H, Eitniear S, Zgodzinski E, Vasko L, Grossman D, Perz JF, Mohr MC. Mycobacterium chelonae eye infections associated with humidifier use in an outpatient LASIK clinic Ohio- 2015. MMWR Morb Mortal Wkly Rep. 2015;64(41):1177.
61. Umer I, Mocherla S, Horvath J, Arora S, Ahmed Y. Mycobacterium abscessus: a rare cause of vascular graft infection. Scand J Infect Dis. 2014;46(11):813–6.
62. Thomas M, D'Silva JA, Borole AJ, Chilgar RM. Periprosthetic atypical mycobacterial infection in breast implants: a new kid on the block! J Plast Reconstr Aesthet Surg. 2013;66(1):e16–9.

Nontuberculous Mycobacterial Disease in Pediatric Populations

Andrea T. Cruz and Jeffrey R. Starke

Introduction

Nontuberculous mycobacterial (NTM) species can cause four major disease morphologies in children. Lymphadenitis is the most common manifestation and is seen most frequently in young (preschool-aged) immunocompetent children. Pulmonary NTM disease is most common in children with underlying pulmonary dysfunction, including children with cystic fibrosis. NTM skin and soft tissue infections (SSTIs) can be caused by direct inoculation of NTM species in immunocompetent children or can be seen as a manifestation of disseminated disease. Disseminated NTM disease, including bacteremia, is predominantly seen in children with underlying immunodeficiencies or children with indwelling vascular catheters. The most common species and host risk factors seen in each form of NTM disease are described in Table 1. Children with disseminated or pulmonary NTM species should be evaluated for immunocompromising conditions. Evaluation for immunocompromise may be performed in collaboration with an immunologist. Preliminary testing may include a complete blood count, testing for human immunodeficiency virus (HIV), and testing for Mendelian susceptibility to mycobacterial disease (MSMD). The latter testing can evaluate for mutations in interferon, STAT, interleukin, and NEMO genes. Children with pulmonary NTM disease should be evaluated for cystic fibrosis (CF); not all mutations resulting in CF can be detected in newborn screens.

The true burden of NTM disease in children is unclear, as it is not a reportable disease to public health authorities. Surveillance studies have demonstrated wide variation in incidence rates in industrialized countries, ranging from 0.6 per 100,000 in Australia to 77 per 100,000 in the Netherlands [1, 2]. Across studies,

A. T. Cruz (✉) · J. R. Starke
Department of Pediatrics, Baylor College of Medicine, Houston, TX, USA
e-mail: acruz@bcm.edu

D. E. Griffith (ed.), *Nontuberculous Mycobacterial Disease*, Respiratory Medicine,
https://doi.org/10.1007/978-3-319-93473-0_16

Table 1 The most common comorbid medical conditions and nontuberculous mycobacterial (NTM) species identified in children

Disease type	Comorbid conditions	Most common NTM species[a]
Lymphadenitis	Usually occurs in previously healthy young (< 5-year-old) children	*Mycobacterium avium* complex (MAC)[b] *M. kansasii* *M. simiae* *M. fortuitum* *M. haemophilum*
Skin/soft tissue	Usually occurs in previously healthy children who sustained cutaneous trauma	*M. fortuitum* *M. abscessus* *M. kansasii* *M. marinum* *M. ulcerans* *M. leprae*
Pulmonary	Cystic fibrosis Interleukin (IL)-12 mutations Polymorphisms in NRAMP1 gene Hematopoietic stem cell transplantation Interferon-gamma receptor mutations	*Mycobacterium avium* complex (MAC)[b] *M. abscessus* *M. kansasii*
Disseminated	HIV infection Chemotherapy Indwelling central venous catheter (CVC) Solid organ or hematopoietic stem cell transplantation Mendelian susceptibility to mycobacterial disease (MSMD) Interleukin (IL)-12 mutations Interferon-gamma receptor mutations STAT 1 mutations NEMO	*Mycobacterium avium* complex (MAC) most common in children without CVCs Several species have been isolated from children with CVCs (*M. scrofulaceum*, *M. haemophilum*, *M. abscessus*, *M. fortuitum*)

CVC central venous catheter, *HIV* human immunodeficiency virus, *NEMO* nuclear factor-kappa B essential modulator deficiency syndrome

[a]Many species have specific growth requirements (e.g., incubation temperature, need for additives to culture media) that may result in lower culture yield with standard cultivation techniques

[b]The most common species isolated in the United States from previously healthy children

however, the most common form of NTM described in children has been lymphadenitis. In industrialized nations, the incidence of NTM infections is far greater than the incidence of tuberculosis.

Lymphadenitis

Epidemiology

Most cases of NTM lymphadenitis occur in previously healthy children, in contrast to the other forms of NTM infection. However, children with certain immunodeficiencies (Table 1) can be predisposed to also having NTM lymphadenitis.

Table 2 Differentiation of nontuberculous mycobacteria (NTM) from tuberculous lymphadenitis

Characteristic	TB	NTM
Age	School aged or adolescent	Toddler or preschool aged
Born in high TB prevalence nation	Yes	No
Contact with adult with TB	Sometimes	Rarely
Abnormal CXR	Up to 50% of cases	Rarely
TST size >10 mm	Almost all cases	Up to 50% of cases
Positive TST in family members	Common	Rare
IGRA result	Often positive	Rarely positive[a]

Adapted in part from [3]
CXR chest radiograph, *TB* tuberculosis, *TST* tuberculin skin test
[a]Exceptions: *M. bovis*, *M. flavescens*, *M. kansasii*, *M. marinum*, and *M. szulgai*

Immunocompetent children with NTM lymphadenitis usually are younger than 5 years of age and most lack risk factors for *Mycobacterium tuberculosis* (MTB) infection [3] (Table 2). However, if an older (school-aged) child were to be diagnosed with NTM lymphadenitis, that child should be evaluated for immunodeficiency.

The most common species isolated from lymph nodes of children in the United States are *Mycobacterium avium* complex (MAC), *M. kansasii*, and *M. simiae*. In European countries, *M. haemophilum* is the most frequently isolated species. One consideration in the United States is *M. bovis*. While a member of the *M. tuberculosis* complex, *M. bovis* is usually transmitted from consumption of unpasteurized dairy products. This species is more common in Latino patients and, due to the mode of transmission, is more associated with extrapulmonary disease than other species [4].

Clinical Presentation

Preschool-aged children usually have palpable cervical lymph nodes even when they are not ill. This examination finding reflects the same immune system maturation process that results in relative tonsillar hypertrophy in children in this age group. Cervical lymph nodes peak in size in otherwise healthy children by 4–7 years of age. However, it is important for clinicians who may care predominantly for adults to be cognizant of this. In contrast to the shotty lymph nodes that can be appreciated in asymptomatic children, children with NTM lymphadenitis present with a slowly enlarging, often painless lymph node in the cervicofacial chain. Usually, a single node is involved, and the most common location is submandibular but can occur in any anatomic location in the head or neck. Preauricular involvement is particularly difficult because of the proximity to the facial nerve. Children usually lack fever or constitutional symptoms. Children may present for subspecialty (including surgical) care after the adenopathy has failed to respond to antistaphylococcal or antistreptococcal antibiotics. The overlying skin may develop violaceous discoloration (Fig. 1). Over time, the nodes increase in size, become tethered to the overlying skin, and become fluctuant. It is the latter two features that are commonly seen before the development of a fistulous sinus tract (Fig. 2). Once

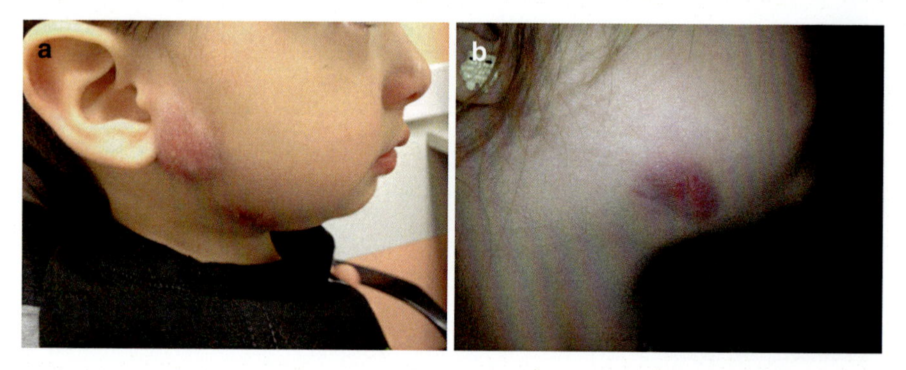

Fig. 1 (**a**) right preauricular and submandibular lymphadenitis, with formation of a fistulous sinus tract over the latter node and fluctuance over the preauricular node. (**b**) right submandibular lymphadenitis with a small draining central fistulous sinus tract

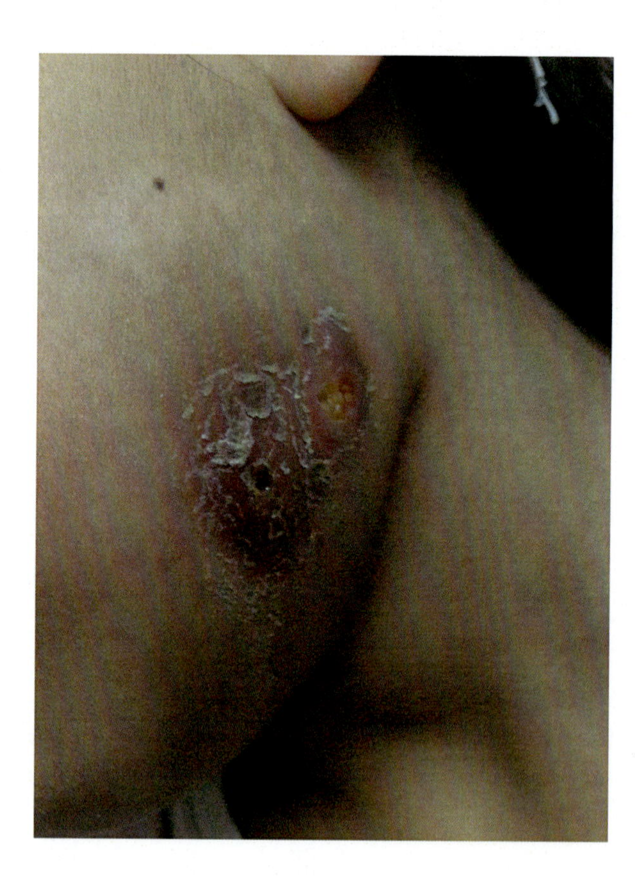

Fig. 2 Fistulous sinus tract caused by nontuberculous mycobacterial lymphadenitis. This child has had the development of several fistulous tracts as her NTM nodes eroded outward. The lesions initially were fluctuant and then became adherent to the overlying skin prior to development of the tract

a sinus tract develops, purulent material may continuously or intermittently drain from the site for weeks or months.

A special case is the child with ipsilateral axillary lymphadenitis following bacille Calmette-Guérin (BCG) administration. BCG can cause a painless bulky axillary lymphadenitis that can occasionally suppurate. This can occur up to 2 years following immunization. Lymphadenitis is estimated to occur in <1/1000 children who receive the BCG vaccine. There are reports of increased cases of lymphadenitis in some countries after a shift to a new strain of the vaccine (Danish 1331 strain) [5]. BCG-related lymphadenitis almost always resolves without treatment in immunocompetent children if suppuration has not occurred. Information on BCG vaccine coverage, the number of doses utilized, and the most common BCG strains used in different countries are available online [6].

Diagnosis

The diagnostic evaluation for a child with suspected NTM lymphadenitis should include a chest radiograph (frontal and lateral) to evaluate for intrathoracic lymphadenopathy that can be seen with tuberculosis or malignancy. As hematologic malignancies are on the differential diagnosis of subacute to chronic lymphadenitis, obtaining a complete blood count and peripheral smear is reasonable if the child has constitutional symptoms or disseminated lymphadenopathy.

Tests for TB infection, either tuberculin skin tests (TSTs) or interferon-gamma release assays (IGRAs), are often performed as part of the evaluation of lymphadenopathy. TST induration can exceed 15 mm in up to 60% of children with NTM lymphadenitis [7], including children with culture-confirmed MAC. Consequently, a large TST induration result per se does not rule out NTM infection, nor does a negative TST result exclude it. IGRAs can help differentiate TB lymphadenitis from NTM, as IGRAs are far more specific than TSTs. However, it is important for clinicians to be aware that several NTM species can yield positive IGRA results. These include *M. bovis* (disease, rather than BCG immunization), *M. flavescens*, *M. kansasii*, *M. marinum*, and *M. szulgai*.

Imaging findings overlap considerably with lymphadenopathy caused by other pathogens. Ultrasound can demonstrate decreased echogenicity, necrosis, matted nodes, and soft tissue edema adjacent to the node(s) [8]. Computed tomography (CT) findings – with intravenous contrast – include asymmetric lymphadenopathy with necrotic, ring-enhancing masses that can involve the subcutaneous fat and skin (Fig. 3). Punctate calcifications can be seen. Unlike lymphadenitis caused by pyogenic bacteria, fat stranding is unusual [9]. While imaging often does not help with the medical management of NTM lymphadenitis, it is useful if surgical intervention is being considered.

If a tissue diagnosis is sought (either by fine needle aspiration (FNA) or excisional biopsy), tissue or purulent material should be sent in a syringe or in sterile container to the laboratory. Purulent material should not be sent on swabs, as this

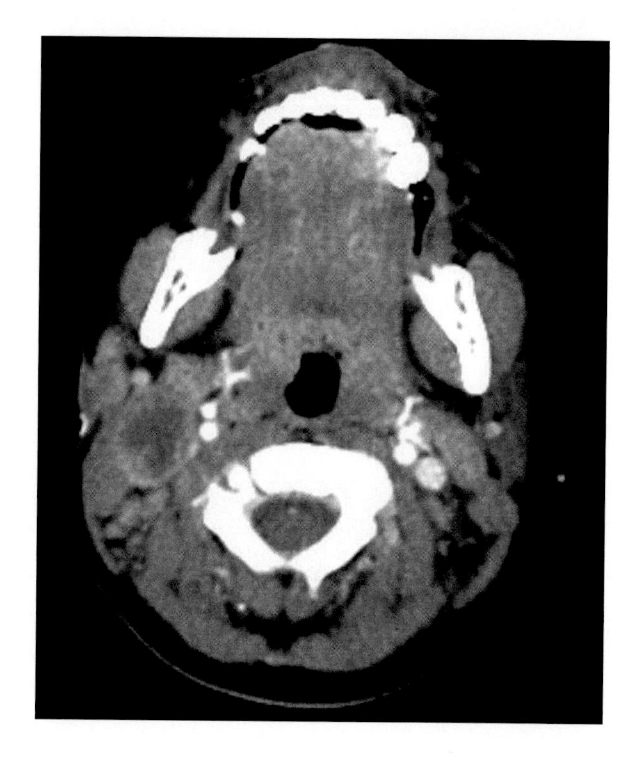

Fig. 3 Computed tomography findings in a toddler with right-sided nontuberculous mycobacterial cervical lymphadenitis. This toddler had several right-sided anterior nodes that were enlarged. The central portion of the nodes appears hypoechoic, and there is only minimal edema in the adjacent fat

will decrease the culture yield. For lymph nodes not amenable to surgical resection (e.g., due to proximity to the facial nerve or nodes that would require a radical neck dissection), microbiologic diagnosis may be feasible through an FNA. If sufficient fluid is aspirated, cultures are positive in up to one-half of cases [10] with a minimal risk of creating a draining sinus tract. Fluid or tissue should be sent for acid-fast stain and culture, histopathology (and flow cytometry if malignancy is suspected), and fungal and routine bacterial cultures. Diagnostic findings are described in Table 3.

Treatment

Treatment of NTM lymphadenitis is controversial (Tables 4 and 5). Observation alone may be a very appropriate option for children with lymph nodes in locations where resection may be difficult (e.g., preauricular nodes adjacent to the facial nerve). While combination of medical and surgical therapy is often utilized, several studies have indicated that time to symptom resolution and cosmetic outcome are no different in children who receive only antibiotics compared to those who do receive combined medical and surgical management [11–13]. Most children had resolution within 6 months, and all children who received no antibiotic or surgical therapy had

Table 3 Diagnostic yield of testing for suspected nontuberculous mycobacterial infection[a]

Site	Test	Finding(s)
Lymph node	AFB smear	Positive in a minority of children
	AFB culture	50–80% yield; can take more than 6 weeks for organisms to be speciated
	MAC PCR	Sensitivity of up to 75%; specificity is close to 100%
	Histopathology	Necrosis, granulomas, microabscesses
Pulmonary	AFB smear	Usually negative except in patients with CF
	AFB culture	Positive in a minority; may require invasive procedures
Skin/soft tissue	AFB smear	Sensitivity <25%
	AFB culture	Sensitivity 40–50%, higher in tissue biopsy than in purulent material
	Histopathology	Granulomas, necrosis; neutrophilic predominance
	M. leprae[b]	AFB smear sensitivity 20–90% by FNA ELISA sensitivity ~ 50% PCR sensitivity 70–90%
	M. ulcerans	AFB smear sensitivity up to 40% Culture sensitivity 20–60% (can take 6–8 weeks to grow) PCR sensitivity over 95% Histopathology sensitivity ~ 90% (should be sent from edge of ulcer)
Disseminated	AFB smear, tissue	Positive in less than 25% of children
	AFB culture, blood	90% sensitivity for MAC; bacteremia may be intermittent and low grade for other NTM species
	Histopathology	Poorly developed granulomas

AFB acid-fast bacilli CF cystic fibrosis, *ELISA* enzyme-linked immunosorbent assay, *FNA* fine needle aspirate, *MAC Mycobacterium avium* complex, *PCR* polymerase chain reaction
[a]Sensitivity of all tests is based on the quality and quantity of the specimen, and higher sensitivity is reported for tissue submitted as part of excisional biopsy versus fine needle aspiration; sending swabs of draining lesions is suboptimal. Some NTM have very specific growth requirements (e.g., *M. haemophilum* requires iron-impregnated media), and since laboratory standards vary by country and laboratory, some of the apparent differences in epidemiology may be related to differences in laboratory protocols
[b]Cultures are not usually attempted for *M. leprae*

resolution by 12 months [13]. However, the long duration of symptoms may make the observation-only approach not palatable for some families.

The American Academy of Pediatrics (AAP), the American Thoracic Society (ATS), and the Infectious Diseases Society of America (IDSA) recommend an excisional biopsy when feasible. Several studies indicate the superiority of surgical versus medical management of NTM lymphadenitis in terms of time to symptom resolution and cosmesis [14, 15]. Curettage of nodes may be an alternative if complete excision is not possible (e.g., proximity to the facial nerve or very bulky nodes in which excision would entail a radical neck dissection). Feasibility of surgical excision can be limited in preauricular and submandibular nodes, which course very near the facial nerve. Transient facial nerve palsies have been noted in up to 30% of

Table 4 Antibiotic dosing for nontuberculous mycobacterial disease in children

Drug Class	Antibiotic	Dosing in mg/kg/day[a] [maximum daily dose]
Macrolides	Azithromycin	10 [500 mg]
	Clarithromycin	15–30 mg/kg/day divided BID [1 g/day]
Rifamycins	Rifampin	10–20 [600 mg]
	Rifabutin	10–20 [300 mg]
Aminoglycosides	Amikacin	15 [1 g]
	Streptomycin	15 [1 g]
Fluoroquinolones	Ciprofloxacin	20–30 mg/kg/day divided BID [1.5 g/day]
	Levofloxacin	0–5 years: 7.5–10 mg/kg BID [500 mg] >5 years: 7.5–10 mg/kg [500 mg]
	Moxifloxacin	10 [400 mg]
Carbapenems	Imipenem	60–100 mg divided every 6–8 h [2–4 g]
	Meropenem	60 mg/kg/day divided TID
Other	Cefoxitin	80–160 divided every 6 h [12 g]
	Clofazimine	1–2 [50 mg]
	Dapsone	1–2 [100 mg]
	Doxycycline	1–2 mg/kg/dose BID [200 mg]
	Ethambutol	15–25 [varies by weight bracket][b]
	Isoniazid (INH)	10–15 [300 mg]
	Linezolid	< 10 years: 10 mg/kg/dose BID [600 mg] ≥ 10 years: 300 mg

[a]All doses are single daily doses unless otherwise specified; the mg/kg dose for children often exceeds that of adults due to children metabolizing medications more rapidly
[b]See Griffith et al. [17]

children in whom excisional biopsy for NTM is performed, and up to 5% may have permanent facial nerve paralysis [16]; facial nerve palsies are more common in children with submandibular or preauricular adenopathy. One important consideration is to avoid incision and drainage of suspected NTM nodes, as this can result in creation of a draining sinus tract. The surgical management of NTM disease is discussed in more detail in chapter "Non-tuberculous Mycobacteria in Cystic Fibrosis".

When complete surgical resection is not feasible and observation is not an attractive option for the family, antibiotic management can be initiated. It is important that, similar to TB, multidrug therapy begun to decrease the probability of selecting for drug-resistant organisms. As the most common isolate in the United States is MAC, empiric therapy can consist of a macrolide combined with either rifampin or ethambutol [17]. Early studies of NTM treatment included clarithromycin (15–30 mg/kg/day in two divided doses; maximum dose, 500 mg/dose). However, the twice-daily treatment schedule and the poor taste of clarithromycin make azithromycin (10 mg/kg/day as a single daily dose; maximum dose, 500 mg/dose) a potentially better option. In addition, azithromycin is less likely to alter the metabolism with other antibiotics used to treat NTM species. The second drug is usually either rifampin (10–20 mg/kg/day as a single daily dose; maximum dose, 600 mg) or ethambutol (15–25 mg/kg/day as a single daily dose; maximum dose, 1600 mg). It is

Table 5 Suggested empiric and definitive treatment regimens for children with nontuberculous mycobacterial disease caused by the most common nontuberculous mycobacterial species seen in the United States

Organism	Site	Most common regimen	Usual duration (months)
MAC	Lymphadenitis	Macrolide + (RIF and/or EMB)	Minimum 3–4
	Pulmonary	Macrolide + EMB + RIF	6–12 months
	SSTI	Macrolide + EMB + RIF	6–12
	Disseminated	Macrolide + EMB + rifabutin	6–12 months after immune restoration
M. abscessus	Pulmonary	Usually 3 drugs based on drug susceptibilities[a]	12 months [non-CF related]; antibiotics often cycled for several months in patients with CF
	SSTI	Macrolide[a] + (amikacin, cefoxitin, or imipenem)	4–6
M. fortuitum	Pulmonary	At least 2 drugs to which isolate has in vitro susceptibility[b]	12 months after sputum conversion
	SSTI	At least 2 drugs to which isolate has in vitro susceptibility[b]	4–6
M. kansasii	Lymphadenitis	INH + RIF + EMB	6–12
	Pulmonary and disseminated	INH + RIF + EMB	12 months after sputum conversion
	SSTI	INH + RIF + EMB	6–12
M. marinum	SSTI	Macrolide + (EMB or RIF); osteomyelitis should be treated with all 3 agents	1–2 months after symptom resolution

EMB ethambutol, *INH* isoniazid, *MAC Mycobacterium avium* complex, *RIF* rifampin, *SSTI* skin/soft tissue infection, CF cystic fibrosis

[a]A macrolide should always be used if the *M. abscessus* isolate is susceptible. However, two of the three subspecies (*M. abscessus abscessus* and *M. a. bolletii*) have inducible macrolide resistance. Other commonly used drugs, depending on drug susceptibility testing, include amikacin, imipenem, cefoxitin, tigecycline, linezolid, and minocycline

[b]*M. fortuitum* isolates usually are susceptible to amikacin, carbapenems, fluoroquinolones, macrolides, and sulphonamides

important to find a dose that is feasible given the available formulations of medications (e.g., ethambutol, in the United States, is available in 100 mg and 400 mg tablets) [18]. In addition, providers need to realize that ethambutol can be very difficult to find at commercial pharmacies. For this reason, rifampin may be more commonly utilized. There have been no randomized controlled trials of different antibiotic regimens for NTM; all treatment outcome data have been anecdotal or from non-controlled case series. Therapy for other NTM species is described in Chapters 11, 12, and 13.

At times, it may not be clear if a child has NTM or tuberculous lymphadenitis. In this instance, one option is to start the child on a combined regimen to treat both, such as isoniazid (10–15 mg/kg/day in a single daily dose; maximum dose, 300 mg/day), rifampin, ethambutol, and azithromycin. With this regimen, the child is receiving three drugs for TB (isoniazid, rifampin, and ethambutol) and three drugs for most NTM species that cause lymphadenitis (azithromycin, rifampin, and ethambutol). In young children, adding an additional drug (such as pyrazinamide, PZA) can result in medication intolerance simply due to the volume of medication required, rather than due to adverse effects. In older children, PZA (30–40 mg/kg/day as a single daily dose; maximum dose, 2 grams) should be added, as MTB is more common than NTM species in the school-aged child and adolescent. Of note, *M. bovis* is inherently resistant to PZA, requiring longer courses of therapy.

The optimal duration of therapy is unknown; a 3–6-month course of therapy results in symptom resolution in most children. It is important for clinicians to distinguish between scar tissue and lymphatic tissue, as well as to warn parents that symptom progression (including fistula formation) can occur even when children are receiving effective antibiotic therapy. Immunocompetent children with isolated lymphadenitis in whom complete surgical excision is performed do not require antibiotic therapy.

A recent systemic review found no benefit in using TB medications to treat BCG lymphadenitis [19]. Needle aspiration of fluctuant nodes can result in more rapid symptom resolution [20]. Excision of suppurative nodes is of unclear benefit. Nonsuppurative nodes usually resolve without medical or surgical intervention.

Pulmonary Disease

Epidemiology

Isolated pulmonary infection caused by NTM is rare in immunocompetent children, and several problems impede its epidemiologic description [1]. The infection is rarely fatal, so mortality statistics are not helpful. NTM infections are not covered by mandatory reporting, and smaller laboratories often do not refer isolates to references laboratories for identification and drug susceptibility testing. Most importantly, isolating a NTM organism from respiratory samples, especially gastric aspirates and induced sputum, is not sufficient to establish a diagnosis of pulmonary disease. Distinguishing among saprophytes, colonizers, and true pathogens requires clinical correlation not available from just laboratory reports. Pseudo-outbreaks of respiratory tract colonization caused by various NTM species have been associated with contaminated ice machines, contaminated bronchoscopes and bronchoscopy supplies, showers, potable water supplies, infected laboratory supplies, contamination of topical anesthesia agents, and tap water in hospitals. Isolation of the same NTM species from bronchoscopy samples from two or more children in a short time period should prompt an investigation to determine possible sources of contamination.

While many adults with NTM pulmonary infection have preexisting lung disease of some kind, the majority of children described with pulmonary NTM infection have not had previous lung disease [2, 21]. In these previously normal children, the most common species to cause pulmonary infection have been MAC, *Mycobacterium kansasii*, a rapidly growing mycobacterium and other rare species that cause disease that mimics pulmonary tuberculosis [22]. Anatomic malformations of the lung may predispose to NTM infection with unusual species [23].

An increasing proportion of NTM pulmonary disease in children occurs among those with certain immunologic abnormalities, either as isolated pulmonary disease but, more commonly, as part of a disseminated infection [24] (Table 1). These conditions include advanced HIV infection (usually with a CD4+ count of less than 50 per mm^3), leukemia and other malignancies, and after organ transplantation [25]. Several mutations cause Mendelian susceptibility to mycobacterial diseases (MSMD). Most of the known defects are involved with interferon-γ signaling [26–29].

Various NTM species have been isolated from the respiratory secretions of children with cystic fibrosis (CF) [30]. The chronic, suppurative lung disease present in patients with long-standing CF provides an ideal environment for NTM. Malnutrition, diabetes mellitus, and frequent use of corticosteroids and antibiotics are risk factors for NTM infections in patients with CF. It is often difficult to compare the various reports and analyses of case series because the methods of ascertainment, culture methods, and definitions of colonization and disease have differed widely [31]. The reported prevalence of NTM isolation has varied widely among pediatric CF centers, from highs above 30% of patients to lows of 3%, with most centers reporting 10–15% [32]. In addition, it is difficult to apply the diagnostic criteria of the ATS/IDSA guidelines to this patient population [17]. The rate of isolation of NTM increases as the children become older, but isolation from younger CF patients is not uncommon. It is often difficult to assess the role of the NTM in individual patients because they tend to already have severe and progressive underlying pulmonary disease. In some patients the initial isolate of NTM from the sputum or a bronchoscopy specimen may be associated with a worsening clinical and radiographic course; however, in other CF patients, the presence of NTM in the sputum may be an incidental finding not associated with clinical deterioration. Historically, MAC was isolated most frequently, but *M. kansasii*, *Mycobacterium gordonae*, and the rapid growers also are encountered.

Special mention must be made of *Mycobacterium abscessus* complex (comprised of *M. a. abscessus*, *M. a. bolletii, and M. a. massiliense*), which is causing an increasing number and severity of NTM pulmonary infections in many CF centers. Its appearance in the sputum is often accompanied by a deteriorating clinical course. While it is accepted that most NTM are not transmitted between patients, this issue is controversial when considering the *M. abscessus* complex. In at least one center, whole genome sequencing of *M. abscessus* complex isolates did not demonstrate clustering [33]. However, in several centers, whole genome sequencing of *M. abscessus* complex isolates has demonstrated clustering among CF patients; as no environmental source for the organism could be found, it has been suggested that person-to-person transmission likely occurred [34]. A more extensive analysis of isolates from geographically diverse CF centers has suggested that many *M. abscessus* complex isolates may be acquired through human transmission, potentially via fomites and aerosols, and that recently emerged dominant circulating clones have spread globally [35].

Clinical Presentation

The clinical manifestations of NTM pulmonary infections in children depend on the species involved and the presence of underlying conditions in the child. This section will present the most common manifestations of pulmonary NTM infection in immunocompetent hosts and then will present special considerations in patients with immunocompromise and patients with CF.

The vast majority of children diagnosed with *M. avium* complex lung infection have been under 5 years of age and have been immunocompetent with no underlying anatomic or physiologic pulmonary abnormalities [36, 37]. The most common clinical and radiographic patterns have been similar to those in children with tuberculosis [38]. Some children with pulmonary NTM infection have been initially misdiagnosed as having pulmonary tuberculosis but were further investigated because of unusual clinical or radiographic findings or failure to respond to standard tuberculosis chemotherapy [39]. Most affected children have come to clinical attention because of fairly mild but persistent symptoms including cough and low-grade fever; severe systemic signs or symptoms such as high fever, night sweats, and significant weight loss have been rare [40, 41]. Localized wheezing has been noted occasionally and the diagnosis of aspirated foreign body considered; this is more common with endobronchial involvement [42, 43]. Rare children with NTM lung abscess may have more severe manifestations including high fever, respiratory distress, and signs of sepsis [44].

The most common radiographic presentation of *M. avium* complex pulmonary disease in children has been enlargement of the hilar or mediastinal lymph nodes, sometimes with accompanying distal atelectasis [45]. Occasionally the lymph node enlargement is so great that the presentation is a mediastinal mass [46]. Some children have experienced repeated episodes of fever and cough, interpreted as recurrent pneumonia or bronchitis. These children tend to have more extensive radiographic involvement that is likely due to partial or complete obstruction of a bronchus caused by an enlarged lymph node and/or endobronchial lesion [47]. Several reported children have had acute onset of pneumonia with extensive pulmonary infiltrates in one or several lobes, with a subsequent prolonged chronic course. Involvement of the pleura is exceedingly rare, and a miliary-like picture has not been described.

Several reported children with *M. avium* complex pulmonary disease have undergone bronchoscopy, which has demonstrated granulation tissue within one of the bronchi. Narrowing of the bronchus because of external compression from an enlarged lymph node can be seen occasionally. It is often the culture of this endobronchial material that leads to the isolation of the *M. avium* complex organism. Histologic examination of this tissue usually reveals caseating granulomas or nonspecific chronic inflammation.

While many cases of *M. kansasii* pulmonary infection in adults occur in individuals with underlying pulmonary disease, most of the reported children and adolescents diagnosed with this infection have been previously normal. In the United

States, this organism occurs mainly in the Midwestern and Southwestern states, especially in Texas. The disease is often clinically similar to tuberculosis. The most common signs and symptoms are cough, mild weight loss, and low-grade fevers. Adolescent patients may experience chest pain or hemoptysis, but these are rare in younger children, who usually experience milder illness. The typical radiographic findings in adolescent patients are similar to those with pulmonary tuberculosis. Most pulmonary findings are in the apical regions, but abnormalities may be found throughout the lungs. About 40% of adolescents have bilateral disease. About one-half of these patients have lung cavitation with or without fibrosis. Pleural scarring and nonspecific pulmonary infiltrates are common. Adenopathy is rare and pleural effusion is exceedingly rare. In contrast, in the few reported cases in young children, the radiographic findings have mimicked those of childhood tuberculosis. Intrathoracic adenopathy is usually present and may lead to collapse-consolidation lesions. Calcification of lymph nodes occurs commonly. It is likely that some cases of disease in young children caused by *M. kansasii* are misdiagnosed as tuberculosis because of an inability to isolate the offending organism, the similarity of the radiographic and clinical presentations, the positivity of both the tuberculin skin test and the interferon-gamma release assays with both infections, and the good response of *M. kansasii* to standard tuberculosis chemotherapy.

Reports of pulmonary disease caused by other NTM in otherwise normal children are exceedingly rare [36, 37, 48]. The most common presenting symptom of pulmonary disease caused by a rapidly growing mycobacterium is usually chronic cough, and significant systemic signs and symptoms are usually absent. The young child may have intrathoracic adenopathy or a reticulonodular appearance to the lungs. Pulmonary disease in children caused by other mycobacteria such as *Mycobacterium xenopi*, *Mycobacterium malmoense*, and *Mycobacterium szulgai* is either exceedingly rare or is not detected because of difficulties isolating the organism from children. In the few existing case reports, the clinical and radiographic presentations in children have been similar to that of tuberculosis.

It appears that the prevalence of disease caused by NTM is increasing in patients with CF. One fundamental problem when a NTM is found in the sputum of a patient with CF is determining the difference between colonization and true disease [30]. The American Thoracic Society and Infectious Disease Society of America have published guidelines for the diagnosis of NTM disease [17]. For individuals without cavitary lesions on chest radiography, the criteria include (1) two or more sputum smears being acid-fast stain positive and/or resulting in moderate to heavy growth on culture of the same NTM species, (2) failure of the sputum cultures to convert to negative with either bronchial hygiene or 2 weeks of antimicrobial therapy, and (3) the exclusion of other reasonable causes of pulmonary disease. Unfortunately, these basic guidelines fail to differentiate among NTM species and are difficult to apply to CF patients with extensive underlying and progressive pulmonary disease. The signs and symptoms that often differentiate NTM colonization from infection in the healthy host include productive cough, dyspnea, hemoptysis, malaise, and fatigue, which are often present for the child with CF without NTM infection. Thus, in the child with CF, clinical indicators of tissue damage cannot adequately differentiate between NTM colonization and disease.

The signs and symptoms of NTM disease in children with CF are usually nonspecific. Many patients experience an increase in productive cough, weight loss, low-grade fevers, and night sweats that cannot be explained by other pathogens associated with CF. The incidence of hemoptysis seems to be the same in patients with CF with and without NTM disease. Many patients will experience deterioration of pulmonary function test results (decline in forced expiratory volume in the first second of expiration – FEV_1) in association with the new onset of NTM in the sputum. Patients who have a positive acid-fast stain of the sputum, indicating a higher bacterial load, are more likely to have NTM disease as are those who demonstrate progression by high-resolution CT scan of typical findings associated with NTM disease, such as cavitary disease, subsegmental parenchymal consolidation, atelectasis, nodules, and tree and bud opacities [32, 49].

Diagnosis

The diagnosis of NTM pulmonary disease in an immunocompetent child is often exceedingly difficult. In general, the ATS/IDSA guidelines for diagnosing pulmonary NTM infection [17] should be used, but this likely will result in missed diagnoses in young children. Nonspecific laboratory tests such as blood counts, inflammatory markers, urinalysis, and serum chemistry tests are usually normal in children and adolescents with NTM pulmonary disease. The key to diagnosis is usually a high index of suspicion based on epidemiologic factors and the clinical and radiographic presentation. Affected children are often suspected clinically of having pulmonary tuberculosis but lack epidemiologic risk factors for that disease. The tuberculin skin testing (TST) has been useful in some cases. Infections caused by the *M. avium* complex organisms commonly are associated with reactions of 0–10 mm, but larger reactions occur occasionally. Other species of NTM that cause pulmonary infections – *M. kansasii* and *M. szulgai* – can cause large TST reactions and positive interferon-gamma release assay results. Of course, similar results can be seen with pulmonary tuberculosis. Investigation of adults in the child's environment may be helpful in determining if a positive reaction in these tests is being caused by *Mycobacterium tuberculosis* complex or one of the NTM.

Acid-fast stains of appropriate respiratory specimens or tissues may give an early clue to the presence of NTM infection, but the number of NTM causing disease is usually small, and acid-fast stains of fluids and tissues are frequently negative when the disease is present. When pulmonary involvement is part of disseminated disease, acid-fast stain and mycobacterial culture of blood and urine and any other affected fluid or tissue should be obtained. Similarly, histologic examination of tissue is helpful when classic granulomatous changes are seen, but many NTM infections in immunocompetent hosts cause only nonspecific acute and/or chronic inflammation without granulomas. Unfortunately, many species of NTM can be isolated from the oral and gastric secretions of healthy children. Repeated isolation of the organism from the sputum or a single isolation from a bronchoscopy specimen in association with an abnormal chest radiograph is suggestive but not necessarily diagnostic of significant disease. Five clinical facts have been found useful in distinguishing NTM colonization from disease: (1) the amount of growth usually increases with disease; (2) repeated isolation of the same organism is associated with invasive disease; (3) a site of origin from a closed anatomic site is more significant; (4) the species is usually a pathogen; and (5) the host has risk factors for NTM disease. Disease in the respiratory tract in adults is usually associated with moderate to heavy growth of NTM, but light growth in specimens from children often indicate invasive disease. Many reported children with intrathoracic NTM disease have required biopsy of lymph nodes with appropriate culture and histologic examination to determine the diagnosis.

Treatment

Pulmonary infections caused by NTM in young immunocompetent children are often similar in clinical and radiographic presentation to tuberculosis. In general, until a NTM is proven as the cause of the disease, initial therapy is directed at *M. tuberculosis*; this is particularly important for infants and toddlers and immunocompromised children of any age for whom untreated pulmonary tuberculosis can have grave consequences. Fortunately, the clinical progression of most NTM pulmonary infections is slow enough that a period of several weeks to months of treatment directed at tuberculosis will not have a significant deleterious effect on the outcome of the NTM infection.

Determining the infecting species of NTM is critical for directing chemotherapy. Unfortunately, there have been no published clinical trials of different regimens for treating any of the NTM species in immunocompetent or immunocompromised children with lung disease. In general, the principles and drug regimens used to treat adults with NTM pulmonary infections are also used in children and adolescents (see chapters "Nontuberculous Mycobacterial Disease Management Principles and Nontuberculous Mycobacterial Disease in Pediatric Populations"). However, compared with immunocompetent children with pulmonary NTM infection, adults tend to have more extensive NTM disease with a larger burden of organisms and underlying pulmonary disease. Regimens that are effective in adults are highly likely to be effective in children, but many children may require shorter durations and, occasionally, fewer drugs.

There are no published guidelines and little clinical experience about the need for resectional lung surgery in children with pulmonary NTM disease. In several case reports, excision of endobronchial lesions or large intrathoracic lymph nodes has been undertaken for both diagnostic and therapeutic reasons (as when extensive bronchial obstruction with severe symptoms has occurred), but resection of lung parenchyma has been performed exceedingly rarely and does not appear to be necessary.

One controversial area is lung transplantation for children with CF who also have lung infection with *M. abscessus* complex. While some centers will not consider transplantation in these patients, other centers have had moderate to good success with posttransplant survival up to 80% using aggressive antimicrobial therapy before and after transplantation [50, 51].

Skin and Soft Tissue Infections (SSTIs)

Epidemiology

The most common NTM species associated with SSTIs in immunocompetent children are described in Table 1. While skin involvement can be seen as part of disseminated disease, it is most frequently observed in previously healthy children in whom the portal of entry was penetrating trauma. This trauma can include innocuous-appearing abrasions when children are in contact with water sources. The classic NTM species associated with SSTIs is *M. marinum*. This endemic fish pathogen has been most commonly reported after children are in contact with fish tanks, fish or shellfish, and bodies of water [52]. While chlorination has reduced the incidence of *M. marinum* in industrialized nations, outbreaks have been reported from taro farmers in Micronesia who farmed in water-filled World War III bomb craters [53].

NTM infections can follow a child stepping on a sharp object but can also include older patients obtaining body piercings or tattoos or young children in whom tympanostomy tubes are placed [54]. Disease caused by penetrating trauma is most associated with the rapidly growing mycobacterial (RGM) species (*M. abscessus*, *M. chelonae*, and *M. fortuitum*). Patients may be predisposed to these deeper infections if open fractures are irrigated with tap water, rather than sterile water. Children with central venous catheters (CVCs) should not let the end of the catheter contact tap water. In contrast to disseminated NTM disease, identification of an NTM SSTI should not automatically prompt an automatic evaluation for a primary or secondary immunodeficiency. However, SSTIs may also be the first manifestation of disseminated disease in immunocompromised patients.

Two NTMs that are infrequently seen in the United States are worth special mention; both have incubation periods of several years. *M. leprae*, the causative agent of leprosy, is most common in Southeast Asia, and India accounts for almost 60% of all leprosy cases [55]. While skin-to-skin contact can result in transmission, discharge of bacilli from nasal mucosa is now thought to be a more important route of transmission. *M. ulcerans*, the species causing Buruli ulcer disease, is most common in tropical areas, with the largest number of cases reported from West Africa. This bacterium is spread through contact with water; person-to-person transmission is rare.

Clinical Presentation

NTM SSTIs can have protean manifestations, and there is no one dermatologic finding that is pathognomonic. One way to think of the findings is to think of "a bump you can feel, a wart that won't peel, an ulcer that won't heal" (Fig. 4). Findings can include folliculitis or furunculosis (e.g., after exposure in a hot tub), as well as papules, nodules, or verrucous lesions (Fig. 5). The lesions, usually painless, often

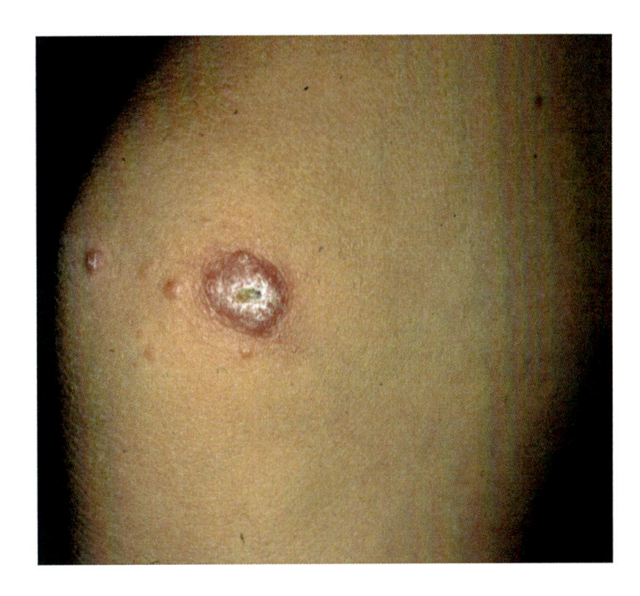

Fig. 4 A verrucous lesion caused by *Mycobacterium marinum*. This lesion over the knee was seen in an adolescent who was kneeling in water while boating. The verrucous lesion is painless and has central ulceration with surrounding satellite lesions

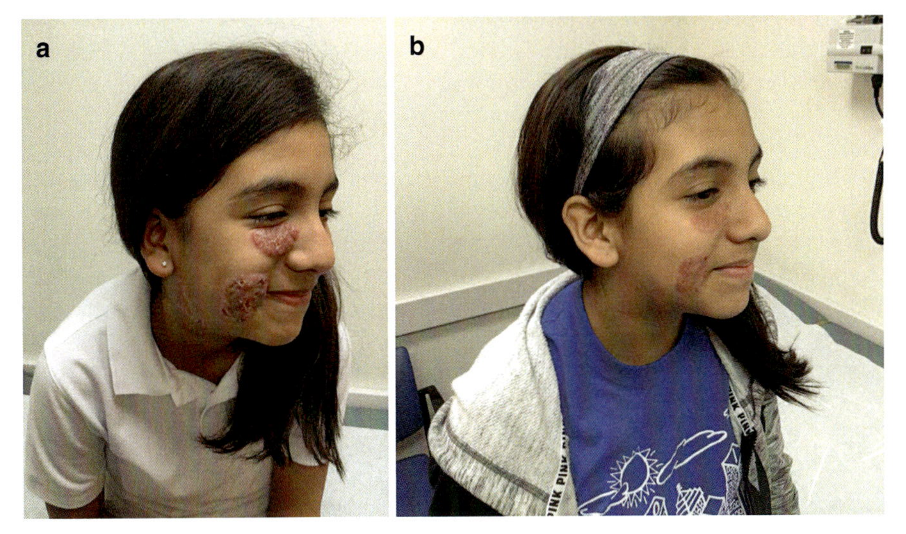

Fig. 5 Cutaneous *Mycobacterium abscessus* infection on the face of an immunocompetent adolescent. The raised, nontender, erythematous, verrucous lesions with some ulceration are noted in this 12-year-old girl prior to initiation of therapy (**a**) and after 3 months of multidrug therapy (**b**)

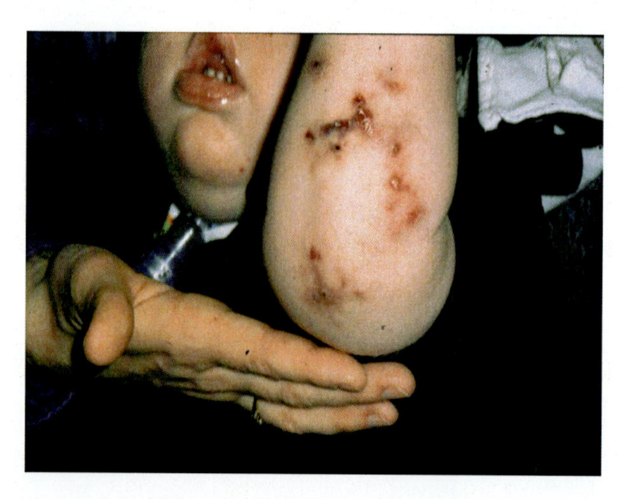

Fig. 6 Forearm lesions in a lymphatic distribution caused by *Mycobacterium marinum*. These nodular lesions began distally on the left arm of this toddler and progressed proximally in a pattern that mimics sporotrichosis

start as indolent, indurated regions and progress to ulcerations [56]. Constitutional symptoms and regional lymphadenopathy are uncommon. Penetrating trauma, including open fractures, can result in subacute or chronic osteomyelitis or pyomyositis. One diagnostic clue is that the child will not have a response to the usual antibiotics used to treat pyogenic SSTIs.

The dreaded complication of *M. marinum* infection is tenosynovitis. Kanavel's signs are the symptoms found in patient with tenosynovitis of the flexor tendons of the hand. Patients hold the affected finger(s) in flexion, there is swelling and pain over the affected tendon, and pain is worsened with passive extension of the fingers. *M. marinum* can also cause septic arthritis, osteomyelitis, or lesions that spread in a centripetal pattern, mimicking sporotrichosis (Fig. 6). *M. marinum* lesions are more common on the fingers and hands.

There are several forms of leprosy, divided into paucibacillary (≤5 skin lesions) and multibacillary forms. The first symptom often is a hypoesthetic area on the skin; this can be seen prior to any dermatologic manifestations being evident. Macules or plaques then develop and peripheral neuropathy symptoms become more evident. The number of lesions and presence of satellite lesions vary based upon the type of leprosy.

Buruli ulcer lesions are classified based upon size (<5 cm, 5–15 cm, >15 cm) and whether lesions are localized (often at the site of antecedent trauma) or disseminated. Similar to other NTM lesions, the morphology of skin findings varies from papules and plaques to nodules and ulcers. Deeper infection can occur, with both contiguous and metastatic osteomyelitis being described. HIV infection has been associated with more aggressive forms of Buruli ulcer disease [57].

Diagnosis

Tissue biopsy is the most sensitive source from which to obtain cultures (Table 3) [58]. Sensitivity of drainage material is lower, particularly if the specimens were obtained by swabs. Some species (e.g., *M. fortuitum*) can grow on routine culture media in addition to growing on Lowenstein-Jensen or other mycobacterial-specific media. It is important for clinicians to differentiate between true infection and contamination. Several pseudo-outbreaks of NTM SSTI and other infections have been reported when laboratory water supplies contain NTM species [59]. Many NTM species (*M. kansasii*, *M. simiae*) are resistant to commonly used disinfecting agents, making laboratory contamination possible.

Surgical procedures often are necessary to diagnose *M. marinum* infections. In one case series, over one-third of patients had biopsy results lacking granulomatous inflammation and had negative acid-fast stains. Instead, biopsy findings can include necrosis, synovitis, or chronic inflammation [52]. Thus, the absence of necrotizing granulomas does not rule out *M. marinum* infection.

Leprosy is predominantly a clinical diagnosis and should be considered in endemic countries in any patient with peripheral neuropathy. The organism cannot be cultured, and smear yield is poor in persons unfamiliar with the techniques required. Skin testing is not recommended, and serologies are insensitive. PCR yield is excellent (Table 3). Similarly, Buruli ulcer disease is primarily diagnosed clinically. The organism is hard to grow in culture due to fastidious temperature requirements; consequently, PCR has become the modality of choice. Tissue should be obtained from the edge of the ulcer, not the center, and should include several skin layers.

Treatment

Management of children with NTM SSTIs often involves combined medical and surgical therapy. Any indwelling foreign bodies (including CVCs) need to be removed, as many species form a biofilm that precludes organism eradication with the foreign body in situ. Surgical debridement is essential for serious localized disease, in order to debulk in addition to draining abscesses into which antibiotic penetration would be poor.

Antibiotic therapy alone is not curative for NTM SSTIs (in contrast to NTM lymphadenitis), but antibiotics decrease time to symptom resolution [17] (Tables 4 and 5). While therapy needs to be organism specific once speciation is available, an empiric regimen may consist of a macrolide in addition to one of the following antibiotics: a fluoroquinolone, doxycycline (for children ≥8 years of age), or trimethoprim-sulfamethoxazole. Parenteral therapy, often including an aminoglycoside (e.g., amikacin) and a carbapenem (e.g., meropenem), should be considered for immunocompromised children or those with serious NTM SSTIs. Once speciation occurs, the child should be changed to a regimen containing at least two drugs to which the isolate is susceptible. As in vitro susceptibilities correlate poorly with in vivo response, a change in antibiotics should be considered if a child does not demonstrate clinical improvement in 1–2 months. Details on the routine antibiotic susceptibilities and treatment of RGM species are described in chapter "Surgical Management of NTM Diseases", and treatment *of M. abscessus* is reviewed in chapter "Management of Lung Diseases Associated with NTM Infection".

M. marinum isolates usually are susceptible to rifamycins, ethambutol, macrolides, and sulphonamides. They are resistant to isoniazid and pyrazinamide, and many isolates are resistant to tetracyclines. ATS guidelines suggest treating with two active agents for 1–2 months following symptom resolution; this often results in a 3–4-month total treatment course [17]. The use of a third antibiotic may be considered for deeper infections. There are no comparative trials on which regimens may be most effective.

Therapy for leprosy is based on the burden of disease. Paucibacillary disease is usually treated for 6–12 months with a combination of rifampin (10–20 mg/kg/day; maximum dose, 600 mg) and dapsone (1–2 mg/kg/day; maximum dose, 100 mg). In contrast, multibacillary disease is treated for 12–24 months with rifampin, dapsone, and clofazimine (1 mg/kg/day; maximum dose, 50 mg) [60]. The WHO recommends treating Buruli ulcer disease with an 8-week course of rifampin (10 mg/kg; maximum dose, 600 mg) and an aminoglycoside, either amikacin or streptomycin (15 mg/kg; maximum dose, 1 g) [61]. Monotherapy is never recommended for any NTM SSTI.

Disseminated Disease (Including Mycobacteremia)

Epidemiology

Disseminated NTM disease is very uncommon in immunocompetent hosts. Most cases occur in children with primary (e.g., Mendelian susceptibility to mycobacterial diseases (MSMD), interferon receptor, or interleukin-12 mutations) or acquired immunodeficiencies, including HIV infection. In the HIV-infected child, MAC used to be the most common pathogen after the CD4+ T cell count dropped below 50. However, with the advent and widespread implementation of highly active antiretroviral therapy (HAART), disseminated NTM has become increasingly rare in HIV-infected children. Now, bacteremia can be caused by direct inoculation into central venous catheters (CVCs) [e.g., when CVCs are flushed with or come into contact with tap water] or from entry into the gastrointestinal and respiratory tracts. Conditions predisposing to disseminated NTM disease are listed in Table 1 [26, 28, 56].

Clinical Presentation

In contrast to children with NTM disease at other sites, children with disseminated NTM often present with constitutional symptoms, including fever, night sweats, myalgia, and weight loss or failure to gain weight. Abdominal pain and diarrhea may also be seen in the initial presentation, particularly in HIV-infected patients in whom the gastrointestinal tract was the portal of entry. Physical examination findings reflect involvement of the reticuloendothelial system (generalized lymphadenopathy, hepatosplenomegaly) or of other affected sites (please see sections on lymphadenitis, pulmonary disease, and SSTIs). Isolated skin findings may be the first sign of disseminated disease. Unmasking immune reconstitution inflammatory syndrome (IRIS), in which NTM disease develops *after* immune system recovery, has been reported in a small percentage of HIV-infected patients in whom HAART was initiated. The main clinical presentations noted included peripheral and visceral lymphadenopathy and pulmonary infiltrates [62].

Diagnosis

Commonly obtained laboratory findings are of minimal utility in children with disseminated NTM disease. Imaging findings can include intrathoracic or mesenteric lymphadenopathy and pulmonary nodules. The diagnosis is made by isolation of an NTM species from sterile sites such as blood, bone marrow, or lymphatic tissue. In affected children with HIV infection, MAC is often isolated from the stool. *M. tuberculosis* PCR and culture should be performed to distinguish between disseminated TB and disseminated NTM disease. While blood cultures are positive in approximately 90% of children with disseminated MAC, mycobacteremia is low grade and more sporadic with other NTM species. As such, multiple mycobacterial blood cultures may be required to document bacteremia. Definitive diagnosis may require more invasive procedures, including biopsies of visceral nodes or bone marrow aspiration. This tissue may then be sent for histopathology in addition to acid-fast stain and culture. As most children with disseminated NTM are immunocompromised, well-developed granulomas may not be observed. AFB smears are positive in less than one-quarter of children. The diagnostic test yield for disseminated NTM is described in Table 3.

Treatment

NTM bacteremia in children with indwelling CVCs should prompt immediate removal of the central line. There are some reports of CVC removal alone being sufficient to resolve the mycobacteremia [63]. However, antibiotic therapy (Tables 4 and 5) is often initiated as NTM bacteremia almost uniformly occurs in immunocompromised children in whom the potential for dissemination is high. Similar to TB, treatment of disseminated NTM is divided into initial and continuation phases. Empiric therapy should target MAC until speciation is known, unless the child has had other NTM species isolated previously. The cornerstone of therapy for MAC is a macrolide (either azithromycin or clarithromycin), to which ethambutol and a rifamycin are added. In HIV-infected patients, use of rifabutin decreases the risk of interaction with HAART. The value of rifabutin over rifampin in HIV-uninfected patients, where drug-drug interactions are less of a concern, is unclear. Addition of an aminoglycoside or a fluoroquinolone should be considered for children who develop disseminated MAC while receiving a macrolide for MAC prophylaxis, given concern for macrolide resistance. For children in whom macrolide resistance is not a concern, either the rifamycin or ethambutol can be stopped after 1–2 months and the child transitioned to two-drug therapy for a total of 6–12 months. Response to treatment can take several weeks to become evident, and treatment failure is defined by persistent mycobacteremia after 8–12 weeks of therapy. Fewer data exist about optimal therapeutic regimens for children with non-MAC disseminated mycobacteremia. Treatment courses for other NTM species causing disseminated disease are less well described. However, once speciation data are available, treatment should be targeted for the species of mycobacteria isolated.

HIV-infected children with disseminated MAC should receive lifelong macrolide prophylaxis unless they are older (> 2 years of age), have completed at least 12 months of MAC therapy and remain asymptomatic, and have had immune restoration (≥ 6 months of CD4+ T cell count recover above age-specific target) on a stable HAART regimen [64].

References

1. Blyth CC, Best EJ, Jones CA, et al. Nontuberculous mycobacterial infection in children: a prospective national study. Pediatr Infect Dis J. 2009;28(9):801–5.
2. Haverkamp MH, Arend SM, Lindeboom JA, et al. Nontuberculous mycobacterial infection in children: a 2-year prospective surveillance study in the Netherlands. Clin Infect Dis. 2004;39:450–6.
3. Carvalho AC, Codecasa L, Pinsi G, et al. Differential diagnosis of cervical mycobacterial lymphadenitis in children. Pediatr Infect Dis J. 2010;29(7):629–33.
4. Gallivan M, Shah N, Flood J. Epidemiology of *Mycobacterium bovis* disease, California, USA, 2003–2011. Emerg Infect Dis. 2015;21(3):435–43.
5. Kuchukhidze G, Kasradze A, Dolakidze T, et al. Increase in lymphadenitis cases after shift in BCG vaccine strain. Emerg Infect Dis. 2015;21(9):1677–9.
6. The BCG World Atlas: A Database of Global BCG Vaccination Policies and Practices. Available online at: www.bcgatlas.org. Accessed 11 Oct 2016.
7. Haimi-Cohen Y, Zeharia A, Mimouni M, et al. Skin indurations in response to tuberculin testing in patients with nontuberculous mycobacterial lymphadenitis. Clin Infect Dis. 2001;33(10):1786–8.
8. Lindeboom JA, Smets AM, Kuijper EJ, et al. The sonographic characteristics of nontuberculous mycobacterial cervicofacial lymphadenitis in children. Pediatr Radiol. 2006;36(10):1063–7.

9. Robson CD. Imaging of granulomatous lesions of the neck in children. Radiol Clin N Am. 2000;38(5):969–77.
10. Ellison E, Lapuerta P, Martin SE. Fine needle aspiration diagnosis of mycobacterial lymphadenitis: sensitivity and predictive value in the United States. Acta Cytol. 1999;32(2):153–7.
11. Lindeboom JA. Conservative wait-and-see therapy versus antibiotic treatment for nontuberculous mycobacterial cervicofacial lymphadenitis in children. Clin Infect Dis. 2011;52(2):180–4.
12. Haimi-Cohen Y, Markus-Eidlitz T, Amir J, Zeharia A. Long-term follow-up of observation-only management of nontuberculous mycobacterial lymphadenitis. Clin Pediatr (Phila). 2016;55(12):1160–4.
13. Zeharia A, Eidlitz-Markus T, Haimi-Cohen Y, et al. Management of nontuberculous mycobacteria-induced cervical lymphadenitis with observation alone. Pediatr Infect Dis J. 2008;27(10):920–2.
14. Lindeboom JA. Surgical treatment for nontuberculous mycobacterial (NTM) cervicofacial lymphadenitis in children. J Oral Maxillofac Surg. 2012;70(2):345–8.
15. Lindeboom JA, Lindeboom R, Bruijnesteijn van Coppenraet ES, et al. Esthetic outcome of surgical excision versus antibiotic therapy for nontuberculous mycobacterial cervicofacial lymphadenitis in children. Pediatr Infect Dis J. 2009;28(11):1028–30.
16. Parker NP, Scott AR, Finkelstein M, et al. Predicting surgical outcomes in pediatric cervicofacial mycobacterial lymphadenitis. Ann Otol Rhinol Laryngol. 2012;121(7):478–84.
17. Griffith DE, Aksamit T, Brown-Elliott BA, et al. An official ATS/IDSA statement: diagnosis, treatment, and prevention of nontuberculous mycobacterial diseases. Am J Respir Crit Care Med. 2007;175(4):367–416.
18. Nahid P, Dorman SE, Alipanah N, et al. Official American Thoracic Society/Centers for Disease Control and Prevention/Infectious Diseases Society of America clinical practice guidelines: treatment of drug-susceptible tuberculosis. Clin Infect Dis. 2016;63(7): e147–95.
19. Cuello-Garcia CA, Perez-Gaxiola G, Jimenez Gutierrez C. Treating BCG-induced disease in children. Cochrane Database Syst Rev. 2013;1:CD008300.
20. Banani SA, Alborzi A. Needle aspiration for suppurative post-BCG adenitis. Arch Dis Child. 1994;71(5):446–7.
21. Tebruegge M, Pantazidou A, MacGregor D, et al. Nontuberculous mycobacterial disease in children – epidemiology, diagnosis & management at a tertiary center. PLoS One. 2016;11(1):e0147513.
22. Asiimwe BB, Bagyenzi GB, Sengooba W, et al. Species and genotypic diversity of nontuberculous mycobacteria isolated from children investigated for pulmonary tuberculosis in rural Uganda. BMC Infect Dis. 2013;13:88.
23. Pham-Huy A, Robinson JL, Tapiero B, et al. Current trends in nontuberculous mycobacteria infections in Canadian children: a pediatric investigators collaborative network on infections in Canada (PICNIC) study. Paediatr Child Health. 2010;15(5):276–82.
24. Stone AB, Schelonka RL, Drehner DM, et al. Disseminated *Mycobacterium avium* complex in non-human immunodeficiency virus-infected pediatric patients. Pediatr Infect Dis J. 1992;11:960–4.
25. Arlotta A, Cefalo MG, Maurizi P, et al. Critical pulmonary infection due to nontuberculous mycobacterium in pediatric leukemia: report of a difficult diagnosis and review of pediatric series. J Pediatr Hematol Oncol. 2014;36(1):66–70.
26. Haverkamp MH, van de Vosse E, van dissel JT. Nontuberculous mycobacterial infections in children with inborn errors of the immune system. J Infect. 2014;68(Suppl 1): S134–50.
27. Safdar A, White DA, Stover D, et al. Profound interferon gamma deficiency in patients with chronic pulmonary nontuberculous mycobacteriosis. Am J Med. 2002;113(9):756–9.
28. Qu H-Q, Fisher-Hoch SP, McCormick JB. Molecular immunity to mycobacteria; knowledge from the mutation and phenotype spectrum analysis of Mendelian susceptibility to mycobacterial diseases. Int J Infect Dis. 2011;15(5):e305–13.

29. Newport MJ, Huxley CM, Huston S, et al. A mutation in the interferon-gamma-receptor gene and susceptibility to mycobacterial infection. N Engl J Med. 1996;335(26):1941–9.
30. Floto RA, Olivier KN, Saiman L, et al. US Cystic Fibrosis Foundation and European Cystic Fibrosis Society consensus recommendations for the management of non-tuberculous mycobacteria in individuals with cystic fibrosis. Thorax. 2016;71(1):1–22.
31. Adjermian J, Olivier KN, Prevots DR. Nontuberculous mycobacteria among patients with cystic fibrosis in the United States. Screening practices and environmental risk. Am J Respir Crit Care Med. 2014;190(5):581–6.
32. Leung JM, Olivier KN. Nontuberculous mycobacteria: the changing epidemiology and treatment challenges in cystic fibrosis. Curr Opin Pulm Med. 2013;19(6):662–9.
33. Harris KA, Underwood A, Kenna DTD, et al. Whole-genome sequencing and epidemiological analysis do not provide evidence for cross-transmission of *Mycobacterium abscessus* in a cohort of pediatric cystic fibrosis patients. Clin Infect Dis. 2015;60(7):1007–16.
34. Bryant JM, Grogono DM, Greaves D, et al. Whole-genome sequencing to identify transmission of *Mycobacterium abscessus* between patients with cystic fibrosis: a retrospective cohort study. Lancet. 2013;381(9877):1551–60.
35. Bryant JM, Grogono DM, Rodriguez-Rincon D, et al. Emergence and spread of a human-transmissible multidrug-resistant nontuberculous mycobacterium. Science. 2016;354(6313):751–7.
36. Nolt D, Michaels MG, Wald ER. Intrathoracic disease from nontuberculous mycobacteria in children: two cases and a review of the literature. Pediatrics. 2003;112(5):e434.
37. Freeman AF, Olivier KN, Rubio TT, et al. Intrathoracic nontuberculous mycobacterial infections in otherwise healthy children. Pediatr Pulmonol. 2009;44(11):1051–6.
38. Gupta SK, Katz BZ. Intrathoracic disease associated with *Mycobacterium avium-intracellulare* complex in otherwise healthy children: diagnostic and therapeutic considerations. Pediatrics. 1994;94(5):741–2.
39. Lopez-Varela E, Garcia-Basteiro AL, Santiago B, et al. Nontuberculous mycobacteria in children: muddying the waters of tuberculosis diagnosis. Lancet Respir Med. 2015;3(3):244–56.
40. Fergie JE, Milligan TW, Henderson BM, et al. Intrathoracic *Mycobacterium avium* complex infection in immunocompetent children: case report and review. Clin Infect Dis. 1997;24(2):250–3.
41. Levelink B, de Vries E, van Dissel JT, et al. Pulmonary *Mycobacterium avium intracellulare* infection in an immunocompetent child. Pediatr Infect Dis J. 2004;23(9):892.
42. Kröner C, Griese M, Kappler M, et al. Endobronchial lesions caused by nontuberculous mycobacteria in apparently healthy pediatric patients. Pediatr Infect Dis J. 2015;34(5):532–5.
43. del Rio Camacho G, Soriano Guillén L, Flandes Aldeyturriaga J, et al. Endobronchial atypical mycobacteria in an immunocompetent child. Pediatr Pulmonol. 2012;45(5):511–3.
44. Glatstein M, Scolnick D, Bensira L, et al. Lung abscess due to non-tuberculous, non-*Mycobacterium fortuitum* in a neonate. Pediatr Pulmonol. 2012;47(10):1034–7.
45. Osorio A, Kessler RM, Guruuprasad H, et al. Isolated intrathoracic presentation of *Mycobacterium avium* complex in an immunocompetent child. Pediatr Radiol. 2001;31(12):848–51.
46. Kelsey DS, Chambers RT, Hudspeth AS. Nontuberculous mycobacterial infection presenting as a mediastinal mass. J Pediatr. 1981;98(3):431–2.
47. Dore ND, LeSouëf PN, Masters B, et al. Atypical mycobacterial pulmonary disease and bronchial obstruction in HIV-negative children. Pediatr Pulmonol. 1998;26(6):380–8.
48. Do PC, Nussbaum E, Moua J, et al. Clinical significance of respiratory isolates for *Mycobacterium abscessus* complex from pediatric patients. Pediatr Pulmonol. 2013;48(5):470–80.
49. Martiniano SL, Nick JA. Nontuberculous mycobacterial infections in cystic fibrosis. Clin Chest Med. 2015;36(1):101–15.
50. Robinson PD, Harris KA, Aurora P, et al. Paediatric lung transplant outcomes vary with *Mycobacterium abscessus* complex species. Eur Respir J. 2013;41(5):1230–2.

51. Koh WJ, Stout JE, Yew W-W. Advances in the management of pulmonary disease due to *Mycobacterium abscessus* complex. Int J Tuberc Lung Dis. 2014;18(10):1141–8.
52. Johnson MG, Stout JE. Twenty-eight cases of *Mycobacterium marinum* infection: retrospective case series and literature review. Infection. 2015;43(6):655–62.
53. Lillis JV, Ansdell VE, Ruben K, et al. Sequelae of World War II: an outbreak of chronic cutaneous nontuberculous mycobacterial infection among Satowanese islanders. Clin Infect Dis. 2009;48(11):1541–6.
54. Franklin DJ, Starke JR, Brady MT, Brown BA, Wallace RJ Jr. Chronic otitis media after tympanostomy tube placement caused by *Mycobacterium abscessus*: a new clinical entity? Am J Otol. 1994;15(3):313–20.
55. World Health Organization. Global leprosy update, 2013: reducing disease burden. Wkly Epidemiol Rec. 2014;89(36):389–400.
56. Kothavade RJ, Dhurat RS, Mishra SN, Kothavade UR. Clinical and laboratory aspects of the diagnosis and management of cutaneous and subcutaneous infections caused by rapidly growing mycobacteria. Eur J Clin Microbiol Infect Dis. 2013;32(2):161–88.
57. Toll A, Gallardo F, Ferran M, et al. Aggressive multifocal Buruli ulcer with associated osteomyelitis in an HIV-positive patient. Clin Exp Dermatol. 2005;30(6):649–51.
58. Sia TY, Taimur S, Blau DM, et al. Clinical and pathological evaluation of *Mycobacterium marinum* group skin infections associated with fish markets in New York City. Clin Infect Dis. 2016;62(5):590–5.
59. El Sahly HM, Septimus E, Soini H, et al. *Mycobacterium simiae* pseudo-outbreak resulting from a contaminated water supply in Houston, Texas. Clin Infect Dis. 2002;35(7):802–7.
60. World Health Organization. Leprosy elimination: WHO recommended MDT regimens. Available online at: http://www.who.int/lep/mdt/regimens/en/. Accessed 10/13/16.
61. World Health Organization. Treatment of *Mycobacterium ulcerans* disease (Buruli Ulcer). 2012. Available online at: http://apps.who.int/iris/bitstream/10665/77771/1/9789241503402_eng.pdf. Accessed 10/13/16.
62. Phillips P, Bonner S, Gataric N, et al. Nontuberculous mycobacterial immune reconstitution syndrome in HIV-infected patients: spectrum of disease and long-term follow-up. Clin Infect Dis. 2005;41(10):1483–97.
63. Shachor-Meyouhas Y, Sprecher H, Eluk O, et al. An outbreak of *Mycobacterium mucogenicum* bacteremia in pediatric hematology-oncology patients. Pediatr Infect Dis J. 2011;30(1):30–2.
64. Department of Health and Human Services. Panel on opportunistic infections in HIV-exposed and HIV-infected children: guidelines for the prevention and treatment of opportunistic infections in HIV-exposed and HIV-infected children. 2013. Available online at: https://aidsinfo.nih.gov/contentfiles/lvguidelines/oi_guidelines_pediatrics.pdf. Accessed 13 Oct 2016.

Non-tuberculous Mycobacteria in Cystic Fibrosis

Robert Burkes and Peadar G. Noone

Introduction

Cystic fibrosis (CF) is a systemic disease caused by autosomal recessive inheritance of defective genes that encode for cystic fibrosis transmembrane conductance regulator (CFTR). Much is now known about the structure and function of CFTR and the clinical phenotypes associated with defective CFTR [1]. The pulmonary aspects are generally responsible for most of the morbidity and mortality associated with the disease. These are thought to relate to abnormal airway host defense, especially reduced mucociliary clearance and retention of airway secretions, leading to a cycle of chronic infection and neutrophilic inflammation of the airways and bronchiectasis [2]. CF most commonly occurs in Caucasians (~1 in 2500 live births), although there is increasing recognition of the occurrence of CF in non-Caucasian or mixed ethnicities [3, 4]. The chronically inflamed, damaged lung in CF provides a rich milieu for chronic infection. Over time patients with cystic fibrosis generally become chronically infected with bacteria such as *Pseudomonas*, other pathogenic gram-negative organisms, and methicillin-resistant *Staphylococcus aureus*. Non-tuberculous mycobacteria (NTM) have been recognized for about 20 years as other important infecting organism in CF patients [5, 6].

R. Burkes · P. G. Noone (✉)
The Division of Pulmonary Medicine, University of North Carolina at Chapel Hill, Chapel Hill, NC, USA
e-mail: Robert.Burkes@unchealth.unc.edu; Peadar_noone@med.unc.edu

D. E. Griffith (ed.), *Nontuberculous Mycobacterial Disease*, Respiratory Medicine,
https://doi.org/10.1007/978-3-319-93473-0_17

Epidemiology

NTM were not commonly recognized in patients with CF prior to the 1990s probably due to multiple factors [7]. Early observations hypothesized that the emergence of NTM in CF patients is likely due to a combination of increased recognition of the role of NTM in the CF microbiome, the aging CF population (as survival improved) with more exposure time to environmental sources of NTM, and improved detection through better culture techniques of airway samples [8]. However, this increase in the prevalence of NTM in the CF population mirrors that of the general population where an increase in NTM infection was noted from 8.7 to 13.9 cases per 100,000 from 2008 to 2013, with NTM observed in older patients compared with non-CF patients [9]. One of the first studies on this issue prospectively and rigorously examined sputum specimens from a large number ($n = 986$) of patients with CF at separate institutions across the United States and found the prevalence of NTM to be 13%, with a range of 7–24% depending on the location of the CF center. *Mycobacterium avium complex* (MAC) was the most commonly encountered NTM (72% of NTM isolates) followed by *Mycobacterium abscessus* (16% of NTM isolates) [5]. A more recent study shows a 20% detection rate of pathogenic NTM in 16,153 persons with CF from 2010-2014. Of the 20% with NTM, 61% had MAC isolated. MAC was three times more likely to be recovered in those who were older and had a lower weight [10]. That study, with its sister “nested cohort” study, both performed during the 1990s and published in 2003, launched what has become a very important issue in monitoring and treating patients with CF and also has a major impact on lung transplant candidacy for patients with advanced end-stage lung disease [6].

The prevalence of positive NTM cultures in patients with CF, and the individual species of NTM that tend to predominate, seems to relate to patient age (even to quite young patients) and geographic locale. A French study of a fairly large number of patients with CF under the age of 24 ($n = 385$, mean age 12) found mycobacterial colonization in 8% of patients. Notably, MAC was not isolated from any patient under 15 years of age, while *Mycobacterium abscessus* was quite prevalent, and recovered across all ages in the cohort. *Mycobacterium gordonae* was also common (18% of mycobacterium-positive isolates) [11]. Environmental factors likely play a role. Areas with higher environmental saturated vapor pressure seem to be associated with a higher prevalence of NTM infection in patients with CF (interestingly, high saturated water vapor pressure also seems to promote *Pseudomonas* colonization). Possibly related to this, NTM have been found to cluster around certain geographic points in the United States. MAC is most common in the upper Midwest, West, and Southeast, and *M. abscessus* is found in the majority of positive NTM cultures in Louisiana, Nebraska, and Delaware [12]. A comparison of NTM culture-positive CF patients ($n = 48$) with non-infected controls ($n = 85$) showed a modest increase in risk of NTM infection in CF patients who participate in indoor swimming, but the total number in the NTM infected group who participated in indoor swimming was low (19%). There was also no difference in the exposure to municipal water sources via showering. The study concluded that common individual risk factors outside of geographic locale need still to be identified [13]. In Scandinavian countries, where microbiologic screening begins between 6 and 10 years of age, the prevalence of CF patients with positive NTM cultures was found to be 11% over a 12-year period. Much like the study from France, *M. abscessus* was found to be the predominant organism in younger patients, followed by MAC. In this cohort, MAC was again found in older patients (median age ~22) when compared to *M. abscessus* (median age ~17). An independent risk factor found in this patient population for the development of NTM colonization was diabetes [14]. In a patient population from Israel (age > 5 years old), where both *M. abscessus* and *M. simiae* were more commonly isolated than MAC, it was noted that CF patients with NTM colonization had significantly lower lung function (forced expiratory volume in 1 s; FEV1); spent more days in the hospital annually, with more antibiotic exposure; and had higher rates of *Pseudomonas* species and *Aspergillus* species compared to non-NTM colonized patients [15]. In contrast, in the United States where MAC is the most predominantly encountered NTM, colonization is often associated with older patients, preserved lung function, and *Staphylococcus aureus* coinfection and appears inversely related to colonization with *Pseudomonas* species [5]. However, patients that develop clinically active (treatment requiring) pulmonary infection with MAC generally develop NTM culture positivity at lower lung function levels, with a greater yearly decline in lung function than those who have only an intermittent or persistent MAC colonization [16]. As an added complication, recent data suggests that common CF therapies may add to the risk of MAC infection. For example, NTM acquisition may be promoted by azithromycin by a mechanism involving reduced macrophage autophagy [17]. Given that azithromycin therapy has become standard in the preventive regimen for CF, this is an important issue for patients and CF teams to watch.

In general, NTM have been considered to be ubiquitous organisms that are acquired by humans from the environment, and not transferred from person-to-person [7]. *M. abscessus* is a hardy organism that can survive desiccation, undergo aerosolization, and may only require a small inoculum to cause infection. There are suggestions in the literature that *M. abscessus* may be capable of person-to-person spread [18]. Studies on *M. abscessus* genomics propose international spread of *M. abscessus* that could be due to a cluster of clones that have spread globally [19]. Studies on MAC have failed to show any convincing person-to-person transmission, but MAC transmission among patients from a common source is conceivable in concept, illustrating the general importance of fomite control in CF centers. It would seem prudent, therefore, to execute rigorous infection control in the form of contact precautions for all patients with CF in the clinic and hospital setting [20, 21].

In summary, the profile of patients with CF infected with NTM is variable worldwide, and regional and global patterns of spread and infection are of increasing concern. In the clinic setting, dry surfaces, medical equipment, and providers' hands can harbor infection such that a program of disinfection and infection control should probably be in place, to prevent transmission of pathogens, including NTM (even if the evidence is still emerging) and especially *M. abscessus* [22].

Pathogenesis

The pathogenesis of NTM infection in a cystic fibrosis patient consists of a complex interaction between the mycobacterium and the dysfunctional airway host defense milieu of the CF airway. Foremost is the inadequacy of mucous transport and reduced airway surface liquid volume. The prevailing hypothesis is that of dehydrated and inspissated mucous leading to reduced mucociliary transport of mucous from the distal airways which causes an enhanced environment for chronic macro-colonization of bacteria in static and remarkably nutrient-rich mucous plugs [23, 24]. Impaired mucociliary clearance is not the only lung defense mechanisms against bacterial colonization that is faulty in cystic fibrosis. Neutrophils are important components of the immune system's initial response to the presence of bacteria. The dehydrated mucous makes neutrophil chemotaxis more difficult than it would be in better hydrated mucous. Viscous mucous is associated with decreased neutrophil uptake of bacteria and killing capacity [25]. Further, neutrophils that enter the airway in cystic fibrosis often have reduced killing capacity due to destruction of the CXCR1 chemokine receptor on the neutrophil surface by unopposed proteolysis in the airway [26]. The acidic microenvironment of the airway surface liquid in CF may play an inhibitory role to innate antimicrobial molecules [27]. Macrophages undergo apoptosis in the airway contributing their DNA to the mucoid sludge of CF, further promoting chronic infection. Taken together, these data support the notion that the abnormal milieu of the CF lung is an ideal environment for organisms such as NTM to flourish.

Non-tuberculous mycobacteria are resilient organisms that can survive in otherwise "sterilized" environments such as water sources, which can escape immune response via biofilm, and have been shown to have the ability to survive within cultured macrophages [28]. The enhanced ability for intracellular survival of NTM is likely magnified in the CF patient who is thought to be deficient in inducible nitric oxide synthetase and may have a suboptimal oxidative burst necessary to eradicate NTM [29, 30]. This effect may be promoted by co-colonizing bacteria and has been shown in studies on *Pseudomonas* and *Burkholderia* species [31]. Further, mouse models of infection with *M. abscessus* suggest that this particular species of NTM has the ability to induce a lethal, tumor necrosis factor alpha (TNFa)-driven inflammatory response in a murine model [32]. Indeed, the ability of NTM to evade, or at least engage in a war of attrition against, the innate immune system is likely magnified in the CF patient with poor airway clearance and cell-mediated killing. The ubiquitous and resilient nature of NTM combined with the inviting nature of the airway environment in CF acts in synergy to promote NTM colonization and overt infection.

Clinical Features

If isolated, and suspected of causing "treatment-requiring" disease, NTM should be assessed across three facets of the condition: the clinical, the microbiological (see below), and radiologic. Also important is exclusion of other causes, which may cause challenges in CF, given the complex nature of the pulmonary phenotype and coexistent infections, all of which may mimic active NTM infection. Patients with clinically significant NTM will usually demonstrate respiratory symptoms comprised of worsening cough, increased sputum production, dyspnea, and possibly hemoptysis. Patients may also complain of insidious onset of fever, chills, and/or night sweats, anorexia, and weight loss. Obviously, this spectrum of symptoms overlaps a good deal with those of a CF exacerbation, hence the need for a comprehensive overview. Chest radiography may show progression of chronic lung disease or findings consistent with NTM infection (nodular changes, tree-in-bud opacities, or cavitary disease), and spirometry may show worsening obstruction, with reductions in the forced expiratory volume in 1 s (FEV1). These overlap with CF lung disease signs and symptoms, generally, such that the presence of NTM after adequate and aggressive treatment of other colonizing gram-negative and gram-positive bacteria (maximal airway clearance, adequate nutrition, control of other possible CF-related issues) should prompt the clinician to suspect clinically relevant NTM infection that may require targeted therapies [33].

Detection and Diagnosis

The overall national prevalence of regular NTM screening among CF patients with acid-fast bacteria (AFB) smear and culture in the United States is estimated to be ~50% [12]. The most recent recommendations for NTM screening come from a consensus document jointly authored by the US Cystic Fibrosis Foundation and the European Cystic Fibrosis Society. This document recommends that patients who spontaneously produce sputum should have yearly smears and cultures performed to detect acid-fast bacteria. These guidelines also suggest that sputum should *not* be induced when evaluating for NTM and oropharyngeal swabs should *not* be employed [34]. Obviously, physician discretion in individual cases remains – the recommendations are the minimum, and for more complex patients, more frequent sampling, including induction of sputum samples, or bronchoscopic samples may be indicated.

Laboratory techniques have improved over the years – while initial efforts to culture mycobacteria from CF sputum samples were hindered by the overgrowth of *Pseudomonas aeruginosa*, since the development of decontamination processes, the yield and accuracy of AFB cultures have improved [35]. Sample decontamination is an important aspect of specimen preparation. Physicians are urged to develop a close liaison with their microbiologic and laboratory colleagues to ensure optimal interpretation of stain and culture results of airway samples.

As with non-CF bronchiectasis, the diagnosis of NTM infection in CF has traditionally followed the guidelines published by the now somewhat dated 2007 American Thoracic Society/Infectious Disease Society of America document on diagnosis, treatment, and prevention of NTM. Microbiologic diagnosis of NTM infection is made by the presence of two or more sputum cultures positive for NTM, a bronchoalveolar lavage culture that is positive for NTM, or a lung biopsy (transbronchial or otherwise) with NTM histopathologic features (granulomata or presence of AFB) and a positive AFB culture for NTM [7]. The advent of culture-independent molecular methods has been timely and should permit more accurate detection of strains of NTM in CF patients – hopefully to ameliorate the problems that have plagued the interpretation of specimens, such as the aforementioned overgrowth by other bacteria. The techniques are still evolving, however, as initial assessment of a 16 s rRNA probes for the detection of NTM in CF sputum samples proved to be disappointing, possibly because of poor mycobacterial lysis during sample preparation. For example, one study assessed a modified lysis protocol for NTM identification; it was found that polymerase chain reaction (PCR) identified NTM in only 3 of 15 culture-positive samples and 16 s rRNA detection was successful in only 4 of 15 culture-positive samples using that institution's standard sample preparation. After the investigators modified preparation protocol, NTM was accurately detected by PCR and 16S rRNA in 6 of 15 culture-positive samples and 8 of 15 culture-positive samples, respectively [36]. Further work on sample preparation protocols need to be undertaken before rRNA probes and DNA PCR can be employed to detect NTM in CF sputum. An attempt to employ a bovine enzyme-linked immuno-absorbent assay was studied in Scandinavia, which proved to be highly sensitive, but was hindered by a prohibitively high false-positive rate [37]. It appears that progress still needs to be made in culture-independent detection of NTM in CF. For the time being, annual smear and culture remain the most prudent way to assess a CF patient for colonization of NTM.

CF patients who are found to persistently harbor NTM species in their airways on routine screening, or during exacerbations, need to be further assessed for the signs and symptoms of active NTM disease, especially if the number of organisms seems heavy (repeatedly strongly stain positive [3–4+ in our laboratory] with heavy culture growth). Most clinicians collect several serial samples of sputum over time, to assess the "burden" of infection as objectively as possible. These data, with, as noted above, persistent constitutional symptoms such as fever, chills, or night sweats, respiratory symptoms of cough with an increase in sputum production, and/or hemoptysis, despite aggressive standard CF therapies should prompt the clinician to consider active "treatment-requiring" NTM infection more strongly. Clearly, radiographic findings that would be indicative of the presence of NTM add to the diagnostic algorithm for active disease [38]. Representative images (Figs. 1, 2, 3 and 4) are included. The accuracy and interobserver agreement in the radiographic diagnosis of NTM disease has been studied in the comparison of chest CT scans between NTM infection, tuberculosis infection, and cystic fibrosis bronchiectasis. The presence of tree-in-bud opacities, consolidation, and atelectasis were associated with the diagnosis of NTM infection [39]. Thus, a patient with CF who is failing to thrive despite aggressive standard treatment including optimal airway clearance and who is persistently stain and culture positive for NTM with compatible radiologic changes, is likely to need treatment for their NTM.

Treatment

As stated above, the consideration to treat NTM in the CF patient should ideally be made after other airway pathogens are addressed with standard CF care, including appropriate systemic antibiotics and airway clearance, and after the comorbidities

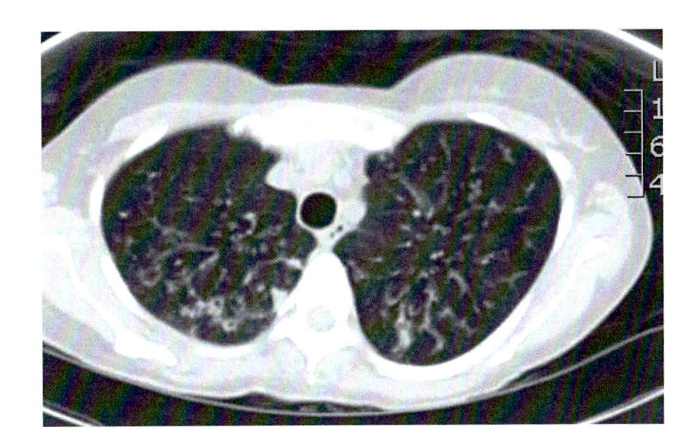

Fig. 1 Computed tomography (CT) cross-sectional image of the upper lung lobes from a patient with cystic fibrosis and MAC infection. These changes (retained airway secretions, dilated airways, and mucus plugging) might be considered to be the same as those of underlying CF

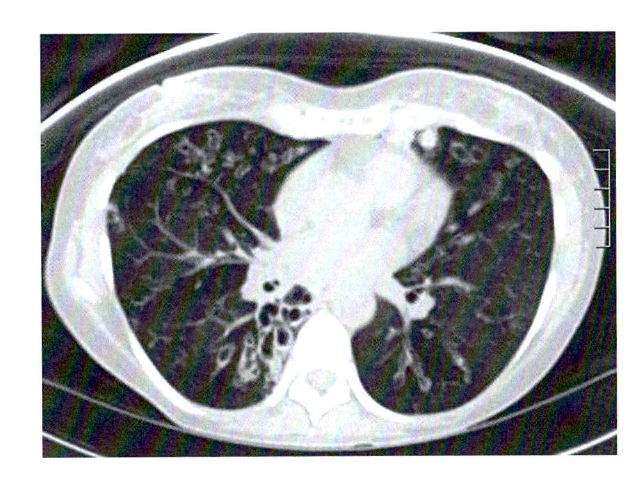

Fig. 2 CT cross-sectional image of the upper lung lobes from the same patient as Fig. 1 showing more pronounced changes (cystic cavitary changes especially) in the dependent areas of the right lower lung

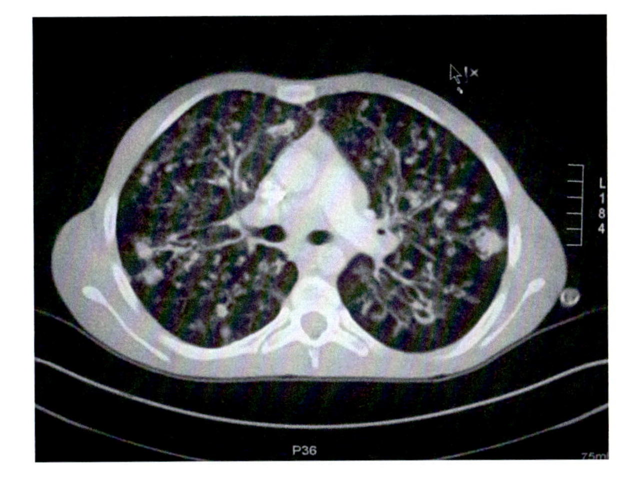

Fig. 3 CT cross-sectional image of the upper lung lobes from a patient with cystic fibrosis and *M. abscessus* infection. The changes are severe, with widespread tree-in-bud, retained secretions and mucus plugging, and early cavitary/cystic abnormalities

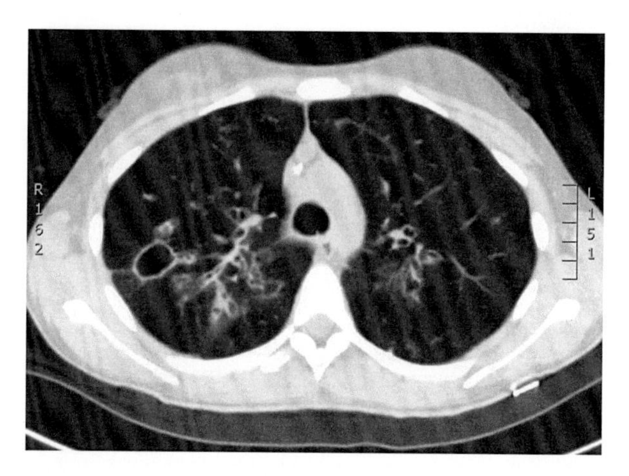

Fig. 4 CT cross-sectional image of the upper lung lobes from a patient with cystic fibrosis and *M. abscessus* infection. While the changes are not as severe as those depicted in Fig. 3, there is a more pronounced cystic cavitary abnormality in the right lower lung and an early cystic change in the left lower lung

of CF are addressed and optimized (allergic bronchopulmonary aspergillosis, diabetes and malnutrition, as examples) [33, 40]. Since the treatment schedules for any of the NTM are onerous, time-consuming, prolonged, and carry the risk of significant adverse events, the decision to treat is not trivial and should be discussed in detail ahead of time with the patient and his or her support system. The placement of long-term indwelling intravenous catheters may be necessary, with all attendant possible complications, as well as the necessity for drug level monitoring. Some patients may simply be unable to handle the rigors of the possible therapies, and some negotiation, for example, on the length of treatment, may be needed and is entirely reasonable. For the purpose of this discussion, we will discuss treatment of MAC and *M. abscessus* separately.

Treatment of MAC in Cystic Fibrosis

At the time of this writing, there is a paucity of randomized, controlled trials concerning the treatment of MAC infection in HIV non-infected patients and none on the treatment of MAC in the CF patient. Treatment of CF patients with MAC infection is generally similar to that of patients without CF, with the exception that patients with CF have altered pharmacokinetics, and thus drug dosing, drug serum levels, and monitoring need extra attention [40].

Determining macrolide sensitivity is a critical first step prior to the initiation of treatment for MAC-associated pulmonary disease in CF. In macrolide-susceptible strains, therapy generally consists of a regimen containing a macrolide, a rifamycin (usually rifampin), and ethambutol [7, 40]. Azithromycin can be considered the preferred macrolide (over clarithromycin) in the treatment of MAC in CF. Not only has azithromycin been shown to have other benefits in CF, pharmacokinetic and dynamic studies on MAC treatment regimens demonstrate azithromycin reaches a therapeutic serum concentration more reliably than does clarithromycin [41, 42].

Those with signs considered to be indicative of a high bacterial "burden," including those with systemic symptoms recalcitrant to standard CF treatments, strongly positive sputum smear, heavy growth on culture, and cavitary lung disease, may garner some benefit from the addition of intravenous amikacin or streptomycin to the aforementioned three-drug oral regimen [7, 40]. In practice, amikacin is more frequently used than streptomycin despite concerns for a moderate increase in ototoxicity. Amikacin via inhalation is one strategy that is frequently employed. Yet, when inhaled, there are concerns that the drug levels at the infected tissues might be lower than with intravenous administration. This is, however, a reasonable strategy during the consolidation phase of *M. abscessus* treatment as described later [40, 43]. One study assessed the role of inhaled liposomal amikacin in a small number of NTM infected patients considered to be refractory to standard treatment (15 with *M. abscessus* and 5 with MAC). Patients were administered inhaled amikacin and followed for a median of 19 months, with 8/20 patients having at least 1 negative culture and 5/20 having persistently negative cultures. Unfortunately, seven patients stopped amikacin in this trial because of toxicity [44]. A larger trial assessed inhaled amikacin in 89 patients finding no statistically significant decrease in a semiquantitative mycobacterial growth scale but did show a greater change of negative culture and improved 6-min walk distance in the treatment group [45]. Further study is needed into the use of inhaled amikacin for treatment of previously treatment-resistant NTM lung infection at this time.

Macrolide resistance due to acquired mutations (not the *erm* mutation of *M. abscessus)*, usually from intentional or unintentional use of azithromycin monotherapy in a MAC-infected patient, can complicate the treatment of the pulmonary disease associated with MAC [38]. Second line of treatment of MAC in the CF infected is not well described. An intravenous aminoglycoside with a rifamycin and ethambutol is a reasonable alternative with the addition of other agent based on clinical experience, a complete list of which is provided in the 2007 ATS guidelines on the treatment on non-tuberculous mycobacteria, and more recently and very comprehensively in the NACF/ESCF guidelines [7, 40]. See Table 1 for treatment strategies of MAC lung disease in CF.

Recently, the anti-leprosy drug clofazimine has shown promise in treating pulmonary MAC infections. In a three-drug regimen with ethambutol and a macrolide in which treatment was carried out for an average of 10 months, cultures became negative after treatment in 26/30 patients, while 20/30 patients continued to have negative cultures at the closing of the study [46]. Clofazimine is therefore an option added to the abovementioned regimen for treatment of macrolide-resistant strains of MAC [40]. As discussed below, surgery can be considered for localized, cavitary disease that is favorable to resection, although these cases are rare in the CF population [40].

Treatment of Mycobacterium Abscessus in Cystic Fibrosis

Subspecies identification seems to be important in predicting the clinical outcome of patients who are treated for *M. abscessus* pulmonary involvement. Patients with the *M. massiliense* subspecies are more likely to have bacterial clearance with treatment and develop less inducible resistance to clarithromycin than subspecies *abscessus* [47]. Juxtaposed to this is subspecies *abscessus* which universally produces a gene, *erm*, that leads to universal macrolide resistance in this particular organism [43]. *Mycobacterium abscessus*, unfortunately, can present with an unpredictable and broad resistance pattern. In the absence of firm evidence-based guidelines, treatment is expert opinion based. Most would consider that the treatment of the infection be considered in phases. The initial "induction" phase, in concept, is an attempt to rapidly decrease the mycobacterial load and usually includes two intravenous antibiotics with favorable activity against the patient's *M. abscessus* strain (usually IV amikacin, with choices of imipenem, cefoxitin, or

Table 1 Suggested regimens for the treatment of MAC and *M. abscessus*

1. First-line treatment in macrolide-sensitive strains of MAC	1. Clarithromycin 500 mg twice daily or azithromycin 500 mg once daily And 2. Rifampin 600 mg once daily (450 mg once daily if less than 50 kg) or rifabutin 300 mg once daily (150 mg if taking CYP3A4 inhibitor or 450–600 mg if taking CYP3A4A inducer) And 3. Ethambutol 15 mg/kg once daily General recommendation is that treatment be continued for 12 months *after* sputum conversion to stain/culture negativity
2. Adjunctive treatment to consider in patients with smear-positive samples, cavitary lung disease on imaging, or signs of systemic illness	Treatment as above with the addition of one of the following to the initial phase (*first 1–3 months*) of treatment: 1. Amikacin 10–30 mg/kg once daily Or 2. Streptomycin 15 mg/kg (max 1000 mg) once daily
3. Second-line treatment of MAC	1. Rifampin 600 mg once daily (450 mg once daily if less than 50 kg) or rifabutin 300 mg once daily (150 mg if taking CYP3A4 inhibitor or 450–600 mg if taking CYP3A4A inducer) And 2. Ethambutol 15 mg/kg once daily And one of 3. Amikacin 10–30 mg once daily Or 4. Moxifloxacin 400 mg once daily Or 5. Clofazimine 50–100 mg daily Treatment duration: similar recommendation as above
M. abscessus:	

(continued)

Table 1 (continued)

1. Initial "intensive" phase	Three months of two intravenous antibiotics with activity against *M. abscessus* and one or more oral medication Two of 1. Amikacin 10–30 mg once daily And/or 2. Tigecycline 100 mg loading dose followed by 50 mg once or twice daily And/or 3. Imipenem 1 g twice daily And/or 4. Cefoxitin 200 mg/kg/day divided into 3 doses (12 g/day maximum) With one or more of 5. Linezolid 600 mg once or twice daily And/or 6. Azithromycin 500 mg daily or clarithromycin 500 mg once daily And/or 7. Moxifloxacin 400 mg daily And/or 8. Minocycline 100 mg twice daily And/or 9. Clofazimine 50–100 mg daily Duration of this phase is controversial, and patients with macrolide-resistant strains may require prolonged courses of IV antibiotics
2. Follow-up "continuation" phase	Continuation phase lasts and indeterminate amount of time and consists of 2–3 of the following, with possible continuation of IV antibiotics in resistant strains. 1. Linezolid 600 mg once or twice daily And/or 2. Azithromycin 500 mg daily or clarithromycin 500 mg once daily And/or 3. Moxifloxacin 400 mg daily And/or 4. Minocycline 100 mg twice daily And/or 5. Clofazimine 50–100 mg daily And/or 6. Inhaled amikacin 250–500 mg once or twice daily

Adapted from Ref. [40]

tigecycline) with concomitant oral medications (linezolid, clofazimine, moxifloxacin, or minocycline) for up to 3 months. An exception that may require prolonged IV antibiotic treatment is infection with the aforementioned *erm*-positive strain based on concerns of efficacy of exclusively oral regimens. Note that the treatment of these particular infections is controversial, and the reader is referred to another chapter of this book, the Rapidly Growing Mycobacterium. The NACF/ESCF guidelines provide a treatment timetable and dosing regimens summarized in the accompanying Table 1. After this, patients may be treated with two oral antibiotics, to include a macrolide with or without inhaled antibiotics (amikacin usually) for up to 16 months of total treatment [40]. There is no evidence-based recommendation for total duration, but most experts and guidelines recommend treatment for 12 months post clearing of sputum (as measured by monthly specimen collection – either expectorated or induced), if "cure" is the goal. Thus, the optimal duration of treatment is not known. Treatment failure is not uncommon and patients may not clear their sputum. Despite this, patients who do not clear their sputum cultures may benefit from repeated or continued treatments, much like repeatedly treating "non-eradicable" gram-negative organisms in the patient with CF [40].

Unfortunately the intravenous antibiotics used in *M. abscessus* infection suffer from a remarkable side effect profile which, together with their unpredictable success rate, adds to a certain nihilistic attitude to its treatment in general. Amikacin is generally considered the anchor of an antibiotic regimen directed at *M. abscessus* and has a side effect profile similar to other aminoglycosides. A study of 52 *M. abscessus* infected patients found that treatment with tigecycline for greater than 1 month led to a 60% clinical improvement rate (10/15 for cystic fibrosis patients) but greater than 90% of patients experienced antibiotic-related side effects, most common being nausea and vomiting [42]. Imipenem has the most favorable side effect profile when used in combination with amikacin, but some would argue its utility. Long-term cefoxitin can cause neutropenia, thrombocytopenia, and hepatitis, and eventually allergy is also very common [48]. While this is problematic, desensitization is possible [49]. The "consolidation" phase of *M. abscessus* treatment usually centers around the use of oral agents, which can be problematic based on the broad resistance patterns of particular strains and subspecies of *M. abscessus*, as noted above. Macrolides are thought to be consistently usable for treatment, but inducible resistance may hinder the use of this class [40]. Despite concerns surrounding clarithromycin and inducible resistance, there may be no difference in the induction of macrolide resistance between clarithromycin and azithromycin [50]. Linezolid is a promising therapy as it has activity against half of *M. abscessus* isolates, but treatment at the highest dose (600 mg twice daily) can lead to hematologic and neurologic side effects [40]. The oral leprosy treatment, clofazimine, is a promising treatment because of its in vitro activity against *M. abscessus* but is not commercially available in the United States at the time of this writing, although it may be obtained for individual patients with permission from the FDA and the manufacturer through forms at the following website: https://www.ntminfo.org/component/k2/item/39-clofazimine [51]. Side effects of this treatment include reddening and dryness of the skin, erythroderma, intestinal obstruction, and retinopathy [52]. Table 1 shows treatment strategies for *M. abscessus* lung disease.

Monitoring Treatment

During the course of treatment, it is important to both monitor for efficacy of treatment and for adverse reaction to medication. The combined US Cystic Fibrosis Foundation and European Cystic Fibrosis Society recommend that during treatment sputum be sampled via the expectorated or hypertonic saline-induced route every 4–8 weeks to assess for microbiological clearance [40]. Likewise, a strategy for monitoring the side effects of treatment in real time needs to be rigorously developed out before beginning treatment; the reader is referred to the comprehensive Table 3 of the NACF/ESCF guidelines on the treatment of NTM in CF [40]. Further, the combined groups recommend high-resolution computed tomography of the chest (HRCT) be performed before and after treatment to monitor for radiographic response [40].

Special Considerations

Lung Transplantation

Because CF patients frequently require lung transplantation as a life-prolonging measure, considering the role of NTM in the lung transplant candidate is important. Many centers regard the presence of mycobacteria in the airways of patients with CF as a contraindication, although there are no data to support that (unlike the situation with *Burkholderia cenocepacia*, as a comparison pretransplant infection, where the effect on posttransplant mortality has been well documented). It is likely team decisions on candidacy are based on local center experience in dealing with mycobacterial infections pre- and posttransplantation. It was the consensus recommendation in the NACF and ESCF summary recently that NTM (both MAC and *M. abscessus*) should *not* preclude transplant candidacy consideration (assuming other candidacy factors are favorable) and are not a major cause of posttransplant mortality [49, 53, 54]. One study looking specifically at *M. abscessus* (the usual cause for concern with lung transplant teams internationally) did not reveal a mortality difference in individuals infected with *M. abscessus* prior to transplantation when compared to the general CF posttransplant population at the same center (though morbidity related to skin and incision site infections may require prolonged posttransplant therapy for mycobacterial disease) [49]. Macrolide-resistant *M. abscessus* is usually regarded as an absolute contraindication to lung transplant at most centers.

Alternate Therapies

Replacement of interferon-gamma in those deficient in the cytokine or as an adjuvant therapy has been evaluated in very small studies in the non-CF population. One study, with only 18 patients in the treatment arm and 14 in placebo, received a course of interferon-gamma for 24 weeks, and there was a significant improvement in a composite endpoint of symptomatic improvement, radiographic improvement, and microbiologic improvement after the treatment phase [55]. The small sample size, lack of similar studies, and use of a composite endpoint need to be noted for this work.

Surgery (non-transplant) is an option to be considered in extreme circumstances. Despite the existence of case series on the matter, it appears it should be reserved for patients with severe, localized, cavitary disease that is nonresponsive to treatment [40, 56]. Therefore, because localized disease is rare in CF patients who have NTM infections, surgery is an option only in rare cases.

Conclusions

Cystic fibrosis patients are at a high risk of developing active NTM pulmonary disease because of their impaired airway defense mechanisms and susceptible lung microenvironment. Annual smear and culture assays for acid-fast bacilli should be performed in CF patients who spontaneously expectorate sputum. When active "treatment-requiring NTM infection is under consideration, the clinician should ensure that there is a plan for optimal management of CF pulmonary and extrapulmonary disease underway." Repeated strongly positive stain and AFB cultures, in the presence of persistent symptoms, and HRCT imaging findings consistent with NTM infection must be assessed in toto, when considering treatment. MAC and M. *abscessus* are treated differently. MAC treatment while certainly not straightforward, and is prolonged, is less complicated than that of *M. abscessus*, which requires organism speciation and drug sensitivity testing, and usually involves an intensive eradication phase for up to 3 months and consolidation treatment with inhaled and oral medications for a very prolonged period of time with all the attendant drug and line adverse event possibilities. Patients being treated for NTM need frequent microbiologic and drug toxicity screening. Interferon-gamma replacement and surgery have been used in rare instances and are not recommended outside of very specific situations. Lung transplant candidacy for patients with NTM remains controversial, but patients with NTM and end-stage disease judged to have limited survival without lung transplant deserve at least a close look to assess for candidacy, as published data suggests no effect on short- or longer-term mortality.

References

1. Elborn JS. Cystic fibrosis. Lancet (London, England). 2016;388(10059):2519–31.
2. Stoltz DA, Meyerholz DK, Welsh MJ. Origins of cystic fibrosis lung disease. N Engl J Med. 2015;372(4):351–62.
3. Stewart CS, Pepper MS. Cystic fibrosis in the African diaspora. Ann Am Thorac Soc. 2016;14(1):1–7.
4. Davies JC, Alton EWFW, Bush A. Cystic fibrosis. BMJ Br Med J. 2007;335(7632):1255–9.
5. Olivier KN, Weber DJ, Wallace RJ Jr, Faiz AR, Lee JH, Zhang Y, et al. Nontuberculous mycobacteria. I: multicenter prevalence study in cystic fibrosis. Am J Respir Crit Care Med. 2003;167(6):828–34.
6. Olivier KN, Weber DJ, Lee JH, Handler A, Tudor G, Molina PL, et al. Nontuberculous mycobacteria. II: nested-cohort study of impact on cystic fibrosis lung disease. Am J Respir Crit Care Med. 2003;167(6):835–40.
7. Griffith DE, Aksamit T, Brown-Elliott BA, Catanzaro A, Daley C, Gordin F, et al. An official ATS/IDSA statement: diagnosis, treatment, and prevention of nontuberculous mycobacterial diseases. Am J Respir Crit Care Med. 2007;175(4):367–416.
8. Olivier KN, Yankaskas JR, Knowles MR. Nontuberculous mycobacterial pulmonary disease in cystic fibrosis. Semin Respir Infect. 1996;11(4):272–84.
9. Donohue MJ, Wymer L. Increasing prevalence rate of nontuberculous mycobacteria infections in five states, 2008–2013. Ann Am Thorac Soc. 2016;13(12):2143–50.
10. Adjemian J, Olivier KN, Prevots DR. Epidemiology of Pulmonary Nontuberculous Mycobacterial Sputum Positivity in Patients with Cystic Fibrosis in the United States, 2010–2014. Annals of the American Thoracic Society. 15(7):2018.
11. Pierre-Audigier C, Ferroni A, Sermet-Gaudelus I, Le Bourgeois M, Offredo C, Vu-Thien H, et al. Age-related prevalence and distribution of nontuberculous mycobacterial species among patients with cystic fibrosis. J Clin Microbiol. 2005;43(7):3467–70.
12. Adjemian J, Olivier KN, Prevots DR. Nontuberculous mycobacteria among patients with cystic fibrosis in the United States: screening practices and environmental risk. Am J Respir Crit Care Med. 2014;190(5):581–6.
13. Prevots DR, Adjemian J, Fernandez AG, Knowles MR, Olivier KN. Environmental risks for nontuberculous mycobacteria. Individual exposures and climatic factors in the cystic fibrosis population. Ann Am Thorac Soc. 2014;11(7):1032–8.
14. Qvist T, Gilljam M, Jonsson B, Taylor-Robinson D, Jensen-Fangel S, Wang M, et al. Epidemiology of nontuberculous mycobacteria among patients with cystic fibrosis in Scandinavia. J Cyst Fibros Off J Eur Cyst Fibros Soc. 2015;14(1):46–52.
15. Levy I, Grisaru-Soen G, Lerner-Geva L, Kerem E, Blau H, Bentur L, et al. Multicenter cross-sectional study of nontuberculous mycobacterial infections among cystic fibrosis patients, Israel. Emerg Infect Dis. 2008;14(3):378–84.
16. Martiniano SL, Sontag MK, Daley CL, Nick JA, Sagel SD. Clinical significance of a first positive nontuberculous mycobacteria culture in cystic fibrosis. Ann Am Thorac Soc. 2014;11(1):36–44.
17. Renna M, Schaffner C, Brown K, Shang S, Tamayo MH, Hegyi K, et al. Azithromycin blocks autophagy and may predispose cystic fibrosis patients to mycobacterial infection. J Clin Invest. 2011;121(9):3554–63.
18. Bryant JM, Grogono DM, Greaves D, Foweraker J, Roddick I, Inns T, et al. Whole-genome sequencing to identify transmission of Mycobacterium abscessus between patients with cystic fibrosis: a retrospective cohort study. Lancet (London, England). 2013;381(9877):1551–60.
19. Bryant JM, Grogono DM, Rodriguez-Rincon D, Everall I, Brown KP, Moreno P, et al. Emergence and spread of a human-transmissible multidrug-resistant nontuberculous mycobacterium. Science (New York, NY). 2016;354(6313):751–7.

20. Kunimoto DY, Peppler MS, Talbot J, Phillips P, Shafran SD. Analysis of Mycobacterium avium complex isolates from blood samples of AIDS patients by pulsed-field gel electrophoresis. J Clin Microbiol. 2003;41(1):498–9.
21. Johnson MM, Odell JA. Nontuberculous mycobacterial pulmonary infections. J Thorac Dis. 2014;6(3):210–20.
22. Foundation CF. Infection prevention and control guidelines for cystic fibrosis: 2013 update. Infect Control Hosp Epidemiol. 2014;35(S1):S1–S67.
23. Martin C, Burgel PR, Lepage P, Andrejak C, de Blic J, Bourdin A, et al. Host-microbe interactions in distal airways: relevance to chronic airway diseases. Eur Respir Rev: Off J Eur Respir Soc. 2015;24(135):78–91.
24. Tang AC, Turvey SE, Alves MP, Regamey N, Tummler B, Hartl D. Current concepts: host-pathogen interactions in cystic fibrosis airways disease. Eur Respir Rev: Off J Eur Respir Soc. 2014;23(133):320–32.
25. Matsui H, Verghese MW, Kesimer M, Schwab UE, Randell SH, Sheehan JK, et al. Reduced three-dimensional motility in dehydrated airway mucus prevents neutrophil capture and killing bacteria on airway epithelial surfaces. J Immunol. 2005;175(2):1090–9.
26. Hartl D, Latzin P, Hordijk P, Marcos V, Rudolph C, Woischnik M, et al. Cleavage of CXCR1 on neutrophils disables bacterial killing in cystic fibrosis lung disease. Nat Med. 2007;13(12):1423–30.
27. Pezzulo AA, Tang XX, Hoegger MJ, Abou Alaiwa MH, Ramachandran S, Moninger TO, et al. Reduced airway surface pH impairs bacterial killing in the porcine cystic fibrosis lung. Nature. 2012;487(7405):109–13.
28. Orme IM, Ordway DJ. Host response to nontuberculous mycobacterial infections of current clinical importance. Infect Immun. 2014;82(9):3516–22.
29. Kelley TJ, Drumm ML. Inducible nitric oxide synthase expression is reduced in cystic fibrosis murine and human airway epithelial cells. J Clin Invest. 1998;102(6):1200–7.
30. SenGupta S, Hittle LE, Ernst RK, Uriarte SM, Mitchell TC. A Pseudomonas aeruginosa hepta-acylated lipid A variant associated with cystic fibrosis selectively activates human neutrophils. J Leukoc Biol. 2016;100(5):1047–59.
31. Assani K, Shrestha CL, Robledo-Avila F, Rajaram MV, Partida-Sanchez S, Schlesinger LS, et al. Human cystic fibrosis macrophages have defective calcium-dependent protein kinase C activation of the NADPH oxidase, an effect augmented by Burkholderia cenocepacia. J Immunol. 2017;198(5):1985–94.
32. Catherinot E, Clarissou J, Etienne G, Ripoll F, Emile JF, Daffe M, et al. Hypervirulence of a rough variant of the Mycobacterium abscessus type strain. Infect Immun. 2007;75(2):1055–8.
33. Nick JA, Pohl K, Martiniano SL. Nontuberculous mycobacterial infections in cystic fibrosis: to treat or not to treat? Curr Opin Pulm Med. 2016;22(6):629–36.
34. Floto RA, Olivier KN, Saiman L, Daley CL, Herrmann JL, Nick JA, et al. US Cystic Fibrosis Foundation and European Cystic Fibrosis Society consensus recommendations for the management of non-tuberculous mycobacteria in individuals with cystic fibrosis: executive summary. Thorax. 2016;71(1):88–90.
35. Whittier S, Olivier K, Gilligan P, Knowles M, Della-Latta P. Proficiency testing of clinical microbiology laboratories using modified decontamination procedures for detection of nontuberculous mycobacteria in sputum samples from cystic fibrosis patients. The Nontuberculous Mycobacteria in Cystic Fibrosis Study Group. J Clin Microbiol. 1997;35(10):2706–8.
36. Caverly LJ, Carmody LA, Haig SJ, Kotlarz N, Kalikin LM, Raskin L, et al. Culture-independent identification of nontuberculous mycobacteria in cystic fibrosis respiratory samples. PLoS One. 2016;11(4):e0153876.
37. Qvist T, Pressler T, Katzenstein TL, Hoiby N, Collins MT. Evaluation of a bovine antibody test for diagnosing Mycobacterium avium complex in patients with cystic fibrosis. Pediatr Pulmonol. 2016;52(1):34–40.
38. Park IK, Olivier KN. Nontuberculous mycobacteria in cystic fibrosis and non-cystic fibrosis bronchiectasis. Semin Respir Crit Care Med. 2015;36(2):217–24.

39. Kwak N, Lee CH, Lee HJ, Kang YA, Lee JH, Han SK, et al. Non-tuberculous mycobacterial lung disease: diagnosis based on computed tomography of the chest. Eur Radiol. 2016;26(12):4449–56.
40. Floto RA, Olivier KN, Saiman L, Daley CL, Herrmann JL, Nick JA, et al. US Cystic Fibrosis Foundation and European Cystic Fibrosis Society consensus recommendations for the management of non-tuberculous mycobacteria in individuals with cystic fibrosis. Thorax. 2016;71(Suppl 1):i1–22.
41. van Ingen J, Egelund EF, Levin A, Totten SE, Boeree MJ, Mouton JW, et al. The pharmacokinetics and pharmacodynamics of pulmonary Mycobacterium avium complex disease treatment. Am J Respir Crit Care Med. 2012;186(6):559–65.
42. Ribeiro CM, Hurd H, Wu Y, Martino ME, Jones L, Brighton B, et al. Azithromycin treatment alters gene expression in inflammatory, lipid metabolism, and cell cycle pathways in well-differentiated human airway epithelia. PLoS One. 2009;4(6):e5806.
43. Chmiel JF, Aksamit TR, Chotirmall SH, Dasenbrook EC, Elborn JS, LiPuma JJ, et al. Antibiotic management of lung infections in cystic fibrosis. II. Nontuberculous mycobacteria, anaerobic bacteria, and fungi. Ann Am Thorac Soc. 2014;11(8):1298–306.
44. Olivier KN, Shaw PA, Glaser TS, Bhattacharyya D, Fleshner M, Brewer CC, et al. Inhaled amikacin for treatment of refractory pulmonary nontuberculous mycobacterial disease. Ann Am Thorac Soc. 2014;11(1):30–5.
45. Olivier KN, Griffith DE, Eagle G, McGinnis Ii JP, Micioni L, Liu K, et al. Randomized trial of liposomal amikacin for inhalation in nontuberculous mycobacterial lung disease. Am J Respir Crit Care Med. 2016;195(6):814–23.
46. Field SK, Cowie RL. Treatment of Mycobacterium avium-intracellulare complex lung disease with a macrolide, ethambutol, and clofazimine. Chest. 2003;124(4):1482–6.
47. Koh WJ, Jeon K, Lee NY, Kim BJ, Kook YH, Lee SH, et al. Clinical significance of differentiation of Mycobacterium massiliense from Mycobacterium abscessus. Am J Respir Crit Care Med. 2011;183(3):405–10.
48. Jeon K, Kwon OJ, Lee NY, Kim BJ, Kook YH, Lee SH, et al. Antibiotic treatment of Mycobacterium abscessus lung disease: a retrospective analysis of 65 patients. Am J Respir Crit Care Med. 2009;180(9):896–902.
49. Lobo LJ, Chang LC, Esther CR Jr, Gilligan PH, Tulu Z, Noone PG. Lung transplant outcomes in cystic fibrosis patients with pre-operative Mycobacterium abscessus respiratory infections. Clin Transpl. 2013;27(4):523–9.
50. Maurer FP, Castelberg C, Quiblier C, Bottger EC, Somoskovi A. Erm(41)-dependent inducible resistance to azithromycin and clarithromycin in clinical isolates of Mycobacterium abscessus. J Antimicrob Chemother. 2014;69(6):1559–63.
51. van Ingen J, Totten SE, Helstrom NK, Heifets LB, Boeree MJ, Daley CL. In vitro synergy between clofazimine and amikacin in treatment of nontuberculous mycobacterial disease. Antimicrob Agents Chemother. 2012;56(12):6324–7.
52. Kar HK, Gupta R. Treatment of leprosy. Clin Dermatol. 2015;33(1):55–65.
53. Malouf MA, Glanville AR. The spectrum of mycobacterial infection after lung transplantation. Am J Respir Crit Care Med. 1999;160(5 Pt 1):1611–6.
54. Skolnik K, Kirkpatrick G, Quon BS. Nontuberculous mycobacteria in cystic fibrosis. Curr Treat Options Infect Dis. 2016;8(4):259–74.
55. Milanes-Virelles MT, Garcia-Garcia I, Santos-Herrera Y, Valdes-Quintana M, Valenzuela-Silva CM, Jimenez-Madrigal G, et al. Adjuvant interferon gamma in patients with pulmonary atypical Mycobacteriosis: a randomized, double-blind, placebo-controlled study. BMC Infect Dis. 2008;8:17.
56. Rolla M, D'Andrilli A, Rendina EA, Diso D, Venuta F. Cystic fibrosis and the thoracic surgeon. Eur J Cardio-thorac Surg: Off J Eur Assoc Cardio-thorac Surg. 2011;39(5):716–25.

Healthcare-Associated Outbreaks and Pseudo-Outbreaks of Nontuberculous Mycobacteria

Barbara A. Brown-Elliott and Richard J. Wallace Jr.

Introduction

Nontuberculous mycobacteria (NTM) are hardy organisms ubiquitous in the environment. They are common microorganisms in municipal and hospital water systems, including hot water and chlorinated systems [1–4]. Not only are the NTM able to survive in harsh environmental conditions, but they are also resistant to many disinfectants commonly used in healthcare facilities [3, 5, 6]. Indeed, some of the species, including *M. abscessus*, *M. fortuitum*, and *M. chelonae*, are among the most commonly encountered and difficult to treat healthcare-associated infections [7]. All species of NTM are hydrophobic, enabling the formation of biofilms, the layer of organisms at the water-solid interfaces of pipes in tubings in hospital and municipal (household) water supplies. Some species, especially *M. abscessus* complex, exhibit two colonial morphotypes, smooth and rough. While the rough morphotype has been associated with increased virulence, the smooth type has an increased production of glycopeptidolipids that is believed to increase their mobility and ability to form biofilms which increase their resistance to antimicrobials and disinfectants [8, 9]. It is hypothesized that up to 90% of all biofilms contain NTM [10–12]. The incidence of these organisms in the healthcare environment is steadily increasing causing a multiplicity of opportunistic infections including postsurgical infection (e.g., cardiopulmonary and cosmetic surgery); central line infections; bacteremia; ophthalmic infections; sternal wound infections; infected augmentation mammaplasty sites; infections following insertion of prosthetic devices including prosthetic

B. A. Brown-Elliott (✉)
Department of Microbiology, Mycobacteria/Nocardia Research Laboratory,
The University of Texas Health Science Center, Tyler, TX, USA
e-mail: Barbara.Elliott@uthct.edu

R. J. Wallace Jr.
Department of Microbiology, Mycobacteria/Nocardia Laboratory,
The University of Texas Health Science Center, Tyler, TX, USA

D. E. Griffith (ed.), *Nontuberculous Mycobacterial Disease*, Respiratory Medicine,
https://doi.org/10.1007/978-3-319-93473-0_18

heart valves, lens implants, artificial knee and hips, and metal rods inserted to stabilize bones following fractures [13–23]; and invasive and disseminated diseases, especially among immunocompromised patients [24]. Additionally, outbreaks of postinjection abscesses associated with the *Mycobacterium abscessus* complex have previously been reported throughout the world including Colombia, China, Korea, and the USA [25–29] (see Table 1) primarily in immunocompetent individuals.

It should be noted that mycobacterial taxonomy in the pre-molecular era (prior to 1990) was not as advanced as currently due to the unavailability of molecular testing during those years. Thus, for example, species that were reported as *M. chelonae* (previously *M. chelonae* subsp. *chelonae*) may be modern technology classified as *M. abscessus* complex (previously *M. chelonae* subsp. *abscessus*), and some species and subspecies (e.g., subsp. *massiliense*) were not recognized at that time.

Waterborne Outbreaks

The majority of hospital- and healthcare-associated NTM outbreaks and pseudo-outbreaks have involved exposure of invasive devices or non-intact skin and soft tissue to tap water and water-based solutions [30, 31] including in tap water contaminated dialysis systems, cardioplegia solution contaminated by tap water, ice machines, surgical and bronchoscopic instruments rinsed in tap water, and most recently heater-cooler units used in cardiac surgery (see Table 1). A 1988 study of hemodialysis centers in the USA by Carson and colleagues reported that 55% of incoming city water contained NTM, the majority of which were rapidly growing mycobacteria (RGM) [32]. More recent studies in the twenty-first century estimate 20–60% of hospital water system contain NTM [11].

A recent systematic review by Li and colleagues reviewed 21 outbreaks associated with NTM from 1982 to 2015 [33], including 2 outbreaks of *M. avium* in a hospital hot water system involving disseminated disease in AIDS patients [33–35]. Nine outbreaks of *M. chelonae*/*abscessus* complex from 1982 to 2009 including infections associated with tap water following the use of tympanostomy tubes in 17 pediatric patients, liposuction in 34 patients, 8 patients who received injections in a podiatry clinic, 22 patients infected post-rhinoplasty, and 6 patients post-sternotomy with *M. fortuitum* and *M. abscessus* complex were detailed [33, 36–40] (see Table 1). Additionally, the authors reviewed three outbreaks associated with hemodialysis and peritoneal dialysis equipment which were improperly disinfected [41–43] and multiple patients with severe subcutaneous infections resulting from rinsing a multiple inoculation device in tap water in a mesotherapy clinic at a tertiary care center in France [44].

The investigators also reviewed two cases (non-outbreaks) with *M. fortuitum* including a laboratory-confirmed case of breast infection in a patient with breast cancer with *M. fortuitum* related to the hospital water supply [45] and a case of nosocomial acquisition of *M. fortuitum* disseminated infection in a patient with leukemia linked to showering with tap water [46].

Table 1 Nontuberculous mycobacteria outbreaks and pseudo-outbreaks

Organism type of infection	References	Year
Rapidly growing mycobacteria		
M. chelonae		
Peritoneal dialysis	Band et al. [43]	1982
Tympanostomy	Lowry et al. [36]	1988
Hemodialysis	Band et al. [43]	1982
	Bolan et al. [42]	1985
	Lowry et al. [41]	1990
Podiatric injections	Wenger et al. [38]	1990
Rhinoplasty	Soto et al. [39]	1991
Liposuction	Meyers et al. [37]	2002
Pulmonary (pseudo-outbreaks associated with contamination of automatic bronchoscopy washer	Chroneou et al. [100]	2008
Mesotherapy	Carbonne et al. [44]	2009
Cutaneous (dermal facial injections)	Rodriguez et al. [54]	2013
Hematopoietic cell transplant	Iroh tam et al. [53]	2014
Ophthalmic infections	Edens et al. [17]	2015
***M. fortuitum* group**		
Sternotomy	Kuritsky et al. [40]	1983
Breast abscess	Kauppinen et al. [46]	1999
Septicemia (disseminated)		
Breast (reconstruction) abscess	Jaubert et al. [45]	2015
Cardiac surgery	Robicsek et al. [81]	1978
Cutaneous (post-nail salon whirlpool baths)	Winthrop et al. [76]	2004
Pulmonary, localized, infected port peritonitis	Brown-Elliott et al. [55]	2011
Wounds (associated with "medical tourism")	Schnabel et al. [68]	2016
Pulmonary pseudo-outbreak associated with contaminated ice machine	Laussucq et al. [21]	1988
***M. mucogenicum* group** (*M. phocaicum, M. mucogenicum*)		
Catheter sepsis (malignancy)	Baird et al. [50]	2011
	Kline et al. [49]	2004
	Cooksey et al. [51]	2008
	Livni et al. [52]	2008
	Tagashira et al. [20]	2015
Sepsis (sickle cell)	Ashraf et al. [18]	2012
M. immunogenum		
Pulmonary (pseudo-outbreak of contaminated bronchoscopes)	Wilson et al. [6]	2001
Hypersensitivity pneumonitis (associated with metal-working fluids)	Wilson et al. [6]	2001
Keratitis	Sampaio et al. [114]	2006
Blepharoplasty	Flesner et al. [47]	2011

(continued)

Table 1 (continued)

Organism type of infection	References	Year
***M. abscessus* complex**		
Wounds (postsurgical)	Chadha et al. [56]	1998
Ophthalmic infections	Hung et al. [118]	2016
M. abscessus subsp. *abscessus* and subsp. *massiliense*		
Pulmonary post (lung transplant and cardiopulmonary bypass)	Baker et al. [61]	2017
Postinjection abscess	Villanueva et al. [25]	1997
	Zhibang et al. [26]	2002
M. abscessus subsp. *massiliense*	Kim et al. [27]	2007
M. abscessus complex		
Pseudo-outbreak associated with bronchoscope washer	Fraser et al. [107]	1992
Pseudo-outbreak, laboratory-distilled water	Lai et al. [110]	1998
Wounds (associated with "medical tourism")	Schnabel et al. [68]	2016
Sternotomy	Kuritsky et al. [40]	1983
M. wolinskyi		
Septicemia (aortic prosthesis associated with heater-cooler units	Dupont et al. [87]	2016
Wound infections (post-cardiothoracic surgery)	Nagpal et al. [86]	2014
M. mageritense		
Cutaneous (post-nail salon whirlpool baths)	Gira et al. [79]	2004
M. neoaurum		
Catheter sepsis	Baird et al. [50]	2011
Slowly growing mycobacteria		
***M. avium* complex**		
Septicemia (HIV/AIDS)	von Reyn et al. [34]	1994
Pulmonary (HIV/AIDS)	Tobin-D'Angelo et al. [35]	2004
M. avium		
Disseminated (AIDS and non-AIDS associated with contaminated water)	Aronson et al. [60]	1999
M. chimaera		
Endocarditis, septicemia ventricular graft infection (associated with heater-cooler units)	Sax et al. [90]	2015
	Kohler et al. [89]	2015
	Perkins et al. [92]	2016
	Marra et al. [91]	2017
	Lyman et al. [88]	2017
M. genavense		
Disseminated in HIV patients on corticosteroids associated with hospital tap water	Hillebrand-Haverkort et al. [58]	1999
M. gordonae		
Pulmonary, pseudo-outbreak associated with contaminated bronchoscope equipment	Scorzolini et al. [101]	2016

Table 1 (continued)

Organism type of infection	References	Year
Pulmonary pseudo-outbreak associated with contaminated tap water supply	Zlojtro et al. [108]	2015
M. kansasii		
Gastric washing contamination with tap water	Lévy-Frébault et al. [22]	1983
M. lentiflavum		
Disseminated, cervical lymphadenitis associated with drinking water	Marshall et al. [23]	2011
M. simiae		
Pulmonary	Conger et al. [59]	2004
M. szulgai		
Keratitis following laser-assisted in situ keratomileusis (LASIK)	Holmes et al. [15]	2002
M. xenopi		
Surgical site infection following discovertebral surgery (associated with contaminated water supply)	Astagneau et al. [48]	2001

Six outbreaks of *Mycobacterium mucogenicum* from 2001 to 2012 from contamination of central venous catheter sites with tap water in multiple immunocompromised patients were also detailed [33]. There was also one case series of *Mycobacterium immunogenum* associated with direct application of ice to eyelids following blepharoplasty [47] involving 3 patients and 49 patients with postsurgical *Mycobacterium xenopi* infections following the use of tap water to rinse discovertebral surgical equipment [48] (see Table 1).

Twelve additional cases and outbreaks from 1997 to 2015 associated with water reservoirs were also summarized by Kanamori and colleagues [24]. These included four cases of *M. mucogenicum* again involving catheter sepsis in a central venous line following exposure to tap water [49–52]. Two outbreaks of *M. chelonae* occurred in hematopoietic all transplant patients exposed to contaminated water and ice [53] and cutaneous infections in patients after cosmetic dermal injections following application of nonsterile ice to the skin [54]. Contaminated water was also incriminated in a large case series of *Mycobacterium porcinum* pulmonary and extrapulmonary infections [55]. Furthermore, inadequate sterilization of surgical instruments resulted in an outbreak of *M. abscessus* in postsurgical wounds [56] (see Table 1).

Most waterborne outbreaks have involved RGM as noted in the above reviews [2, 24, 33, 57]. However, Kanamori also noted four unusual outbreaks with slowly growing mycobacterial species including *Mycobacterium genavense* disseminated disease in three HIV-infected patients following ingestion of contaminated water [58] and a second outbreak involving pulmonary infections with *Mycobacterium simiae* in patients with lung cancer and chronic pulmonary disease [59] (see Table 1). A third unusual outbreak involved 49 patients with *M. xenopi* traced to the use of tap water to rinse surgical instruments following vertebral disk surgeries [48].

Another outbreak involved another slowly growing NTM, *M. szulgai*, which caused keratitis following exposure of medication syringes to contaminated ice following laser-assisted in situ keratomileusis (LASIK) [57]. *M. avium* has also been associated with waterborne outbreaks [34, 35]. A 1999 study of the genetic relatedness of isolates of *M. avium* in potable water in Los Angeles, California, showed a close relationship between clinical (patient) and water isolates and thus proposed potable water as a presumptive source of nosocomial infection [60].

Recently a large biphasic outbreak of *M. abscessus* complex (*M. abscessus* subsp. *abscessus* and *M. abscessus* subsp. *massiliense*) among lung transplant and cardiopulmonary surgical patients at a tertiary care hospital in the southeastern USA was reported. The investigators initiated a multidisciplinary epidemiologic field and laboratory investigation. Phase 1 of the outbreak occurred from August 2014 to May 2014 [61]. Thirty-six of 71 (51%) Phase 1 cases were lung transplant patients with respiratory cultures positive for *M. abscessus* complex. Seventeen patients (27%) died within 60 days of the first positive culture. After strict control measures including elimination of tap water to the aerodigestive tract among high-risk patients were implemented, the incidence rate decreased from 3.0 cases per 10,000 patient days to baseline (0.7 cases/10,000 patient days). In Phase 2 (December 2014 to June 2015), 12 of 24 (50%) cases occurred in cardiac surgery patients with invasive infections. Twenty-one patients required cardiopulmonary bypass during surgery. All patients had extensive perioperative comorbidities and complicated postoperative courses. Nine patients died within 60 days of the first positive cultures. After implementation of an intensified disinfection protocol and the use of sterile water in the heater-cooler unit (HCU) of the cardiopulmonary bypass machines, Phase 2 of the outbreak was also resolved. Additional water engineering mitigation intervention to decrease improve water flow and increase disinfectant levels also contributed to resolution of the outbreak. Complete eradication of *M. abscessus* complex and biofilms from tap water and piping was not possible given the mycobacterial burden in the hospital water supply and environmental persistence of NTM. However, a sustained decrease in cases of *M. abscessus* complex was achieved following institution of mitigation strategies [61].

Cosmetic Procedure Outbreaks

Cosmetic surgical procedures including liposuction, breast augmentation, and mesotherapy have also been associated with exposure to tap water, resulting in outbreaks of *Mycobacterium chelonae*, *Mycobacterium fortuitum*, and *Mycobacterium immunogenum* [16, 37, 62–66]. In 2008, outbreaks involving *M. abscessus* complex and *M. fortuitum* among US tourists ("lipotourists") who underwent abdominoplasty in the Dominican Republic (DR) were reported [67]. Schnabel and coworkers reported 21 case patients from 6 states who had surgery in 1 of 5 DR clinics from 2013 to 2014. Ninety-two percent of the patients were culture positive for *M. abscessus* complex [68]. Understanding the role of "medical tourism" in risk assessment for disease and improving patient protection in this context require continual monitoring by the international public health agencies and medical institutions and heightened vigilance among immigrants and travelers. Additional reports of NTM outbreaks have been described following tattooing and fractionated CO_2 laser cosmetic resurfacing procedures [69, 70].

Biomesotherapy, a new therapeutic procedure for pain management and general well-being, combines homo-toxicology, mesotherapy, and acupuncture with injections of homeopathic formulations at specific body points with simultaneous oral treatment of homeopathic formulations which has also been associated with exposure to tap water and with outbreaks of *M. chelonae* including 27 cases in a single practitioner's office in Australia [71]. Additional outbreaks of NTM including alcohol-resistant strains of NTM, following acupuncture, have also been reported including in two patients with *M. chelonae* and two patients with *M. nonchromogenicum* [72].

A single strain of 2% glutaraldehyde disinfectant-resistant *M. abscessus* subsp. *massiliense* recently was associated with a widespread epidemic of more than 150 patients following laparoscopic surgeries and cosmetic procedures in Brazil from 2006 to 2007. The strain was also recovered from pulmonary samples and urine from non-epidemic patients. After 2008, reports of the surgically related infections decreased. However, in 2010 more cases surfaced in other geographic areas of Brazil emphasizing the epidemiological importance of this strain [73, 74]. Other RGM postsurgical outbreaks in 62 hospitals in Brazil including laparoscopic, arthroscopic plastic surgery or cosmetic procedures have been continuing since 2004 [75].

Several case reports and outbreaks of lower extremity furunculosis involving *M. fortuitum*, *M. abscessus*, and *M. mageritense* associated with contaminated water baths at nail salons have also been reported as described in Table 1 [76–79].

Cardiothoracic Surgery Outbreaks

Outbreaks in cardiothoracic surgical sites following exposure to tap water have occurred since the 1970s [80]. One of the first outbreaks associated with cardiac surgery was reported in 1978. Nineteen patients with *M. fortuitum* complex (later identified at CDC as *M. fortuitum* and *M. abscessus* complex) infections following open-heart surgery were described [81–83].

In 1983, Kuritsky et al. reported sternal wound *M. fortuitum* and *M. abscessus* complex infections, endocarditis, and saphenous graft site infections due to a contaminated cardioplegia solution. The outbreak strain was later found in the cold tap water and ice used to cool the cardioplegia solution [40]. Later in 2010, another *M. fortuitum* outbreak related to the use of a single contaminated patch for septal defect repair was described in three children in Serbia [84] who recovered from the infections following antimicrobial treatment. Again in 2010, contamination during the manufacturing of biological prosthetic valves in Brazil resulted in multiple cases of

M. chelonae endocarditis (identified several years following the patient diagnoses in 2002) [85].

Outbreaks of NTM in patients following cardiothoracic surgery have continued and include other NTM species. In 2014, a cluster of six cases of *Mycobacterium wolinskyi*, a rarely encountered RGM, was reported following cardiothoracic surgery in a large academic medical center. Although no single point source of infection or environmental source was established, when two potential sources including a self-contained water system used in heart-lung machines for cardiothoracic surgeries and a cold-air blaster were removed, no active cases occurred [86]. A similar case in France in 2016 recovered multiple NTM species, but not *M. wolinskyi*, from the HCU and tap water sources [87].

In 2015, outbreaks of invasive infections with *Mycobacterium chimaera*, a species within the *Mycobacterium avium* complex, associated with contaminated Stöckert 3 T HCU (LivaNova, formerly Sorin Group Deutschland, GmbH) HCU devices used during cardiopulmonary bypass for cardiac surgery were reported in three European hospitals [88, 89]. In Zurich, Switzerland, six cases of disseminated *M. chimaera* cardiac infection were identified from August 2008 to the end of May 2012. Following the Swiss cases, healthcare authorities in Germany and the Netherlands reported similar cases [90]. In total, nine cases (eight adults and one infant plus one presumptive case) were described. Initially, the adults complained of fever, shortness of breath, fatigue, and weight loss. All patients exhibited anemia, pronounced lymphocytopenia, and thrombocytopenia; creatinine levels, C-reactive protein, lactate dehydrogenase, and transaminases were also elevated. In the infected infant, clinical findings included episodes of fever and failure to thrive. The presumptive case presented with fever of unknown origin. The patient had been treated for presumptive sarcoidosis due to granulomatous hepatitis. *M. chimaera* was detected in the sternoclavicular joint, bone marrow, liver, and blood; however no endocarditis was detected [89]. Despite extensive treatment, five patients died. Following investigation, public health authorities in Europe recommended placing the Sorin 3-T HCU outside the operating room in an area with independent air flow. Additionally, they recommended that the water reservoir and piping should be made airtight and/or reliably disinfected. Numerous reports in Europe detailed the epidemiological risk assessment and investigations of the outbreaks: Investigation of *Mycobacterium chimaera* infection associated with cardiopulmonary bypass. Public Health England. https://www.gov.uk/government/publications/health-protection-report-volume-9-2015/hpr-volume-9-issue-15-news-30-april; Risk assessment on *Mycobacterium chimaera* infections associated with heater-cooler units. ECDC. http://ecdc.europa.eu/en/publications/Publications/myocbacterium-chimaera-infection-associated-with-heater-cooler-units-rapid-risk-assessment-30-April-2015.pdf.

Investigation of *Mycobacterium chimaera* infection associated with cardiopulmonary bypass: an update. Public Health England. https://www.gov.uk/government/publications/health-protection-report-volume-9-2015/hpr-news-volume-9-issue-18-21-may; Epidemiological update: invasive infections with *Mycobacterium chimaera* potentially associated with heater-cooler units used during cardiac surgery. ECDC. http://ecdc.europa.eu/en/press/news/_layouts/forms/News_DispForm.

aspx?ID=1223&LIst=8db7286c-fe2d-476c-9133-18ff4cb1b568&Source=http%3A%2F%2Fecdc.europa.eu%2Fen%2Fpress%2Fnews%2FPages%2FNews.aspx.

Ongoing whole-genome sequencing indicated a match between patient isolates and air samples collected near the HCUs. The outbreak was subsequently found to result from contamination at the factory in Germany in units built before 2014. Of note, these patients were not severely immunodeficient, but manifestation of bacteremic embolization and manifestations similar to other disseminated NTM disease including osteomyelitis or other bone lesions, cholestatic hepatitis or granulomatous nephritis, and chorioretinitis or vasculitis of unknown origin preceded cardiac infection symptoms. Biopsy samples of affected organs and tissues and cardiac valve samples for mycobacterial culture and/or molecular analysis were important in the investigation. As with all foreign body infections, the primary treatment resulting in resolution of the infections was the removal of the prosthetic material and surgical debridement [89].

Following the European reports of the recovery of *M. chimaera* from multiple HCU devices, Lyman and colleagues from the Pennsylvania Department of Public Health notified the CDC in Atlanta, Georgia, about a prolonged cluster of eight NTM patients following cardiothoracic surgery and exposure to HCUs from January 2010 to July 2015 who exhibited clinical signs of NTM infection [88]. Only three case isolates were available and all were identified as *M. chimaera*. Water and biofilm samples from the HCUs also grew *M. chimaera*. Additionally, *M. chimaera* was also cultured from air samples near the HCU exhaust vent. Subsequent whole-genome sequencing and pulsed-field gel electrophoresis for DNA strain typing proved the clinical HCU and air sample isolates of *M. chimaera* were highly related [88]. These results strongly suggested transmission of *M. chimaera* from the HCUs through aerosolization. Other cases were subsequently identified in 2016, including the University of Iowa Hospitals and Clinics [91]. More than 250,000 cardiothoracic surgeries are performed per year in the USA alone, and it has been estimated that approximately 60% of them have been associated with these HCUs [92]. The field investigation by the Pennsylvania Department of Health with the CDC investigation identified approximately 1300 potentially exposed patients. This investigation emphasized the global nature of this ongoing outbreak and prompted the issuance of the following FDA/CDC guidelines [92].

Centers for Disease Control (CDC) and Prevention Guidelines

In October 2015 (and updated in 2016), as a result of the increasing reports of invasive NTM infection with *M. chimaera* following cardiac surgical patients exposed to HCUs, the CDC issued recommendations for "increased vigilance for NTM infections by health departments, healthcare facilities, and individual providers." Moreover, the Food and Drug Administration (FDA) also issued a "Safety Communication on Nontuberculous Mycobacterium Infections Associated with HCUs" regarding the proper use and maintenance of HCUs [93, 94].

In summary, for health departments, the main recommendations included provisions that [1] local and state health departments should communicate with healthcare facilities who provide surgeries that use Sorin 3-T HCUs to ensure that all HCUs are in proper working condition and clinical staff should be instructed how to identify NTM infections potentially associated with HCUs and [2] health departments should track/monitor potential HCU-related infections and encourage healthcare facilities to report events to the FDA and be prepared for further investigations as needed. Thus far, no cases of disease with any HCU manufacturer other than the LivaNova (previously Sorin) have been reported.

For healthcare facilities (HCFs):

1. Immediate assessment of HCUs for safe and proper use and vigilance of possible NTM infection.
2. Ensure HCFs are following the FDA's recommendations for maintenance, cleaning, disinfection, and monitoring HUCs.
3. Microbiology laboratory data and records of surgeries should be reviewed to identify any patients with NTM within 4 years following cardiac surgery if their HCU tests positive for NTM or if there is concern for patient infection related to the HCU.
4. If the HCU is suspected to have caused an NTM infection or tested positive for NTM, the HCF should promptly notify the local health department, submit report to the FDA, and assess means for notifying patients.

Healthcare providers (HCPs) should have increased suspicion for NTM infection among patients with signs of infection and history of cardiac surgery:

1. The HCP should assess patients who report signs or symptoms of NTM infection within 4 years following cardiac surgery.
2. The HCP should assess patients for exposure to cardiac surgery, exposure to the LivaNova (previously Sorin 3-T) HCUs. Healthcare exposures such as injections, plastic surgery, and dialysis may also be associated with NTM infection and may warrant consultations with FDA or public health authorities.
3. The HCP should order mycobacterial cultures in patients who show signs and symptoms of NTM infection who have undergone cardiac procedures within 4 years.

Finally, CDC advised that patients who have recently had cardiac surgery should contact their HCP if they have signs or symptoms of NTM infection including fever, pain, erythema, heat, pus around the incision, night sweats, joint and/or muscle pain, weight loss, and/or fatigue. In infants failure to thrive is an important sentinel event and should be reported. The patient should also contact the HCP if they have any questions about exposure to HCUs and keep in communication with their clinician for epidemiological and clinical evaluation/tracking purposes.

The above recommendations are designed to increase awareness of the possibility of infection/outbreaks among HCFs and patients to facilitate earlier diagnosis and treatment in cases [88]. Recognition and reporting of such cases are crucial for evaluation and management of potentially infected individuals and identification of options to mitigate risk of infection/outbreaks [93].

M. abscessus Complex Outbreak in Cystic Fibrosis Clinic

One of the most publicized recent NTM respiratory outbreaks involved patients with *M. abscessus* subsp. *massiliense* among five patients in a cystic fibrosis clinic in Seattle [95]. The index case was a 22-year-old man who was diagnosed at another center several years prior to the outbreak. Eight months after his transition to the Seattle clinic, four additional patients were found to have the same organism as confirmed by DNA strain typing. No environmental source of the strain was found despite efforts to recover the strain. The current hypothesis, largely based on current whole genomic sequence data, is that transmission may have occurred by patient-to-patient spread or by contact with contaminated areas or equipment within the clinic [96, 97]. Additional reports have identified the same strain in the United Kingdom and in South America triggering massive ongoing investigations and efforts in both the USA and Europe to determine the mode of transmissibility of *M. abscessus* [97]. Moreover, this global outbreak has provided strong impetus for rigorous prevention and control measures in cystic fibrosis clinics in the USA and Europe [98, 99].

Pseudo-Outbreaks

NTM pseudo-outbreaks are characterized by positive NTM culture results from patients in the absence of true infection. Typical pseudo-outbreaks or clusters of pseudo-infections, known as pseudo-outbreaks, usually result from contamination during specimen processing/handling. Like outbreaks, NTM colonization of potable water systems has resulted in nosocomial pseudo-outbreaks. An increase in the frequency of an NTM species or numbers of organisms isolated in patients without signs of infection should raise high suspicion of an NTM pseudo-outbreak. Recognition of pseudo-outbreaks of NTM may be delayed due to the length of time it takes to culture and identify some species of NTM.

As seen in Table 1, the most common type of NTM pseudo-outbreaks has involved respiratory samples. Tap water contamination of bronchoscopes or bronchoscopy equipment are frequent causes of NTM pseudo-outbreaks [100, 101]. NTM pseudo-outbreaks related to ice machines (including one outbreak involving an unusual NTM, *M. paraffinicum*), contaminated water, and bronchoscope reprocessing have also been described [101–103]. Early studies revealed that contamination occurred when the terminal rinse for the bronchoscopy equipment was found to be tap or distilled water, not sterile water [104–107].

A 2015 study of 135 patients with positive cultures for *M. gordonae*, a typical nonpathogenic NTM, was associated with tap water contamination of sputum cultures [108].

A pseudo-outbreak of *M. abscessus* subsp. *massiliense* (previously subsp. *bolletii*) was described in Brazil. The implicated strain was the same clone as the one which was identified in the previously described large laparoscopic surgical outbreak [109].

Laboratory contamination has also resulted in pseudo-outbreaks. Contamination of laboratory-distilled water was previously associated (as cited in Table 1) with a pseudo-outbreak of *M. abscessus* complex [110]. Another laboratory-related pseudo-outbreak was related to NTM contamination of the sampling needle of an automated radiometric mycobacterial culture detection system [111].

An initial report of a 2016 pseudo-outbreak was reported as *M. fortuitum* in 12 patients from 2 hospitals in less than 1 week. Treatment was instituted for several patients. However, subsequent investigation proved the organisms to be *M. terrae*, a common laboratory/environmental contaminant which is not typically considered pathogenic [112]. This issue not only emphasizes the problem of unnecessary treatment but also the necessity for prompt and accurate identification of isolates to species level, optimally using molecular methods.

Pseudo-outbreaks can be problematic for healthcare personnel as they can cause additional expense and time to investigate and unnecessary administration of potentially toxic medications to patients.

Surveillance and Control of Outbreaks

Hospital and community water supplies and water-related devices are the most common sources of healthcare-associated outbreaks and pseudo-outbreaks involving NTM. This chapter is not meant to be an exhaustive review of all of the published NTM outbreaks and pseudo-outbreaks, yet it does provide a review of many of the most important cases.

Ongoing active surveillance and education of healthcare personnel should help facilitate the identification of healthcare-associated infections and pseudo-infections and thus help to limit the extent of problems associated with outbreaks and pseudo-outbreaks. It is likely that many outbreaks go unrecognized. Clinical laboratories and infection control and prevention personnel should be vigilant to recognize any surgical or device-related infection (e.g., central line infection, sternal wound infection, etc.) or any increase in the number of NTM bronchoalveolar lavage infection reported (as they do for bacterial species). Large numbers of reports of one species of NTM should be investigated often requiring medical chart review or correspondence with the clinical staff carefully to determine whether the increase represents contamination or a possible outbreak. Advances in species strain identification methods have provided better understanding of reservoirs and transmission of NTM pathogens. The understanding of pathways of NTM transmission is essential for development of risk assessments, prevention strategies, and control measures of healthcare outbreaks and pseudo-outbreaks. Laboratories that do not have capability of molecular identification should send their isolates in question to a qualified reference laboratory with experience in DNA strain typing. Strain variation among NTM is well recognized and requires expertise in performance of the strain typing method and interpretation of results. Strain typing is an essential component to an NTM epidemiological investigation, especially if a single clone can be associated to the source of the outbreaks or pseudo-outbreaks. Pulsed-field gel electrophoresis (PFGE) is the standard method for most NTM species DNA strain typing [83, 113, 114]. Other techniques including repetitive sequence-based PCR (rep-PCR, DiversiLab; bioMerieux; Marcy-l'Etoile, France), randomly amplified polymorphic DNA PCR (RAPD), and variable number tandem repeat (VNTR) have also been useful to show strain differentiation in some species [2, 16, 61, 115, 116]. Whole genomic sequencing may soon be available as another technology. Molecular techniques to identify specific strains of NTM should be instituted as soon as possible in the investigation.

The premise of the investigation of outbreaks and pseudo-outbreaks is to characterize the exposed individuals and identify the sources of these increasingly recognized microorganisms with the major aim to institute control measures to mitigate the effects of the outbreak and prevent future outbreaks.

Control of healthcare-associated outbreaks of NTM requires thorough investigation by infection control personnel, prompt identification of the source of infection by application of molecular DNA strain typing in the laboratory, and institution of effective control measures designed to prevent and/or mitigate the outbreak. Records of routine cleaning of equipment should be reviewed along with clinical and laboratory policies and procedures to assure strict adherence to protocols. In some cases direct observation of the performance of the procedure may be necessary to provide evidence of sources of contamination. Autoclaving of equipment or in the case of plumbing fixtures, flushing with appropriate disinfectants that are mycobactericidal may be necessary. Single-use devices are preferable to multi-use devices, including medication vials. Tap water should not be used in the collection and processing of NTM samples. Careful monitoring of glutaraldehyde concentration in disinfectant and close adherence to the shelf life of disinfectant activity are essential for the prevention of healthcare-associated infections and pseudo-infections [117].

Prevention of Outbreaks

The use of an alcohol terminal rinse of equipment used for invasive procedures (e.g., bronchoscopes) may aid in preventing NTM outbreaks and pseudo-outbreaks. Increasing the temperature for probe sterilization between samples in automated culture systems may also help to prevent cross-contamination and decrease the incidence of pseudo-outbreaks related to laboratory procedures. Overall, strict adherence to manufacturer guidelines for all equipment and reagents is essential for the prevention of outbreaks and pseudo-outbreaks due to NTM.

Although no controlled studies have been performed, a suggestion to increase the temperature of the hot water may be a reasonable and effective control measure for the prevention of NTM outbreaks and pseudo-outbreaks. However, most states regulate the allowable temperature (mean 116 °F) because of the risk of scald injuries. Analysis of risk-benefit and cost-effectiveness is necessary to compare the risk of scald injury to inhibition of NTM contamination [117].

Education of clinical and laboratory staff to recognize increases in numbers of NTM isolates and development of multiple types of control measures including cleaning, disinfection, and engineering approaches as well as ongoing surveillance and clinical management can help to reduce the risk for healthcare-associated infections and pseudo-infections. Communication between healthcare personnel and the laboratory is essential to the investigation and the development of control and prevention strategies.

References

1. Wallace RJ Jr, Brown BA, Griffith DE. Nosocomial outbreaks/pseudo-outbreaks caused by nontuberculous mycobacteria. Annu Rev Microbiol. 1998;52:453–90.
2. Brown-Elliott BA, Philley JV. Rapidly growing mycobacteria. Microbiol Spectr. 2017;5:TNM17-0027-2016.
3. Covert TC, Rodgers MR, Reyes AL, Stelma GN Jr. Occurrence of nontuberculous mycobacteria in environmental samples. Appl Environ Microbiol. 1999;65:2492–6.
4. Williams MM, Chen T-H, Keane T, Toney N, Toney S, Armbruster CR, Butler WR, Arduino MJ. Point-of-use membrane filtration and hyperchlorination to prevent patient exposure to rapidly growing mycobacteria in the potable water supply of a skilled nursing facility. Infect Control Hosp Epidemiol. 2011;32:837–44.
5. van Ingen J, Boeree MJ, Dekhuijzen PNR, van Soolingen D. Environmental sources of rapid growing nontuberculous mycobacteria causing disease in humans. Clin Microbiol Infect. 2009;15:888–93.
6. Wilson RW, Steingrube VA, Böttger EC, Springer B, Brown-Elliott BA, Vincent V, Jost KC Jr, Zhang Y, Garcia MJ, Chiu SH, Onyi GO, Rossmoore H, Nash DR, Wallace RJ Jr. *Mycobacterium immunogenum* sp. nov., a novel species related to *Mycobacterium abscessus* and associated with clinical disease, pseudo-outbreaks, and contaminated metalworking fluids: an international cooperative study on mycobacterial taxonomy. Int J Syst Evol Microbiol. 2001;51:1751–64.
7. Cortesia C, Lopez GJ, de Waard JH, Takiff HE. The use of quaternary ammonium disinfectants selects for persisters at high frequency from some species of non-tuberculous mycobacteria and may be associated with outbreaks of soft tissue infections. J Antimicrob Chemother. 2010;65:2574–81.
8. Greendyke R, Byrd TF. Differential antibiotics susceptibility of *Mycobacterium abscessus* variants in biofilms and macrophages compared to that of planktonic bacteria. Antimicrob Agents Chemother. 2008;52:2019–26.
9. Howard ST, Rhoades E, Recht J, Pang X, Alsup A, Kolter R, Lyons CR, Byrd TF. Spontaneous reversion of *Mycobacterium abscessus* from a smooth to a rough morphotype is associated with reduced expression of glycopeptidolipid and reacquisition of an invasive phenotype. Microbiology. 2006;152:1581–90.
10. Schulze-Röbbecke R, Janning B, Fischeder R. Occurrence of mycobacteria in biofilm samples. Tuberc Lung Dis. 1992;73:141–4.
11. Sood G, Parrish N. Outbreaks of nontuberculous mycobacteria. Curr Opin Infect Dis. 2017;30:404–9.
12. Galassi L, Donato R, Tortoli E, Burrini D, Santianni D, Dei R. Nontuberculous mycobacteria in hospital water systems: application of HPLC for identification of environmental mycobacteria. J Water Health. 2003;1:133–9.
13. Brown-Elliott BA, Wallace RJ Jr. Clinical and taxonomic status of pathogenic nonpigmented or late-pigmenting rapidly growing mycobacteria. Clin Microbiol Rev. 2002;15:716–46.

14. Eid AJ, Bergari F, Sia IG, Wengenack NL, Osmon DR, Razonable RR. Prosthetic joint infection due to rapidly growing mycobacteria: report of 8 cases and review of the literature. Clin Infect Dis. 2007;45:687–94.
15. Holmes GP, Bond GB, Fader RC, Fulcher SF. A cluster of cases of *Mycobacterium szulgai* keratitis that occurred after laser-assisted in situ keratomileusis. Clin Infect Dis. 2002;34:1039–46.
16. Sampaio JL, Chimara E, Ferrazoli L, da Silva Telles MA, Del Guercio VM, Jericó ZV, Miyashiro K, Fortaleza CM, Padoveze MC, Leão SC. Application of four molecular typing methods for analysis of *Mycobacterium fortuitum* group strains causing post-mammaplasty infections. Clin Microbiol Infect. 2006;12:142–9.
17. Edens C, Liebich L, Halpin AL, Moulton-Meissner H, Eitniear S, Zgodzinski E, Vasko L, Grossman D, Perz JF, Mohr MC. *Mycobacterium chelonae* eye infections associated with humidifier use in an outpatient LASIK clinic – Ohio. Morb Mortal Wkly Rep. 2015;64:1177.
18. Ashraf MS, Swinker M, Augustino KL, Nobles D, Knupp C, Liles D, Ramsey KM. Outbreak of *Mycobacterium mucogenicum* bloodstream infections among patients with sickle cell disease in an outpatient setting. Infect Control Hosp Epidemiol. 2012;33:1132–6.
19. Shachor-Meyouhas Y, Sprecher H, Eluk O, Ben-Barak A, Kassis I. An outbreak of *Mycobacterium mucogenicum* bacteremia in pediatric hematology-oncology patients. Pediatr Infect Dis J. 2011;30:30–2.
20. Tagashira Y, Kozai Y, Yamasa H, Sakurada M, Kashiyama T, Honda H. A cluster of central line-associated bloodstream infections due to rapidly growing nontuberculous mycobacteria in patients with hematologic disorders at a Japanese tertiary care center: an outbreak investigation and review of the literature. Infect Control Hosp Epidemiol. 2015;36:76–80.
21. Laussucq S, Baltch AL, Smith RP, Smithwick RW, Davis BJ, Desjardin EK, Silcox VA, Spellacy AB, Zeimis RT, Gruft HM, Good RC, Cohen ML. Nosocomial *Mycobacterium fortuitum* colonization from a contaminated ice machine. Am Rev Respir Dis. 1988; 138:891–4.
22. Lévy-Frébault V, David HL. *Mycobacterium kansasii*: drinking water contaminant of a hospital. Rev Epidemiol Sante Publique. 1983;31:11–20.
23. Marshall HM, Carter R, Torbey MJ, Minion S, Tolson C, Sidjabat HE, Huygens F, Hargreaves M, Thomson RM. *Mycobacterium lentiflavum* in drinking water supplies, Australia. Emerg Infect Dis. 2011;17:395–402.
24. Kanamori H, Weber DJ, Rutala WA. Healthcare outbreaks associated with a water reservoir and infection prevention strategies. Clin Infect Dis. 2016;62:1423–35.
25. Villanueva A, Calderon RV, Vargas BA, Ruiz F, Aguero S, Zhang Y, Brown BA, Wallace RJ Jr. Report on an outbreak of post-injection abscesses due to *Mycobacterium abscessus*, including management with surgery and clarithromycin therapy and comparison of strains by random amplified polymorphic DNA polymerase chain reaction. Clin Infect Dis. 1997;24:1147–53.
26. Zhibang Y, Bixia Z, Qishan L, Lihao C, Xiangquan L, Huaping L. Large-scale outbreak of infection with *Mycobacterium chelonae* subsp. *abscessus* after penicillin injection. J Clin Microbiol. 2002;40:2626–8.
27. Kim H-Y, Yun Y-J, Park CG, Lee DH, Cho YK, Park BJ, Joo S-I, Kim E-C, Hur YJ, Kim B-J, Kook YH. Outbreak of *Mycobacterium massiliense* infection associated with intramuscular injections. J Clin Microbiol. 2007;45:3127–30.
28. Tiwari TS, Ray B, Jost KC Jr, Rathod MK, Zhang Y, Brown-Elliott BA, Hendricks K, Wallace RJ Jr. Forty years of disinfectant failure: outbreak of postinjection *Mycobacterium abscessus* infection caused by contamination of benzalkonium chloride. Clin Infect Dis. 2003;36:954–62.
29. Galil K, Miller LA, Yakrus MA, Wallace RJ Jr, Mosley DG, England B, Huitt G, McNeill MM, Perkins BA. Abscesses due to *Mycobacterium abscessus* linked to injection of unapproved alternative medication. Emerg Infect Dis. 1999;5:681–7.
30. Phillips MS, von Reyn CF. Nosocomial infections due to nontuberculous mycobacteria. Clin Infect Dis. 2001;33:1363–74.

31. Decker BK, Palmore TN. The role of water in healthcare-associated infections. Curr Opin Infect Dis. 2013;26:345–51.
32. Carson LA, Bland LA, Cusick LB, Favero MS, Bolan GA, Reingold AL, Good RC. Prevalence of nontuberculous mycobacteria in water supplies of hemodialysis centers. Appl Environ Microbiol. 1988;54:3122–5.
33. Li T, Abebe LS, Cronk R, Bartram J. A systematic review of waterborne infections from nontuberculous mycobacteria in health care facility water systems. Int J Hyg Environ Health. 2017;229:611–20.
34. von Reyn CF, Maslow JN, Barber TW, Falkinham JO III, Arbeit RD. Persistent colonization of potable water as a source of *Mycobacterium avium* infection in AIDS. Lancet. 1994;343:1137–41.
35. Tobin-D'Angelo MJ, Blass MA, del Rio C, Halvosa JS, Blumberg HM, Horsburgh CR Jr. Hospital water as a source of *Mycobacterium avium* complex isolates in respiratory specimens. J Infect Dis. 2004;189:98–104.
36. Lowry PW, Jarvis WR, Oberle AD, Bland LA, Silberman R, Bocchini JA Jr, Dean HD, Swenson JM, Wallace RJ Jr. *Mycobacterium chelonae* causing otitis media in an ear-nose-and-throat practice. N Engl J Med. 1988;319:978–82.
37. Meyers H, Brown-Elliott BA, Moore D, Curry J, Truong C, Zhang Y, Wallace RJ Jr. An outbreak of *Mycobacterium chelonae* infection following liposuction. Clin Infect Dis. 2002;34:1500–7.
38. Wenger JD, Spika JS, Smithwick RW, Pryor V, Dodson DW, Carden GA, Klontz KC. Outbreak of *Mycobacterium chelonae* infection associated with use of jet injectors. JAMA. 1990;264:373–6.
39. Soto LE, Bobadilla M, Villalobos Y, Sifuentes J, Avelar J, Arrieta M, Ponce de Leon S. Post-surgical nasal cellulitis outbreak due to *Mycobacterium chelonae*. J Hosp Infect. 1991;19:99–106.
40. Kuritsky JN, Bullen MG, Broome CV, Silcox VA, Good RC, Wallace RJ Jr. Sternal wound infections and endocarditis due to organisms of the *Mycobacterium fortuitum* complex. Ann Intern Med. 1983;98:938–9.
41. Lowry PW, Beck-Sague CM, Bland LA, Aguero SM, Arduino MJ, Minuth AN, Murray RA, Swenson JM, Jarvis WR. *Mycobacterium chelonae* infection among patients receiving high-flux dialysis in a hemodialysis clinic in California. J Infect Dis. 1990;161:85–90.
42. Bolan G, Reingold AL, Carson LA, Silcox VA, Woodley CL, Hayes PS, Hightower AW, McFarland L, Brown JW III, Petersen NJ, Favero MS, Good RC, Broome CV. Infections with *Mycobacterium chelonei* in patients receiving dialysis and using processed hemodialyzers. J Infect Dis. 1985;152:1013–9.
43. Band JD, Ward JI, Fraser DW, Peterson NJ, Silcox VA, Good RC, Ostrey PR, Kennedy J. Peritonitis due to a *Mycobacterium chelonei*-like organism associated with intermittent chronic peritoneal dialysis. J Infect Dis. 1982;145:9–17.
44. Carbonne A, Brossier F, Arnaud I, Bougmiza I, Caumes E, Meningaud J-P, Dubrou S, Jarlier V, Cambau E, Astagneau P. Outbreak of nontuberculous mycobacterial subcutaneous infections related to multiple mesotherapy injections. J Clin Microbiol. 2009;47:1961–4.
45. Jaubert J, Mougari F, Picot S, Boukerrou M, Barau G, Ali Ahmed S-A, Cambau E. A case of postoperative breast infection by *Mycobacterium fortuitum* associated with the hospital water supply. Am J Infect Control. 2015;43:406–8.
46. Kauppinen J, Nousiainen T, Jantunen E, Mattila R, Katila ML. Hospital water supply as a source of disseminated *Mycobacterium fortuitum* infection in a leukemia patient. Infect Control Hosp Epidemiol. 1999;20:343–5.
47. Flesner C, Deresinski S. Investigation of an outbreak of SSI due to *Mycobacterium immunogenum* after blepharoplasty. Am J Infect Control. 2011;39:e204–5.
48. Astagneau P, Desplaces N, Vincent V, Chicheportiche V, Botherel A, Maugat S, Brucker G. *Mycobacterium xenopi* spinal infections after discovertebral surgery: investigation and screening of a large outbreak. Lancet. 2001;358:741–51.
49. Kline S, Cameron S, Streifel A, Yakrus MA, Kairis F, Peacock K, Cooksey RC. An outbreak of bacteremias associated with *Mycobacterium mucogenicum* in a hospital water supply. Infect Control Hosp Epidemiol. 2004;25:1042–9.

50. Baird SF, Taori SK, Dave J, Willocks LJ, Roddie H, Hanson M. Cluster of non-tuberculous mycobacteraemia associated with water supply in a haemato-oncology unit. J Hosp Infect. 2011;79:339–43.
51. Cooksey RC, Jhung MA, Yakrus MA, Butler WR, Adékambi T, Morlock GP, Williams M, Shams AM, Jensen BJ, Morey RE, Charles N, Toney SR, Jost KC Jr, Dunbar DF, Bennett V, Kuan M, Srinivasan A. Multiphasic approach reveals genetic diversity of environmental and patient isolates of *Mycobacterium mucogenicum* and *Mycobacterium phocaicum* associated with an outbreak of bacteremias at a Texas Hospital. Appl Environ Microbiol. 2008;74:2480–7.
52. Livni G, Yaniv I, Samra Z, Kaufman L, Solter E, Ashkenazi S, Levy I. Outbreak of *Mycobacterium mucogenicum* bacteraemia due to contaminated water supply in a paediatric haematology-oncology department. J Hosp Infect. 2008;70:253–8.
53. Iroh Tam P-Y, Kline S, Wagner JE, Guspiel A, Streifel A, Ward G, Messinger K, Ferrieri P. Rapidly growing mycobacteria among pediatric hematopoietic cell transplant patients traced to the hospital water supply. Pediatr Infect Dis J. 2014;33:1043–6.
54. Rodriguez JM, Xie YL, Winthrop KL, Schafer S, Sehdev P, Solomon J, Jensen B, Toney NC, Lewis PF. *Mycobacterium chelonae* facial infections following injection of dermal filler. Asethet Surg J. 2013;33:265–9.
55. Brown-Elliott BA, Wallace RJ Jr, Tichindelean C, Sarria JC, McNulty S, Vasireddy R, Bridge L, Mayhall CG, Turenne C, Loeffelholz M. Five year outbreak of community- and hospital-acquired *Mycobacterium porcinum* infections related to public water supplies. J Clin Microbiol. 2011;49:4231–8.
56. Chadha R, Grover M, Shama A, Lakshmi A, D M, Kumar S, Mehta G. An outbreak of post-surgical wound infections due to *Mycobacterium abscessus*. Pediatr Surg Int. 1998;13: 406–10.
57. Merlani GM, Francioli P. Established and emerging waterborne nosocomial infections. Curr Opin Infect Dis. 2003;16:343–7.
58. Hillebrand-Haverkort ME, Kolk AH, Kox LF, Ten Velden JJ, Ten Veen JH. Generalized *Mycobacterium genavense* infection in HIV-infected patients: detection of the mycobacterium in hospital tap water. Scand J Infect Dis. 1999;31:63–8.
59. Conger NA, O'Connell R, Laurel V, Olivier K, Graviss EA, Williams-Bouyer N, Zhang Y, Brown-Elliott BA, Wallace RJ Jr. *Mycobacterium simiae* outbreak associated with a hospital water supply. Infect Control Hosp Epidemiol. 2004;25:1050–5.
60. Aronson T, Holtzman A, Glover N, Boian M, Froman S, Berlin OGW, Hill H, Stelma G Jr. Comparison of large restriction fragments of *Mycobacterium avium* isolates recovered from AIDs and non-AIDs patients with those of isolates from potable water. J Clin Microbiol. 1999;37:1008–12.
61. Baker AW, Lewis SS, Alexander BD, Chen LF, Wallace RJ Jr, Brown-Elliott BA, Isaacs PJ, Pickett LC, Patel CB, Smith PK, Reynolds JM, Engel J, Wolfe CR, Milano CA, Schroder JN, Davis RD, Hartwig MG, Stout JE, Strittholt N, Maziarz EK, Saullo JH, Hazen KC, Walczak RJ Jr, Vasireddy R, Vasireddy S, McKnight CM, Anderson DJ, Sexton DJ. Two-phase hospital-associated outbreak of *Mycobacterium abscessus*: investigation and mitigation. Clin Infect Dis. 2017;64:902–11.
62. Centers for Disease Control and Prevention. Rapidly growing mycobacterial infection following liposuction and liposculpture—Caracas, Venezuela, 1996-1998. Morb Mortal Wkly Rep. 1998;47:1065–7.
63. Centers for Disease Control and Prevention. Nontuberculous mycobacterial infections after cosmetic surgery—Santo Domingo, Dominican Republic, 2003-2004. Morb Mortal Wkly Rep. 2004;53:509.
64. del Castillo M, Palmero DJ, Lopez B, Paul R, Ritacco V, Bonvehi P, Clara L, Ambroggi M, Barrera L, Vay C. Mesotherapy-associated outbreak caused by *Mycobacterium immunogenum*. Emerg Infect Dis. 2009;15:357–8.
65. Regnier S, Cambau EM, Meningaud JP, et al. Clinical management of rapidly growing mycobacterial cutaneous infections in patients after mesotherapy. Clin Infect Dis. 2009;49:1358–64.
66. van Dissel JT, Kuijper EJ. Rapidly growing mycobacteria: emerging pathogens in cosmetic procedures of the skin. Clin Infect Dis. 2009;49:1365–8.

67. Furuya EY, Paez A, Srinivasan A, Cooksey R, Augenbraun M, Baron M, Brudney K, Della-Latta P, Estivariz C, Fischer S, Flood M, Kellner P, Roman C, Yakrus M, Weiss D, Granowitz EV. Outbreak of *Mycobacterium abscessus* wound infections among "lipotourists" from the United States who underwent abdominoplasty in the Dominican Republic. Clin Infect Dis. 2008;46:1181–8.
68. Schnabel D, Esposito DH, Gaines J, Ridpath A, Barry MA, Feldman KA, Mullins J, Burns R, Ahmad N, Nyangoma EN, Nguyen DB, Perz JF, Moulton-Meissner HA, Jensen BJ, Lin Y, Posivak-Khouly L, Jani N, Morgan OW, Brunette GW, Pritchard PS, Greenbaum AH, Rhee SM, Blythe D, Sotir M. Multistate US outbreak of rapidly growing mycobacterial infections associated with medical tourism to the Dominican Republic, 2013-2014. Emerg Infect Dis. 2016;22:1340–7.
69. Falsey RR, Kinzer MH, Hurst S, Kalus A, Pottinger PS, Duchin JS, Zhang J, Noble-Wang J, Shinohara MM. Cutaneous inoculation of nontuberculous mycobacteria during professional tattooing: a case series and epidemiologic study. Clin Infect Dis. 2013;57:e143–7.
70. Culton DA, Lachiewicz AM, Miller BA, Miller MB, MacKuen C, Groben P, White B, Cox GM, Stout JE. Nontuberculous mycobacterial infection after fractionated CO_2 laser resurfacing. Emerg Infect Dis. 2013;19:365–70.
71. Ivan M, Dancer C, Koehler AP, Hobby M, Lease C. *Mycobacterium chelonae* abscesses associated with biomesotherapy, Australia, 2008. Emerg Infect Dis. 2013;19:1493–5.
72. Woo PCY, Leung K-W, Wong SSY, Chong KTK, Cheung EYL, Yuen K-Y. Relatively alcohol-resistant mycobacteria are emerging pathogens in patients receiving acupuncture treatment. J Clin Microbiol. 2002;40:1219–24.
73. Duarte RS, Lourenço MC, Fonseca Lde S, Leão SC, Amorim Ede L, Rocha IL, Coelho FS, Viana-Niero C, Gomes KM, da Silva MG, Lorena NS, Pitombo MB, Ferreira RM, Garcia MH, de Oliveira GP, Lupi O, Vilaça BR, Serradas LR, Chebato A, Marques EA, Teixeira LM, Dalcolmo M, Senna SG, Sampaio JL. Epidemic of postsurgical infections caused by *Mycobacterium massiliense*. J Clin Microbiol. 2009;47:2149–55.
74. Matsumoto CK, Chimara E, Ramos JP, Campos CED, De Suza Caldas PC, Lima KVB, Lopes ML, Duarte RS, Leão SC. Rapid tests for the detection of the *Mycobacterium abscessus* subsp. *bolletii* strain responsible for an epidemic of surgical-site infections in Brazil. Mem Inst Oswaldo Cruz, Rio de Janeiro. 2012;107:969–70.
75. Viana-Niero C, Lima KVB, Lopes ML, da Silva Rabello MC, Marsola LR, Brilhante VCR, Durham AM, Leão SC. Molecular characterization of *Mycobacterium massiliense* and *Mycobacterium bolletii* in isolates collected from outbreaks of infections after laparoscopic surgeries and cosmetic procedures. J Clin Microbiol. 2008;46:850–5.
76. Winthrop KL, Albridge K, South D, Albrecht P, Abrams M, Samuel MC, Leonard W, Wagner J, Vugia DJ. The clinical management and outcome of nail salon-acquired *Mycobacterium fortuitum* skin infection. Clin Infect Dis. 2004;38:38–44.
77. Vugia DJ, Jang Y, Zizek C, Ely J, Winthrop KL, Desmond E. Mycobacteria in nail salon whirlpool footbaths, California. Emerg Infect Dis. 2005;11:616–8.
78. Winthrop KL, Abrams M, Yakrus M, Swartz I, Ely J, Gillies D, Vugia DJ. An outbreak of mycobacterial furunculosis associated with footbaths at a nail salon. N Engl J Med. 2002;346:1366–71.
79. Gira AK, Reisenauer H, Hammock L, Nadiminti U, Macy JT, Reeves A, Burnett C, Yakrus MA, Toney S, Jensen BJ, Blumberg HM, Caughman SW, Nolte FS. Furunculosis due to *Mycobacterium mageritense* associated with footbaths at a nail salon. J Clin Microbiol. 2004;42:1813–7.
80. Wallace RJ Jr, Musser JM, Hull SI, Silcox VA, Steele LC, Forrester GD, Labidi A, Selander RK. Diversity and sources of rapidly growing mycobacteria associated with infections following cardiac surgery. J Infect Dis. 1989;159:708–16.
81. Robicsek F, Daugherty HK, Cook JW, Selle JG, Masters TN, O'Bar PR, Fernandez CR, Mauney CU, Calhoun DM. *Mycobacterium fortuitum* epidemics after open-heart surgery. J Thor Cardiovasc Surg. 1978;75:91–6.
82. Hoffman PC, Fraser DW, Robiesek F, O'Bar PR, Mauney CU. Two outbreaks of sternal wound infections due to organisms of the *Mycobacterium fortuitum* complex. J Infect Dis. 1981;143:533–42.

83. Hector JSR, Pang Y, Mazurek GH, Zhang Y, Brown BA, Wallace RJ Jr. Large restriction fragment patterns of genomic *Mycobacterium fortuitum* DNA as strain-specific markers and their use in epidemiologic investigation of four nosocomial outbreaks. J Clin Microbiol. 1992;30:1250–5.
84. Vukovic D, Parezanovic V, Savid B, Dakic I, Laban-Nestorovic S, Ilic S, Cirkovic I, Stepanovic S. *Mycobacterium fortuitum* endocarditis associated with cardiac surgery, Serbia. Emerg Infect Dis. 2013;19:517–9.
85. Strabelli TMV, Siciliano RF, Castelli JB, Demarchi LMMF, Cardoso Leão S, Viana-Niero C, Miyashiro K, Sampaio RO, Grinberg M, Uip DE. *Mycobacterium chelonae* valve endocarditis resulting from contaminated biological prostheses. J Infect. 2010;60:467–73.
86. Nagpal A, Wentink JE, Berbari EF, Aronhalt KC, Wright AJ, Krageschmidt DA, Wengenack NL, Thompson RI, Tosh PK. A cluster of *Mycobacterium wolinskyi* surgical site infection at an academic medical center. Infect Control Hosp Epidemiol. 2014;35:1169–75.
87. Dupont C, Terru D, Aguilhon S, Frapier J-M, Paquis M-P, Morquin D, Lamy B, Godreuil S, Parer S, Lotthé A, Jumas-Bilak E, Romano-Bertrand S. Source-case investigation of *Mycobacterium wolinskyi* cardiac surgical site infection. J Hosp Infect. 2016;93:235–9.
88. Lyman MM, Grigg C, Kinsey CB, Keckler MS, Moulton-Meissner H, Cooper E, Soe MM, Noble-Wang J, Longenberger A, Walker SR, Miller JR, Perz JF, Perkins KM. Invasive nontuberculous mycobacterial infections among cardiothoracic surgical patients exposed to heater-cooler devices. Emerg Infect Dis. 2017;23:796–805.
89. Kohler P, Kuster SP, Bloemberg G, Schulthess B, Frank M, Tanner FC, Rössle M, Böni C, Falk V, Wilhelm MJ, Sommerstein R, Achermann Y, Ten Oever J, Debast SB, Wolfhagen MJHM, Bravo Bruinsma GJB, Vos MC, Bogers A, Serr A, Beyersdorf F, Sax H, Böttger EC, Weber R, van Ingen J, Wagner D, Hasse B. Healthcare-associated prosthetic heart valve, aortic vascular graft, and disseminated *Mycobacterium chimaera* infections subsequent to open heart surgery. Eur Heart J. 2015;36:2745–53.
90. Sax H, Bloemberg G, Hasse B, Sommerstein R, Kohler P, Achermann Y, Rössle M, Falk V, Kuster SP, Böttger EC, Weber R. Prolonged outbreak of *Mycobacterium chimaera* infection after open-chest heart surgery. Clin Infect Dis. 2015;61:67–75.
91. Marra AR, Diekema DJ, Edmond MB. *Mycobacterium chimaera* infections associated with contaminated heater-cooler devices for cardiac surgery: outbreak management. Clin Infect Dis. 2017;65:669–74.
92. Perkins KM, Lawsin A, Hasan NA, Strong M, Halpin AL, Rodger RR, Moulton-Meissner H, Crist MB, Schwartz S, Mardens J, Daley CL, Salfinger M, Perz JF. Notes from the field: *Mycobacterium chimaera* contamination of heater-cooler devices used in cardiac surgery – United States. MMWR Morb Mortal Wkly Rep. 2016;65:1117–8.
93. Centers for Disease Control and Prevention. CDC advises hospitals to alert patients at risk for contaminated heater-cooler devices used during cardiac surgery. https://emergency.cec.gov/han/han00397.asp. 2016.
94. Centers for Disease Control and Prevention. Non-tuberculous *Mycobacterium* (NTM) infections and heater-cooler devices interim practical guidance: Updated October 27, 2015. In: Services HaH, editor; 2015. Department of Health and Human Services, Centers for Disease Control and Prevention. https://www.cdc.gov/HAI/pdfs/outbreaks/CDC-Notice-Heater-Cooler-Units-final-clean.pdf
95. Aitken ML, Limaye A, Pottinger P, Whimbey E, Goss GH, Tonelli MR, Cangelosi GA, Ashworth M, Olivier KN, Brown-Elliott BA, Wallace RJ Jr. Respiratory outbreak of *Mycobacterium abscessus* subspecies *massiliense* in a lung transplant and cystic fibrosis center. Letter to the Editor. Am J Respir Crit Care Med. 2012;185:231–3.
96. Bryant JM, Grogono DM, Rodriguez-Rincon D, Everall I, Brown KP, Moreno P, Verma D, Hill E, Drijkoningen J, Gilligan P, Esther CR, Noone PG, Giddings O, Bell SC, Thomson R, Wainwright CE, Coulter C, Pandey S, Wood ME, Stockwell RE, Ramsay KA, Sherrard LJ, Kidd TJ, Jabbour N, Johnson GR, Knibbs LD, Morawska L, Sly PD, Jones A, Bilton D, Laurenson I, Ruddy M, Bourke S, Bowler ICJW, Chapman SJ, Clayton A, Cullen M, Daniels T, Dempsey O, Denton M, Desai M, Drew RJ, Edenborough F, Evans J, Folb J, Humphrey H,

Isalska B, Jensen-Fangel S, Jönsson B, Jones AM, Katzenstein TL, Lillebaek T, MacGregor G, Mayell S, Millar M, Modha D, Nash EF, O'Brien C, O'Brien D, Ohri C, Pao CS, Peckham D, Perrin F, Perry A, Pressler T, Prtak L, Qvist T, Robb A, Rodgers H, Schaffer K, Shafi N, van Ingen J, Walshaw M, Watson D, West N, Whitehouse J, Haworth CS, Harris SR, Ordway D, Parkhill J, Floto RA. Emergence and spread of a human-transmissible multidrug-resistant nontuberculous mycobacterium. Science. 2016;354:751–7.

97. Bryant JM, Grogono DM, Greaves D, Foweraker J, Roddick I, Inns T, Reacher M, Haworth CS, Curran MD, Harris SR, Peacock SJ, Parkhill J, Floto RA. Whole-genome sequencing to identify transmission of *Mycobacterium abscessus* between patients with cystic fibrosis: a retrospective cohort study. Lancet. 2013;381:1551–60.
98. Tettelin H, Davidson RM, Agrawal S, Aitken ML, Shallom S, Hasan NA, Strong M, de Moura VCN, De Groote MA, Duarte RS, Hine E, Parankush S, Su Q, Daugherty SC, Fraser CM, Brown-Elliott BA, Wallace RJ Jr, Holland SM, Sampaio EP, Olivier KN, Jackson M, Zelazny AM. High-level relatedness among *Mycobacterium abscessus* subsp. *massiliense* strains from widely separated outbreaks. Emerg Infect Dis. 2014;20:364–71.
99. Floto RA, Olivier KN, Saiman L, Daley CL, Herrmann J-L, Nick JA, Noone PG, Bilton D, Corris P, Gibson RL, Hempstead SE, Koetz K, Sabadosa KA, Sermet-Gaudelus I, Smyth AR, van Ingen J, Wallace RJ, Winthrop KL, Marshall BC, Haworth CS. US Cystic Fibrosis Foundation and European Cystic Fibrosis Society consensus recommendations for the management of non-tuberculous mycobacteria in individuals with cystic fibrosis. Thorax. 2016;71:i1–i22.
100. Chroneou A, Zimmerman SK, Cook S, Willey S, Eyre-Kelly J, Zias N, Shapiro DS, Beamis JF, Craven DE. Molecular typing of *Mycobacterium chelonae* isolates from a pseudo-outbreak involving an automated bronchoscope washer. Infect Control Hosp Epidemiol. 2008;29:1088–90.
101. Scorzolini L, Mengoni F, Mastroianni CM, Baldan R, Cirillo DM, De Giusti M, Vullo V. Pseudo-outbreak of *Mycobacterium gordonae* in a teaching hospital: importance of strictly following decontamination procedures and emerging issues concerning sterilization. New Microbiol. 2016;39:25–34.
102. LaBombardi VJ, O'Brien AM, Kislak JW. Pseudo-outbreak of *Mycobacterium fortuitum* due to contaminated ice machines. Am J Infect Control. 2002;30:184–6.
103. Wang S-H, Mangino JE, Stevenson K, Yakrus MA, Cooksey R, Butler WR, Healy M, Wise MG, Schlesinger LS, Pancholi P. Characterization of "*Mycobacterium paraffinicum*" associated with a pseudo-outbreak. J Clin Microbiol. 2008;46:1850–3.
104. Maloney S, Welbel S, Daves B, Adams K, Becker S, Bland L, Arduino M, Wallace RJ Jr, Zhang Y, Buck G, Risch P, Jarvis W. *Mycobacterium abscessus* pseudoinfection traced to an automated endoscope washer: utility of epidemiologic and laboratory investigation. J Infect Dis. 1994;169:1166–9.
105. Fraser V, Wallace RJ Jr. Nontuberculous mycobacteria. In: Mayhall CG, editor. Hospital epidemiology and infection control. Baltimore, MD: Williams & Wilkins; 1996. p. 1224–37.
106. Fraser V, Zuckerman G, Clouse RE, O'Rourke S, Jones M, Klasner J, Murray P. A prospective randomised trial comparing manual and automated endoscope disinfection methods. Infect Control Hosp Epidemiol. 1993;14:353–89.
107. Fraser VJ, Jones M, Murray PR, Medoff G, Zhang Y, Wallace RJ Jr. Contamination of flexible fiberoptic bronchoscopes with *Mycobacterium chelonae* linked to an automated bronchoscope disinfection machine. Am Rev Respir Dis. 1992;145:853–5.
108. Zlojtro M, Jankovic M, Samarzija M, Zmak L, Jankovic VK, Obrovac M, Jakopovic M. Nosocomial pseudo-outbreak of *Mycobacterium gordonae* associated with a hospital's water supply contamination: a case series of 135 patients. J Water Health. 2015;13:125–30.
109. Guimarães T, Chimara E, do Prado GV, Ferrazoli L, Carvalho NG, Simeão FC, de Souza AR, Costa CA, Viana Niero C, Brianesi UA, di Gioia TR, Gomes LM, Spadão FS, Silva MD, de Moura EG, Levin AS. Pseudooutbreak of rapidly growing mycobacteria due to

Mycobacterium abscessus subsp. *bolletii* in a digestive and respiratory endoscopy unit caused by the same clone as that of a countrywide outbreak. Am J Infect Control. 2016;44:e221–6.
110. Lai KK, Brown BA, Westerling JA, Fontecchio SA, Zhang Y, Wallace RJ Jr. Long-term laboratory contamination by *Mycobacterium abscessus* resulting in two pseudo-outbreaks: recognition with use of random amplified polymorphic DNA (RAPD) polymerase chain reaction. Clin Infect Dis. 1998;27:169–75.
111. Bignardi GE, Barrett SP, Hinkins R, Jenkins PA, Rebec MP. False-positive *Mycobacterium avium-intracellulare* cultures with the Bactec 460 TB system. J Hosp Infect. 1994;26:203–10.
112. Bettiker RL, Axelrod PI, Fekete T, St. John K, Truant A, Toney S, Yakrus MA. Delayed recognition of a pseudo-outbreak of *Mycobacterium terrae*. Am J Infect Control. 2006;34:343–7.
113. Zhang Y, Yakrus MA, Graviss EA, Williams-Bouyer N, Turenne C, Kabani A, Wallace RJ Jr. Pulsed-field gel electrophoresis study of *Mycobacterium abscessus* isolates previously affected by DNA degradation. J Clin Microbiol. 2004;42:5582–7.
114. Sampaio JLM, Junior DN, de Freitas D, Höfling-Lima AL, Miyashiro K, Alberto FL, Leão SC. An outbreak of keratitis caused by *Mycobacterium immunogenum*. J Clin Microbiol. 2006;44:3201–7.
115. Zhang Y, Rajagopalan M, Brown BA, Wallace RJ Jr. Randomly amplified polymorphic DNA PCR for comparison of *Mycobacterium abscessus* strains from nosocomial outbreaks. J Clin Microbiol. 1997;35:3132–9.
116. Sampaio JL, Viana-Niero C, de Freitas D, Höfling-Lima AL, Leão SC. Enterobacterial repetitive intergenic consensus PCR is a useful tool for typing *Mycobacterium chelonae* and *Mycobacterium abscessus* isolates. Diagn Microbiol Infect Dis. 2006;55:107–18.
117. Brown-Elliott BA, Wallace RJ Jr. Nontuberculous mycobacteria. In: Mayhall CG, editor. Hospital epidemiology and infection control. 4th ed. Philadelphia: Lippincott Williams & Wilkins; 2012. p. 593–608.
118. Hung JH, Huang YH, Chang TC, Tseng SH, Shih MH, Wu JJ, Huang FC. A cluster of endophthalmitis caused by *Mycobacterium abscessus* after cataract surgery. J Microbiol Immunol Infect. 2016;49:799–803.

Index

D. E. Griffith (ed.), *Nontuberculous Mycobacterial Disease*, Respiratory Medicine,
https://doi.org/10.1007/978-3-319-93473-0

Q

R

S

PGSTL